INFORMATION TECHNOLOGY AND THE NETWORKED ECONOMY

SECOND EDITION

Patrick G. McKeown, Ph.D.

University of Georgia

THOMSON
COURSE TECHNOLOGY

Australia • Canada • Mexico • Singapore • Spain • United Kingdom • United States • Japan

Information Technology and the Networked Economy, Second Edition

by Patrick G. McKeown, Ph.D.

Senior Vice President, Publisher
Kristen Duerr

Managing Editor
Jennifer Locke

Product Manager
Tricia Boyle

Developmental Editor
Marilyn R. Freedman

Contributing Developmental Editor
Deborah Kaufman

Production Editor
Melissa Panagos

Contributing Production Editor
Brooke Albright

Marketing Manager
Jason Sakos

Associate Product Manager
Janet Aras

Editorial Assistant
Christy Urban

Text Designer
Ann Small - Wills

Cover Designer
Rakefet Kenaan

Photo Researcher
Abby Reip

Manufacturing Manager
Denise Powers

Printed in the United States
1 2 3 4 5 6 7 8 9 BM 06 05 04 03 02

For more information, contact Course Technology, 25 Thomson Place, Boston, MA 02210; or find us on the World Wide Web at www.course.com.

ISBN 0-030-34851-X

To my family—my wife, Carolyn; my daughter,
Ashley, and her husband, Todd; and my son, Christopher

BRIEF CONTENTS

CONTENTS

PART 5
Issues in the Networked Economy 361

PREFACE

The beginning of the 21st century has seen a rapid change from the industrial economy to a networked economy built on computers, connectivity, and human knowledge. The networked economy is characterized by rapidly changing market conditions and methods of commerce. Instead of leveraging human strength with machines as was done in the industrial economy, the networked economy leverages human knowledge with computers and connectivity to produce goods and services. The networked economy requires that organizations concentrate on improving their organizational productivity rather than worrying about personal productivity. To meet the needs of tomorrow's organizations, colleges and universities must immediately begin to prepare students to work in the networked economy. The second edition of *Information Technology and the Networked Economy* is aimed at providing today's business students with the knowledge of the networked economy necessary to be successful employees and managers in the 21st century.

New in the Second Edition

As in the first edition, *Information Technology and the Networked Economy, Second Edition* is divided into sections. With the addition of a new chapter on electronic commerce technology and the refocusing of the existing electronic commerce chapter to concentrate on strategy, the book now includes a new section on electronic commerce. This combination of electronic commerce strategy and technology provides students with an understanding of the "what" and "how" of electronic commerce. In addition to this new chapter and section, every chapter has been extensively updated to take into account the continuing changes in the global networked economy brought about by advances in information technology. This updating includes new information on existing topics, extensive revisions to entire sections to include the latest information, and, in several cases, sections on new topics. For example, the chapter on electronic commerce strategy includes the new topics of threats to business from the Internet and strategies for countering these threats. All key terms are now defined in the margin as they occur for easy reference by the reader.

In addition, all boxed features are either new or have been extensively updated from the first edition. They have also been renamed to better reflect the focus of each type of box. Many of the quick review questions within chapters have been rewritten, as have the review questions at the end of chapters. In addition, the number of review questions has been doubled from 10 to 20. The discussion questions have been rewritten and expanded, and a new type of end-of-chapter exercise, Research Questions, has been added. These questions ask readers to carry out research, either on the Web or in person, and to write a paper or create a presentation on their findings. The WildOutfitters.com case at the end of each chapter has similarly been updated.

Learning Objectives

The second edition of *Information Technology and the Networked Economy* is built around achieving the following six key learning objectives. In so doing, it ensures that students will be prepared to be successful employees and managers in the networked economy. After reading this book, the student will be able to:

1. Understand how information technology has created the networked economy and discuss the implications of this transformation.
2. Describe how people use information technology to process data into information and share data, information, and resources.
3. Discuss how information technology enables organizations to handle the present, remember the past, and prepare for the future through the use of information systems.
4. Describe how firms use electronic commerce strategy and technology to transform the way they carry out operations.
5. Discuss the processes involved in developing and acquiring information systems.
6. Understand the effects that information technology and the networked economy are having on crime, security, and ethics and the social issues created by the networked economy and information technology.

Achieving these learning objectives will go a long way toward providing the student with an understanding of the networked economy, information technology, information systems, and their impact on society.

Organization

To achieve the learning objectives, the second edition of *Information Technology and the Networked Economy* is divided into five parts, as shown in the table below. In general, after coverage of Part 1, any of the other parts can be covered in any order.

Part	Topical Coverage
1	Introduction to Information Technology and the Networked Economy
2	Information Systems in Organizations
3	Electronic Commerce: Strategy and Technology
4	Development of Information Systems
5	Issues in the Networked Economy

Part 1 provides information about the networked economy and information technology. This part includes chapters on the networked economy, elements of information technology, and networks for sharing data, information, and resources. This section introduces the risks facing all organizations and the use of information systems to address those risks.

Part 2 covers the effects of information technology on organizations and includes chapters on transaction processing systems for handling the present, organizational memory for remembering the past, and decision support systems for preparing for the future. This section provides the student with a complete discussion of information systems as they enable organizations in the networked economy not just to survive, but to grow.

Part 3 discusses electronic commerce strategy and technology. The chapter on electronic commerce strategy provides the student with a basis for understanding the benefits and threats associated with the Internet and Web and strategies for dealing with those threats. The chapter on electronic commerce technology discusses a number of the technologies used to make electronic commerce a viable revenue stream for businesses.

Part 4 considers the issues involved in developing or acquiring information systems. This includes topics on designing new information systems and deciding whether to acquire, outsource, or internally develop the new system. This section also covers the process of developing an information system. These chapters provide the student with an understanding of the systems development process including the structured systems development approach, rapid application development (RAD), outsourcing, and acquisition.

Finally, part 5 covers the impact of information technology and the networked economy on society in the areas of security, crime, privacy, ethics, health, and societal issues. This section includes a chapter on crime and security in organizations, a chapter on privacy and ethical issues, and a chapter on the societal issues associated with information technology and the networked economy.

Approach

To prepare readers to be successful in the networked economy by understanding the impact of information technology and information systems, *Information Technology and the Networked Economy, Second Edition* uses a variety of pedagogical elements, including two running cases, quick review questions after each major section, boxed features, review questions, discussion questions, and research questions.

FarEast Foods, Inc. Running Case

An important element is the FarEast Foods, Inc. case that runs through all of the chapters. This case is based on a fictitious company that distributes Asian foods via retail stores, catalogs, and the Internet. It provides students with a look at the ways in which companies use information technology and information systems to transact business in the networked economy. FarEast Foods takes orders over the Internet that it fulfills by ordering individual items from wholesalers. The company combines individual food items to create a shipment that a package delivery company picks up and delivers to the customer. Students can simulate the purchase process by visiting the FarEast Foods Web site at www.fareastfoods.com. As students move through the book, the various aspects of information technology and information systems are applied to the company. For example, in the chapters on systems development (Chapters 9 and 10), the running case describes how FarEast Foods enhances its information system, and students can experience that enhancement at the Web site. Throughout the book, whenever the FarEast Foods running case is discussed, the material is highlighted by the FarEast Foods logo in the margin, similar to what you see here.

Quick Review Questions

The Quick Review questions enable students to check their understanding of the material immediately after reading it. Answers to these questions are available at www.course.com so students can gauge their comprehension of the material.

Marginal Glossary

To help readers with the terminology that is so much a part of information systems and information technology, we have added a new feature to this edition—marginal glossary definitions for each key term in the text. With these definitions prominently displayed in the margin, the reader can easily determine the meaning of a word, phrase, or acronym. Readers will also find a traditional glossary at the end of the book, as in the first edition.

Boxed Features

The five boxed features in each chapter include an opening case that focuses on management issues, a "Technology on the Edge" box that provides information on new technologies, an "Internet in Action" box that provides an interesting example of using the Internet, an "IT Innovators" box that discusses one of the pioneers in IT and the networked economy, and an end-of-chapter management case with associated critical thinking questions. These boxes provide interesting information about elements of the networked economy beyond the material covered in the body of the chapter. The chapter opening and closing case boxes pay particular attention to information technology and cutting edge management issues in the networked economy at a wide variety of companies that students will recognize, including Lands' End, General Electric, Dreamworks, Home Depot, Dell Computers, and more.

Learning Objectives, Summary, and Review, Discussion, and Research Questions

Each chapter opens with a series of questions that the student will be able to answer after reading the chapter. The material in the chapter is summarized by providing answers to these questions. The review, discussion, and research questions at the end of the chapter provide readers with an opportunity to review what they have learned from the chapter and to research and discuss issues associated with the material. The learning objectives, summary, review questions, discussion questions, and research questions all work in concert to guide students to mastery of the material.

WildOutfitters.com Running Case

The running case at the end of each chapter, WildOutfitters.com, introduces the reader to Alex and Claire Campagne, owners of a small shop specializing in equipment and provisions for outdoor recreation located near the New River Gorge of West Virginia. The Campagnes are moving their business onto the Internet, and the case asks students to apply what they have learned in the chapter to the development of the company. The WildOutfitters.com cases also request that readers use Microsoft Office (or equivalent software) to solve problems associated with the situation described in the case.

Instructional Resources

In addition to this textbook, a variety of instructional resource items are a part of the teaching tools for *Information Technology and the Networked Economy, Second Edition*.

Instructor's Manual

An Instructor's Manual can also be found at www.course.com and on the Instructor's Resource CD. The Instructor's Manual contains a variety of items to assist the instructor. These items include the following: sample syllabi, learning objectives, chapter outlines, detailed lecture notes, quick quizzes, and solutions to end-of-chapter material.

PowerPoint Presentations

These include a complete set of PowerPoint slides, created by Mark Huber and Craig Piercy of the University of Georgia. The authors of this very useful teaching aid have extensive experience teaching an introductory information systems course using the first edition of this textbook since its release in Spring 2000. They have class-tested the new slides using pre-release chapters of the book with almost one thousand students. The slides are available for download at www.course.com (password protected).

ExamView

This textbook is accompanied by ExamView, a powerful testing software package that allows instructors to create and administer printed, computer (LAN-based), and Internet exams. ExamView includes more than 1000 test questions, which were also created by Mark Huber and Craig Piercy. Students can use ExamView to generate detailed study guides that include page references for further review. The computer-based and Internet testing components allow students to take exams at their computers and also save the instructor time by grading each exam automatically.

FarEast Foods, Inc. Demonstration Web Site

Users may also access a Web site, www.fareastfoods.com, which accompanies the FarEast Foods, Inc. case that runs throughout the text. This site allows students to interact with a simulated electronic commerce company. Although students cannot actually receive goods from www.fareastfoods.com, they can carry out all of the other activities described in the text.

Student and Solution Files for WildOutfitters.com Case

Student and Solution Files can be found at www.course.com and on the Instructor's Resource CD. Solution files are password protected and available only to instructors. Student files include any files necessary for the user to work with the WildOutfitters.com case.

Acknowledgments

Anyone familiar with writing a textbook such as this knows that the final product is not just the work of the author, but the result of a team effort. The team for *Information Technology and the Networked Economy, Second Edition* includes many talented people, and I am appreciative of their efforts. First, I want to thank Richard T. Watson of the Terry College of Business at the University of Georgia for his work advising me on the text. In his role as advisor, Rick discussed the topical coverage of each chapter with me, made suggestions for material to be included, and acted as first reviewer of each chapter. Several other colleagues at the University of Georgia including Hugh Watson, Dale Goodhue, and Mark Huber were available for many helpful suggestions during the writing of the text and I am grateful to them. I also want to thank Craig Piercy of the University of Georgia for writing the WildOutfitters.com cases that appear at the end of each chapter. He brought to this part of the book a special talent for making the cases interesting as well as useful in the learning process. Dave Preston of the University of Georgia did a fine job of writing answers to all of the exercises, both within the chapters and at the end of the chapters. The health section of Chapter 13 is primarily the result of work carried out by my wife, Carolyn McKeown, RN, BSN, and I want to express special thanks to her for that effort.

I also want to express my appreciation for those who reviewed one or more chapters of the manuscript. The final text reflects many of their ideas. These reviewers include Paul M. Bauer, University of Denver; Sonny Butler, Eastern Kentucky University; Charmayne Cullom, University of Northern Colorado; George Heilman, University of Northern Colorado; Brent Hussin, University of Wisconsin at Green Bay; Jack Klag, Colorado Technical University; Terence M. Waterman, Golden Gate University; Richard T. Watson, University of Georgia; and Dennis L. Williams, California Polytechnic State University.

I would like to thank those at Course Technology who were involved in editing and producing this textbook. Marilyn Freedman did an outstanding job as Developmental Editor in working with me to create the final text. Her analyses of the reviews were extremely helpful as was her attention to every detail of the project. Tricia Boyle worked as Product Manager and I appreciate her efforts to keep the project on schedule. Thanks also go to the Production Editor, Melissa Panagos; the Photo

Researcher Abby Reip; and to the Designer, Ann Small, who designed the book. I also want to express my appreciation to Jennifer Locke, Managing Editor, for overseeing the project to a successful completion.

Finally, my acknowledgements would be incomplete without again mentioning my wife of 35 years, Carolyn. Without her love, support, and work on the project, I would not have been able to complete it on time and it would not have been worth doing at all.

About the Author

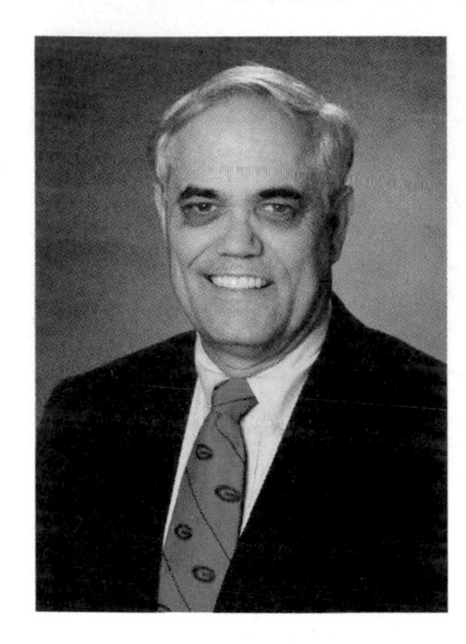

Dr. Patrick G. McKeown, head of the top-ten ranked Department of Management Information Systems in the Terry College of Business, has been at the University of Georgia since 1976. He received his bachelor's and master's degrees at the Georgia Institute of Technology and his Ph.D. from the University of North Carolina—Chapel Hill. McKeown has published more than 30 textbooks and close to 50 articles in the areas of management science and information systems. He was the 1997 recipient of the Terry College of Business Distinguished Service Award for his work in enhancing computer literacy through computer projects in non-computer courses. McKeown was a 1998 Fulbright scholar in Portugal, where he taught an MBA course in electronic commerce, and has served as a reviewer for various Fulbright programs since that time. He has also taught in France at Lyon III University, in Finland with the Helsinki School of Economics, and at the Graduate School of Business Leadership at the University of South Africa.

PART 1

INTRODUCTION TO INFORMATION TECHNOLOGY AND THE NETWORKED ECONOMY

The dramatic growth of the Internet and the World Wide Web are changing the way we live, work, and play in many ways. One important change has been the transition from the industrial economy to the networked economy. The networked economy is based on computers, connectivity, and human knowledge and will involve changes in the way goods and services are created, produced, sold, and distributed. Like the industrial and agricultural economies, the networked economy must have an underlying infrastructure. Its infrastructure is known as information technology (IT), and the primary components of IT are the computer and computer networks, which make connectivity possible.

Part 1 introduces you to the networked economy and information technology. This introduction begins with a broad discussion of both areas, followed by chapters on computer hardware and software and computer networks. After reading this first part of the book, you should have a good grasp of the networked economy and information technology.

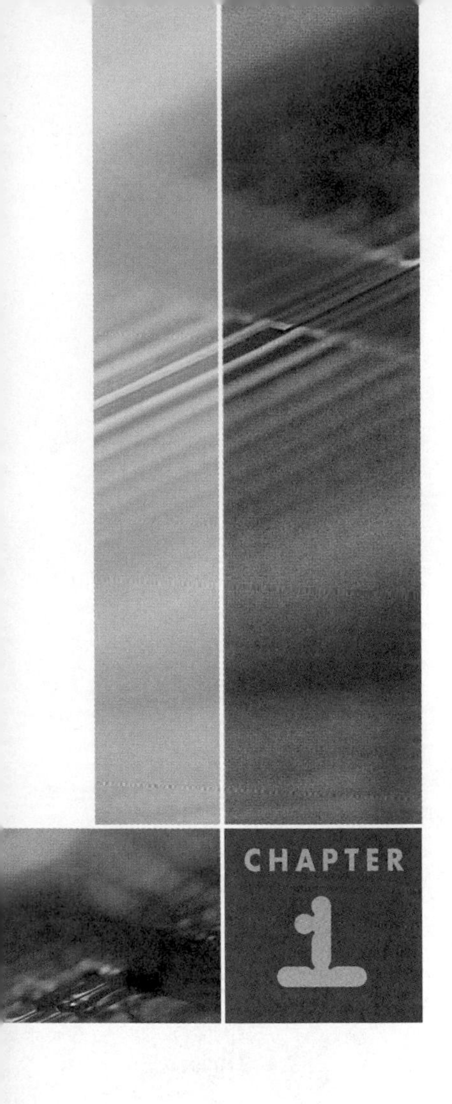

CHAPTER 1

THE NETWORKED ECONOMY

LEARNING OBJECTIVES

After reading this chapter, you will be able to answer the following questions:

- How has the networked economy evolved?
- What are the key elements of the networked economy?
- What is the role of information systems in organizations?
- What is the IS cycle, and what are the three IS functions?
- How do the IS functions relate to business risks?

Getting Personal at Lands' End

Lands' End is a direct clothing merchant that has a long history of customer service. For example, it was the first catalog company to set up toll-free lines to take orders and one of the first to accept credit cards. In 1995, it was the first cataloger to offer a Web site to its customers. From the beginning of its Web-based service, Lands' End was dedicated to continue its long-standing customer service commitment through e-mail responses to customer queries. While other retailers did not bother to respond to e-mails, Lands' End first set a goal of responding within 24 hours and then pushed for a three-hour turnaround. In a 1999 e-mail test of retailers during a holiday weekend conducted by an independent group, Lands' End responded in 30 minutes and was rated as one of the best "e-tailers" on the Web for customer service.

Lands' End has now taken its customer service efforts even further by adding a variety of technology-based services. For example, its Ask Me service allows users to use the telephone or text-based chat to talk with a customer service representative while using the Web. With this system, both customer and representative can see the same Web page; the representative can even put items in the customer's shopping basket for him or her. Other Lands' End services include the following:

> Shop with a Friend, where two customers can view the same Web page and chat about items
> Virtual Model, where a customer can put clothes on a model that is identical in dimensions to him or her
> Personal Shopper, where a Lands' End system suggests items based on customers answering a series of questions
> Wish List, where customers can create a list of items from which family and friends can select purchases

With these services, Lands' End plans to stay at the top of catalogers in terms of customer service.

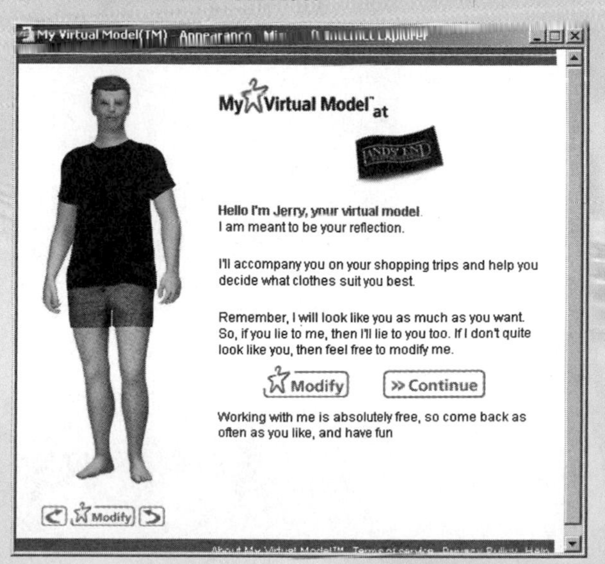

Lands' End has added a number of features to their Web site, including a Virtual Model like this one.

Source: http://landsend.com.

A Connected World

The end of the twentieth century and the first years of the twenty-first century have been exciting times. The rise of networked technology as an important form of communications and commerce has profoundly influenced the way people communicate and the way companies and organizations do business. High-speed communication links connect technology and people. Networked computers share trillions of characters of information daily, enabling us to withdraw money from ATMs anywhere in the world, check the weather in distant locations, and buy airline tickets anywhere at any time. Networked technology is also having a dramatic effect on the relationships between companies and their customers, suppliers, and employees. Television networks using satellite technology make it possible to watch hundreds of channels from locations anywhere in the world. Telephone networks, both wired and wireless, make real-time voice and text communications possible to nearly anywhere in the world. Global positioning satellite (GPS) systems enable pinpoint navigation anywhere in the world and are frequently found today in the dashboards of high-end automobiles. And this list includes only a few of the growing number of network applications. As noted in the opening case, technology is also becoming an important part of providing customer service at companies such as Lands' End. For example, Figure 1.1 shows how a car rental agency can provide customer service by using three of 24 GPS satellites combined with computized technology to determine the position of an automobile stranded near Denver, Colorado.

Figure 1.1 Using GPS to find a stranded car

Credit: HowStuffWorks at http://www.howstuffworks.com

All networks depend on computers for direction. In fact, the computer is one of the most important machines in use in the world today—if not *the* most important. A **computer** is a device that accepts data and manipulates it into information based on a sequence of instructions. That is, computers process raw data into useful information and then send this information over the network to other computers. Without computers, many of the developed world's factories, transportation systems, and other infrastructure components would quickly grind to a halt. This issue forced the massive upgrade of the world's computer systems in preparation for the change from the twentieth to the twenty-first century. The computer has been deemed one of the 10 most important inventions of the last 2000 years, along with items such as eyeglasses and the printing press.

computer
A device that accepts data and manipulates it into information based on a sequence of instructions.

The Impact of Computer Networks on Business

A widely used form of networked technology is the **computer network**, which consists of two or more connected computers. Such computer networks form the nervous system of modern companies and organizations by enabling stakeholders—managers, employees, suppliers, and customers—to interact electronically. In fact, all networked technologies are at their heart computer networks because they are run by computers. The largest and most widely used computer network is the **Internet**, which connects thousands of smaller computer networks, thereby linking together millions of computers around the world. The Internet enables worldwide communications and makes available a virtually infinite amount of information to its more than 500 million users worldwide, including more than 165 million users in the United States. Figure 1.2 shows a conceptual computer network with several computers connected by high-speed telephone wire, television cable, and wireless connections.

computer network
A combination of two or more computers with a communications system that allows exchange of data, information, and resources between the computers.

Internet
A worldwide network of computer networks in private organizations, government institutions, and universities, over which people share files, send electronic messages, and have access to vast quantities of information.

Figure 1.2 Conceptual computer network

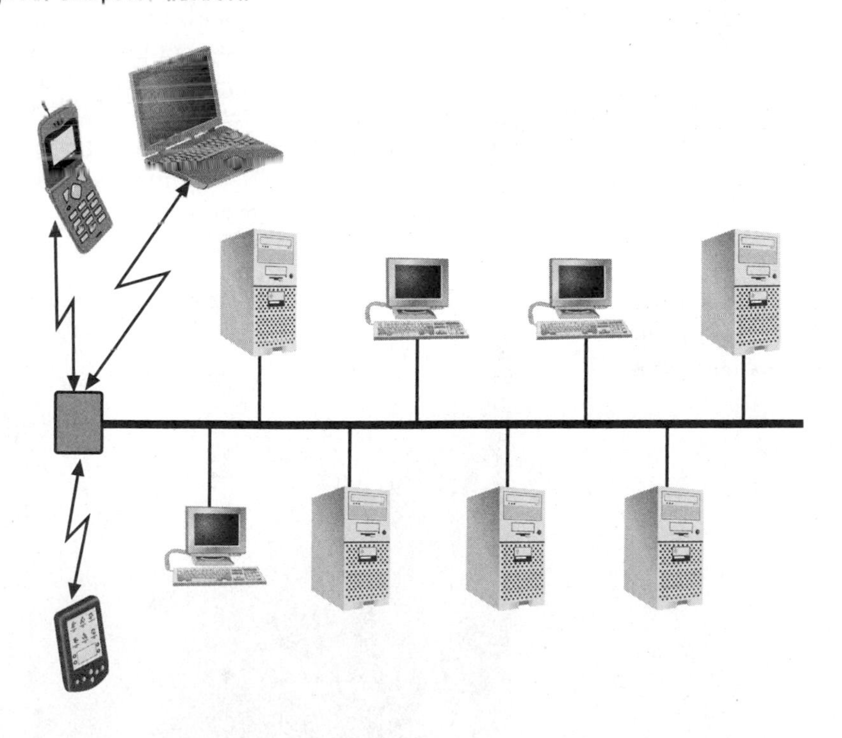

It would be impossible to list every way that networked technology has changed our lives. Just as the printing press made books available to a large proportion of the population, leading to widespread education, networked technology is making it possible to carry out a virtually unlimited number of activities from our homes or offices. Much of this activity is carried out via the Internet, and one of its most popular applications, the World Wide Web. The size of the Web is estimated to have passed *6 billion* pages of information in early 2002, with more than 7 *million* new pages being added each day. In addition to the Internet, other networks are beginning to have significant effects on us. For example, mobile phone networks are growing very rapidly, as are new applications for them. A current trend is to combine mobile phones with the Internet to provide the best features of both—that is, wireless voice communication and access to virtually unlimited amounts of information. Today, it is virtually impossible to find a company of any size in the developed countries that does not have an internal computer network that enables employees to share information and ideas.

As you are surely aware, the Internet has made it possible for individuals to find information on purchased items from catalog services such as Lands' End, check out features and prices before actually buying a car, communicate with friends and relatives with e-mail or instant messaging, join a chat group discussing computer applications, listen to distant radio stations live, download software for a free trial, and buy and sell items on an online auction. Surely you can think of many more ways to use the Internet.

Companies are also finding that the Internet can be very useful—commercial use of the Internet has grown dramatically since 1995, with 2001 sales to consumers reaching $61.8 billion.[1] Although still only a small fraction of the multitrillion-dollar U.S. economy, Internet sales are expected to continue to grow as more consumers find items of interest available online.

electronic commerce
The activity of carrying out business transactions over computer networks.

In addition to **electronic commerce**—the activity of carrying out business transactions over computer networks—companies are using the Internet to communicate with other companies or to extend their existing business models instead of trying to create entirely new ones. In 2002, sales between businesses—sometimes called business-to-business (B2B) commerce—were expected to be $830 billion.[2] The Gartner Group, an independent research and consulting group, has predicted that, by 2005, $8.5 *trillion* in orders will be placed by companies over computer networks. In addition, many companies are finding that a "clicks and mortar" combination approach can be successful because customers can visit the physical location to return items purchased over the Internet or to discuss problems with the purchase.

Mobile telephones are now capable of sending live video to other mobile telephones.

1. Amy Winn, "Cyber Fraud Rampant." *Atlanta Journal/Constitiution*, March 5, 2002, p.D2.
2. http://www.ecommercetimes.com/perl/story/16314.html

Movement to the Networked Economy

networked economy
Enhanced, transformed, or new economic relationships based on computer networks and human knowledge.

Computer networks are not just a new way of handling business transactions or searching for information; they also provide a better way of doing business. Computer networks are the basis of a new type of economy—a networked economy. For 200 years, people have lived and worked in the *industrial economy*, which was built on the existence of capital, in the form of factories and machines, and labor, in the form of employees. In contrast, the **networked economy** combines enhanced, transformed, or new economic relationships based on computers, connectivity, and human knowledge.

Because all economies are built on the premise of meeting the needs and desires of humans by carrying out transactions, the networked economy encompasses a wide variety of economic relationships between people, including those between firms and customers, between firms and employees, and so on. Although you may not buy an automobile over the Internet, you are very likely to use the Internet to conduct research prior to making a purchase at a dealership.

Steam engines like the one that powers this locomotive were an important part of the infrastructure of the industrial economy.

A primary result of the movement to the networked economy is that organizations of all types need to learn how to use the new combination of computers, connectivity, and human knowledge to remain competitive and to survive. Organizations can no longer hope to do the same thing they have been doing for years; instead, they must change or risk being driven out of business by competitors they did not even know existed. The essence of the networked economy is not just change; it is change at an accelerating rate of *speed*. Companies must continually scan their environment for new ways to serve their customers or else face the prospect that another company will serve their customers instead. This requirement may mean radically changing the way these companies have done business or it actually may spur companies to move into new businesses.

The transition to the networked economy affects you both as an individual and as an employee. As an individual, you need to learn how to take advantage of the new opportunities for information, employment, and entertainment that are constantly becoming available in this new economy. As an employee, you need to look for ways to help your organization take advantage of the opportunities for new markets afforded by the networked economy. You need to do so regardless of the type of position you hold in your organization, because opportunities are not restricted to those elements of the company normally associated with technology. This textbook aims to help business students prepare to take advantage of the almost unlimited possibilities associated with the networked economy.

Impact on Businesses

Although the economic rules of supply and demand remain valid, the networked economy's use of computers, connectivity, and human knowledge has resulted in a *better* way of doing business. Because these computer networks enable humans to use their unique intelligence to share ideas, workers can find more efficient ways of carrying out business operations as well as find totally new ways of doing things. In fact, this innovation is much more important than efficiency, because it leads to competitive advantages that seldom come from efficiency alone.

All companies and organizations face three major problems or risks: low demand for products or services, inefficient handling of transactions, and a lack of innovation to stay ahead of competitors. Known as demand, efficiency, and innovation risks, these three problems must be solved by every company or organization if it hopes to be successful. In the industrial economy, a great deal of effort went into solving the efficiency risk. In fact, the production worker's job was to find out how to do his or her job better, thereby increasing the productivity of the factory. Entire fields of study have been built around the concept of optimizing the known. However, because they failed to address the demand and innovation risks—through marketing and research and development efforts—many companies have been left behind and gone out of business. In the fully networked economy, in which most repetitive production tasks will be handled by robots controlled by chips and humans will be freer to use their unique knowledge to develop innovations, companies that do not innonvate will disappear. As noted management author Peter Drucker said, "Business has only two basic functions—marketing and innovation."[3] These three risks will be addressed again later in this chapter and again throughout the book.

It is important to understand that *success in a networked economy is nonlinear*—by the time you realize something is a success, it has probably gone beyond the point where you can take advantage of it. Therefore, individuals and organizations must constantly scan their environment to find new ideas and technologies *before* they become successes. Unlike the industrial economy, in which companies had the luxury of taking a wait-and-see attitude about new technology, companies in the networked economy cannot afford to delay. Although there is no longer a panic-driven approach to trying new solutions to business problems, good companies are always pushing the envelope to find better ways to serve their customers.

Continuing innovations in computer networks will undoubtedly result in goods and services such as software, movies, music, books, financial services, and so on, which already are either electronic in nature or can be converted to an electronic form, being delivered directly to the home. In this way, the networked economy is generating the benefits promised by the many dot-com companies that have gone out of business over the past few years.

Creative Destruction

creative destruction
A concept emphasizing that the most important part of the change process for a business is not what remains after the change but rather what has been destroyed.

The speed of change in the networked economy means that organizations must constantly reinvent themselves if they are to survive. Termed **creative destruction** by the European-born U.S. economist Joseph Schumpeter (1883-1950), this concept emphasizes that the most important part of the change process for a business is not what *remains* after the change but rather what has been *destroyed*.

Austrian-born U.S economist Joseph Schumpeter developed the theory of creative destruction.

Without the destruction of the old ways of carrying out business, organizations cannot create the new ones. Creative destruction often requires an entirely new way of thinking about the problems facing a business. Executives may need to redefine the problems or reframe the questions; simply doing business as usual will not suffice. For example, Reuben Mattus decided that he needed to creatively destroy his existing approach to marketing ice cream products to achieve success. Rather than try to compete with large dairy products companies by selling cheap ice cream, he decided to use traditional fresh ingredients in his ice cream, changed the product name to Haagen-Däzs, and raised the price. It did not matter that the ice cream's name had no meaning in any language or that the same product now cost more—Mattus had successfully redefined his approach to business.

The need to carry out creative destruction is never more important or more difficult than when a company is at the top of its industry. Often, its employees share a feeling of having done well and have little desire to rock the boat. It is at this time that looking for new opportunities becomes especially critical; otherwise, the company can easily be surpassed by competitors as they take advantage of newly emerging markets. There are numerous examples of companies that were industry leaders one year, only to find themselves outstripped by the competition during the next year. At one time, Lotus Development Corporation (now owned by IBM) held the dominant position in the spreadsheet market, selling more copies of Lotus 1-2-3 than all of its competitors combined. When Lotus failed to take full advantage of the transition to the Windows operating system, however, it quickly lost much of its market share to Microsoft Excel.

3. "Quote Disk 1,2,3," DBUG, 1991.

Other companies have been able to creatively destroy themselves and become successful in the face of severe competition. For example, consider Edmunds.com, Inc. This company was founded in 1966 as Edmunds, Inc., for the purpose of publishing new and used vehicle guides in magazine and book form to assist automotive buyers in making informed decisions.

Edmunds was making a nice profit selling its guides, but the owners decided that they needed to move to an electronic form for distribution of the same information. In 1995, Edmunds became the first company providing such vehicle information to establish a site on the World Wide Web. The amazing thing about this approach was that Edmunds gave away on the Web the same information it sold in its magazines. This decision has turned out to be the right one—the site draws thousands of visitors each day. Edmunds makes its money by linking visitors to other automobile-related sites that sell financing, insurance, and other services. These other companies pay Edmunds a commission on each sale made to someone coming to them through the Edmunds Web site. In 2000, Edmunds became the first source for vehicle pricing information for users of wireless Web-enabled devices.

Even when it is not necessary to creatively destroy the current organization, the networked economy means that almost all companies must look for ways to extend their business to take advantage of the Internet and other businesses. They must practice *creative extension*. For example, as noted in the case at the beginning of the chapter, Lands' End was able to extend its traditional catalog clothing business by using the Web as a new front end for the company's ordering system.

1. How does the networked economy differ from the industrial economy?
2. What are the three risks that every company faces?

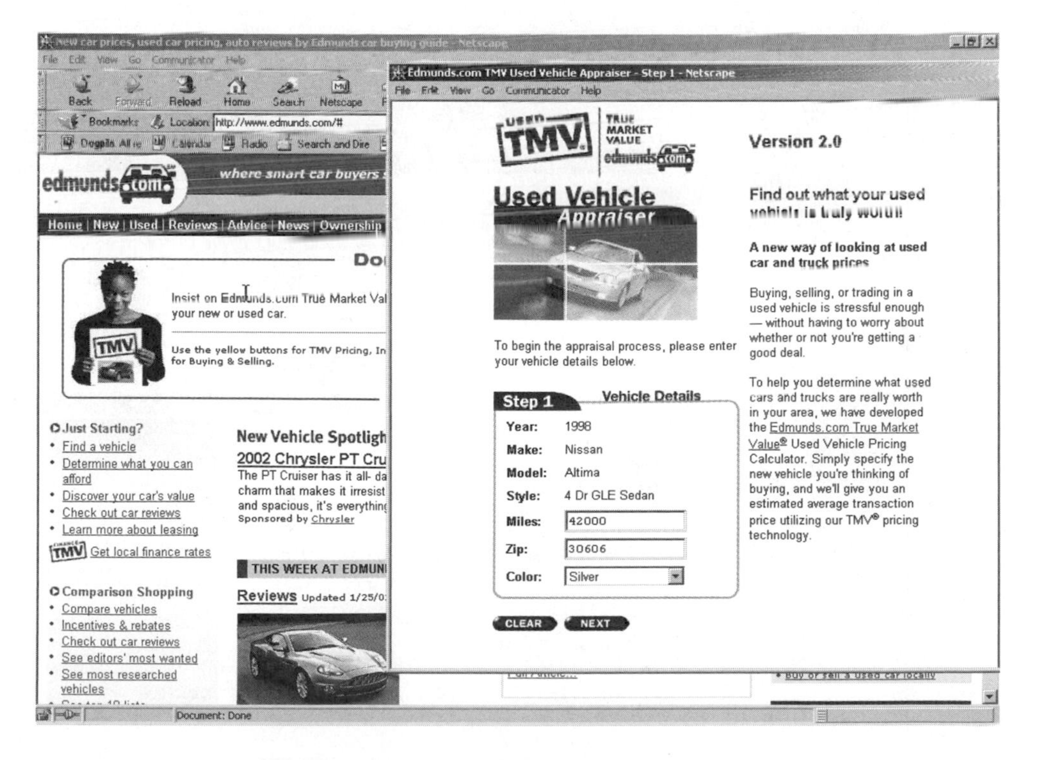

The Web site for Edmunds.com provides a great deal of information about both new and used automobiles.

INTERNET IN ACTION

Recession-Resistant Dot-Coms?

As you are undoubtedly aware, many of the so-called dot-com companies, which sprang to life during the heady final years of the twentieth century, are no longer in existence. In fact, by the end of 2001, more than 750 dot-coms had folded since January 2000.[4] However, some dot-com companies have actually prospered in this day and time of failed technology-based companies: those that deal with job placements. Companies such as Monster.com, JobsOnline.com, and Headhunter.net are doing quite well by providing recruiters and job-seekers with an online way to reach one another. In fact, the largest of these companies, Monster.com, purchased two smaller online placement companies in mid-2001—Jobline.com and Hotjobs.com—for a combined total of more than $575 million. Monster.com claims that a total of 30,000 new résumés per day are added to its listings and that (in mid-July 2001) it had more than 415,000 job listings—down from a high of more than 500,000 listings. One interpretation of the reduced numbers is that employers are hiring, but are seeking more experienced workers.

How do job placement companies make their money? They charge employers a fee to post a job online and to search the résumé database for matches to their needs. For example, to list one job on Monster.com costs a company $295, but companies can pay a bulk rate to list more than one job. Job-hunters pay nothing for this service, resulting in many of them posting resumes to more than one online job site. However, most people agree that Monster.com is the go-there-first site, because it has spent a great deal of effort in educating people to the idea of using an online job search. These sites are not without their competitors. For example, some of the biggest newspapers have taken their long-standing classified sections and put them online.

4. http://www.webmergers.com.

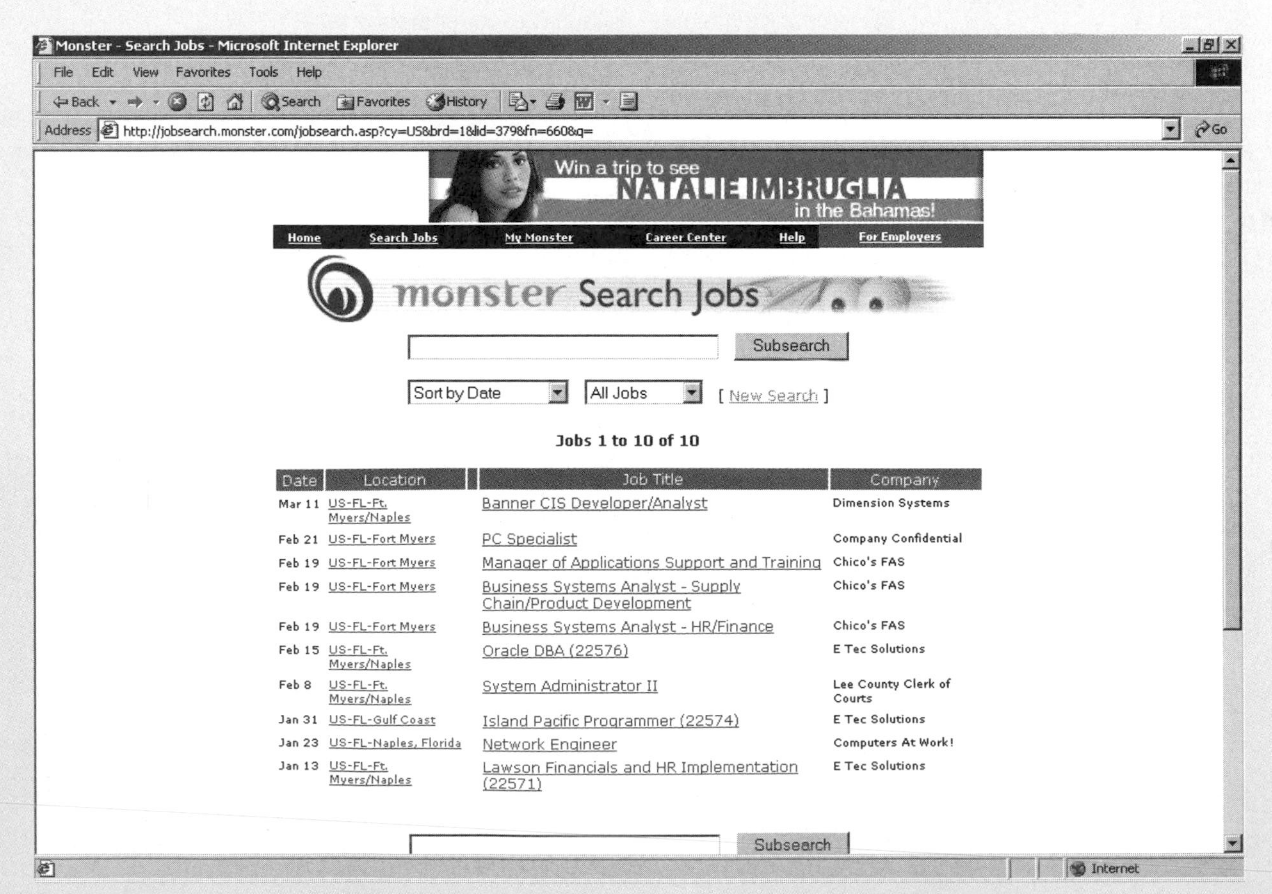

Monster.com is one of the companies that was able to survive the dot-com crash of 2000-2001.

Source: Frances Katz, "Recession resistant: Online job-placement services riding the waves." *Atlanta Journal/Constitution*, June 6, 2001, p. G6.

Elements of the Networked Economy

By now, you should be convinced that the networked economy is bringing about great changes in our personal lives as well as in business and industry. Let's now take a closer look at the new economic relationships and new jobs being spawned by the networked economy and the elements of the networked economy—computers, connectivity, and knowledge. These three elements work together so that each element *multiplies* the effects of the other elements, thereby enhancing, transforming, and creating new economic relationships, as shown in Figure 1.3.

Economic Relationships

Traditionally, management primarily has been concerned with three stakeholder relationships: relationships with customers, relationships with employees, and relationships with suppliers. However, in the networked economy, a whole host of different relationships are now possible. Consumers are not always just customers; in some cases, they take on the role of employees. For example, when Microsoft was preparing the Windows 95 operating system, it had more than 1 million consumers acting as testers using advance copies. While performing the testing, were these testers considered customers or unpaid employees of Microsoft who benefited from being among the first to use the new operating system?

In addition to modifying existing relationships, the networked economy can bring new relationships into being. For example, relationships among customers have become more important to firms because those customers may form user groups that provide important feedback to the firms about their products. In some cases, employees within the firms actually form direct relationships among themselves and customers, as they become special service representatives. For example, a customer might work directly with a technician on a problem or communicate with other customers though an Internet newsgroup. The firms that benefit most from these changes will be those that take the greatest advantage of new and different relationships by thinking "outside the box" about ways to improve customer service. Figure 1.4 shows the various relationships that are possible in the networked economy.

Figure 1.3 Elements of the networked economy

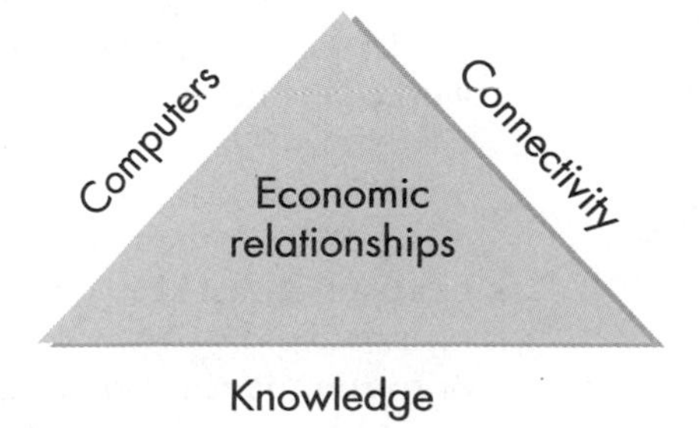

Figure 1.4 Possible relationships in the networked economy

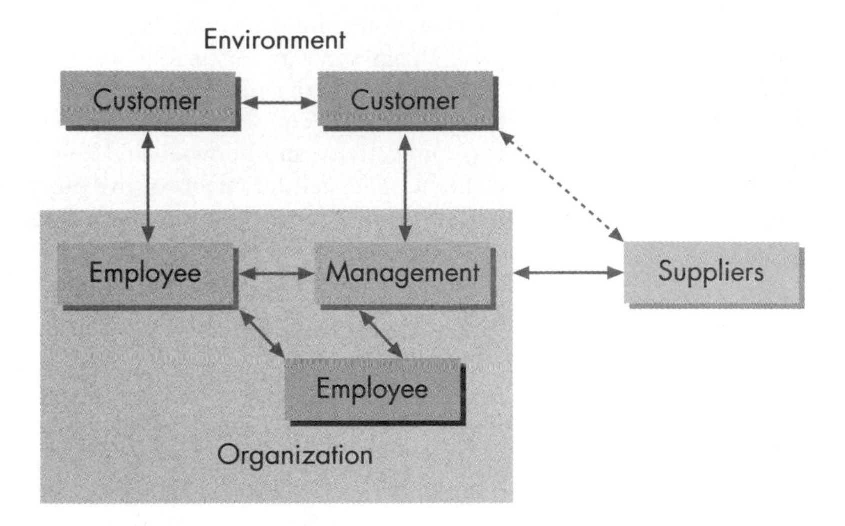

Computers

infrastructure
The underlying foundation or basic framework of a system or organization.

information technology (IT)
Technology that is used to create, store, exchange, and use information in its various forms.

chip
A tiny piece of silicon consisting of millions of electronic elements that can carry out processing activities.

The industrial economy and the agricultural economy that preceded it were both supported by infrastructures. An **infrastructure** is the underlying foundation or basic framework of a system or organization. The infrastructure of the industrial economy included canals, roads, railroads, power plants, factories, and so on; these components enabled companies to bring in raw materials, produce finished goods, and transport them to the customer. The infrastructure of the networked economy is based on computers and communication networks, and it is commonly referred to as information technology. **Information technology (IT)** refers to technology that is used to create, store, exchange, and use information in its various forms. The most obvious IT device is the computer found on millions of desks in offices around the world, but many other IT devices help run the networked economy as well. They include routers that control the Internet, all types of office machines, personal digital assistants (PDAs) like the Palm, mobile phones, manufacturing robots, contactless toll cards, and global positioning satellite (GPS) systems among many, many other IT devices.

Computers provide the processing and communications capabilities for the networked economy. When you think of a computer, you probably immediately envision a desktop or laptop computer. In fact, computers are present virtually everywhere, in every aspect of daily life. Computers handle the millions of transactions that occur every day over the Internet, at local grocery stores, or at shops in the mall, for instance. Computers that keep track of inventory at all stages of production and distribution are included in the networked economy, as are computers that are used to design products or to run the network. Today, close to 600 million computers are in use worldwide.

All of information technology is built around the **chip**, the ever-shrinking marvel of miniaturized electronic circuitry that carries out instructions from the user or the manufacturer.

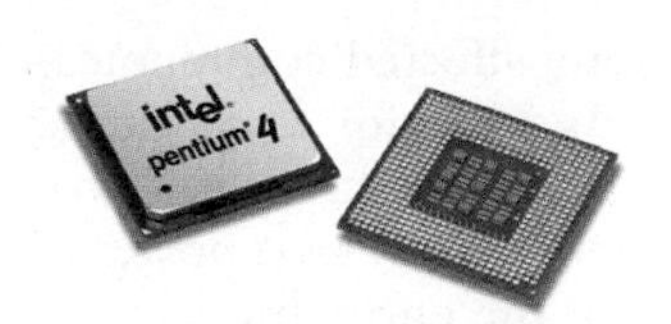

The Intel Pentium 4 chip is the most recent in a long line of processing chips that have powered personal computers.

Some chips contain more than 20 million transistors—electronic on/off switches—that are connected on the chip in such a way as to carry out an almost infinite variety of operations. If you think for a moment, you'll recognize that fax machines, mobile phones, pagers, handheld games, digital clocks, and other electronic devices all have at least one chip in them. However, the chip's use is far more widespread than those devices and computers. Chips are used in many other devices to measure and report data to computers, where the data are processed or used to control operations of some machine. For example, there are more chips in today's automobile than there are in a personal computer.

Since their invention in the early 1970s, chips that power today's computers and other information technology have followed Moore's law, which states that they will double in power every 18 months. This evolution has resulted in faster, easier-to-use computers as well as in cheaper, more useful machines of all types. In fact, it now appears that chips are doubling in speed even faster than predicted by Moore's law. Chips are also becoming so cheap to manufacture that they have become part of our throw-away economy. For example, chips are used in one-time usage smart cards and radio frequency identification (RFID) devices, which can be used to identify almost anything, even items in a box.

Connectivity

connectivity
The availability of high speed communications links that enable the transmission of data and information between computers and conversations between people.

Although computers are a key element in our changing world, another development—connectivity—has magnified their potential. **Connectivity** refers to the availability of high-speed communications links that enable the transmission of data and information among computers and conversations between people. This communication has involved the use of both wired and wireless media. On the wired side, connectivity includes the use of fiber-optic cable, new ways of using traditional copper wire to send voice and data over telephone lines, and increasing use of TV cable as a two-way communications medium. On the wireless side, a large variety of approaches are being used for primarily short-distance (line-of-sight) communications.

Downloading Digital Video

TECHNOLOGY ON THE EDGE

In the fall of 1998, a company named Diamond Multimedia changed the face of music when it released what would be the first of many devices that could play MP3 files. An **MP3** file is a digital file that has been compressed to one-twelfth its original size with little loss of sound quality. MP3 stands for MPEG-1, Audio Layer 3, whereas MPEG is a standard compression method sponsored by the Motion Picture Experts Group and one of several standards for compressing audio and video. MP3 changed the face of music by allowing huge files on music CDs to be compressed to a size that allowed them to be stored internally on a computer hard disk. As you are well aware, this development led to the creation of Napster and its unsuccessful fight over its legality with the Recording Industry Association of America (RIAA).

Today, a similar fight is brewing over digital video files compressed using the MPEG-4 compression standard for video. Named DivX by its anonymous creators (no relation to the defunct DVD player of the same name), MP4 video files are small enough to be stored on a single CD-ROM or downloaded over the Internet via a high-speed connection. Although still in its infancy compared to MP3 music, DivX video files are already causing concerns in the film industry, which, because of their size, previously was not worried about pirated movie files. To further complicate matters, hackers have come up with a way to break the security code protecting DVDs, allowing them to pull the video files off and then convert them into DivX format. The Motion Picture Association of America (MPAA) is leading the fight against illegal copying of DVDs and will most likely be moving against DivX movies soon.

Source: John Borland, "Hacker's video technology goes open source." *CNET News,* January 16, 2001.

bandwidth
The capacity of a communication channel, expressed in bits per second.

The ever-increasing power of chips has also significantly affected communications, leading to dramatically increased capacity, or **bandwidth**, for transmitting data and information between computers. Author George Gilder has noted that bandwidth is increasing at a much faster rate than computer power—by a power of 10 every two years. Because communication is the basis for the entire human culture, increased communications capabilities will have far-reaching implications. In the year 2000, the volume of data (computer-to-computer) traffic over telecommunications networks surpassed that of voice traffic. Experts predict that, in the next few years, voice traffic will make up only five percent of the total traffic volume.[5] In a sense, the era of computers as computational devices may soon be over because their greatest use will become that of communication devices.

Knowledge

knowledge
A human capacity to request, structure, and use information.

Although computers and connectivity are necessary elements of the networked economy, without human knowledge, they would be worthless. **Knowledge** can be defined as the capacity to request, structure, and use information. For example, it takes knowledge to understand the meaning of the numbers generated by a networked computer, say, in a departmental payroll, and to know if these numbers are within acceptable values. Although a variety of computer programs have attempted to incorporate human knowledge, they cannot make the decisions that require hunches, intuition, and leaps into totally unrelated areas, which humans make every day without a second thought.

data
Facts, numbers, or symbols that can be processed by humans or computers into information.

information
Data that have been processed into a form that is useful to the user.

To fully grasp the concept of knowledge, you need to understand two other terms: data and information. Together with knowledge, data and information are widely used in discussing the networked economy. **Data** consist of facts, numbers, or symbols that can be processed by humans or computers into information. Data on its own has no meaning and must be interpreted in some way before it can be useful. Although this interpretation can be accomplished by humans, today it is more commonly achieved by inputting the data into a computer and processing it into a meaningful form known as **information**. Information comes in many forms, including documents, reports, tables, charts, and so on, all of which are meaningful to humans. For example, 1247.93 is one piece of data that has no meaning by itself, but when it is combined with other numbers on your bank statement, it becomes information. Processing data into information takes on many forms. Figure 1.5 shows the process of converting data into information.

Examples of transforming data into information include the following:

- Combining grade reports from many professors, each of whom submits his or her grades for numerous students, into an individual grade report for each student
- Creating the summary of gross pay, deductions, year-to-date amounts, and net pay included with paychecks for individual employees based on the number of hours submitted by each employee to his or her division or department
- Transforming the numbers and formulas in a spreadsheet into a chart
- Creating a model of a hurricane from observations, assumptions, and formulas

There are literally millions of other situations in which data are processed into information; you can probably come up with quite a few after just a moment's reflection.

Figure 1.5 Converting data into information

Data → Processing → Information

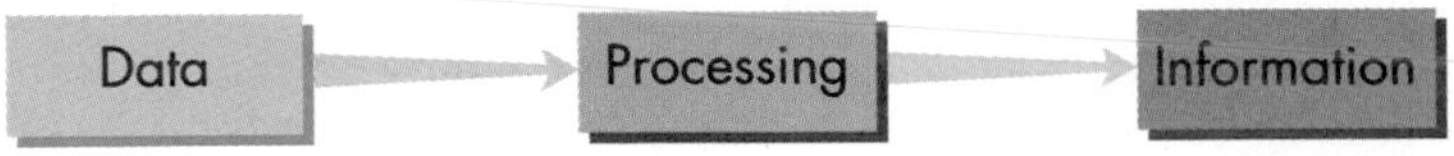

5. Kagan, Jeffrey, "It's for you." Atlanta Journal/Constitution, October 15, 2000, p. P1.

In thinking about data and information, you should realize that the difference between data and information lies in the eye of the beholder; that is, what is *data* to one person or organization is *information* to another. For example, an employee's payroll statement is information to the employee, but is a data item when the departmental manager must combine it with payroll values for all other employees in the department to determine the departmental payroll. The departmental payroll is information to the department manager, but the comptroller of the company seesdepartmental payroll as data that must be combined with payrolls for all other departments to determine the overall company payroll. Figure 1.6 illustrates how the same thing can be data to one person, but information to another.

At one time, converting data into information was the primary purpose of computers and software, and the term *data processing* came about to describe this use of computers. Early thinking about the use of computers was that processing sufficient data into information would almost automatically improve the management of organizations. Over time, it has become quite clear that information by itself is almost as useless as data and that, without the uniquely human capability of knowledge, it would be impossible to improve the management of organizations based solely on information. One of the key uses of knowledge in a business environment is to make decisions that will determine the future well-being of the organization. These decisions can focus on which people to hire, which products to develop, or which strategies to use to increase profits. As noted by Drucker in his book *Post-Capitalist Society,* knowledge is the "only meaningful resource." This analysis has been extended by others to say that knowledge is the only means of sustainable competitive advantage for an organization.

In transforming data into information, the goal is to combine it with knowledge to make decisions in organizations. The relationships among data, knowledge, information, and decisions in information systems are shown in Figure 1.7. In this figure, a request for information from a knowledgeable person results in data being processed into information. Personal knowledge is then used to interpret the information, reach conclusions, and make decisions that lead to action. This cycle can be repeated as often as necessary to acquire sufficient information to make a decision. Be aware, however, that if managers fail to have the wisdom to use their knowledge to interpret the information available to them, the whole system breaks down. For example, in one case, the CEO of a technology company noticed that it was taking longer to close deals but failed to take action to slow production. As a result, his company was stuck with a large volume of unsold inventory, resulting in layoffs and significantly reduced profits.

Figure 1.6 Different views of data and information

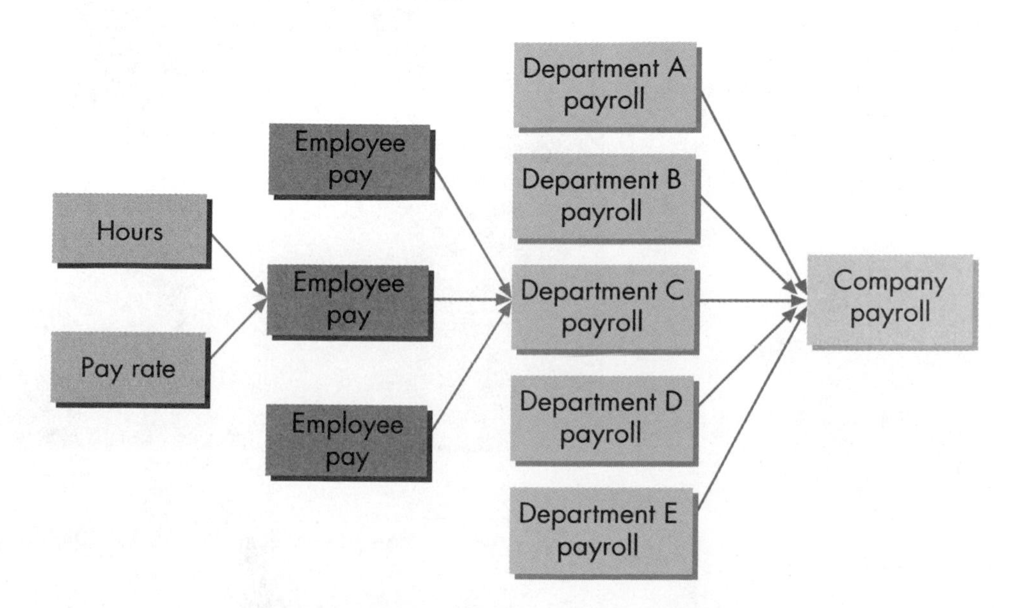

Figure 1.7 Information, knowledge, and decisions

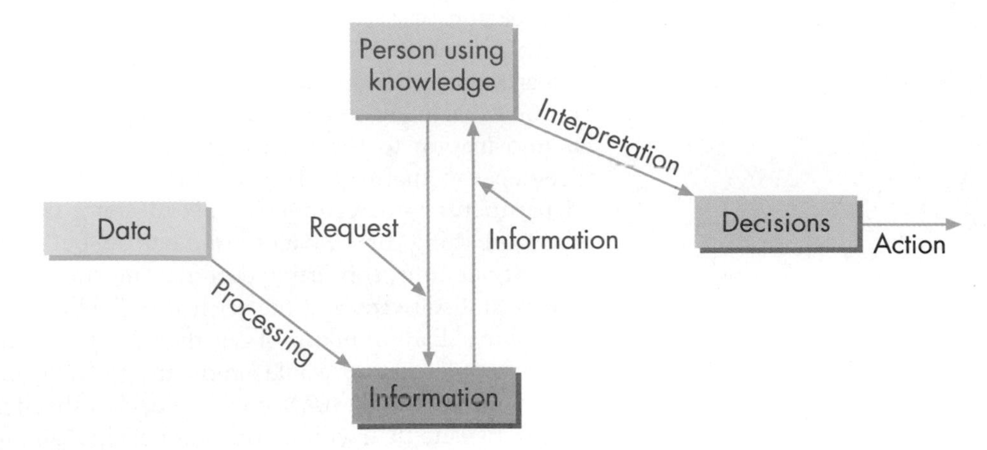

Source: R.T. Watson, *Organizational Memory* (New York: John Wiley, 1999), p.27

Knowledge Workers

The networked economy is referred to in the popular media by a variety of terms, including the *Information Age* and the *digital economy*. However, the use of the terms *information* and *digital* do not clearly describe the importance of knowledge used by people in the networked economy. Workers in organizations who use their knowledge to work with information have been referred to as **knowledge workers**. This term was coined by Drucker as early as 1959. Many of the 75 percent of U.S. workers in the service sector are actually knowledge workers rather than workers in industries typically associated with service. In fact, in addition to the traditional economic sectors (manufacturing, agriculture, and service), there should probably be a fourth sector of the economy to reflect this emphasis on knowledge. Drucker later extended this concept to define today's environment as the "knowledge society" in which, he argues, knowledge is not just another resource of production, but rather the only meaningful resource.

knowledge workers
Workers in organizations who use their knowledge to work with information.

To understand the importance of knowledge workers, consider Microsoft Corporation. In 1998, Microsoft became the most valuable company in the world, surpassing General Electric.[6]

Although not the biggest company in terms of number of employees or revenue, Microsoft is the second most valuable company in the world.

6. http://www.singapore.cnet.com/news/2001/04/05/20010405v.html.

The value of a company is measured by multiplying the number of shares of stock outstanding times the value of a single share of stock. At the end of 2001, even though it was not ranked among the top 50 firms in terms of revenues, Microsoft was still the second most valuable company, despite the drop in the value of its stock resulting from the antitrust decision against it and the Setember 11, 2001 terrorist attacks.

What makes Microsoft so valuable is not the small amount of land that it owns or the few factories that make copies of its software. With fewer than 40,000 employees, it is far from the *biggest* firm in the world. However, each employee has a great deal of knowledge that goes into the software created and distributed by Microsoft; it is this knowledge that makes Microsoft so *valuable*.

This textbook is aimed at making you a better knowledge worker by providing you with an understanding of the skills and capabilities necessary to be successful in the networked economy. It is only a beginning—the networked economy is changing so fast that you will need to constantly scan the environment to update your skills.

1. Identify at least three devices in your environment that contain chips. How many of these devices are connected in some way?
2. Give an example of a type of knowledge that is unique to you or to a member of your family.

Peter F. Drucker

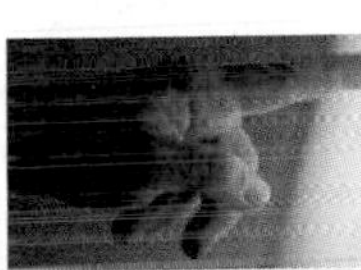

IT INNOVATORS

Even before he coined the term "knowledge worker" to describe individuals who use their minds instead of their muscles in their work, Peter Drucker was generating ground breaking ideas for business. Born in Vienna in 1909, Drucker migrated to the United States in 1937 and began teaching in the area of management. In addition to the concept of the knowledge worker, he has originated other important management concepts, such as management by objectives, the transition from assembly lines to flexible production, and the notion of empowerment. Drucker has written a large number of articles and has been a heavy contributor to the *Harvard Business Review* and the *Wall Street Journal*. He has also written 31 books, with the latest being *Management Challenges for the 21st Century*. Drucker has received numerous honors, including having the Claremont University Management School named after him. *Business Week* called him the "most enduring management thinker of our time," and *Forbes* featured him on its cover in 1997 under the headline "Still the Youngest Mind."

Even after he has passed his ninetieth birthday, Drucker continues to think about the knowledge worker. In an October 1999 article in *Atlantic Monthly*, he writes about the future of these individuals. He notes that, in our Internet-driven world, knowledge workers will demand more than just the financial rewards associated with being employees; they will only stay with an organization that can satisfy their values and give them social recognition and social power. Instead of being employees, they will demand to be well-paid partners in the organization.

Peter F. Drucker has been a leading innovator in the world of management for over 50 years.

Sources: *Peter F. Drucker biography*, http://www.pdf.org/leaderbooks/drucker/bio.htm, and Peter F. Drucker, "Beyond the Information Revolution." *Atlantic Monthly*, October, 1999, pp. 47-57.

Information Systems: IT in Organizations

Organizations have used IT for many years in an effort to reduce their labor costs by automating many of their operations. Many organizations have also used IT to ensure that the correct information is made available to the correct person so that workers can make better decisions. The most successful companies have found that their goal must be to use IT to serve their customers better. Using IT to better serve customers increasingly will become the norm in the networked economy as prices and profit margins associated with manufactured goods drop. As a student of business, knowledge of information technology is critical to your success in the workplace. As recently as 10 years ago, it may have been possible to get by without knowledge of information technology if it was not your major area of interest. Today, regardless of your major, you will need to know how to use information technology to perform your job better and to provide better service to your organization's customers.

You will often hear the terms *information technology* and *information systems* used interchangeably, sometimes leading to some confusion. In this book, the term **information systems (IS)** refers to systems that develop the information that managers and other employees combine with knowledge to make decisions. Basically, you can think of information systems as information technology applied to organizations. To fully understand information systems, you first need to understand the concept of a *system*.

information systems (IS)
Systems that develop the information that managers and other employees combine with knowledge to make decisions.

Systems

A **system** is a group of elements (people, machines, cells, and so forth) organized for the purpose of achieving a particular goal. Almost everything around you is a part of one type of system or another. Examples of systems include your college, a business, or a computer system. Your college is a system composed of the faculty, staff, and students in addition to the buildings and equipment; this system is organized for the purpose of providing a college-level education to the students. A business is composed of a variety of elements, including production, marketing, distribution, accounting, and so on, with a goal of making a profit and increasing the value of its owners' investment. A computer system is composed of input, output, processing, and secondary storage elements. The goal of a computer system is to process data into information.

system
A group of elements (people, machines, cells, and so forth) organized for the purpose of achieving a particular goal.

All systems have input, processing, output, and feedback. Anything that enters the system is classified as **input**, which is transformed in some way by **processing** and then **output** in some form. **Feedback** is information about the output that may cause the system to change its operation.

input
Receiving the data to be manipulated and the instructions for performing that manipulation.

processing
Converting data into information.

output
The result of processing as displayed or printed for the user.

feedback
Information about output that may cause a system to change its operation.

As an example of a system, consider a business like Lands' End. This business has numerous types of input, processing, output, and feedback going on simultaneously. One important type of input consists of orders from customers. These orders must be processed, resulting in output in the form of shipments to customers. An important type of feedback takes the form of complaints from customers regarding the timeliness of deliveries or quality of clothing items. If items are delivered late, Lands' End might choose a different package delivery company. Similarly, poor quality should result in a change to different suppliers. If feedback is ignored, the company could quickly go out of business. Obviously, almost all businesses have many other systems besides those of fulfilling orders, but this example should give you an idea of how you can view a business as a system. Figure 1.8 shows how the process of carrying out a transaction can be viewed as a system.

IS Functions

Information systems are used in organizations for three primary purposes or functions: handling the present, remembering the past, and preparing for the future. Handling the present means that the organization must be able to take care of its day-to-day business, primarily by processing transactions that involve customers, suppliers, and employees. These transactions must be stored if the organization is to

Figure 1.8 Transaction viewed as a system

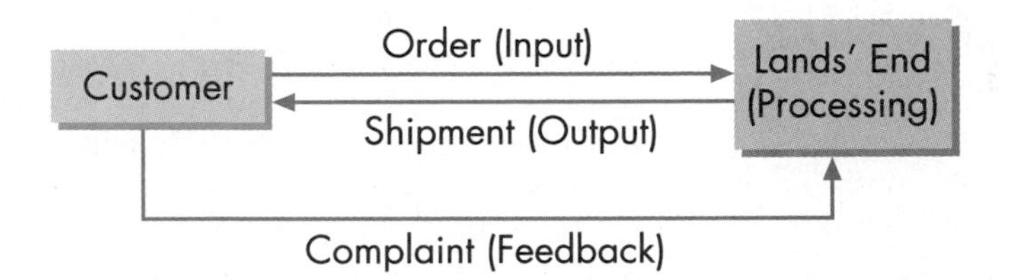

remember its past. Data on transactions are then used to prepare for the future, which results in decisions that determine the way transactions are handled in the future. Because each function provides input to the next function, with the third function leading back to the first, they can collectively be thought of as a cycle of functions, called the **IS cycle**, shown in Figure 1.9. Note that there is no break in the cycle and that it continues from one operation to the next. The three primary purposes will be discussed here; they will be related to the three risks facing every business in the next section.

IS cycle
Information systems for handling the present, remembering the past, and preparing for the future.

Handling the Present

In many cases, the original purpose of many types of IS was to process data into information. This operation continues to be extremely important for all organizations because it handles the present by processing transactions. Originally, transactions were limited to physical purchases between buyer and seller, in which the buyer might be the end consumer, another company, or another unit within the same company. In this case, the transaction process includes ordering a product, transferring the funds between buyer and seller to pay for the product, and delivering the product to the buyer. However, in the networked economy, because so many more events can be handled as electronic transfers of information between computers or other IT devices, the definition of a **transaction** needs to be broadened to include any event that involves the digital transfer of money *or* information between entities. Transactions now include payments to employees, employee payments to pension funds or tax-deferred savings accounts, payments of tuition, transfers of funds between accounts, queries to a Web page, and so on.

transaction
Any event that involves the digital transfer of money or information between entities

Figure 1.9 The IS cycle

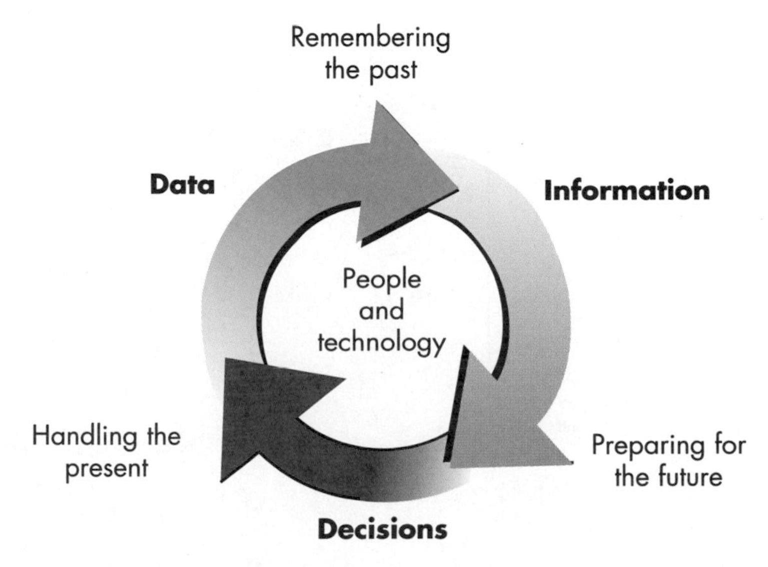

Source: The concepts underlying this figure were developed in conjunction with Richard T. Watson

The magnetic strip on the back of a credit card has sufficient information storage capacity to enable transactions to take place.

If an organization fails to process transactions accurately and in a timely manner, it will not continue to exist for very long. If customer transactions are not handled, revenue will not come in. If transactions involving suppliers are not handled appropriately by the organization ordering and paying for raw materials, the firm will have nothing to offer its customers. Finally, the organization must ensure that its employees are paid in a timely manner or else they will cease to work for that company.

The transaction process in the networked economy is quite different from that in the industrial economy because of the capability to carry out all of these events electronically. For example, if you use any of the many electronic bill-payment services, the process is quite different than when you write a paper check or visit the payee to pay with cash. Table 1.1 compares the payment of a bill using a networked economy method—the electronic bill payment system—and an industrial economy method—the paper check. Using electronic bill payment, it is much easier to pay bills.

At the end of the process, you have an instant record of paid bills without having to deal with paper bills and returned checks. Money can easily be transferred between accounts, and it is even possible to set up automatic transfers and payments. Transaction processing systems will be covered in detail in Chapter 4.

Remembering the Past

Handling the present through transaction processing involves checking existing data (for example, to determine whether the product is in stock and whether the buyer has good credit) and storing new data (for example, the number of units of which item was sold to whom and for how much). Checking existing data and storing new data require an extensive amount of **organizational memory** in the form of databases, data warehouses, information management systems, knowledge management systems, and so on. The most common form of organizational memory is the **database**, which is a collection of information that is arranged for easy manipulation and retrieval. Organizational memory, in turn, enables an organization to perform the vital function of remembering the past. To see how this process works, think about what happens

organizational memory
Remembering the past through data, information, and knowledge management.

database
A collection of information that is arranged for easy manipulation and retrieval.

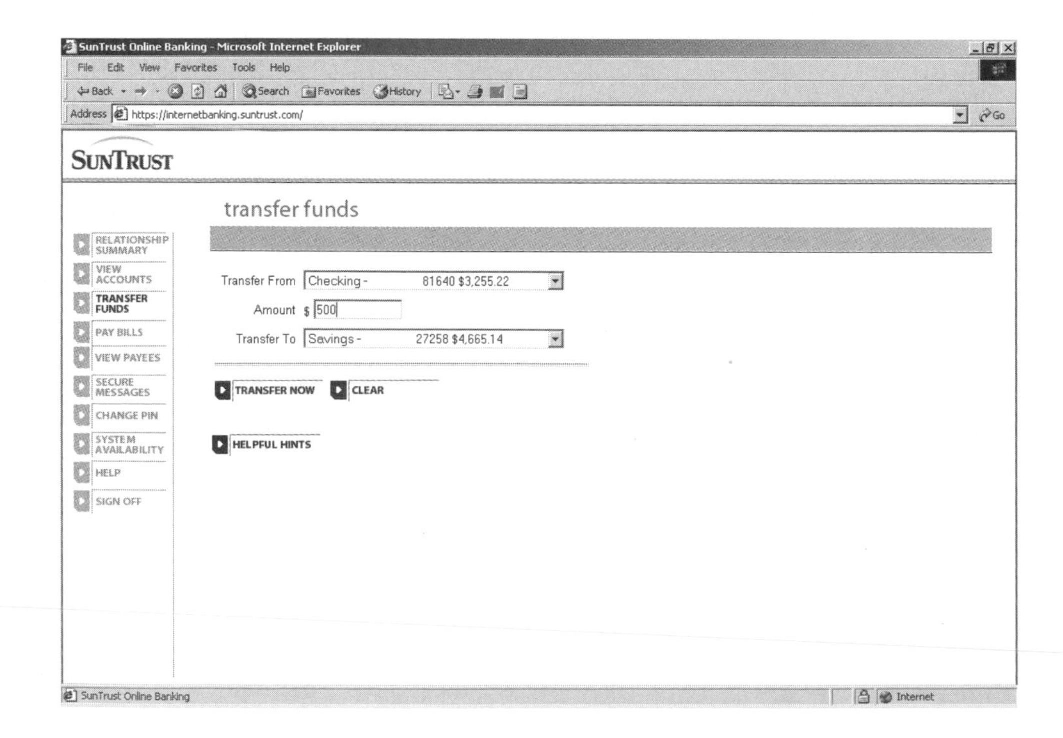

With an electronic payment system, such as the one shown here, it is easy to pay bills or transfer money between accounts.

Table 1.1 Comparison of Bill Payment Systems

Step	Electronic Bill Payment	Paper Check Payment
1.	Bill is received via e-mail or Web site	Paper bill is received through postal system
2.	Electronic bill-payment software is accessed via computer; payee and amount are entered, and instructions are given to pay bill	Check is manually written and sent to payee through postal system
3.	Software electronically transfers funds between payer's bank account and that of payee	Check is deposited by payee in its bank for processing; check goes though banking system to payer's bank for processing, and payer's account is reduced by amount of check
4.	Electronic bill-payment system shows transaction immediately and updates payer's account balance; bill-payment system stores record of transaction, which can be accessed as needed	Paid check is sent back to payer via postal system to be manually compared to payer's checkbook balance and physically stored for later reference

when you purchase a product from Lands' End (www.landsend.com). The transaction processing includes looking up the customer's personal information such as your address, credit card number, and previous purchases. It also includes checking the availability of the items you are ordering or determining whether they must be back-ordered and sent at a later date. When the transaction is completed, all of these data must be updated or, if you are a new customer, new data must be created. Organizational memory systems will be covered in detail in Chapter 5.

Preparing for the Future

Although handling the present and storing the data and information generated by this process are critical to the efficient operation of almost any type of organization, another important operation is extracting the data and information and combining it with human knowledge to prepare the organization for the future. Preparing for the future involves using information technology to improve the capability of employees to understand, respond to, manage, and create value from information. Often this endeavor involves using a variety of information systems to help employees make better decisions. Examples of the use of information systems to prepare for the future include using data mining to find needed data in data warehouses, combining data with models of the firm to predict the result of various courses of action, and presenting information to executives in a form that they can use to make decisions. Data warehouses, data mining, and firm modeling are all **decision support systems**, which have been in use for almost a quarter of a century and have become important to the financial health of today's firms. Decision support systems will be covered in detail in Chapter 6.

decision support systems
Information systems aimed at helping an organization prepare for the future by making good decisions.

Transforming the Organization

In addition to helping employees make better decisions, information technology is having another effect on organizations—it is *transforming* them into new types of organizations. As noted in the discussion of the networked economy, organizations must constantly undergo the process of creative destruction if they want to remain in existence. Today, they can implement creative destruction through electronic commerce and by taking advantage of the opportunities offered by the networked economy. For example, many firms, such as Edmunds.com as discussed earlier, are transforming the way they serve their customers by using the Internet and World Wide Web to engage in electronic commerce. Whereas many firms are simply replicating their traditional methods of doing business, others are actually changing the processes by which they do business. Edmunds did not merely try to replicate its paper magazine in an electronic form; it changed the entire way it approached the distribution of information. Companies that are changing the way they do business will be able to take full advantage of the new information technology by transforming themselves into different entities.

Quick Review

1. Give another example of a system and list the input, output, processing, and feedback for it.
2. What three functions must information systems carry out for any organization?

Using IS Functions to Deal with Business Risks

Earlier, three risks that every organization faces were mentioned: demand risk, efficiency risk, and innovation risk. Failure to successfully deal with any of these risks can result in an organization's demise. Many of the now defunct Internet-based companies such as eToys.com and Pets.com found that the cost and effort involved in selecting, packing, and shipping products to their customers made their operations inefficient and unprofitable. Even if they had sufficient demand, they could not efficiently meet it and make a profit. Finally, failure to innovate in terms of developing new or improved products, services, or operations has resulted in many well-known companies being forced out of business by their competitors. Reducing these risks can be closely linked to the IS functions discussed in the preceding section.

Reducing Demand Risk

Every company faces the risk of lowered demand, which can lead to unsold inventory and lower profits, which could in turn result in layoffs. Even well-known and well-established companies with a history of success are vulnerable to the risk of lowered demand. One way to address this risk is through improved customer service. If customers perceive that they are being served by one company better than by another, they will probably return for repeat business—and bring their friends and associates with them. Information technology in the networked economy can deliver improved customer service in a number of ways, including the following:

- Providing self-service to customers on a 24 hours a day, seven days a week (24/7), basis by making products and services available on the Web
- Providing product information to customers on the Web in a variety of ways
- Enabling customers to have contact with other customers, resulting in customers adding value to customers

You can understand how IT helps reduce demand risk by examining the underlying IS functions that play a role in improved customer service. For example, the first approach to improved customer service—self-service checkout—uses IT to handle the present more efficiently. The second approach—making information available to customers—uses all of the memory of the organization, its customers, and, potentially, its suppliers so as to improve the level of service to customers. The IT involved also provides improved search technologies that enable people to find information on products and services quickly and easily. Finally, by enabling customers to add value to one another, the organization prepares for the future by locking customers into the organization.

Improved customer service also helps people conserve their scarcest resource in the networked economy—time. With the world moving at an ever faster pace, any way to reduce the time necessary to carry out an activity will be seen as valuable. Instead of being limited by business hours, during which traditional merchants are open, banks, stores, and other providers can be open on a 24/7 basis in the networked economy, meeting customer needs rather than their own needs. This trend does not mean that people will stop sleeping, only that Internet technology will provide services around the clock.

Reducing Efficiency Risk

If companies fail to carry out transactions efficiently, they run the risk of making stakeholders unhappy, which results in lower profits and lost sales. For example, customers become unhappy when transactions fail to go through or are lost. They also become unhappy when returns are not credited. Similarly, suppliers and employees become unhappy when they are not paid in a timely manner. Failure to deal with stockholders in an efficient manner can lead to a stockholder revolt and removal of the current management team. Finally, failure to meet tax deadlines will usually result in the taxing agency placing a lien on the company or worse.

The IS function of handling the present is key to reducing efficiency risk. It does so by enabling a company to use a variety of networks to lower its transaction costs. A recent study at the University of California, Berkeley, showed that moving traditional business functions to the Internet is saving U.S. companies tens of billions of dollars each year and significantly boosting the growth rate of U.S. productivity.[7] Many IT elements are involved in this process, and how companies handle the present is constantly being improved. For example:

- Ordering products and services over the Web leads to much lower transaction costs
- Specialized networks enable companies and their suppliers to conduct business
- Companies are paying employees in a more efficient manner by electronically delivering paychecks directly to the bank and sending paystubs to employees via e-mail
- The U.S. Internal Revenue Service is developing more and more ways for individuals and companies to file and pay for taxes electronically

Reducing Innovation Risk

Of all the risks, innovation risk is the greatest in the networked economy. If companies and organizations fail to innovate, they face extinction as their rivals find better ways to provide the same or better products and services. Innovation risk is reduced through an appropriate application of the IS function of preparing for the future. IT can implement this IS function in a number of ways, including providing access to ideas from customers, employees, suppliers, and Web search engines.

Customers are often the best source of new ideas for a company if an easy way is provided for them to send in those ideas. Electronic methods for customers to share ideas include the company Web page, e-mail, bulletin boards, and chat groups. Employees can also be a great source of innovations, as they often discover better ways to do their jobs if they are given proper means to share their discoveries with other employees. Likewise, suppliers can be a fine source of new ideas, as they want the company to prosper and remain a customer. Finally, a search of the Web can often generate new ideas.

7. http://interactive.wsj.com/articles/SB1007419980707882280.htm.

many-to-many communication
A form of communication in which many people can communicate with many other people.

One of the results of the World Wide Web is an increase in **many-to-many communication**, in which any person on the Internet can easily communicate with a large number of people also using the Internet. Many-to-many communication represents a dramatic change from traditional forms of communication that have been one-to-one (for example, telephone calls) or one-to-many (for example, newspapers, radio, and television). With many-to-many communication, a tremendous amount of information flows around the Web each day. Companies taking advantage of many-to-many communications will find many new ideas about how to carry out their business.

1. Suggest some ways that an organization with which you are familiar can use IT to reduce its demand risks.
2. What is many-to-many communication?

FarEast Foods, Inc.

To gain a better understanding of a networked economy company, throughout this book you will interact with a fictional distributor of Asian food items named FarEast Foods, Inc. This company started out as an Asian food store in New York City and then branched out to San Francisco and Los Angeles. Ten years ago, the company opened a mail-order distribution center in Denver and moved its corporate headquarters there. Several years ago, the company began marketing its packaged products (no fresh or frozen food) over the World Wide Web at www.fareastfoods.com. The firm's Web page is shown in Figure 1.10.

FarEast Foods markets packaged goods, but is not involved in their production, storage, or distribution other than to combine orders from wholesalers to be picked up by a package delivery company. Prospective buyers can view and order a wide variety of Asian foods, paying by credit card, and receiving their choices via package delivery. The business process for the company is shown in Figure 1.11. Note that there are five elements in the FarEast Foods Web-based business process: the Web site, the food wholesalers, the bank, the package delivery company, and the customer. Note also that a variety of electronic and physical flows occur between the elements. FarEast Foods and its Web-based business will be used in all the chapters of this book as a continuing example of a networked economy company.

Figure 1.10 Web page for FarEast Foods

Figure 1.11 Organization of FarEast Food

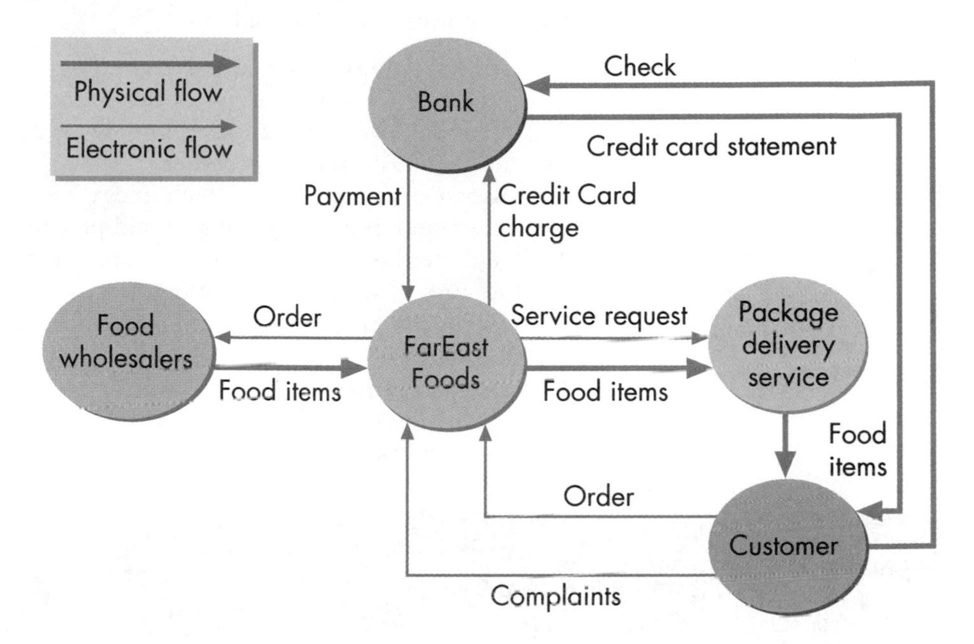

In terms of the IS cycle, FarEast Foods handles the present in a number of ways, including face-to-face transactions in its three stores, voice transactions from the mail-order system, and Web-based transactions on its Web site. It remembers the past by storing information on all prior orders, regardless of the source of the order, in a computer system that will bring up a customer's prior orders upon entry of his or her telephone number. This information is used to suggest other items that the customer might wish to purchase or order based on his or her prior purchase history and on purchases made by other customers buying similar items. By letting FarEast Foods know which products are in demand in which areas and which items need to be discounted to increase sales, this same information provides the basis for FarEast Foods to prepare for the future.

1. Visit the Web site for FarEast Foods at www.fareastfoods.com and "purchase" some food items.
2. Describe the transaction process at www.fareastfoods.com.

A Look Ahead

Chapter 1 introduced the networked economy and information technology. This textbook is divided into five parts, with Chapter 1 opening the first part. The remaining chapters in Part 1 provide additional information about the information technology infrastructure for the networked economy, including a chapter on the hardware and software components of information technology and networks for sharing data and information. Part 2 describes the use of information systems in organizations to handle the present, remember the past, and prepare for the future. Part 3 focuses on the use of the Internet for commercial purposes and includes chapters on both electronic commerce strategy and technology. Part 4 addresses the processes that organizations use in designing and then developing or acquiring information technology. Finally, Part 5 addresses the implications of the networked economy and information technology for crime, security, privacy, and ethics and discusses issues arising from the growth of the networked economy.

Learning Objectives

This textbook will develop your skills to master six learning objectives. After reading this book, you will be able to:

1. Understand how information technology has created the networked economy and discuss the implications of this transformation
2. Describe how information technology is used to process data into information and share data and information
3. Discuss how information technology can be used in organizations to handle the present, remember the past, and prepare for the future
4. Describe the use of electronic commerce strategy and technology to transform the ways in which organizations carry out their operations
5. Discuss the processes involved in the development and acquisition of information systems
6. Understand the effects that information technology and the networked economy are having on crime, security, and ethics and the issues created by the networked economy and information technology

Case*Study*

For GE, Business Slowdown Spells Opportunity

For a 110-year-old company that sprang from the Industrial Revolution and whose former chairman, John F. (Jack) Welch, once described himself as a "Neanderthal" when it came to the Internet, the business slowdown in 2001 just meant more opportunity for General Electric (GE) to push ahead with its plans to digitize every aspect of the company. To achieve this goal, GE's 7000 tech people are creating dozens of new applications that will generate demand for hardware and software to support them. GE planned to spend $3 billion on IT in 2001, up 12 percent from 2000 and three times the overall rise in IT spending across the economy. Welch told company employees that the slowing economy was not the time to slow down digitization efforts, but rather a period in which they should seek to widen their lead over competitors. With hardware prices being down and good IT people available, Welch encouraged GE employees to push ahead.

GE is pursuing the goal of digitizing everything because it feels this change will lead to both lower costs and higher revenues. On the cost side, GE believes it can save $1.6 billion annually through its digital efforts by using online auctions and ordering systems as well as digital invoices. In 2001, GE planned to carry out 30 percent of its annual buying through e-auctions, up from only 15 percent in 2000. By reducing the number of paper invoices from the 3.1 million handled in 2000 to zero by the end of 2001, GE hoped to save millions of dollars. Another initiative hopes to enable divisions to compare costs of various items (say, stationary) and then use the cheapest suppliers. The software to do so cost only $18,000, but will save $500,000 on business cards alone.

From a revenue point of view, GE sold $7 billion of goods and services online in 2000, a mere 5 percent of its total $129 billion in revenue. However, it hoped to sell $15 billion in 2001 and eventually generate 30 percent of its revenues online. The company has also discovered that the best way to generate new revenue is to develop new products that involve information. For example, GE believes it can collect data on its products in the field and then sell the data back to customers to help them with their planning. For example, how often does a CT scanner break down?

Former CEO Jack Welch pushed GE to digitize as much of its business as possible.

Source: "Lion in winter." *Forbes.com*, April 30, 2001.

Think About It

1. How is GE using information technology to help it achieve success in the networked economy?
2. Discuss GE's situation in terms of business risks and IS functions.
3. How is GE taking advantage of the business downturn to increase its competitive advantage?

SUMMARY

To summarize this chapter, let's answer the questions posed at the beginning of the chapter.

How has the networked economy evolved? Computers combined with high-speed communications media form computer networks that are changing the way we live and work. These networks enable individuals, organizations, and businesses to transmit huge amounts of voice and data anywhere in the world almost instantaneously. They are rapidly changing the way business is carried out, as more and more people go online to make purchases from businesses, 24 hours a day, seven days a week. The networked economy is an economy based on enhanced, transformed, or new economic relationships resulting from a combination of computers, connectivity, and human knowledge. A result of the movement to the networked economy is that organizations of all types need to learn how to use the combination of computers, connectivity, and knowledge to remain competitive if they want to remain viable. The essence of the networked economy is not just change, rather, it is change at an accelerating rate of speed.

What are the key elements of the networked economy? The networked economy is built on economic relationships based on computers, connectivity, and knowledge. Relationships are changing in the networked economy to include not just business-to-customer but also customer-to-business, employee-to-customer, and customer-to-customer relationships. It is also necessary for companies to practice creative destruction or extension to avoid being left behind as well as to deal with the three business risks: demand, efficiency, and innovation.

Computers are essential to the networked economy because they provide the required processing capabilities and are the most readily visible form of information technology. IT is often referred to as the infrastructure of the networked economy. Connectivity is required to share data and information between computers. Human knowledge is necessary to use the information that results from processing raw data into information. Without knowledge, none of the other elements would be of any use.

What is the role of information systems in organizations? Information technology in organizations is often referred to as information systems. A system is a group of elements (people, machines, cells, and so forth) organized for the purpose of achieving a particular goal. All systems have input, processing, output, and feedback. Information systems have the goal of transforming data into information that can be combined with knowledge to make decisions in organizations. They are aimed at handling the present by carrying out transactions, remembering the past with organizational memory, and preparing for the future with decision support systems. Information systems are also critical to transforming the organization.

What is the IS cycle, and what are the three IS functions? The IS functions involve handling the present, remembering the past, and preparing for the future. *Handling the present* refers to taking care of all the transactions with which an organization must deal. *Remembering the past* refers to creating and using organizational memory. *Preparing for the future* refers to extracting data and information and combining it with human knowledge to help make better decisions that will determine the future of the firm. Information technology is also important in transforming organizations into new forms. In so doing, it addresses the need for creative destruction.

How do the IS functions relate to business risks? IT in the organization in the form of the IS functions can be applied in many ways to reduce the three business risks—demand, efficiency, and innovation. Handling the present is most important in dealing with the efficiency risk by making sure that the organization processes transactions efficiently. Remembering the past is important in dealing with all risks, but is most closely related to the demand and innovation risks by remembering what customers want and providing ideas for the future based on past experience. Preparing for the future is closely related to all three risks, because a company must make decisions on future ways of handling demand, becoming more efficient, and creating new products and processes through innovation.

REVIEW QUESTIONS

1. What is a computer network? What is the relationship between the Internet and computer networks?
2. Give three examples of computer networks other than those mentioned in the text.
3. Why do we say that "success is nonlinear in the networked economy"? What does this statement mean for businesses?
4. What is creative destruction, and why is it important in the networked economy?
5. What are the three stakeholder groups with which management has traditionally been concerned? How has this changed?
6. What are the three elements of the networked economy?
7. Why is information technology referred to as the *infrastructure of the networked economy*? What is the most evident form of IT?
8. What is Moore's law, and what does it have to do with the networked economy?
9. How does information differ from data? How does knowledge differ from information?
10. Give two examples of converting data into information other than those discussed in the text.
11. What is a knowledge worker? Give two examples of knowledge workers other than those given in the text.
12. How does information technology relate to information systems? What is the difference?
13. What are the four parts of a system? Which part helps it correct for past errors?
14. How do transactions in the networked economy differ from those in the industrial economy?
15. In what ways does information technology help an organization handle the present?
16. How is the past remembered in an organization? Where are the data typically stored?
17. In what ways does information technology help an organization prepare for the future?
18. What types of systems are used to help an organization prepare for the future? Give two examples of how these systems accomplish this goal.
19. In what other way does information technology help organizations beyond handling the present, remembering the past, and preparing for the future?

DISCUSSION QUESTIONS

1. Discuss how Lands' End is using information technology to avoid demand, efficiency, or innovation risks.
2. Discuss a company that has gone out of business because it failed to practice creative destruction, or describe a company that did practice creative destruction or extension.
3. Discuss how Monster.com is using information technology to be successful in the networked economy.
4. Discuss some of the economic relationships of which you have been a part as either a purchaser, user, or person returning some product.
5. Discuss ways in which FarEast Foods could take advantage of information technology in setting up www.fareastfoods.com.

RESEARCH QUESTIONS

1. Research the history of Haagen-Däz ice cream or Edmunds.com. Write a two-page paper on your findings, concentrating on the concept of creative destruction or extension.
2. Research the demise of a dot-com company with which you are familiar. Write a two-page paper describing its failure to handle demand, efficiency, or innovation risks.
3. Look into both Moore's and Gilder's laws and determine how well they have predicted the actual growth in chip speed and bandwidth since they were first suggested.

Claire stomped the mud off her boots and shrugged off her backpack before entering through the screen door to the kitchen. The aroma of stew cooking had made the walk up the hill to the house seem longer than any of the mountain trails she had hiked over the last three days. A warm shower and one of Alex's home-cooked meals would be the perfect way to wind down, she thought.

Turning from the stove, Alex greeted her with a kiss: "Hi dear, I didn't expect you so soon. How was your adventure?"

"Not bad at all. I caught a glimpse of a black bear on the bald above Panther Creek Falls," Claire replied. "And the trip was just long enough to give the new equipment a good field test. The new Featherlite 2000 performed better than most other backpacking tents I've tried. I'll be touting its virtues in my next review."

Claire and Alex Campagne are the owners of a small shop specializing in equipment and provisions for outdoor recreation. Their location near the New River Gorge of West Virginia places them in the heart of an area rich with the possibility of outdoor adventure. Hiking, camping, rock climbing, and whitewater rafting combine with the rugged beauty of the region to entice visitors from all over the East Coast. Having arrived in the area to sample the world class white-water of the Gauley and New Rivers six years ago, the couple fell in love with the place and decided that here they could begin their dream of owning their own business. They left two fairly lucrative jobs, pulled up stakes, and found a home/store just off the main highway near the New River bridge.

Since starting their shop almost five years ago, the Campagnes have become well known for their knowledge of the region and all manner of outdoor equipment. Spending most of their personal time in the outdoors, they try to personally use most of the equipment that they sell. Their well-earned reputation has spread slowly and mostly by word of mouth. In addition to their often sought-after information about the region and equipment, the hot muffins that the Campagnes bake on weekend mornings help to provide a homey ambience and have made the shop a favorite starting point for the many adventurers who stop by for provisions and a chat with the owners.

Not satisfied with the small success they have enjoyed in starting and maintaining Wild Outfitters, Claire has been interested in finding some way to increase sales. Due to their location near the action and the nature of their business, most transactions occur on the weekend. Thus, the couple feels that some business is lost to the big sports stores in the city, such as Eastern Mountain Sports and REI. One problem they foresee with increased sales would be a need to maintain more and longer hours at the store. Currently, they have been opening from Thursday to Sunday. This schedule leaves the rest of the week to enjoy the outdoors and test equipment. More hours serving customers would leave less time for these pursuits.

Another idea the Campagnes have been toying with is developing their own line of products under the Wild Outfitters brand name. Although they haven't yet worked out the details, this product line could include anything from guide books to outdoor clothing to camp cuisine. Claire and Alex have tried a few items in the store that sold well. To be successful, the product line would need to have better advertising than the word-of-mouth process that they have relied on so far.

Lately, Alex has been playing around with the Internet and posting a personal home page. Besides a brief background of the store, the site includes reports describing the outdoor sports that he and his wife have enjoyed. These accounts come complete with photos, maps, descriptions of the area, and brief reviews of the equipment used. The couple has been wondering how they might turn this new hobby into an asset for their business.

As she began to wash for dinner, Claire thought more about what she would write about the equipment in her next trip report. Yes, she thought, the tent worked wonderfully, but the ache in her shoulders told her that the review of the backpack would not be so kind. Perhaps a few well-timed moans and groans will convince Alex to help her with a nice backrub while she types.

Think About It

1. How could electronic commerce benefit the Campagnes?
2. What special knowledge do the Campagnes have that may help them to succeed in the networked economy?
3. Which features would you include on WildOutfitters.com to help customers and to make the site more efficient while also increasing opportunities for the Campagnes to receive innovative suggestions?
4. Can Wild Outfitters be thought of as a system? If so, identify the inputs, outputs, processes, and feedback of the current business.
5. What steps of creative destruction might you take with Wild Outfitters?
6. Look over your answers to the previous questions and develop business goals for the Campagnes. Make sure that your goals are measurable, and include a tentative time for achieving each goal.

Hands On

7. Think about the type of site you would like to start on the Web. Provide a general description of your site. What product or service would you sell? What would you call it?

Hands On

8. Check one of the domain registration sites on the Web to see if your name is available. Research domain registration and briefly describe what you would need to do to register your site's name.

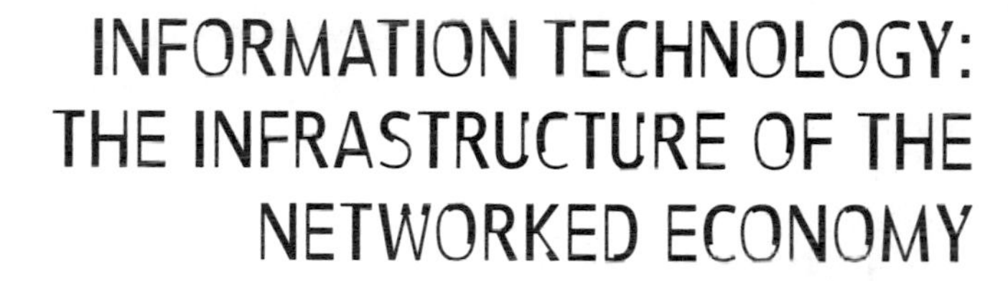

INFORMATION TECHNOLOGY: THE INFRASTRUCTURE OF THE NETWORKED ECONOMY

LEARNING OBJECTIVES

After reading this chapter, you will be able to answer the following questions:

How does information technology support the networked economy?

What are the two primary parts of any computer, and what roles do they play?

What are the four major hardware elements of the computer, and how do they process data into information?

Why is software necessary to use a computer, and what are the two primary types of software?

How has computer processing evolved over time, and what are the three types in use today?

Dreamworks Animates with Linux

In the animation wars between Disney and Dreamworks, Dreamworks is depending on Linux-based computers to give it a leg up. At each of the three animation production centers that Dreamworks SKG operates, computers running the Linux operating system are crucial to the animation process. These centers are specialized for the type of animation being created—Bristol, United Kingdom for claymation products such as *Chicken Run*; Palo Alto, California, for computer graphics projects such as *Shrek* and *Antz*; and Glendale, California, for traditional animation. Currently, Dreamworks has more than 600 computers running Linux, partially because of the dramatically reduced cost associated with it. However, the main reason for using this operating system is the increased productivity animators are finding with Linux-based computers. In fact, they are replacing five-year old special-purpose animation workstations with PCs running Linux because of the noticeable increase in speed.

An animator's computer is quite different from the PCs that most people use: An animator needs a high-performance workstation with two monitors, a high-performance graphics system, and specialized software for motion picture production. To typical computer users, the animator's software tools may be unfamiliar. For example, Dreamworks' IT staff created a special-purpose tool called ToonShooter to run under Linux. ToonShooter is used to scan the paper sketches created during the production process into the animator's computer. Dreamworks now has 60 versions of ToonShooter in use at its three production facilitites.

Workstations such as this one were used to create the animation in the movie, *Shrek*.

Source: Robin Lowe, "Dreamworks feature Linux and animation." *Linux Journal.com*, July 10, 2001, http://www.linuxjournal.com.

Elements of Information Technology

Without the help of information technology, the Dreamworks animation studios would not be able to create their popular animated motion pictures. In fact, most companies today rely on information technology to help produce their products, deliver their services, or carry out interal functions. Recall from Chapter 1 that *information technology (IT)* is the *infrastructure* of the networked economy; it consists of the elements necessary for the networked economy to exist and for transactions to take place. IT is also necessary for accessing data and information and for sharing data, information, and knowledge with other people in the networked economy. Because the computer serves as the basis for most of information technology, let's look at the computer more closely.

A Closer Look at Computers

hardware
The electronic part of the computer that stores and manipulates symbols under the direction of the computer software.

software
One or more programs that direct the activity of the computer.

program
A series of instructions to the computer.

All computer-based information systems are combinations of two elements: hardware and software. The computer's machinery is referred to as **hardware** and consists of chips and other electronic devices. Typical hardware includes system boxes, disk drives, monitors, keyboards, printers, and so on. Although computer hardware has continuously become smaller, faster, and cheaper, it can do *nothing* without the instructions commonly referred to as **software**. A common saying helps to differentiate hardware from software: "If you bump into it, then it's hardware!" Software is composed of one or more sets of instructions called **programs**. The many types of computer systems in use in modern organizations are composed of a multitude of programs that work alone or in groups to meet the needs of the organization. Machines with embedded chips, such as autombiles, microwave ovens, and mobile telephones, have programs already built into the chips.

Two features have strongly contributed to the rapid growth of computer use: speed and value. The speed of a computer is limited only by the speed at which electrical signals can be transmitted between chips, enabling desired operations to be carried out in a fraction of the time needed to do them manually. This ability enables even personal computers to execute millions of operations per second in carrying out software instructions. In terms of value, the price of computers has dropped dramatically at the same time as their power and speed have increased. Computers are even more valuable when they are connected via networks.

The opening case described how Dreamworks uses sophisticated hardware in the form of personal computers with multiple monitors and high-performance graphics capabilities. The company also uses a special-purpose hardware item—the ToonShooter—to transfer paper sketches to the computer. In terms of software, Dreamworks has moved to the Linux operating system to run its animation software, thereby increasing its productivity.

A Closer Look at Hardware

mainframe
A very large and fast computer that requires a special support staff and a special physical environment.

Computers are the backbone of the networked economy, and they come in all shapes and sizes. They are roughly classified by size, ranging from large computers usually referred to as mainframes that handle many users simultaneously to personal computers (either desktop or laptop) to handheld personal digital assistants. **Mainframes** are large computers that have been around since the dawn of computing in the 1950s and provide the heavy-duty processing required by many corporate activities.

This mainframe handles very large and complex pro cessing tasks.

personal computer (PC)
A small, one-user computer that is relatively inexpensive to own and does not require a special environment or special knowledge for its use.

minicomputer
A computer that is between a mainframe and a personal computer in size.

personal digital assistant (PDA)
A form of handheld computer that does not require a separate keyboard or monitor.

nanosecond
One-billionth of a second.

Personal computers (PCs) are relatively newer, smaller computers that have been around for a little more than 20 years and are responsible for making computing available to individuals. In organizations, they are typically connected over local area networks (LANs). Between the mainframe and the PC is the **minicomputer**, which is used to provide centralized computer access to multiple users but without the level of support required by a mainframe. Handheld computers, or **personal digital assistants (PDAs)**, and mobile telephones are the smallest computers in use today and are becoming an increasingly important part of the overall technology used in organizations. They are widely employed as electronic personal information managers (PIMs), replacements for calculators, paper calendars, address books, telephone books, and tools for keeping notes on meetings, events, contacts, and so on. When combined with a modem, these devices can be used to check e-mail and surf the Web. Regardless of their size, all computers operate via electronic circuitry, have similar input and output devices, and need software to run. Mainframes and PCs will be discussed in more detail in a later section in this chapter.

Most operations performed by hardware are electronic—that is, there are few moving parts. Instead, the operations are accomplished by millions of transistors that act as electronic switches. Because the operations are electronic, hardware is both fast and accurate when carrying out the instructions of the software. The speed of many operations performed within the computer is limited only by the speed at which transistors can change from "on" to "off" and is measured in billionths of a second **(nanoseconds)**. These operations obey physical laws that do not change from operation to operation, which produce the computer's high degree of accuracy. Nevertheless, if the software instructions are incorrect—even in a single place—incorrect results will occur.

All IT hardware—whether in a large computer, PC, handheld device, mobile phone, or other chip-based device—consists of three elements: *input, processing,* and *output*. Most IT hardware has a fourth element known as *secondary storage*. Figure 2.1 shows a typical computer with a system unit, a keyboard and mouse for input, and a monitor, printer, and speakers for output.

Figure 2.1 Typical computer system

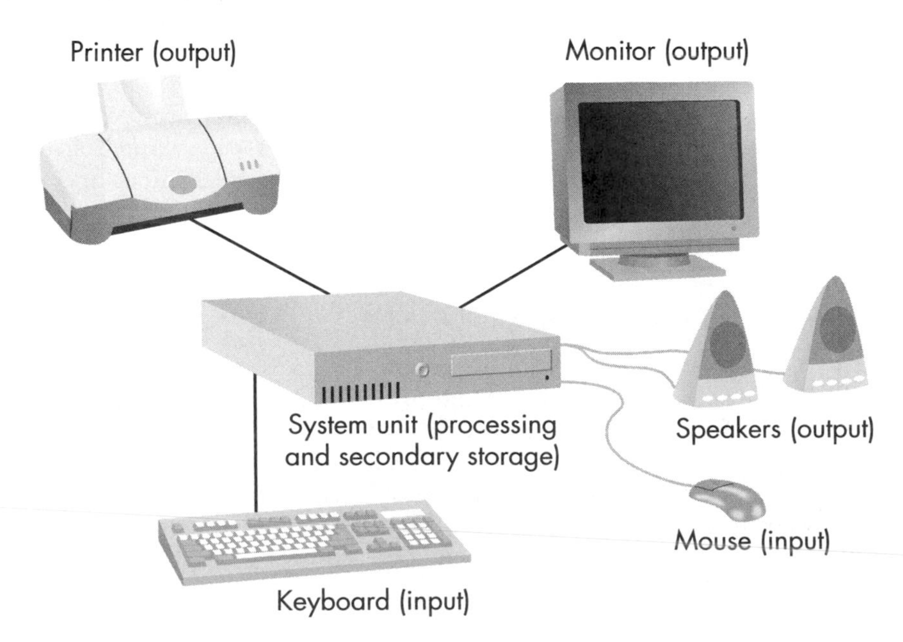

system unit
The main case of the PC housing the processing unit, internal memory, secondary storage devices, and modems.

peripherals
Devices such as printers and speakers that are attached to a PC or a computer network.

Housed within the **system unit** (and often invisible from the outside) are the processing unit, internal memory, secondary storage disk drives, and a modem for telecommunications input and output. Some systems require additional hardware devices to generate the video displayed on the monitor, and many now come with built-in connections for LANs. All system units include a number of connections for the keyboard, mouse, monitor, printer, and other devices known as **peripherals**. All new systems come with special types of connections that can be used for a variety of peripherals, which can be swapped in or out as needed. You have probably used a computer system similar to the one shown in Figure 2.1 and are aware of the terminology used to describe it.

Now, let's take a closer look at the four key elements of computer hardware: input, processing, output, and secondary storage. At the same time, let's look at how computer hardware is used to process data into information.

Input

ainput
Receiving the data to be manipulated and the instructions for performing that manipulation.

keyboard
An input device made up of keys that allow for input of alphanumeric and punctuation characters.

mouse
An input device—about the size of a mouse and connected to the computer by a long cord or wireless signal—that allows input through movement over a flat surface.

bar codes
Combinations of light and dark bars that are coded to contain information.

smart card
A card, containing memory and a microprocessor, that can serve as personal identification, a credit card, an ATM card, a telephone credit card, a critical medical information record, and cash for small transactions.

touch screen
A type of input device that enables the user to touch parts of the computer screen to select commands.

Smart cards such as this one have a chip, instead of a magnetic stripe, on which to store data.

Input enables hardware to receive data to be processed and the software instructions for processing those data. Input is critical to handling the present because it represents the first step in dealing with virtually every transaction that an organization carries out, whether it be making a sale, accepting a shipment into inventory, paying an employee, or transmitting payroll taxes to the government. If the organization hopes to adequately handle the present, input must be handled efficiently. The importance of input was shown dramatically in the November 2000 U.S. presidential election, during which questions about punch card input from the state of Florida delayed a final decision on the winner into early December. As a result of problems with punch card input, both the states of Florida and Georgia are moving to touch screen input for the 2004 elections.

As shown in Figure 2.1, the most common input devices for a computer are a **keyboard** for entering data and instructions, a **mouse** for pointing to elements on the computer screen, a modem, and some forms of secondary storage. The keyboard is a data input device widely used for creating documents, spreadsheets, databases, and so on. It is also the primary method of entering the instructions to create the programs that make up software. Although not shown in Figure 2.1, a microphone, scanner, or digital camera may also be connected to the computer for input.

Bar codes are far and away the most common and accurate form of data input for transactions in business and industry. In this system, a bar code reader converts data in the form of a bar code into a form the computer that is running the transaction system can use. When all items in an order have been processed, a common form of payment is the credit card. Credit card input comes from a magnetic strip on the card that is used to charge a customer's account for the amount of a purchase. Bar code and credit card readers have become so commonplace that some stores have set up self-service centers where customers can scan their items and swipe their cards with no help from an attendant. Likewise, self-service gas pumps that include a credit card reader are quite common.

Whereas magnetic strip credit cards are the predominant form of input for consumer transactions in the United States, much of the rest of the developed world has already moved to smart cards. **Smart cards** use chips to store a great deal more information than is possible with the magnetic strips on credit cards. They can also store funds, making it possible to use them without contacting a bank for verification. And, because they are programmable, it is possible to add funds to smart cards at ATMs.

Touch screens are another popular form of input device, especially when selections must be made. This type of input is widely used with ATMs in Europe, in contrast with the keypad used in much of the United States. As noted earlier, touch screen input will soon replace punch card input in elections in at least two states. Input from mobile phones is also becoming popular in Europe. Users there can access their bank and investment accounts and even pay for a car wash using their mobile phones.

Europeans frequently use mobile phones to carry out financial transactions.

modem
A communications device that converts computer characters into outgoing signals and converts incoming signals into computer characters.

The **modem** is a telecommunications device that can be used for both input and output. It is essential for connecting the computer to the networked economy because it is often used to link the computer to the Internet. In business settings, modems are widely used to connect credit card scanners to bank computers so as to obtain authorization for the use of the credit card for a purchase. Different types of modems are used to connect the computer to a telephone line or to a broadband television cable.

An important form of input (and output), especially in a business setting, is the Ethernet LAN connection. This connection enables office workers to communicate with co-workers also connected to the LAN as well as to connect to the Internet. Newer PC systems come with built-in Ethernet LAN connections that look like large telephone jacks into which a LAN cable is inserted. Ethernet connections are also used with the cables originating at high-speed connections.

Processing

Once input has taken place, the hardware must use programmed instructions to convert the data into information. Although all other elements of hardware are important, it is the processing step that separates the computer and computer-based systems from other systems. Converting data into information is the second step in the process through which computers handle the present. For example, every transaction involving a bar code reader must process the data stored in the bar code into the item name and price information that is then displayed at the checkout stand and entered into an inventory database. In addition to ensuring that transactions are handled efficiently in the present, the processing component of hardware must carry out myriad other tasks to enable people to use information technology so as to remember the past and prepare for the future. Processing is crucial to ensuring that data and information are stored in such a way that they can be found when needed or used to develop forecasts of future conditions.

central processing unit (CPU)
The part of the computer that handles the actual processing of data into information.

random access memory (RAM)
The section of memory that is available for storing the instructions to the computer and the data to be manipulated.

Processing of data into information in computers is accomplished using two types of chips: processing chips and memory chips. The actual processing of data into information takes place on the processing chips. Because the instructions to process data items must be carried out sequentially, it is necessary to store the instructions and data as well as the resulting information internally on memory chips. In computers, the processing chip or chips make up the **central processing unit (CPU)**, and the internal storage is carried out on **random access memory (RAM)** chips. More than 90 percent of CPU chips in use today are based on a standard created by Intel. Although not all such machines use Intel chips, they use chips that will run the same software as Intel chips.

This is the main or motherboard for an Intel Pentium 4-based PC.

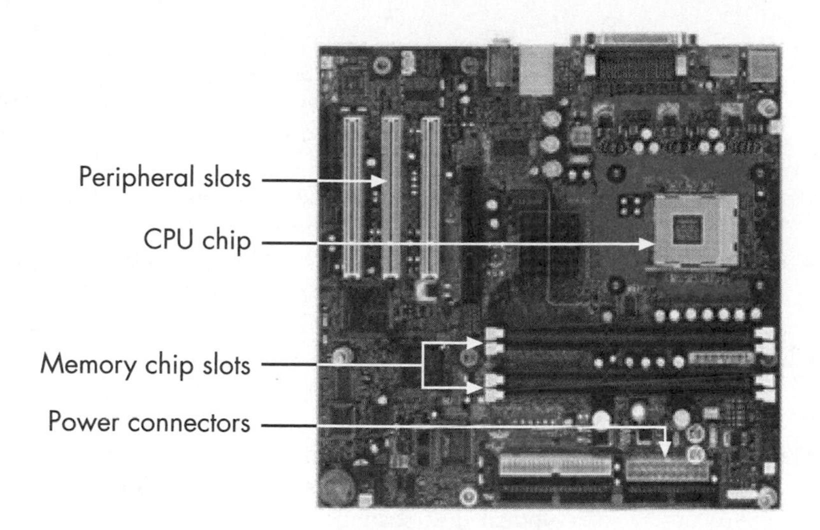

read-only memory (ROM)
The section of memory that is placed in the computer during the manufacturing process and remains there even after the computer is turned off.

firmware
Instructions on a ROM chip.

In addition to CPU chips and RAM chips, an important type of memory chip is known as read-only memory. **Read-only memory (ROM)** chips have instructions built into them at the factory and are an important component of many types of information technology. In computers, ROM chips carry out essential operations such as starting the computer and storing information on how the computer interacts with other hardware elements—keyboard, monitor, printer, secondary storage, and so on. In other types of information technology, in which there is no need for RAM or secondary storage, the ROM chips contain all of the software needed by the device to operate. ROM chips are used in machines such as mobile phones, PDAs, automobiles, and CD and DVD players. The software stored on ROM chips in these devices is often referred to as **firmware**.

Output

output unit
A unit that provides the result of processing for the user.

monitor
A cathode ray tube or flat-panel output device that displays output.

printer
An output device that places words, symbols, and graphics on paper.

speakers
Devices designed to broadcast sounds.

Because processed information is worthless unless it is provided to the user in some way, the computer must have an output unit. The **output unit** provides the result of processing in some useful form. Common forms of computer output include information displayed on a **monitor**, printed on paper by a **printer**, or broadcast from audio speakers. In addition, output is often sent to secondary storage for permanent storage or transmitted over a network using a modem. Output to the small screen on a mobile phone is also popular, especially in Europe.

A monitor is required for almost any computer system for two reasons. First, the data or instructions input from the keyboard or other input device are shown on the monitor. Second, the monitor is an almost instantaneous outlet for the results of processing. Two types of monitors are especially popular today: cathode ray tube (CRT) and flat-panel monitors. The latter monitors use the same liquid crystal display technology as is employed in laptop computers.

Printers are necessary to obtain a hard copy of a document or the screen contents. Two types of printers commonly used with PCs are ink-jet printers and laser printers. The ink-jet printer sprays tiny droplets of ink to create letters and graphic images. The laser printer works much like a copier except that, instead of photocopying an existing document, it converts output from a computer into a printed form.

Speakers are useful for listening to the audio track while watching movies on DVD or over the Internet, for providing sound to games, for listening to audio CDs, for listening to talks broadcast over the Internet, or for conducting telephone calls over the Internet.

Secondary Storage

secondary storage
Storage area that is used to save instructions, data, and information when the computer is turned off.

disk drive
A device that writes information to or reads information from a magnetic disk.

hard disk
A type of magnetic disk that is fixed in the computer.

CD-ROM disks
A form of read-only optical storage using compact disks.

DVDs
Digital versatile disks used for storing data as well as audio and video programs.

CD-RW drives
Optical disk drives that can write to a CD-ROM as well as read from it.

Because the internal memory of a computer is both limited and nonpermanent, most computers include some form of secondary storage. **Secondary storage** is permanent storage, usually on a magnetic disk or an optical disk device. With either of these devices, internal memory can access stored data, information, or instructions when the CPU deems it necessary. Because the secondary storage unit must locate the needed material, read it, and then transfer it to internal memory, secondary storage is a much slower form of memory than internal memory is; this slow transfer of information, however, is balanced by the virtually unlimited storage capacity.

Magnetic-disk secondary storage is composed of metal or plastic covered with an iron oxide whose magnetic direction can be arranged to represent symbols. This magnetic arrangement is carried out on a **disk drive**, which spins the disk while reading from it and writing information to it. Magnetic disks can be either fixed (also known as **hard disks**) or portable (like floppy disks and Zip disks).

Optical-disk secondary storage comes in the form of compact disk-read only memory (**CD-ROM disks**) and digital versatile disks (**DVDs**), both of which are portable and primarily *read-only* storage devices. Many computers sold today come with CD drives that can read both types of disks. Others come with **CD-RW drives** that allow users to burn their own CD-ROMs. Such drives are useful for storing large amounts of data and information. In addition, they enable users to create personalized music CDs from commercial CDs or from other sources that allow computer users to share music.

Conceptual Computer

To understand how input, processing, output, and secondary storage work together to process data into information, consider the conceptual computer shown in Figure 2.2. In addition to the hardware elements, this conceptual computer shows the flows into, within, and out of the computer. The processing of data into information begins with the input unit sending data and instructions to the internal memory unit. Next, the instructions and data are sent to the processing unit as they are needed. If necessary, the processing unit may request that additional data or instructions be retrieved from secondary storage. Once the processing is completed, the resulting information is sent to internal memory from the processing unit. The internal memory unit then transfers the information to the output unit for display, printing, or listening. The information may also be stored in secondary storage for later use or additional processing.

Hard drives like this one provide the primary secondary storage for PCs.

Figure 2.2 Conceptual computer

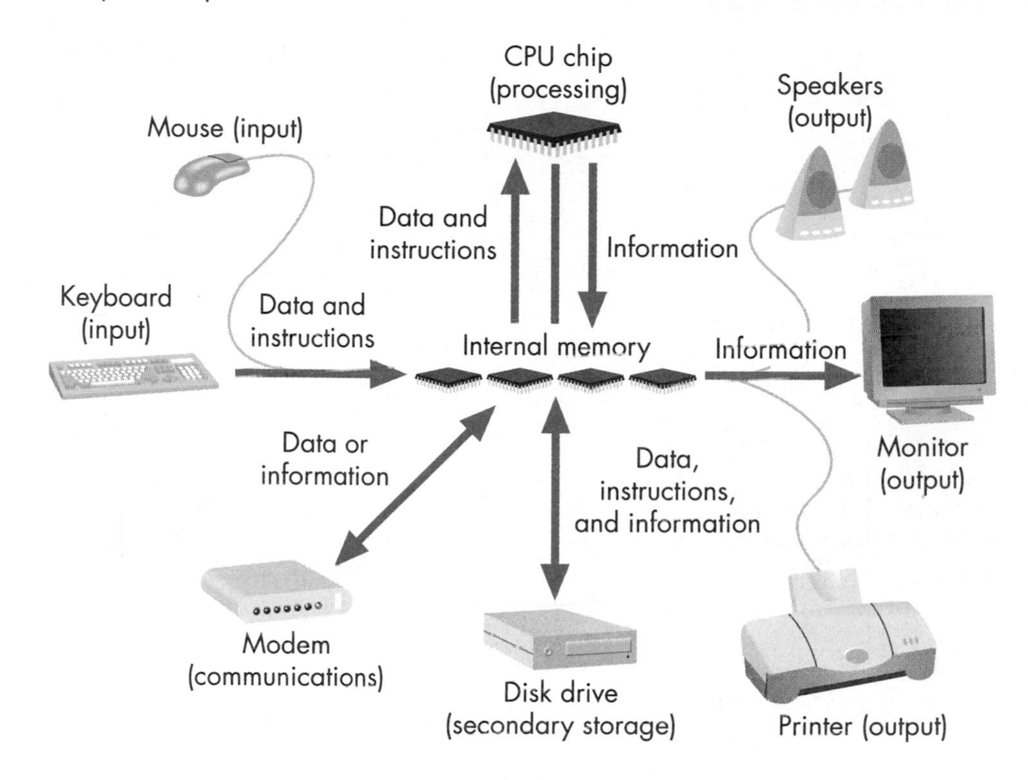

Bits, Bytes, and Binary Numbers

binary number system
Base-2 number system based on zeros and ones.

bit
The basic unit of measure in a computer; contraction of **BI**nary and digi**T**.

byte
A group of eight bits—equivalent to a single character.

ASCII
(pronounced "as-key") An acronym for *American Standard Code for Information Interchange.*

Because humans have 10 fingers, we use a base-10 number system to perform calculations. Similarly, we use written symbols to communicate ideas to other people. Instead of using written characters, information technology hardware employs transistorized on/off switches on chips to store and proceess numbers and symbols. Because these switches have two states—on or off—they carry out calculations using base-2, or the **binary number system**, instead of the base-10 system that humans use. Each transistorized switch corresponds to one **bit** (**BI**nary digi**T**) of storage. For example, the decimal number 20 is 10100 as a binary number. Just as base-10 numbers can be added, subtracted, multiplied, divided, and so on, so, too, can binary numbers.

In contrast, to represent and store letters, punctuation marks, and special symbols internally and on secondary storage, standardized representations were created. A standard representation involves using a group of eight bits, called a **byte**, in which each bit pattern represents a given symbol. Why was eight bits chosen? Possibly because 256 characters can be achieved with this number, which was thought to be a sufficient set. Although several patterns of bits have been suggested, the standard code for personal computers is **ASCII** (pronounced "as-key"), which is an acronym for *American Standard Code for Information Interchange*. For example, the letter *A* is coded as 01000001 in ASCII. Because each character can be represented by one byte in ASCII, the terms *byte* and *character* are used interchangeably.

ASCII codes are used to transmit information between the keyboard and internal memory and between internal memory and the display screen. They are also used for storing information in internal memory and secondary storage on personal computers. Table 2.1 shows the ASCII codes representing several letters and symbols. Note that the ASCII representation of a number is different from the corresponding binary form of that number and that the lowercase and uppercase letters have different ASCII representations.

Table 2.1 ASCII Code

Character	ASCII Code	Character	ASCII Code
0	00110000	a	01100001
1	00110001	b	01100010
2	00110010	y	01111001
A	01000001	z	01111010
B	01000010	!	00100001
Y	01011001	;	00100001
Z	01011010		

Unicode Worldwide Character Standard (Unicode)
An international coding scheme that can process and display written texts in many languages.

kilobyte (KB)
A measure of computer storage equal to 1024 bytes or approximately one-half page of text.

megabyte (MB)
A measure of computer storage equal to approximately 1 million bytes or 500 pages of text.

gigabyte (GB)
A commonly used measure of computer storage, approximately equal to approximately 1 billion bytes of storage or 500,000 pages of text.

terabyte
A measure of computer storage equal to approximately 1 trillion bytes or 500 million pages of text.

ASCII was developed to encode the English language, and many other encoding schemes were developed for other languages. To address the need for a single, consistent encoding scheme, the **Unicode Worldwide Character Standard (Unicode)** was created as a system to process and display written texts of the many languages of the modern world. Currently, Unicode has 34,168 distinct coded characters derived from 24 supported language scripts that cover the principal written languages of the world. Additional work is under way to add the few modern languages not yet included in this standard. The Unicode standard has been adopted by industry leaders such as Apple, Hewlett-Packard, IBM, and Microsoft. It is supported by many types of computers, all modern Web browsers, and many other products.

As discussed earlier, data, information, and instructions must be stored in a computer's internal memory as well as its secondary storage. The amount of storage in both is measured in **kilobytes (KB)**, **megabytes (MB)**, and **gigabytes (GB)**. A kilobyte is approximately 1000 bytes, a megabyte is approximately 1 million bytes, and a gigabyte is approximately 1 billion bytes. Trillions of bytes **(terabytes)** of data are now commonly being stored in many commercial situations. Internal memory (RAM) in PCs is typically measured in megabytes, whereas most hard drives today store data in tens of gigabytes. Table 2.2 shows the relationships between bits, bytes, and the various amounts of memory and secondary storage (where the number of pages refers to text pages and is approximate).

Table 2.2 Memory Relationships

Memory Term	Amount of Memory	Amount Represented
1 bit	1 transistor on/off switch	-
1 byte	8 bits	One character
1 kilobyte	1024 bytes	One-half double-spaced page
1 megabyte	1024 KB (1,048,576 bytes)	500 pages
1 gigabyte	1024 MB (1,073,741,824 bytes)	500, 000 pages
1 terabyte	1024 GB (1,099,511,627,778 bytes)	500,000,000 pages

To read an analog thermometer, you need to estimate the level of the mercury against the scale; reading a digital thermometer is much easier.

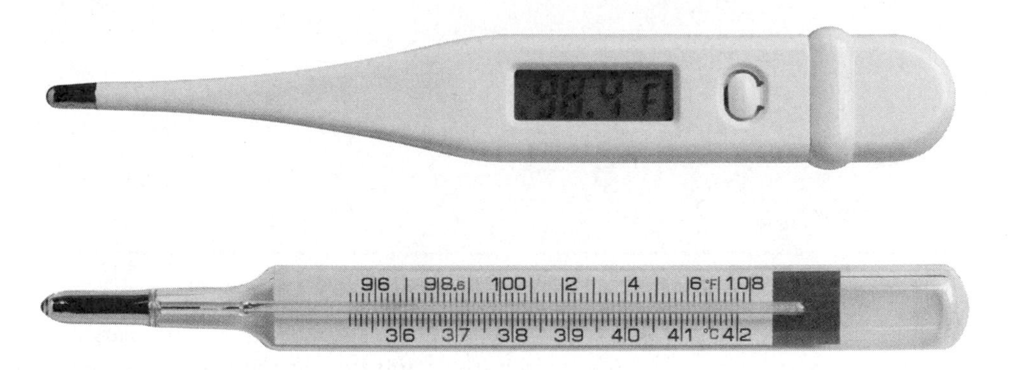

Digital Storage

digital devices
Devices that process and store data in a binary (0-1) form.

analog devices
Devices that convert conditions, such as movement, temperature, and sound, into analogous electronic or mechanical patterns.

Computers are **digital devices** based on the binary (0–1) system. Over time, there has been a concomitant move from paper storage to digital storage of data and information. Digital storage of information is much more compact than paper storage and is less subject to deterioration. In addition, it is possible to create backup versions of the information. To understand the benefits of digital storage, it is helpful to compare it to storage using **analog devices**, which convert conditions, such as movement, temperature, and sound, into analogous electronic or mechanical patterns.

You may have used both analog and digital devices to play music. For example, an analog audiocassette uses physical changes in the magnetic field on a tape to represent the music or other sounds that have been recorded. In contrast, a compact disk or minidisk contains all sounds in a digital form that is played using a laser light.

Although analog machines can capture the subtle nature of the real world, they cannot produce repeated copies of output without showing marked signs of deterioration. In contrast, digital devices provide a level of consistency in manipulation, storage, and transmission that is not possible with analog devices. Also, a compact disk or DVD can be played thousands of times without loss of quality, something that is not possible with an audiocassette tape. Like the trend seen in information storage, a rapid movement is occurring from analog to digital technology in many other devices, including mobile telephones and cameras. Table 2.3 compares analog and digital services for the same needs.

Table 2.3 Analog versus Digital Storage Methods

Purpose	Analog	Digital
Reproduce music or speech	Cassette tapes or vinyl records	Compact disk, minidisk, or MP3 devices
Reproduce video	Analog video tape	Digital video tape, DVD, or DivX CDs
Send audio	Plain old telephone service (POTS), radio, and analog mobile telephones	Digital mobile telephones, DSL service
Send audio and video	Broadcast and analog television	Direct broadcast satellite service, digital cable service
Photography	Film cameras, analog video recorders	Digital cameras, digital video recorders
Make appointments	Paper calendar or appointment book	Personal digital assistants

1. List the hardware components of a conceptual computer.
2. Why are a byte and a character essentially the same thing?

A Closer Look at Software

Although hardware advances in recent years have been mind-boggling in terms of increased speed and storage and decreased cost, without software, information technology would be nothing more than a well-constructed combination of silicon chips and electronic circuitry. Recall that software is the general term for all of the instructions given to the computer or other chip-based devices by the user or the manufacturer. The idea of information technology without software has been described as being similar to a car without a driver and a camera without film. Regardless of your favorite analogy, hardware must have software to direct it.

The two major types of software are operating system software and application software. Operating system software controls the hardware and other software, and

Using Face Recognition to Catch Cheaters in Casinos

TECHNOLOGY ON THE EDGE

A major problem for any casino is catching the cheaters who like to take advantage of them. The average casino is very unsophisticated in its efforts to watch out for known cheaters. They typically depend on VCRs, TV monitors, and mugbooks to try to identify the known cheaters who mingle with thousands of honest visitors in the casino. Often, cheaters are able to make a hit on the casino and disappear before a manual search of the mugbooks can turn up a match.

Some casinos, however, have begun attacking the cheating problem by using **facial recognition technology** to look for known cheaters. This technology uses a system known as biometrics that focuses on the eyes of people in the casino; it translates the image of a face into a numerical code that can be compared to faces stored on a database. The same technology was used at Super Bowl XXXV in Tampa in January 2001 to scan the faces of visitors looking for wanted criminals attending the game. Although some civil liberties experts considered this application to be an invasion of privacy, when used in combination with casino surveillance cameras and databases of known cheaters, the system can be used to quickly recognize people who would like to take advantage of a casino.

The Trump Marina Casino in Atlantic City has been using a facial recognition system since 1997. Three days after it was installed, it helped the casino catch a gang of eight baccarat cheats. The Trump Marina Casino has a database of approximately 10,000 people who have been arrested, evicted, or thrown out of it or other casinos to which it compares the faces of players at various games. According to a surveillance supervisor at the Trump Marina, "Speedwise, it's unbelievable." This speed makes it possible to catch cheaters before they have a chance to steal from the casino.

Casinos aren't the only places where this technology is useful. Facial recognition technology systems are likely to become an important part of airport security systems designed to protect people from terrorist attacks.

Source: "Casinos use facial recognition technology." http://www.*CNN.com/Sci-Tech*, February 26, 2001.

Figure 2.3 Relationship between types of software

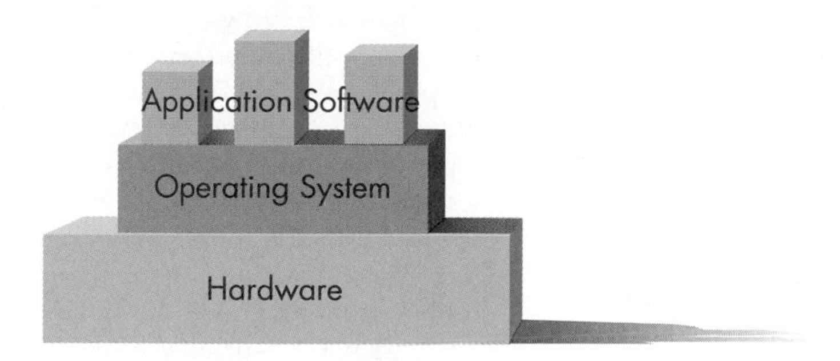

application software carries out the processing tasks. Application software includes word processing, spreadsheet processing, transaction processing, and other programs. Application software also handles such tasks as calculating payrolls, generating grade reports, running electronic commerce Web sites, forecasting sales demand, running the Internet, and controlling robots, among many other things. Except in special situations, both types of software are at work in a computer at the same time, each serving a different purpose. Figure 2.3 shows the relationships among the various types of software within the computer. Note that the operating system handles interactions between the application software and the hardware. Both types of software will be discussed after you learn some necessary software terminology.

Software Terminology

user interface
What the user sees on the screen.

command-driven interface
An interface in which the software responds when the user enters the appropriate command or data.

prompt
An indicator on the computer screen that data or commands should be entered.

Every type of software has its own method for entering data and commands called its **user interface**. The two most common user interfaces are command-driven and graphical. In software with a **command-driven interface**, the software responds when the user enters the appropriate command or data. A **prompt** ususally indicates to the user where on the screen a command or data is expected, and users must enter commands or data based on their knowledge of the software. The MS-DOS operating system was the predecessor to Windows and remains available in it for working with commands. In Figure 2.4, the first line shows the dir/w command entered at a prompt (the C:\> symbol), which you use to provide a listing of the files and folders on the hard disk.

Figure 2.4 Use of command-driven software

```
C:\>dir/w
 Volume in drive C has no label.
 Volume Serial Number is 1E2D-1EE2

 Directory of C:\

[WINDOWS]                   [FOUND.000]                 LFNBK.DAT
RESETLOG.TXT                [FOUND.001]                 [FOUND.002]
[dos]                       HALLOC.EXE                  [TOSHIBA]
[MOUSE]                     [Temp]                      [Program Files]
[Lotus]                     SCANDISK.LOG                [Acrobat3]
full_fe24.exe               [mhcappt]                   [MHCOLD]
[newmhcappt]                MHCAPPT.EXE                 README.TXT
VSVIEW.VBX                  [Spreadsheets]              [My Downloads]
[Documents and Settings]    [MSDOS7]                    MSDOS.SYS
CONFIG.SYS                  [My Documents]              [I386]
f1Scheduler.dat             [ADOBEAPP]                  AdobeWeb.log
[QUICKENW]                  [PICDISK]                   [TOC4]
[PSFONTS]                   [My Music]                  WS_FTP.LOG
salaries.txt                Copy of salaries.txt        FINES.TXT
              16 File(s)        677,907 bytes
              26 Dir(s)   7,613,521,920 bytes free

C:\>
```

graphical user interface (GUI)
An interface that uses pictures and graphic symbols to represent commands, choices, or actions.

icons
Graphical figures that represent operations in a GUI.

The type of user interface that is standard for today's personal computers is the **graphical user interface**, or **GUI** (pronounced "gooey"). A GUI is also standard for almost all Web browsers regardless of the type of computer. In a GUI, **icons** (pictures) on the monitor represent the functions to be performed. A user executes a function by positioning a mouse pointer over the corresponding icon and clicking a mouse button. Selecting an icon can cause a command to execute or a new screen of icons representing other options to be displayed. For example, if an icon of a file folder represents a file and an icon of a trashcan represents the erasure of a file, the user can erase a file by pointing to the file folder icon and dragging it to the trashcan icon. The GUI first became popular upon its inclusion on the Apple Macintosh in 1984; it was brought to Intel-based PCs in 1990 with the introduction of Microsoft Windows 3.0. Since then, GUI-based Windows has become the most widely used operating system for PCs. Similarly, all software written to run with the Macintosh and Windows operating systems uses a GUI. There is even a GUI for Linux-based computers. Figure 2.5 shows the use of a GUI on a Windows-based machine and an Apple Macintosh running the OS-X operating system to display the same application.

Operating System Software

operating system software
The software that manages the many tasks going on concurrently within a computer.

As shown in Figure 2.3, the software that controls the hardware is the **operating system software**. Operating system software is extremely important because it controls the operations of all other software, including application software, in addition to controlling the hardware.

Operating system software manages the many tasks occurring concurrently within a computer. These tasks include starting application software, supervising multiple applications running simultaneously, allocating memory to the applications, and managing the storage of programs, data, and information on secondary storage. Operating system software must manage various input and output devices as well as the frequent transfer of data and programs between internal memory and secondary storage, too. On computers that serve multiple users, the operating system may also manage the allocation of memory and processing time to each of the multiple users and provide security to each user and to the system as a whole.

Figure 2.5 Example of a Windows GUI (left) and a Macintosh GUI (right)

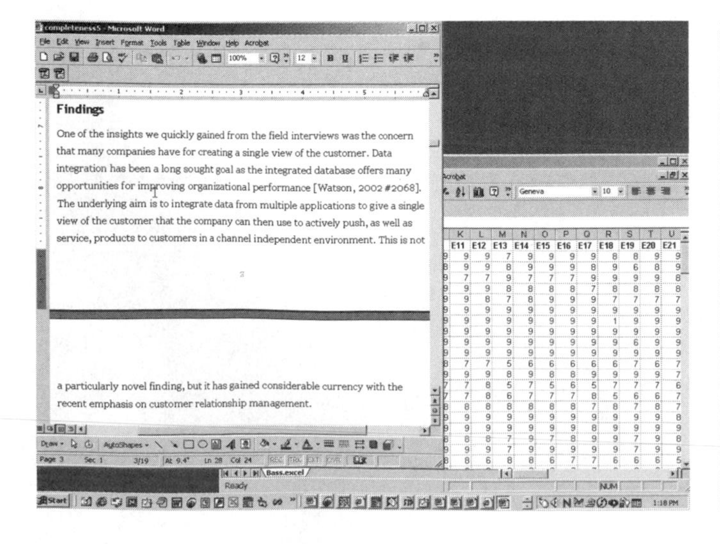

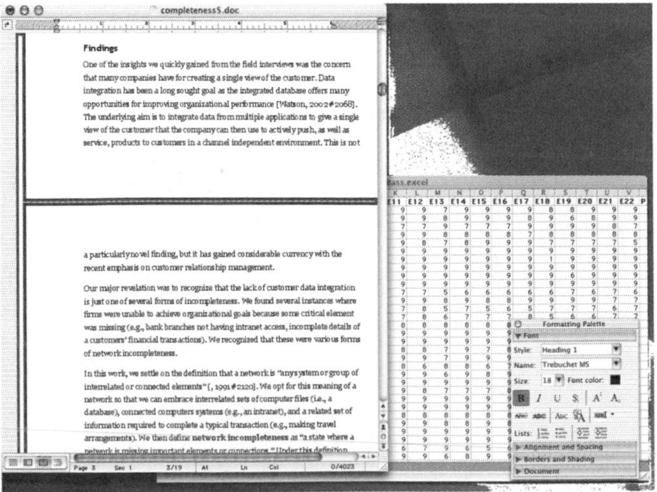

For PCs, the most widely used operating system is Microsoft Windows (95, 98, ME, 2000, or XP). More than 90 percent of all PCs use some form of the Windows operating system. Macintosh computers use their own operating system; the most current version is OS-X (for operating system, version 10). For business computers, the most popular operating systems are Windows NT or 2000, Windows XP, and UNIX/Linux.

Windows NT and 2000 are similar in operation to Windows 95, 98, and ME, but offer increased security features and the capability to manage multiple computers on a network. Windows XP, the newest version of Windows, includes many additional features.

UNIX, the oldest operating system, was originally developed by AT&T in the early 1970s for its internal computers. It is widely used for directing networks with many Internet computers running it. Linux is a freely available version of UNIX that has been deployed on a growing number of business computers as well as some home computers. For example, as noted in the opening case, Dreamworks switched to the Linux operating system because of the increased productivity it provided for PCs relative to specialized computers. The latest Macintosh operating system, OS-X, is actually a version of UNIX that has been adapted to the Macintosh hardware.

files
Programs, data, or information to which the user or software assigns a name.

Operating systems manage the storage of programs, data, and information on secondary storage and the many transfers from secondary storage to internal memory. Programs, data, and information are stored in structures called **files**, to which the user or software assigns a name. Application software carries out its designated purpose by working with files, and the operating system is responsible for making files available when they are needed by application software.

On personal computers, the types of files generated by application software have become interchangeable through standardization. For example, most word processing, spreadsheet, database, and presentation graphics software can now read files created by multiple applications. The standard for each of these applications is the type of file generated by the Microsoft Office Suite: .doc files for word processing, .xls files for spreadsheets, .mdb files for database software, and .ppt files for presentation graphics software. In addition, Web files use an .htm or .html extension and can be read from any browser regardless of the type of computer.

Comparison of Operating Systems

The operating system functions discussed previously apply to all computers regardless of size. However, some important differences exist among the operating systems for mainframes, networks, and personal computers. The primary differences between the three types of operating systems relate to the numbers of users and the complexity of the peripheral devices to be managed. Mainframes and network operating systems manage multiuser systems and a large number of peripheral devices, and both require a high degree of security. By comparison, most personal computer operating systems manage a single user and a handful of peripheral devices.

Mainframes must manage a large number of storage, input, and output devices, and a network operating system must manage numerous hard drives, backup devices, and printers. In contrast, a personal computer system usually has at most three or four storage devices, one keyboard and mouse, one printer, one monitor, and a set of speakers. Because of these differences, mainframe operating systems are extremely large programs that require the maintenance services of a staff of systems programmers. Network operating systems may also be quite large, requiring support from individuals with special training and certification on the particular network operating system. Personal computer operating systems are much less complex and must be able to operate without day-to-day maintenance. Finally, all three types of operating systems today are multitasking systems enabling the computer to work on more than one job or program concurrently.

Table 2.4 compares mainframe, network, and personal computer operating systems in terms of number of users, security, number of peripherals, number of tasks performed, support required, and gives an example of each.

Table 2.4 Comparison of Operating Systems Based on Computer Type

Feature	Mainframe	Network	Personal Computer
Number of users	Multiple	Multiple	One
Security	Sophisticated	Sophisticated	Minimal or none
Peripherals	Complex	Numerous	Few
Number of tasks	Many	Many	Many
Support	Systems programmers	Network-certified personnel	User
Example	OS390	Novell NOS	Windows XP

Application Software

By far, the largest amount of software available to the computer user is available in the area of application software. The applications for which software has been written cover the entire range of human activities. In fact, it is quite difficult to think of a single area of human endeavor for which application software has not been developed. This is especially true for business, industry, government, and personal uses. Just as our daily lives would be changed dramatically if no chips had been developed, most organizations would quickly grind to a halt if their application software ceased working. Various types of application software have been and are being written to carry out the three IS functions to deal with the business risks facing all organizations.

Application software for organizations comes in two primary forms: software developed commercially and software developed within an organization. Commercially developed software is usually purchased as **commercial off-the-shelf (COTS) software**. COTS is also sometimes referred to as **shrink-wrapped software** because it typically comes in a package containing the disk, instructions, and documentation all wrapped together. COTS is used for a wide variety of tasks in the modern firm, ranging from clerical and personal productivity to accounting and inventory control applications. The goal of many software companies is to market their software as COTS. Such software is also widely available over the Internet as downloads from Web sites.

commercial off-the-shelf (COTS) software
Commercially prepared software on disk(s), with instructions and documentation all wrapped together.

shrink-wrapped software
See *commercial off-the-shelf software.*

All of these software packages are commercial off-the-shelf (COTS) software.

The most widely used COTS software are word processing programs to create documents, spreadsheets to carry out financial and other quantitative analyses, database programs to manage lists and tables of data, and presentation programs to create presentations. A virtually endless variety of other application software exists, including Web browsers for accessing the World Wide Web, utilities such as virus protection programs and CD-ROM burners, games, and so on.

Software developed by a company specifically for its own use or developed for it by a contract software firm often has important competitive implications for the company. Although companies often purchase COTS software, such as word processing programs or spreadsheets for many tasks in their businesses, in many cases other software must be custom-developed to meet their particular needs. As a consequence, no matter how good off-the-shelf software may become, the demand for programmers to work for contract software firms or within organizations to develop software for their special needs will always exist and almost certainly grow as we move further into the networked economy.

Douglas Engelbart

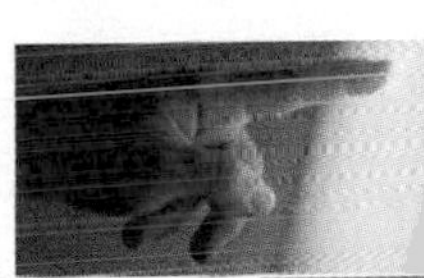

IT INNOVATORS

To find the origins of many of the GUI and network concepts to which users have become accustomed, you have to go back to the almost prehistoric (in computer terms) year of 1963, when Douglas Engelbart published a paper describing how to augment human intellect through the use of the computer. Almost 30 years ahead of its time, this paper described unknown concepts such as personal workstations, networks of users communicating with one another, and word processing. On the strength of this paper, Engelbart received a grant to try out some of his ideas.

Five years later, on December 5, 1968, Englebart made a 90-minute presentation to the American Federation of Information Processing Societies' Fall Joint Computer Conference that stunned the computing world. In this presentation, Engelbart and his visionary colleagues from the Stanford Research Institute (SRI) demonstrated a mouse-driven interface (his rolling pointer prototype was made of wood), hypermedia like that now used with the Web, multiple windows, outline processing, display editing, context-sensitive help, and many of the other computing functions now found at users' fingertips. Considering that computing in 1968 was completely command-driven and text-based on mainframes, this presentation was a world-changing event.

Thirty years later on December 9, 1998, the 73-year old Engelbart was honored at a symposium at Stanford University. Called *Engelbart's Unfinished Revolution*, the conference recalled that fateful day in 1968 when Doug gave what is now referred to as the "greatest demo of all time." Although personal computers have implemented his ideas on GUI, Englebart now believes that the networking of computers will have the greatest impact on the computing world.

Douglas Engelbart was a visionary who first created much of the information technology in use today.

Computer Programs and Languages

programming
The process of writing a series of instructions for the computer to follow in performing some specific task.

computer language
A language used by humans to give instructions to computers.

Earlier in this chapter, the idea that a computer manipulates and stores symbols by turning switches on and off was discussed. For the computer to know which switches should be on and which should be off, it must use a very specific set of rules called a program. All software—whether built into the machine by the manufacturer, purchased from a software development firm, or developed within the organization—is created by one or more persons using a process known as **programming**. Because the hardware knows nothing other than what it is told by the program, the program must be quite specific about the process that converts data into information. A software development team uses a step-by-step approach to creating programs and assumes that the computer does not know any of the steps. These steps are then converted into a program that is written in some **computer language**, which is a language used by humans to give instructions to a computer.

Programs may be written in any of several different computer languages. Some of the more commonly used languages and their uses are listed in Table 2.5 along with their full names (if different from the commonly used name). Each computer language, like a human language, has its own vocabulary and grammatical rules. Most, however, share a similar logical approach to communication with the computer.

Java is a new type of computer language that was originally developed to run on networks but is now being used in many different ways, particularly in e-commerce applications. As an example of a popular computer language, consider Figure 2.6, which shows an application in VB .NET for a video store. In the figure, you can see both the interface for this application and some of the corresponding programming instructions (code) necessary to implement the logic behind it.

Table 2.5 Commonly Used Computer Languages

Language	Full Name	Common Use
COBOL	Common Business Oriented Language	Writing business software, usually for mainframes
Fortran	Formula Translator	Writing scientific and engineering software
VB .NET	Visual Basic .NET	Writing software for PCs
C++	C++	Writing software for PCs or network servers
Java	Java	Writing software for all types of computers and applets to run on browsers

Figure 2.6 Computer program in Visual Basic

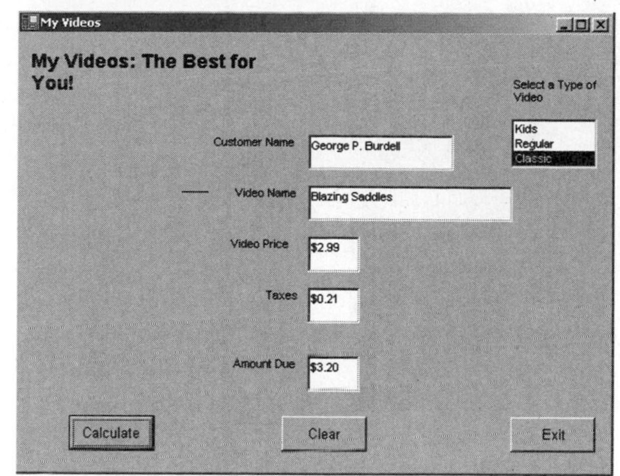

Application screen

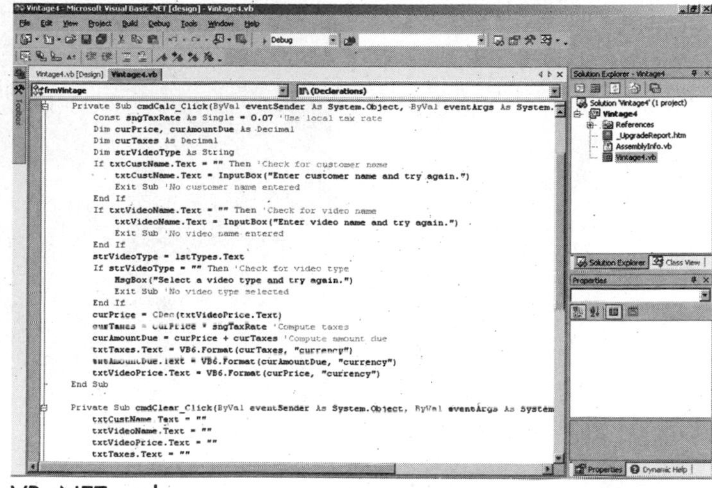

VB .NET code

1. Why is software an essential element of every computer? What are the two main types of software?
2. Why is programming necessary for software development?

Organizational Computing

To understand computing in the organizational environment, you must look at the three types of computing that coexist in the corporate world—mainframes, PCs connected by a LAN, and client/server computing.

Mainframes

The original mainframe computers were huge machines that took up large rooms. Today, however, they have shrunk in size to the point that some of the smaller mainframes resemble a large PC.

For years, some have predicted that mainframe computers would go the way of dinosaurs, but this prediction has not proved true. In fact, just the reverse is occurring as the need to process and store large amounts of information has grown. Customers of these large computers include any organization that requires large-scale transaction processing; production processing; or scientific computations; including airlines, manufacturers, retailers, finance, banking and insurance companies, telecommunications and travel service providers, and energy companies, many of which are upgrading from older mainframe systems. Although mainframes (and their smaller cousins the mini-computers) are larger and more powerful than desktop PCs, they work on exactly the same principles as the smaller computers, with operating systems running application software to accomplish the tasks required of them. As mentioned earlier, the operating systems for these machines must be able to handle requests from multiple users to accomplish multiple tasks.

Early mainframes like these required huge amounts of electricity to power their vacuum tubes.

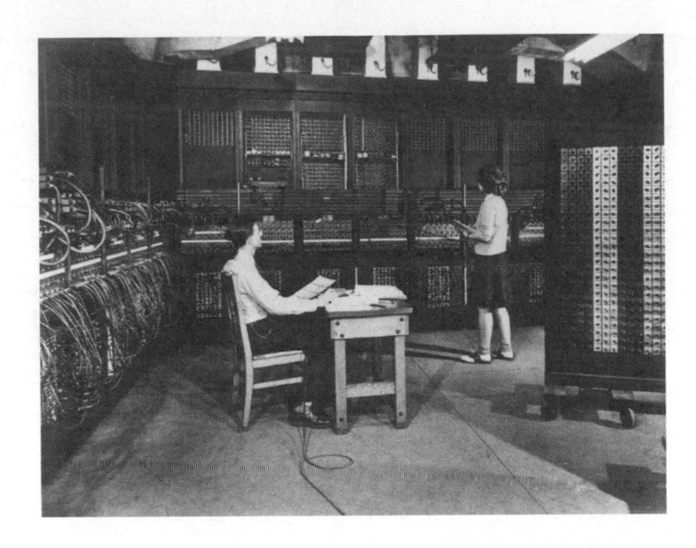

supercomputer
The biggest, fastest computer used today.

parallel processing
Processing that uses multiple CPU chips to perform multiple processing operations at the same time.

dumb terminal
A computer with no CPU or secondary storage. Its sole purpose is to serve as an input/output device for a mainframe.

remote job entry (RJE)
A site where data are stored locally on a PC and then submitted to the mainframe or supercomputer for manipulation.

host
A computer to which other computers are connected.

business-critical application
A software application that is critical to the continued existence of the organization.

Surpassing mainframes in terms of their computing capabilities are **supercomputers**, which specialize in high-speed processing and often use **parallel processing** to take advantage of the capabilities of multiple processors. Weather forecasters and scientists trying to model global warming trends, for example, tend to use supercomputers because the processing tasks are divided among the supercomputer's multple processors and run in parallel.

Users typically interact with mainframe computers through dumb terminals or PCs. A **dumb terminal** is a low-cost device that consists of a keyboard, a monitor, and a connection to the mainframe, but no processing chip, internal memory, or secondary storage. It is restricted to character input and output and cannot work with GUIs or pointing devices. Even when a PC is used as an interface to a mainframe, it typically emulates a dumb terminal and uses only a fraction of its computing power. Figure 2.7 shows the use of a mainframe with dumb terminals and a remote job entry site. At a **remote job entry (RJE)** site, data are stored locally on a PC and then transferred to the mainframe or supercomputer for computation.

This type of computing offers a number of advantages. The mainframe—or, as it is often called, **host**—provides massive centralized computing power and handles all of the data processing and storage. The mainframe has the capability to run applications that are critical to the continued profitability and existence of the organization—so-called **business-critical applications**. To access the mainframe, a user must have a user identification number and a password, resulting in a high level of security.

Mainframe computing also has a number of disadvantages. Mainframes are typically restricted to command-driven applications that involve only text and numbers, are very expensive to purchase and maintain, can stop all processing when they fail (or go down), and can be a bottleneck to processing when the demand is extremely high. Table 2.6 summarizes the advantages and disadvantages of mainframe computing.

Figure 2.7 Use of mainframe with dumb terminals and remote job entry

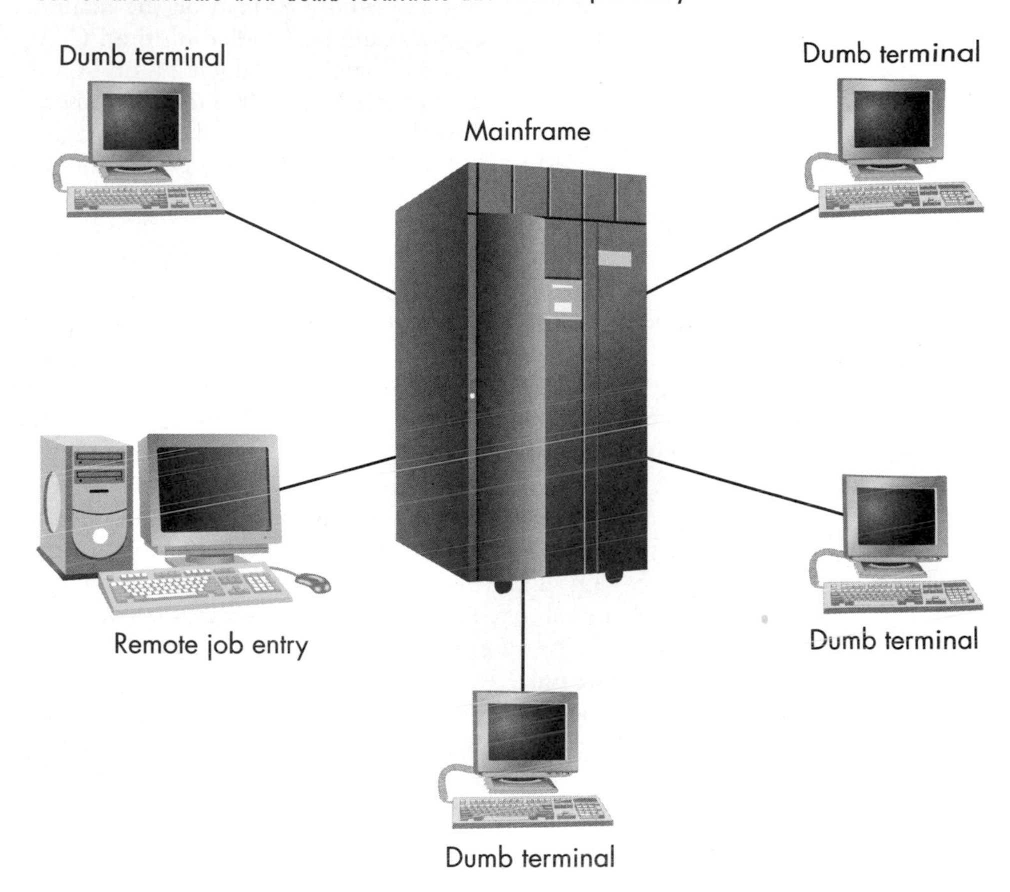

Personal Computers

The primary reason for the dramatic growth in the use of computers over the last 25 years has been the personal computer (PC). The PC has become a fixture in homes, offices, factories, and schools around the world. In addition, the PC introduced the GUI in the workplace, leading to increased employee productivity in a more user-friendly environment. Similarly, it introduced distributed computing to the workplace, extending organizations' computing power from the machine room, where the mainframe lives, to the desktops of employees. With the advent of the

Table 2.6 Advantages and Disadvantages of Mainframe Computing

Advantages	Disadvantages
Centralized computing power, including management and backup	Character-based applications
High levels of security	High initial cost
Capability to run enterprise applications	Problems with failure of centralized computing
	Inability to keep up with high demand
	High cost of upgrading machines and software

PC, users were no longer restricted to text-based dumb terminals or forced to depend on the availability or load on the mainframe because many of the mainframe's functions could be handled on the PC. An additional benefit is that computing power and storage cost much less on PCs than they do on mainframes, with PC computing power being 1000 times cheaper. In summary, the PC has revolutionized the world of organizational computing.

Most PCs in the organization are now connected via a LAN to facilitate sharing of data, information, and resources. In fact, it is very rare to find an unconnected computer in an organization today.

Much of this tremendous growth of the PC can be attributed to the standardization of both hardware and software. By having a single operating system (Windows) in more than 90 percent of all PCs and a standard CPU chip (the Intel standard) in 95 percent of the machines, users and developers can depend on a single **Wintel (Windows + Intel)** standard. In addition, the standardization of Web browsers has boosted the popularity of PCs—it is possible to visit almost any Web site with a PC regardless of which browser you use.

Wintel
A combination of Windows and Intel, in which a PC with an Intel chip runs a version of the Windows operating system.

Although an individual PC is initially *much* less expensive to purchase than a mainframe, taken collectively, the total cost of ownership of PCs in an organization is much *more* than the total cost of a mainframe due to the higher cost of managing and supporting those PCs. This higher cost includes the need for a technical staff to install new hardware and software as it becomes available or to repair machines that fail. Another element of the higher cost of management and maintenance of PCs is the purchase cost of new software or new versions of existing software. In fact, support for an organization's personal computers makes up about 70 percent of the cost of owning them. A PeopleSoft advertisement in the *Wall Street Journal* claimed that computer software on PCs cost businesses $500 *billion* in 2000.[1] In addition, according to the Gartner Group, whatever a company pays for a PC, the total cost will be 3⅓ times higher than the purchase price over the three-year life of the machine. For example, if a company pays $1500 for a PC, the total lifetime cost will be $5000.[2]

In addition to these costs, PCs present other problems, including a lack of centralized management and backup and security risks, including physical security, data security, and virus protection. Table 2.7 shows the advantages and disadvantages of PCs in organizational computing.

Table 2.7 Advantages and Disadvantages of PCs

Advantages	Disadvantages
Standardized hardware and software	High cost of management and maintenance
Ease of use (GUI)	Lack of centralized control
Low processing costs	Security risks
Distributing computing	Data duplication across machines
High user productivity	

1. PeopleSoft advertisement, *Wall Street Journal*, April 25, 2001, p. B2.
2. Matt Hamblen, "Kodak touts TCO success amid glodal PC, laptop rollout." *Computerworld*, February 4, 2002, http://www.computerworld.com.

Client/Server Computing

client/server computing
A combination of clients and servers that provides the framework for distributing files and applications across a network.

client
A computer running an application that can access and display information from a server.

server
A computer on a network running an application that provides services to client computers.

workstation
A client computer that allows the use of specialized applications requiring high-speed processing of data into information.

network computer (NC)
A computer that can be used only when connected to a client/server network.

In mainframes and PCs, you see the two extremes of corporate computing—centralized hosts and distributed computing. Mainframes store and process data on a central machine that is remote from the user unless the data are stored at a RJE site. PCs store and process data locally to the user. Obviously, in many situations, both the power of a mainframe and the ease of use of a PC are needed. Even when PCs are connected via a LAN, they still may not meet the needs of many business-critical applications. To meet these needs, **client/server computing** was developed. Combinations of local and remote storage processing leads to four basic computing architectures, shown in Figure 2.8. Note that the client/server system overlaps both local and remote storage and processing.

In client/server computing, the operation of processing of data into information is shared between multiple small computers known as **clients** and a host computer known as a **server**. The client computers are linked to the server computer via a network. Although the terms *client* and *server* often are used to refer to the machines being used, it is actually the *client software* and *server software* that carry out the processing. In fact, it is possible to run several types of client and server software on the same machine. For example, you probably have both Web browser and e-mail client software running simultaneously on your computer.

Clients on the network are typically PCs, a type of high-powered small computer built for specialized applications called a workstation, or a new type of computer called a network computer. A **workstation** is often used for tasks that require processing capabilities beyond those available on a standard PC, such as simulating complex business situations. In contrast, a **network computer (NC)** looks much like a PC without a hard drive, CD-ROM or DVD drives, or internal sockets for expanding the capabilities of the computer. The NC is dedicated to working on a network; it does not work as a stand-alone computer.

Figure 2.8 Basic client/server architectures

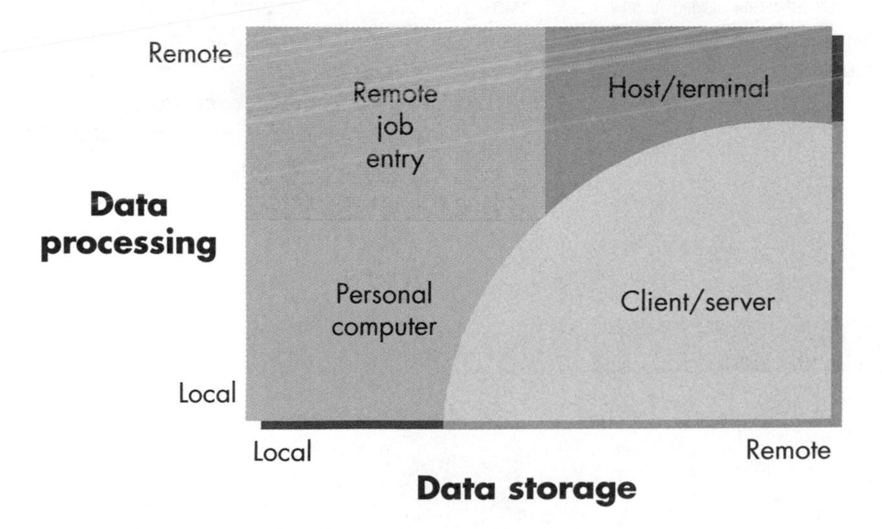

Source: R. T. Watson, *Organizational Memory*, 3rd ed. (New York: John Wiley, 2002), p. 356.

thin client
A client computer on a network that cannot be used in a stand-alone mode.

fat client
A client computer, usually a PC, that can also be used as a stand-alone computer.

NCs offer a number of advantages over PCs in organizations, mainly in terms of the total cost of ownership due to lower purchase, setup, and software installation costs. In a client/server environment, NCs are often referred to as **thin clients** and PCs and workstations, are called **fat clients**, in which *thin* and *fat* refer to client capabilities. Thin clients have little or no secordary storage and are meant to be used only as clients, depending on the server for software and secondary storage. Fat clients, in contrast, are full-blown computers in their own right, having secondary storage and the capability to be used independently of the server. The evolution of computing has gone from a strictly host-based system with dumb terminals (extremely thin clients), to a system of distributed computing with fat clients, to a client/server system with either fat or thin clients.

file server
A server computer with a large amount of secondary storage that provides users of a network with access to files.

The server in a client/server network is typically dedicated to a specific type of processing, such as providing files with a file server, responding to database queries with a database server, or handling high-speed processing with an application server. Because both clients and servers are capable of processing, processing is shared between the two computers depending on the capability of each. In the simplest form of client/server computing, a **file server** manages the network. The file server does not handle any of the computing load—rather, its primary role is to control access to the network, manage communications between fat clients in the form of PCs, and make data and program files available to the individual PCs. In this case, the processing load is still distributed among the individual PCs.

Three-Tiered Architecture

three-tiered client/server architecture
A client/server architecture in which an intermediate computer exists between the server and the client.

A widely used client/server setup involves the use of a client, an application server, and a database server. In this client/server environment, a user working at a GUI-based client PC or workstation sends a request for data or processing to an application server, which decides which data are needed and sends a query to the database server to retrieve those data. The database server processes the query and returns the matching data to the application server, which in turn processes the data into the form required by the user. Because one client and two servers are involved in this process, this setup is commonly referred to as a **three-tiered client/server architecture**. Note that a three-tiered client/server architecture is different from the situation in which a file server supplies the entire database file, with the local PC still handling all of the processing. Figure 2.9 shows a three-tiered client/server architecture.

In three-tiered client/server architecture, requests for processing on a server often come from another server, which in turn may be processing a request from a client. For example, a client may make a request to a Web server, which in turn makes a request to an application server, which then makes a request to a database server. One notable strength of client/server computing is the capability to string a series of servers together to respond to a client's request.

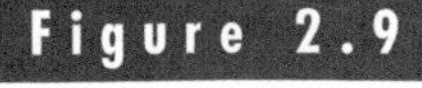

Three-tiered client/server architecture

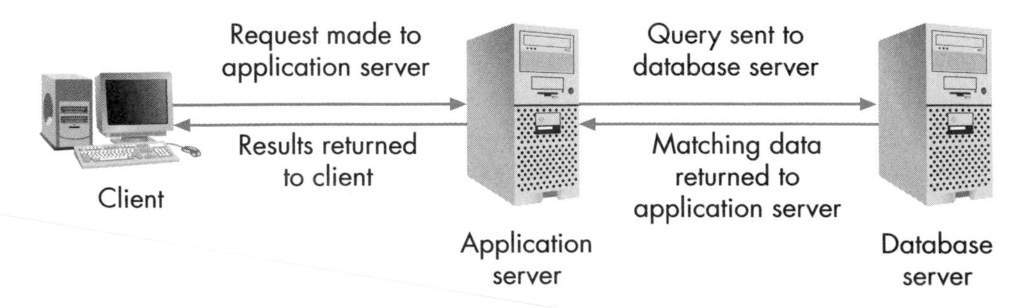

Table 2.8 Servers on a Network

Server Type	Purpose
File	Provides both software and data files to users
Database	Handles queries to a large database and returns matching records
Application	Handles high-speed processing
Web	Handles requests for Web pages
Mail	Sends and receives e-mail for the entire organization
Fax	Sends and receives faxes for the entire organization

Network computers (NCs) like this one have no secondary storage and are always connected to a network.

A familiar example of a three-tiered architecture client/server application involves the use of client software in the form of a bill-paying program, like Quicken, that communicates with the user's bank to handle personal financial affairs. The bill-paying program on the client PC enables the user to send electronic checks, keep an up-to-date register, and track income and expenditures. It does so by sending messages to the application server at the bank, which accesses a database server that stores information on the user's account and generates electronic checks that are sent to the payees.

Today, a variety of servers are found in client/server environments. In addition to file, database, and application servers, there are Web, e-mail, and fax servers, as shown in Table 2.8. Each of these servers has application software that is specialized for the task that the server carries out on the network. For example, Web server software is specialized to handle requests for Web pages, whereas e-mail server software is specialized to send and receive e-mail. It is worth noting that in many organizations an application server is replacing a mainframe for handling large-scale processing tasks. In fact, IBM refers to its large computers as *enterprise servers* to emphasize their use in client/server computing as application servers.

Figure 2.10 shows a typical client/server network that includes file, fax, mail, database, application, and Web servers and a variety of clients.

Figure 2.10 Servers on a network

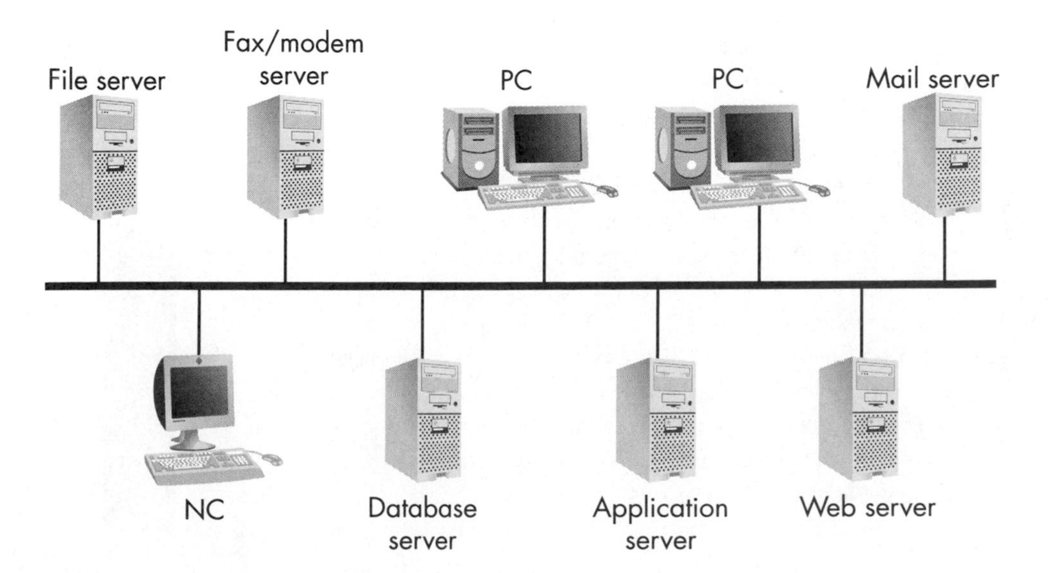

Table 2.9 Advantages and Disadvantages of Client/Server Systems

Advantages	Disadvantages
Computing burden can be shared among servers and clients	Programming relationship between clients and servers is more complex
Servers can be specialized to one particular type of task	System upgrades require that all clients and servers be upgraded regardless of location
Upgrading system can be done in small steps	
Loss of client does not stop other clients from accessing server	

Like the other types of computing, client/server computing has its own advantages and disadvantages, as shown in Table 2.9. The primary advantage of client/server computing lies in its capability to share processing and data storage responsibilities among multiple machines and to use specialized servers to meet specific needs. The primary disadvantage is the complexity inherent in sharing responsibilities among multiple machines. Overall, however, the trend in corporate computing is toward wider use of client/server computing because of its increased flexibility.

1. List the types of computing that are in use in organizations.
2. List the types of servers that one might find in an organization.

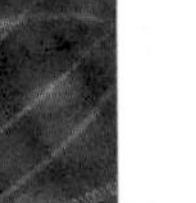

INTERNET IN ACTION

Dave's Guides

Numerous Web sites provide a wealth of information on information technology infrastructure, inlcuding buying a PC, setting it up, and troubleshooting. You can find one of the most complete sources of such information at the PC-Guide Web site created by Dave Strauss of Michigan State University, www.css.msu.edu/PC-Guide.html. This site provides access to a series of guides as well as to two Web-based courses that Dave teaches. The guides include *Buying a Home PC, Buying a Used PC, Setup and Upgrade, Troubleshooting,* and *Frequently Asked Questions.*

In the section on buying a home PC, the Web site includes a great deal of information and advice on a variety of topics, including operating systems, processing chips, internal memory, modems, and graphics. For example, if you are unsure of what is meant by some term, such as *USB*, you can look it up in *Dave's Guides to Buying a Home PC*. There you will find that *USB* means Universal Serial Bus and is a way of connecting devices to your computer that provides a much higher degree of flexibility than previous methods. If you want to know more, Dave provides a link to Intel's Web site as a part of the discussion. This site provides one of the best discussions around on processing chips under the heading "Processor Paranoia," where types and speeds of chips are compared. In summary, this a good site to bookmark for future reference on anything to do with Windows-based PCs. Its only shortcoming is a lack of information on Apple computers.

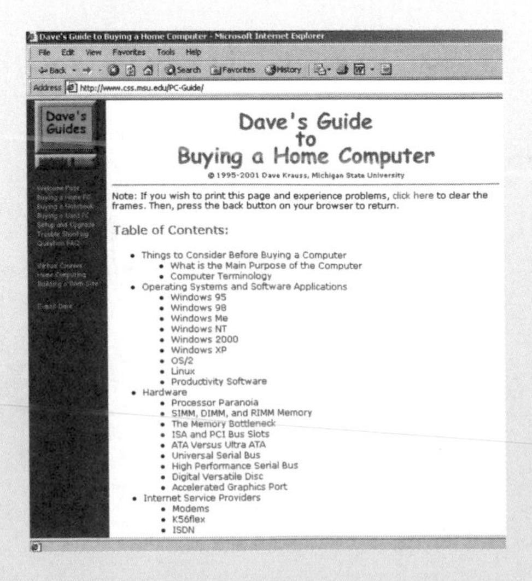

Hardware and Software at FarEast Foods

Service is what distinguishes FarEast Foods from its competitors. The customer service representatives use a client/server network to handle telephone orders for Asian foods. They also deal with customer requests to return food items that were incorrectly ordered and customer complaints about poor delivery service or poor food quality. Currently, each customer service representative has a full-featured PC (a fat client), which is linked to a database server and has its own hard drive and printer. Any queries to the database server and subsequent processing are carried out using software residing on the PC. The data required might include data about the requested food product or previous transactions retrieved from the database server. The PCs run the latest version of Windows, whereas the database server runs the Windows 2000 server operating system and database software compatible with Windows 2000. Much of the processing on the client PCs is handled via applications written in Visual Basic and a spreadsheet application written in Excel. Excel is also used for ad hoc calculations required to handle returns or give credit for problem orders. This setup requires that the Windows operating system software, the VB .NET applications, and at least Excel (plus possibly other Microsoft Office applications) be resident on the PC at all times. Web browser and e-mail clients are also necessary for the client computers. This setup is shown in Figure 2.11.

Figure 2.11 Use of fat clients to access a database server

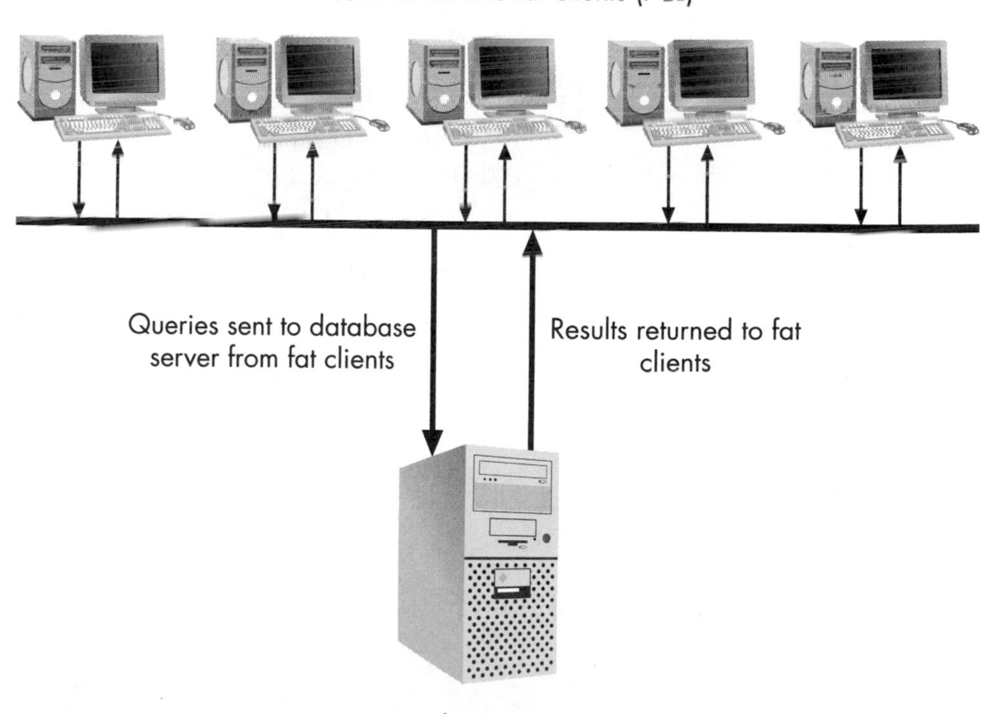

The high cost of frequently updating the software on the fat PC clients has prompted FarEast Foods to consider moving to a client/server setup using thin clients running the Linux operating system. An application server will be added between the thin client and the database server to process requests from the thin client. With such a system, when the customer service agent needs to carry out processing to handle an order, return, or complaint, the thin client sends the appropriate request to the application server, which then queries the database server for data and processes it into information, which is returned to the client. This setup is a classic case of a three-tiered client/server application. In situations for which processing is required on the client, the client sends a request to the application server for the appropriate application software, along with a request for the necessary data from the database server.

Because the thin clients will run the Linux operating system, both the application servers and the database servers must be switched to the Linux operating system. In this new setup, the application server will carry out all processing for an application, and the thin client's main task will be presenting information to the customer service representative. Figure 2.12 shows the proposed client/server system.

1. List some advantages of FarEast Foods' current system over the proposed system. List some advantages of changing to the new system.
2. Where might the company spend less, and where might it spend more on this solution?

Figure 2.12 Proposed customer service system

CaseStudy

Propping Up Aging Mainframes

Even though the personal computer and Internet have received the majority of the publicity over the last 10 years, organizations of all sizes have continued to use mainframes to carry out their truly important back-office operations. As a result, 70 percent of all data are estimated to still be sitting on these machines, even though many people might imagine that they went out with the first Bush administration in the early 1990s. Because of their huge investment in mainframe information technology infrastructure, companies are looking for ways to extend the life and effectiveness of these machines.

One increasingly popular approach uses software developed by Atlanta-based Jacada, Ltd., to link the mainframes and their data to other forms of IT. One type of Jacada software makes data on mainframes directly available to stakeholders over the Web or through wireless devices. Other software from the company is being employed to help IT workers do their jobs better without extensive retraining.

For example, Delta Airlines is using Jacada software to make operating information, flight schedules, and daily changes mandated by weather that is stored on company mainframes available over the Internet to its 27,000 employees who have been given laptops. Previously, Delta's pilots and flight attendants had to go to a special kiosk or use the telephone to access this type of information. Delta hopes that speedy availability of this type of information will improve service in ways that the public can notice and appreciate.

Jacada is also finding success by creating Web-based front ends for university registration systems. Universities often built their registration systems around mainframes, which typically required text-based systems. These text-based systems are difficult to modify so that they can run on the Web. However, the Jacada solution creates Web-based front ends that work easily with the already-existing mainframe systems. For example, the University of Georgia used a Jacada solution to update its existing Online Accredited Student Information System (OASIS) from a clunky text-based system that was only available from machines on campus to one that is Web-based and available to anyone with Internet access. Youngstown State University carried out a similar conversion.

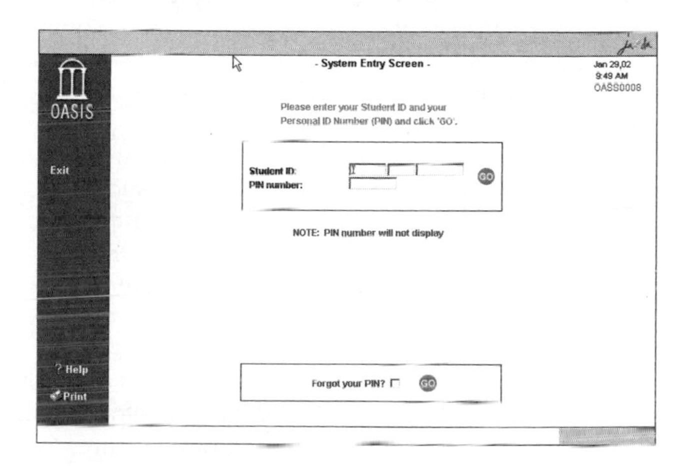

Sources: Ernest Holsendolph, "To the rescue: Jacada's software props up aging mainframes." *Atlanta Journal/Constitution*, January 17, 2001, p. C6; Melanie Horton, "University looks to expand OASIS." *Red and Black*, February 21, 2001, p. 1.

Think About It

1. Why is a product like Jacada necessary?
2. How might Delta Airlines benefit from the Jacada upgrade?
3. Go to the UGA OASIS home page and run through the turorial. Compare this registration system to the one you are using at your school.

SUMMARY

To summarize this chapter, let's answer the questions posed at the beginning of the chapter.

How does information technology support the networked economy? Information technology is the infrastructure of the networked economy—it provides the elements necessary for the networked economy to exist. This infrastructure allows raw data to be processed into information and is necessary for transactions to take place in the networked economy. It is also necessary for accessing existing data and information and sharing data, information, and knowledge with other people in the networked economy.

What are the two primary parts of any computer, and what roles do they play? The two parts of the computer are hardware and software. Hardware is the physical, electronic part of information technology that carries out various operations under the direction of software. Computers come in a variety of sizes, ranging from large mainframe computers to handheld devices such as the Palm or Pocket PC. Between these extremes are minicomputers, which are essentially smaller versions of mainframes, and personal computers, which are found in virtually every office in the developed world.

What are the four major hardware elements of the computer, and how do they process data into information? Information technology hardware consists of three elements: input, processing, and output. Secondary storage is also included with most computers. The input element enables the hardware to receive the data to be processed and the software instructions for processing those data. Once the data are input, the hardware must be able to process the data into information using software instructions. This task is accomplished using two types of chips: processing chips and memory chips. The actual processing of data into information takes place on the processing chips. Data, information, and instructions are stored on memory chips. Because the processed information is worthless unless it is provided to the user in some way, the computer has an output unit. Secondary storage units are necessary because memory chips are both limited in space and nonpermanent. Letters, numbers, punctuation marks, and special symbols are stored internally and on secondary storage using a standardized representation that uses groups of eight bits, called bytes. Each bit pattern represents a given symbol.

Why is software necessary to use a computer, and what are the two primary types of software? Without software, information technology would be nothing more than a well-constructed combination of silicon chips and electronic circuitry. *Software* is the general term for the instructions given to the computer or other chip-based device by the user or the manufacturer. The two major types of software are operating systems software and application software. Both work in the computer at the same time, with each serving a different purpose.

The two most common user interfaces are command driven and graphical. With a command-driven interface, software simply waits for the user to enter the appropriate command or data. The standard interface for personal computers is the graphical user interface (GUI). In a GUI, the user selects icons with a pointer; the icons represent the functions to be performed.

Operating system software enables application software to run and controls the computer itself. In this process it has a number of responsibilities, including starting the computer, managing hardware, controlling access to the computer, providing an interface for the user, and ensuring efficient use of the CPU. When running application software, the operating system must determine the order in which programs will be run, carry out file and disk management, and manage memory.

Application software performs all of the tasks that make a computer useful. It includes games, personal productivity software, Web browsers used on personal computers, and specialized programs that process grades, compute payrolls, and handle transactions on larger computers.

All software—whether built into the machine, purchased from a software development firm, or developed within an organization—is created by one or more people through a process known as programming. Programming entails the use of one of the many available computer languages.

How has computer processing evolved over time, and what are the three types in use today? The three types of computing that coexist within the corporate world are mainframe computing, PCs on a LAN, and client/server computing. Mainframes handle all of the processing and storage, leading to centralized massive computing power and the capability to run enterprise-level business-critical applications. PCs are cheaper to purchase, easier to use, and result in more user productivity than mainframe computers. In organizations, they are typically connected over an internal network called a local area network (LAN).

Client/server computing is the newest form of corporate computing; it attempts to combine the best features of mainframes and PCs. With client/server computing, processing is shared between multiple small computers known as clients and a host computer known as a server. Clients are typically PCs, workstations, or network computers. In most cases, the server is dedicated to a specific type of processing. A variety of servers are employed in today's client/server environment, including database servers, application servers, Web servers, e-mail servers, fax servers, and modem servers. Requests for processing often come from another server, which in turn may be processing a request from a client.

REVIEW QUESTIONS

1. Why is information technology a necessary element of the networked economy?
2. What are the two parts of a computer? What is the purpose of each part?
3. What are the four elements of computer hardware? Give an example of each part.
4. What is the purpose of each part of the computer? Which parts are typically found in the system unit?
5. List three types of input and three types of output for a computer.
6. Which computer element is used for both input and output? Which computer element is necessary for connection to the Internet?
7. What unit is typically used to measure RAM? To measure hard drive capacity?
8. Why is digital storage rapidly replacing analog storage?
9. Which two types of software were discussed in the text? Which one is used to actually process transactions in the networked economy?
10. Which two types of user interfaces are in common use today? Which one uses icons, toolbars, and menu bars to carry out operations?
11. Which type of software is used to manage the operations of the hardware and software? List the operations carried out by this type of software.
12. How are data, information, and instructions organized within a computer?
13. Compare mainframe, network, and PC operating systems in terms of security and technical support.
14. What are the three types of computing used in the corporate or organizational environment? Which is the oldest type of computing used in the corporate or organizational environment?
15. Which type of organizational computing is initially the cheapest but is the most expensive in the long run? Why?
16. Which is the newest type of computing used in the corporate or organizational environment?
17. What is the difference between a client and a server? What is the difference between a fat client and a thin client?
18. List five types of servers. Describe the purpose of each.
19. Which type of server has taken over many of the tasks previously handled by mainframes?
20. What are the disadvantages of client/server computing?

DISCUSSION QUESTIONS

1. Briefly describe and discuss two examples of application software with which you are familiar. Include in your discussion the types of data files with which the application software works.
2. If you own a computer, list the various hardware elements in it. Also, list the amount of RAM and the size of the hard disk(s). What types of secondary storage does your computer have? If you don't own a computer, visit the Web site for a computer retailer, say, Dell Computer, and create a list for a computer offered for sale there.
3. Discuss the transition from analog to digital technology. Describe two other devices that have made or are making the analog-to-digital transition.
4. Discuss the strengths and weaknesses of mainframes and PCs in a corporate or organizational environment. How does the client/server system address some of the weaknesses of both systems?
5. Discuss the difference between a thin client and a fat client. Which would be most useful as a standalone computer in case the server was out of operation? Why?

RESEARCH QUESTIONS

1. Visit the Web site for the *Linux Journal* and read the complete article on Dreamworks' use of Linux-based computers for animation. In particular, consider Table 1, which lists the steps in the animation process. Write a two-page paper discussing the process and the role that computers play in it.
2. Visit the site for a computer retailer such as Dell, Gateway, or IBM. Configure your dream computer on all three sites and compare the prices. Write a two-page paper on the elements in your dream computer and the differences between the configurations and prices at the three companies.
3. Facial recognition systems, like those discussed in the *Technology on the Edge* box, are being suggested as a way of providing a higher degree of security at U.S. airports. Research this suggested use of facial recognition systems and compare it with other suggested security methods. Describe your comparison in a two-page paper.

"I forgot. Is this rock the Web server or the printer?" asked Claire.

Alex replied, "That's the server, this stick is the printer."

After a long day of paddling, Alex and Claire were discussing their plans for WildOutfitters.com at their campsite by the river. Alex was attempting to explain the IT system with a diagram made up of whatever was at hand. By the light of the campfire, he had placed a few stones and sticks to represent various hardware devices, and he had drawn lines in the dirt to represent the connections between them. If left as it was, future archeologists might be puzzled by the strange patterns, but Alex seemed to know exactly what they meant. Now, he only hoped that he could explain it adequately to Claire.

The Campagnes have decided to set up a Web storefront to sell outdoor sport products—of their own design and from other companies—over the Internet. For this purpose, they have obtained the license for the URL WildOutfitters.com, named after their current business. They hope that a Web storefront will not only increase their sales with orders from the Web, but also serve as a form of advertising for their physical location.

The couple plans to start small with a minimum of features but retaining the flexibility to add more functionality as desired. The initial site will consist primarily of a catalog of products along with product reviews. A shopping cart feature will be provided for accumulating items to purchase. Payment for online orders will be by credit card only. Shipping and handling charges will be calculated automatically and added to the purchase amount, based on the total cost of the products ordered.

Claire and Alex figure that their site, while small compared to other Web stores, will require a number of pages. A purchaser will enter the site through a home page offering background information about the store and its products. This page will include contact information, the store location, hours for the store, and appropriate links to the online catalog. In addition to the catalog links, the home page will include links to pages describing the shipping and privacy policies.

With the online catalog, the user will be able to browse the products by category or search by keyword. The search engine will return pages listing matching products, photos, and prices. A customer may then click the photo of a product to get more information and a product review written by Claire or Alex. Various forms and message pages will be needed to process a customer's order, verify the customer's credit card information, and notify him or her of any problems that occur.

Although the initial storefront will be limited primarily to the catalog, Alex and Claire plan to purchase hardware and software for the system that will allow them to expand their Web offerings in the future.

The couple discussed these plans over a warm meal and late into the night by the fire. After moving a stone here and a stick there, they brewed a pot of coffee and looked on their system design with satisfaction.

Finally, after surveying the crude diagram laid out before her, Claire announced in her best caveman voice: "Ugh, computer good! Tomorrow make wheel."

With a groan, Alex decided to call it a night and headed for the tent.

Think About It

1. Which type of corporate computing system would you recommend for the Campagnes' business to support both the store and their future Web site?
2. Research Web hosting services that are available for a small business such as Wild Outfitters. After comparing the available services for price as well as the services offered, would you recommend that the Campagnes use such a service for their site or host the site on their own system? Why?
3. What are the procedures for obtaining and using a Web address such as WildOutfitters.com?
4. What hardware devices should be purchased for WildOutfitters.com? Why?
5. What application software should be purchased for the system? Why?

Hands On

6. Search newspaper and magazine advertisements or Web sites for vendors of the hardware and software that you discussed in Questions 4 and 5. Prepare a spreadsheet listing the desired items and their price from each vendor. Calculate the total cost of the systems from each vendor. Be sure to include any extra charges, such as shipping and maintenance or technical support contracts.

Appendix to Chapter 2: A Short History of Computers

The development of computers is usually described as occurring in generations. The first generation is considered to span the period 1946–1959. This generation of computers was characterized by the use of vacuum tubes in the CPU and internal memory units, the first commercial computers, and many fundamental advances in computing. The first commercial computer was the UNIVAC 1 (**UNIV**ersal **A**utomatic **C**omputer), a machine sold to the Census Bureau in 1951.

In the second generation of computers, developed during the period 1959–1964, the vacuum tube was replaced by the transistor. The transistor, a solid-state device, was the major breakthrough that allowed computers to have reasonable size and power. The materials of which a solid-state device is composed allow it to be instructed to permit or not permit a flow of current. Because solid-state devices did not use the hot filament found in vacuum tubes, the use of transistors reduced the computer's heat output and power requirement. Transistors also increased the reliability of the computer because they did not burn out as often as vacuum tubes did. This development reduced the cost of owning and operating a computer. The second generation saw tremendous growth in the use of computers by government, business, and industry.

The introduction of the integrated circuit in 1965 signaled the onset of the third generation of computers. With this technological advance, an entire circuit board containing transistors and connecting wires could be placed on a single chip. The result was greater reliability and compactness combined with lower cost and power requirements. During this period, IBM controlled the mainframe market with its 360 (later to be 370) series of computers. This series was so well designed and built that its successors remain in heavy use today.

The fourth and current generation of computers began in 1971 with the introduction of the microprocessor—a central processing unit on a chip. This generation featured the introduction of supercomputers. These "monster computers" are in heavy demand for military and meteorological applications that require high-speed operation. Another important advance of this generation has been the introduction of the personal computer, which has made the power of the computer available to anybody who wishes to use one.

The Evolution of the PC

The term *PC* was coined by a computer scientist, Alan Kay, in a 1972 paper entitled "A PC for Children of All Ages." As a result of Kay's work in this area, Xerox built a PC called the Alto, although the company never put it on the market. Other established computer companies also considered the concept of a PC but decided that no market for such a machine existed. Not until 1975 did an Albuquerque, New Mexico, company called MITS release the first PC in kit form. This machine, named Altair after a planet in the *Star Trek* TV series, had just 1 KB of memory and was very slow by today's standards. MITS received 5000 orders for the Altair after it was pictured on the cover of *Popular Electronics*. A pioneer in the field, now-computer science publisher Rodney Zaks, remarked that "never before had such a powerful tool been invented and so few people realized what it could do."

Although MITS was the first company to sell a PC, it was up to Apple, Radio Shack, and Commodore to popularize its use. They were among almost 100 companies that rushed to put out PCs in the years immediately after the MITS machine's introduction. An amazing success story of this period is that of the Apple Company, formed by two young Californians, Steve Jobs and Steve Wozniak. The duo built the first Apple computer in their garage.

With all of these infant companies competing for the emerging computer market, Apple made a real breakthrough in 1978 when it offered a disk drive to go along with the original Apple II. This key addition, along with the VisiCalc spreadsheet software package offered only on the Apple, allowed Apple to leapfrog over Radio Shack and Commodore into first place among the pioneer companies.

The next breakthrough arrived in 1981, when IBM offered its PC, which used an Intel CPU chip. Although not a technological innovation, the IBM PC almost immediately became an industry standard and legitimized the concept of a PC. It was followed by the Apple Macintosh in 1984 and the short-lived IBM PS/2 line in 1987.

The last 20 years have seen a continued movement toward faster and less expensive computers, to the point that the under-$500 PC is a reality. Although the vast majority of PCs follow the Wintel standard, Apple has made recent inroads into this market with its iMac and PowerMac lines of computers. At the mainframe level, the introduction of air-cooled instead of water-cooled computers has led to a reduction in their size to the point that they are often indistinguishable from a standard personal computer.

The other dramatic change over the last 10 years has been the rapid growth in use of the Internet, e-mail, and the World Wide Web for communication, research, commerce, entertainment, and a whole host of other purposes. These developments have driven hardware manufacturers to make these applications available to an ever-widening audience at faster and faster connection speeds. They have also influenced the development of software applications with Web browser interfaces.

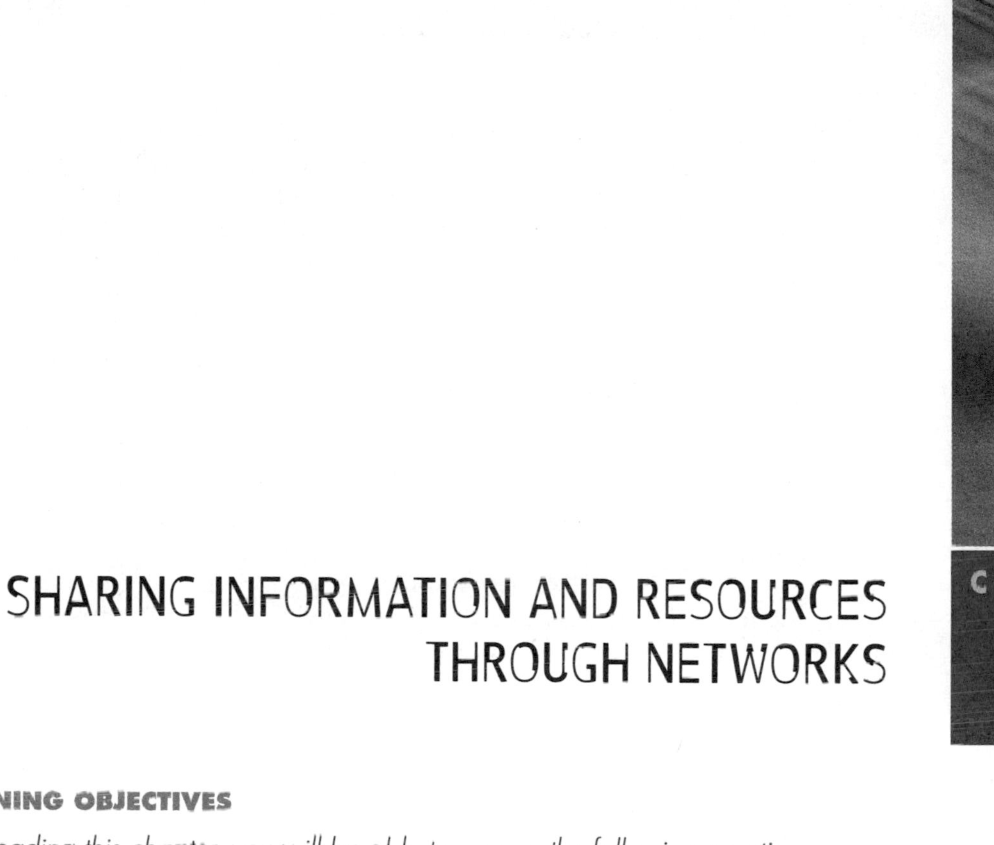

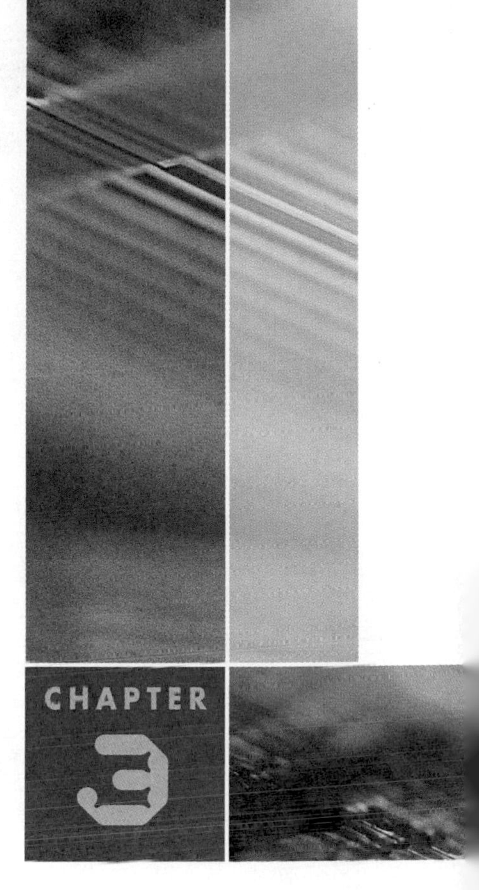

SHARING INFORMATION AND RESOURCES THROUGH NETWORKS

LEARNING OBJECTIVES

After reading this chapter, you will be able to answer the following questions:

- What are computer networks, and what is their role in the networked economy?
- What is the network layer model, and how does it work for a wide area network?
- What are the parts of a local area network?
- What are wireless networks, and why have they become so popular?
- What is the Internet, and how is it used?
- What is the World Wide Web, and how does it work?

Using Networks to Turn Paper into Cardboard

If there is a poster child for an old-economy company, it's Corrugated Supplies, located in Bedford Park, Illinois, an industrial suburb of Chicago. It has a fairly simple business: turn big rolls of paper into flat corrugated cardboard sheets. This cardboard, in turn, is used as raw material for other manufacturers to produce finished products such as boxes, displays, inserts, and other packaging needs. The paper comes in 8000-pound rolls that are stacked almost to the ceiling of Corrugated's 100,000-square-foot manufacturing building. These rolls must be moved by forklifts to production machinery that slits, trims, and corrugates it into almost 1000 feet of cardboard per minute in 200 combinations of paper strengths and styles.

Corrugated Supplies competes in a small-margin, niche market where service is critical as the company tries to meet the widely varying needs of its customers. In looking for ways to improve its systems, the head of the IT group discovered that the forklifts were the bottleneck, because the drivers had to figure out what material to pull and where it needed to go on the floor for the manufacturing operation. When an order was received, the forklift operators had to scurry around to find the paperwork that would tell them which orders to load first and in which tractor trailer the order would be leaving. To solve this problem, PCs connected to a wireless network were mounted in the forklifts. Using the PCs, drivers can receive the needed shipping information by using touchscreen input over the network. This change has resulted in the ship-through rate being cut from 3 days to 24 hours, making the company's customers much happier.

Machines like these at Corrugated Systems convert sheets of paper into various types of cardboard.

Source: Alan Joch, "Business case: Thinking outside the boxes." *http://www.networkmagazine.com/article/NMG20010521S006*, June 5, 2001.

Computer Networks

In Chapter 2, you learned about the use of hardware and software for processing data into information. For information to be useful, however, it must be shared with others over a computer network. A computer network exists whenever two or more computers are linked through some type of communications medium. In the chapter-opening case, Corrugated Supplies found ways to use networking technology to better serve its customers. The best-known computer network is the Internet, which hundreds of millions of people use to access the World Wide Web and send e-mail.

Computer networks have many uses in addition to sharing information. For example, when you use an ATM either near your home or in another part of the world, information technology reads the magnetic strip on your card, compares it to the personal identification number (PIN) keyed in, and decides whether you are a valid user of that card. When you attempt to withdraw cash from the ATM, it makes a connection with a local computer, which in turn connects to your home bank's computer to verify that your account balance is sufficient to cover the amount to be withdrawn. If you are in a foreign country, this process involves a currency conversion between your home currency and that of the country in which you are withdrawing the money. Through computer networks, this entire process usually takes less than a minute.

Computer networks are also important for sharing resources, usually other computers. As noted in Chapter 2, computing is moving toward a client/server model in which client computers make requests of server computers for files, database records, results of processing data, or Web pages over the Internet. In the client/server model, sharing resources is as important as sharing information.

Chapter 2 discussed information technology's role as the infrastructure for the networked economy. This chapter will extend that discussion and explain how networks support the networked economy. Although networks can link a variety of information technology devices, the focus here will be on networks that connect computers. Computer networks can be classified in several ways, but the most common classification scheme is based on size. Network types can be defined by their geographic scope, with the two primary types being local area networks and wide area networks.

Network Types

local area network (LAN)
A computer network composed of at least one client and one server that is restricted to a single geographical area.

A **local area network (LAN)** is a computer network composed of at least one client and one server that is confined to a single geographical area—for example, an office, a building, or a group of buildings. LANs are a very important part of business information systems because they allow workers to communicate with one another and to share information, software, and hardware. Figure 3.1 shows a typical LAN for a small business with multiple clients and servers.

backbone
A transmission medium created to connect networks.

Because even a small organization can have multiple LANs, it is necessary to link them together. The most common way of linking multiple LANs is through a **backbone**, which typically is a transmission medium created to connect networks. Backbones for connecting LANs can span a mile or more and offer high-speed communications between the LANs. Backbones can also connect LANs to the Internet. As an example of multiple LANs in a company, consider the network at FarEast Foods, shown in Figure 3.2. At its corporate headquarters in Denver, several LANs are connected by a backbone that, in turn, connects to the Internet. These networks include the human resources LAN that handles issues such as payroll, employment records, meeting equal opportunity hiring requirements, and so on. Another LAN in the distribution center connects the individuals who put together the packages of food orders to be shipped. Finally, a LAN connects all customer service representatives. Although some managers have access to all these networks, employees in each area generally have access to only their own LAN. This setup provides security for the various functions within the company.

Figure 3.1 Typical LAN for small business

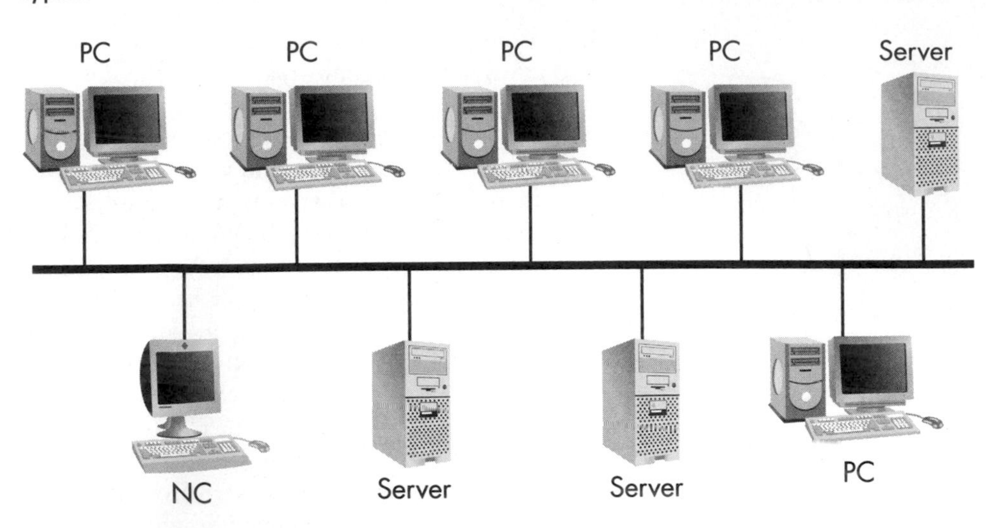

wide area network (WAN)
A network covering more than a single geographic area.

Computers that are connected over a region, country, or the world, are part of a **wide area network (WAN)**. WANs are typically connected by high-speed communication links that enable the sending of messages and files between geographically distant computers. Figure 3.3 shows a wide area network that connects the FarEast Foods corporate offices in Denver to six other cities in the United States.

Figure 3.2 LANs at FarEast Foods corporate headquarters

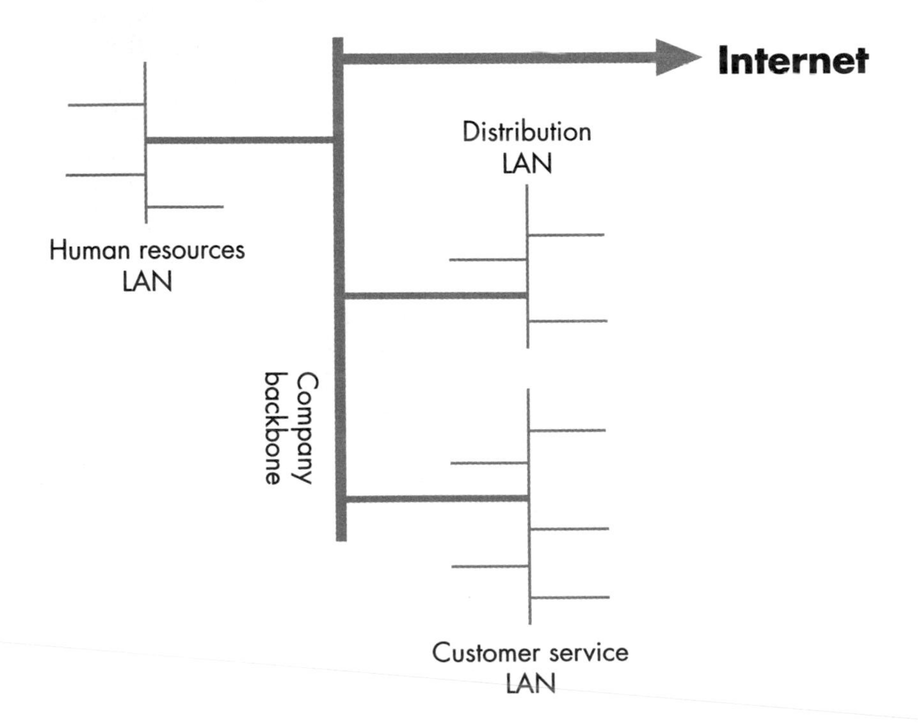

Figure 3.3 WAN connecting Denver to other cities

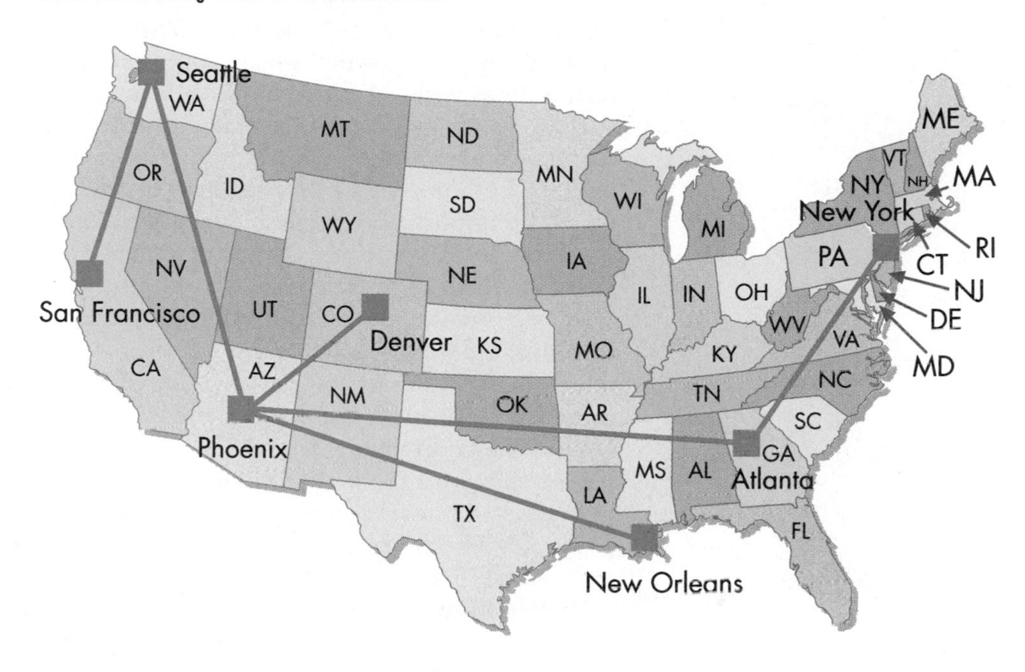

The Internet is a WAN that consists of many interconnected networks—a network of networks. Individuals cannot connect directly to the Internet; instead, they must connect to a network that is, in turn, connected to the Internet. Typically, individuals make this connection through a mainframe computer, a LAN, or a private **Internet Service Provider (ISP)**. An Internet Service Provider is a company that provides access to the Internet to individuals and organizations. You are probably able to access the Internet from your school computer lab over a LAN, whereas you probably connect your home computer over a telephone line or cable modem to an ISP.

Internet Service Provider (ISP) A company that provides access to the Internet to individuals and organizations.

Each network or ISP that connects to the Internet pays for the privilege. Payments go not to the Internet, but rather to some telecommunications company that has created one of the many communication links that make up the Internet. If you access the Internet through your college computer connected to a LAN, then the college has paid for this link to the Internet. The fee is typically a flat rate per month rather than a fee based on usage. Individuals, who usually access the Internet through a local or national ISP or a national information service, may pay a flat fee, an hourly fee, or a combination of both. The ISP or information provider then pays another larger network or a telecommunications company for its connection to the Internet.

The next sections describe WANs in detail first and then LANs. WANs are covered first because of their importance to electronic commerce. The fast-growing world of wireless networks in both the corporate and personal world will also be covered. Finally, the Internet and World Wide Web will be discussed as applications of WANs.

1. In what ways can a computer network be classified according to the geographical area it covers?
2. What is the difference between a backbone and a wide area network?

Understanding Wide Area Networks

Wide area networks provide the infrastructure for the networked economy by making it possible for individuals, companies, and organizations to form economic relationships without regard to place or time. With a WAN, for a company such as FarEast Foods in Denver, Colorado, can deal with suppliers in Hong Kong and Bangkok and customers in Toronto, Canada, and Ft. Myers, Florida, without anyone leaving the home office. Because having a basic understanding of how networks operate is important, a simple model is provided here.

How a WAN operates is often described by using a model that involves *network layers,* in which each layer handles part of the communications between computers. The original version of this model was created by the International Standards Organization (ISO) and consists of seven layers that define the standards with which networks must comply. Our simplified version of this model contains only three layers: the *application software layer,* the *networking software layer,* and the *physical layer,* as shown in Figure 3.4.

As shown in the figure, at the sender end of the network, a message traverses the layers by moving from the application layer, to the networking layer, to the physical layer. The reverse process occurs at the receiver end of the network, with the message first traversing the physical layer, then the networking layer, and finally the application layer. The postal system provides a good way to think about this model. In the postal system, you write a letter, put it in an envelope with a a friend's address on it, and then place the envelope in a mail box. You can think of this activity as the application layer. A postal worker picks up the envelope and takes it to the post office (the networking layer), which decides how to send it to your friend (the physical layer). At the other end, this process is reversed.

Figure 3.4 Network layer model

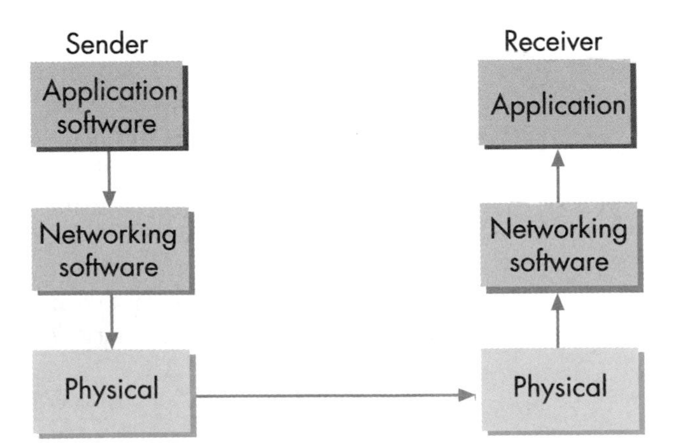

Figure 3.5 Submitting a food order form at the FarEast Foods Web site

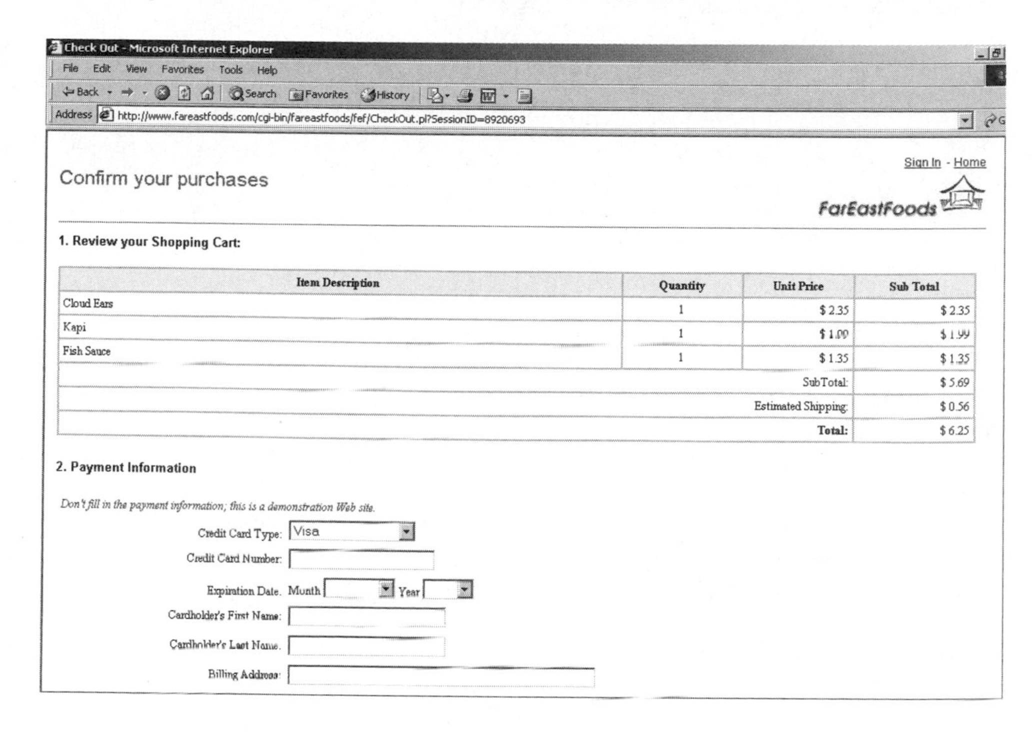

Application Software Layer

protocol
A formal set of rules for specifying the format and relationships when exchanging information between communicating devices.

Simple Mail Transfer Protocol (SMTP)
A communication protocol for transferring mail messages over the Internet.

Hypertext Transfer Protocol (HTTP)
The communication protocol for moving hypertext files across the Internet.

Electronic Data Interchange (EDI)
A communication protocol that allows computers to exchange data and information electronically, thereby automating routine business between retail stores, distributors, and manufacturers.

At the top of the network layer model, as shown in Figure 3.4, is the application software layer. Application software is the software on each computer on the network that the user sees and uses to send and receive messages and data between computers. At the sender end, the application layer includes well-known software applications such as Web browsers and e-mail. This software formats (places in a special form) a message by adding important information to make it conform to a specific standard or **protocol**, which is a special set of rules for communicating. Typical protocols for the application layer of the Internet are **Simple Mail Transfer Protocol (SMTP)** for e-mail, **Hypertext Transfer Protocol (HTTP)** for Web pages, and **Electronic Data Interchange (EDI)** for large-scale exchange of data between organizations. The resulting message comprises a combination of the message generated by the application software and the protocol. The message may also be encrypted (placed in a secure, unreadable form) to protect it from unauthorized readers. For example, suppose a customer goes to the FarEast Foods Web site at www.fareastfoods.com, fills out an order form (see Figure 3.5), and submits that form. In this case, the application is a Web browser, the message is the contents of the food order form, and the protocol is HTTP. In addition, the message is encrypted. The combination of protocol and encrypted message for the food order is shown in Figure 3.6.

Figure 3.6 Application layer message

Food order	Encryption	HTTP

Figure 3.7 Converting data into packets

11001101010111001110011 →	11001101010	198.137.240.92
	11100110011	198.137.240.92

Networking Software Layer

Transmission Control Protocol/Internet Protocol (TCP/IP)
The basic communication language or protocol of the Internet.

ANSI X 12
An EDI protocol used in the United States.

EDIFACT
An EDI protocol used in Europe.

IP address
A numeric address for a server on the Internet consisting of four groups of four digits.

packets
Data that have been grouped for transmission over a network.

In the networking software layer, the message from the application software layer is formatted according to whatever protocol will actually be used to send it over the network. Commonly used protocols for WANs are **Transmission Control Protocol/Internet Protocol (TCP/IP)** for the Internet and **ANSI X 12** or **EDIFACT** for EDI. For now, the discussion here will be restricted to the Internet; EDI will be covered later.

With TCP/IP, the networking software layer carries out a series of operations to prepare the message to be sent across the Internet to a destination computer. It must first convert the address of the server at the destination from a text form (for example, somecomputer.somewhere.org) to an **IP address**, which consists of four groups of numbers in the range 0 to 255 separated by periods. The address is converted by using a conversion table stored either on the user's computer or on a computer with which the local computer can communicate. For example, if you were sending an e-mail message to the president of the United States, whose address is President@Whitehouse.gov, then Whitehouse.gov, which is the name of the e-mail server at the White House, would be converted into the IP address 198.137.240.92.

Next, the message is divided into smaller digital units called **packets**, each of which contains a specific number of bytes. At this step, each packet is given a sequence number, and the destination address is added to it. This process is shown in Figure 3.7, where a series of binary digits has been converted to packets and the IP address is added to them.

Physical Layer

Once the message has passed through the application software layer and the networking software layer, it is ready to be sent out over the the network by the physical layer of the network model. Data and information transmitted over networks travel over various media, including twisted-pair wire, coaxial cable, fiber-optic cable, and microwave and satellite transmission.

twisted pair
A medium for data transfer that is made of pairs of copper wire twisted together.

coaxial cable
A medium for data transfer composed of a center wire, an insulating material, and an outer set of wires. Similar to that used to transmit cable television signals into your home.

fiber-optic cable
The newest type of data transfer medium that consists of thousands of glass fiber strands that transmit information over networks.

Twisted pair, which consists of twisted pairs of copper wires, is similar to the wiring used in much of the existing telephone system. It is used widely in many types of networks, both within and between locations.

Coaxial cable is used to transmit cable television signals into your home. It is also widely used in networks. In many areas, television cable is being converted to a type of cable capable of handling two-way signals instead of the one-way signals associated with television transmissions. This two-way cable enables the connection of home computers to ISPs at much faster speeds than those available with traditional telephone connections.

Fiber-optic cable, the newest medium, consists of hundreds of glass fiber strands that can transmit a large number of signals at extremely high rates of speed. The glass fiber strands also reduce the size of the cable required. However, individual computers are not set up to connect directly to fiber-optic cable, so it is often necessary to use twisted pair or coaxial cable for the last few feet to the computer. Figure 3.8 compares copper wire and fiber-optic cables as means for transmitting the same volume of information.

Fiber optics use strands of glass to transmit signals at high speeds. Satellite dishes, such as the one shown, receive data from orbiting satellites. A pair of copper wires, such as this, is the most common form of communication media. Coaxial cable is the backbone of the television cable system and is now being used for high-speed Internet connections.

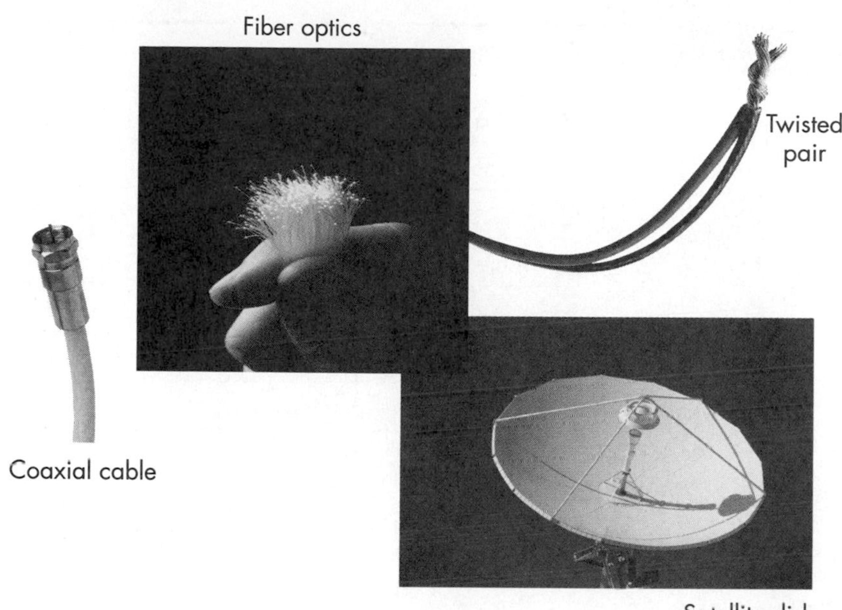

microwaves
High-frequency radio transmissions that can be transmitted between two earth stations or between earth stations and communications satellites, which are commonly used to transmit television signals.

satellite transmission
The use of direct broadcast, which uses microwaves for one-way downloads of data to homes and offices.

Microwaves are high-frequency radio transmissions that can be sent between two earth stations or between earth stations and communications satellites, which is the method commonly used to transmit television and mobile telephone signals. The use of direct broadcast **satellite transmisson**, which use microwaves for one-way downloads of data to homes and offices, is a new way of carrying out transmission of data *to* a user, especially where land lines do not exist or are difficult to install.

A variety of wireless technologies are also becoming popular as a way to provide mobile users with connections regardless of where they may be. The two most popular wireless methods of sending data involve the use of infrared light and radio. Infrared light wireless transmissions, like microwaves, require a line of sight. On the other hand, radio transmissions can pass through walls, so a line of sight is not required. However, security is a problem due to the radio waves going in all directions. Wireless networks will be discussed in more detail in a later section. Table 3.1 compares the various communications media.

Figure 3.8 Comparison of coaxial and fiber-optic cables

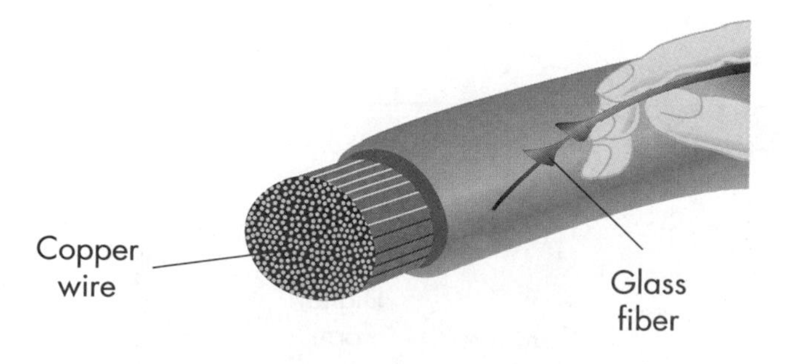

Table 3.1 Comparison of Media

Media	Cost	Error Rates	Speed
Twisted pair	Low	Low	Low-high
Coaxial cable	Moderate	Low	Low-high
Fiber optics	High	Very low	High-very high
Radio	Low	Moderate	Low
Infrared	Low	Moderate	Low
Microwave	Moderate	Low-moderate	Moderate
Satellite	Moderate	Low-moderate	Moderate

Source: Jerry Fitzgerald and Alan Dennis, *Business Data Communications and Networking*, 7th ed., p. 81, New York: Wiley, 2002.

signal type
The type of signal—digital or analog—being used to transmit bits between computers.

Cable modems such as this one are used for high-speed Internet connections.

Other aspects of the physical layer include the signal type and the data rate. The **signal type** is the manner in which data are sent over the network and can be either digital or analog. With digital transmission, bits being transmitted are sent as high and low electronic pulses. With analog signals, the bits are transmitted as wave patterns. A computer can transmit over a digital communications link without changing the data; to transmit over an analog channel, however, the data must be modified. Because most telephone and cable systems use analog signals, modems are used to convert digital signals from the computer into analog signals for transmission over the communications link. However, telephone and cable modems convert between analog and digital forms of data using quite different methods. In a telephone modem, at the sender's end, the modem modulates the digital computer data or information into an analog form that can be sent over standard telephone lines. At the receiver's end, a modem demodulates the analog signal back into a digital form that can be understood by the receiver's computer. The modulation/demodulation process for a telephone modem is shown in Figure 3.9.

Just as telephone modems are necessary to connect two computers over a telephone line, connecting two computers over a cable line requires a cable modem to modulate and demodulate the cable signal into a stream of data. In addition, cable modems incorporate a variety of other functions to allow the PC to be linked to a network.

Figure 3.9 Use of a telephone modem

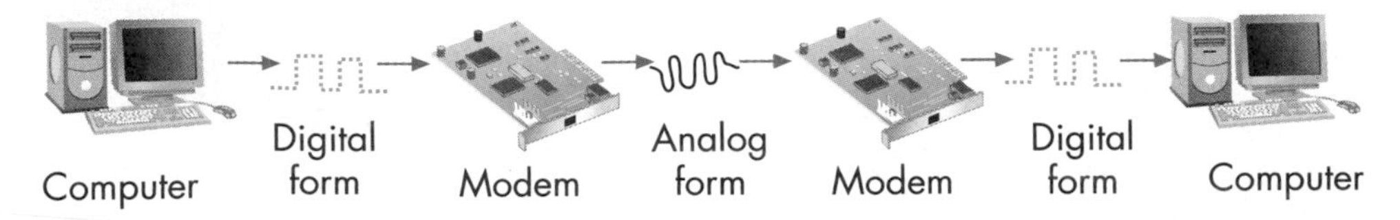

Table 3.2 Maximum Data Rates

Tranmission Method	Maximum Data Rate	Comments
Standard telephone service	56 Kbps	Available everywhere.
Digital subscriber line (DSL)	6 Mbps in; 640 Kbps out	Becoming more widely available. Does not require special equipment. Does not slow down as more people sign up.
Cable	As high as 55 Mbps but averages between 200 Kbps and 2 Mbps	Cable must support two-way communication. Available in many locations but slows down as more people use it in a specific location.
T-1 to T-4	1.544 Mbps-274 Mbps	Leased lines used for commercial telecommunication.

Source: Jerry Fitzgerald and Alan Dennis, *Business Data Communications and Networking*, 7th ed., New York: Wiley, 2002.

data rate
The number of bits per second transmitted between computers.

bits per second (bps)
A measure of the data rate.

digital subscriber line (DSL)
A digital method of data transmission using existing telephone lines.

T-carrier circuit
A digital method of data transmission over dedicated telephone lines.

bandwidth
A measure of the capacity of a communication channel, expressed in bits per second.

baseband
A classification of digital transmission in which the full capacity of the transmission medium is used and multiple sets of data are transmitted by mixing them on a single channel.

broadband
Simultaneous analog transmission of large amounts and types of data, including audio, video, and other multimedia, using different frequencies.

The **data rate** is measured in **bits per second (bps)**. For example, with a telephone modem, the maximum data rate is 56 Kbps. Even when using a 56 Kbps modem, the actual data rate can vary depending on the quality of the telephone line. Other methods of transferring data over telephone lines or cable have higher data rates. They include digital subscriber lines and the various T-carrier circuits. A **digital subscriber line (DSL)** transmits computer data in a digital form via the same telephone line that is used for analog voice communications. **T-carrier circuits** are dedicated digital lines that are leased from a telecommunications company to carry data between specific points. Table 3.2 shows the maximum data rates for various methods of transmitting data.

The term **bandwidth** is often used in relationship to the data rate and meaures how fast data flows on a transmission path. With the increasing demand from users in developed countries for the capability to view high-quality photos, graphics, and full-motion video on their computers, the competition to provide higher bandwidth access is becoming ever more keen among telecommunications providers. The two extremes of bandwidth are **baseband**, in which only a single digital signal is carried through the media, and **broadband**, in which a variety of analog signals are transmitted. Figure 3.10 shows the difference between baseband and broadband transmissions.

When someone orders food from www.fareastfoods.com, the physical layer includes the customer's modem and telephone line or cable modem and cable. When the message reaches the ISP, it is handled by the ISP's modem and hardware connections.

Receiving the Message

At the receiving end, the process followed in sending a message is reversed. That is, the message enters the physical layer and is passed up through the networking software layer, where the packets are put back together using their sequence numbers, the various identification bits are stripped off, and the message is decrypted. The complete message is then sent to the application software layer of the receiving computer, which interprets it. If it is an e-mail message, it is displayed by the e-mail software; if it is a Web page, it is displayed by the Web browser. For the food order to FarEast Foods, the message containing the data is delivered to a Web server, where software interprets the order and starts the process of fulfilling the order.

Figure 3.10 Difference between baseband and broadband transmissions

Baseband transmission

Data, voice, video

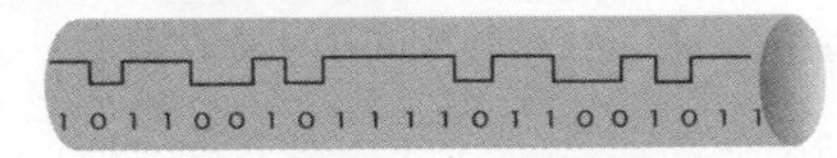

Broadband transmission

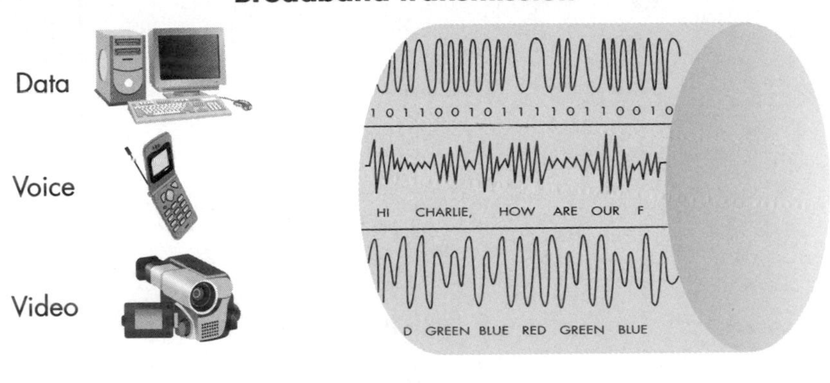

Packet Switching

Recall that the networking software layer divides an Internet message into groups of bytes called packets. When the physical layer sends these packets over telephone lines, it uses an approach that differs from the approach used to send voice and fax telephone calls. For voice and fax calls, a complete path from the caller to the receiver is created and kept open during the duration of the call. Because computers send large amounts of data quickly and then do not send any data for a while, such an approach would be very inefficient, tying up telephone lines when none is needed. Instead, using a technology called **packet switching**, individual packets are routed through the network based on the destination address contained in each packet. With packet switching, many computers in the network can share the same data path, and if a computer on the network becomes inoperable, the packet can find another way to reach its destination. Packet switching has been the key technology that has made the Internet work as well as it does.

packet switching
In a wide area network, dividing long messages into smaller data units that can be transmitted more easily through a network.

router
A computer that determines the path that a message will take from the sending computer to the receiving computer.

Routers, such as the one shown here, are essential to sending data over the Internet.

When a group of data packets, like the food order or an e-mail message, is sent to a computer with an IP address, software on the sending computer transmits the packets to the *nearest* router for retransmission to other routers on the network. A **router** is a special type of computer that has one purpose: accepting packets and determining the *best* way to send them to the destination computer. That is, the router specializes in *switching the packets.*

Note that in this context, the terms *nearest* and *best* do not have the same meanings as they would for someone taking a trip from Charleston, South Carolina, to Seattle, Washington. Instead, they refer to the least-congested network path to the eventual goal. Because packets travel so fast over the Internet, delay time is more important than distance in determining total time to deliver a message. Very sophisticated software has been written to carry out this process on the routers and speed the data packets to their destination. Figure 3.11 shows the process of sending data from one computer to another on the Internet. Note that the packets don't flow through all routers—just the ones that are necessary to get the message to the destination.

Because a packetized message can be reconstructed using the sequence order that is attached to each packet, all packets do not necessarily need to follow the same path through the network or arrive in the order in which they were sent. As they arrive, the destination computer acknowledges received packets. If the sending

Figure 3.11 Sending data between computers on the Internet

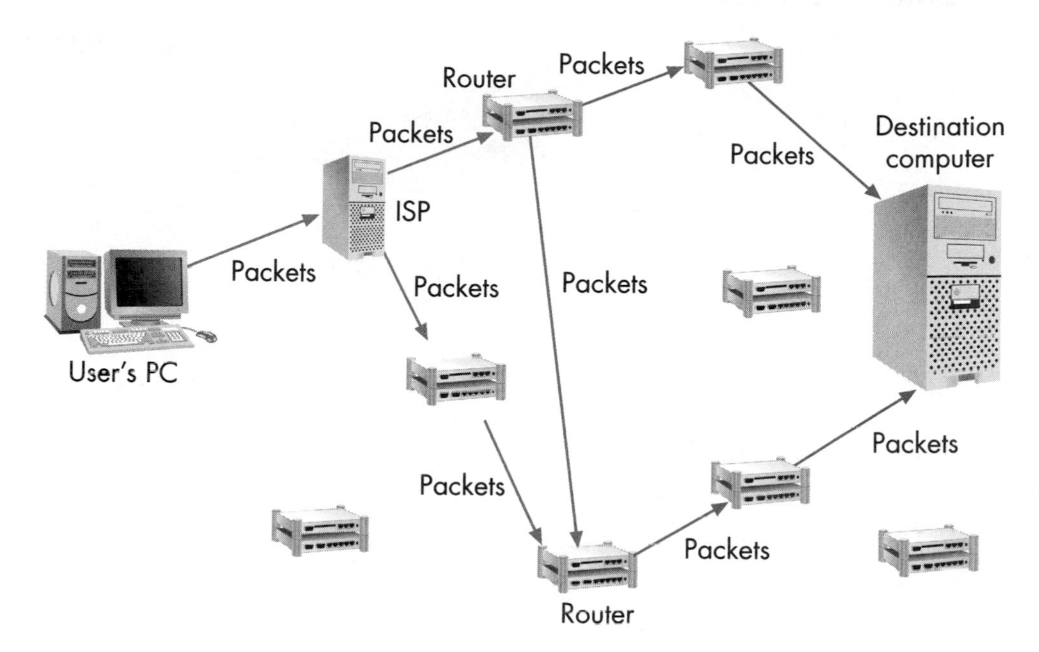

computer does not get an acknowledgment that a packet has arrived within a certain time frame, it automatically resends the unacknowledged packets. This practice ensures that all packets are eventually received.

Electronic Data Interchange

Although networks have been discussed largely in the context of the Internet in this chapter, it is important to be aware that EDI is a heavily used protocol when businesses must exchange data and information, automating much routine business among retail stores, distributors, and manufacturers. Instead of sending paper documents, such as purchase orders, invoices, bills of lading, and shipping slips, back and forth through traditional communications channels, EDI allows companies to transmit the same information electronically between their computers.

Electronic Data Interchange (EDI) is a popular method of sending data between trading partners.

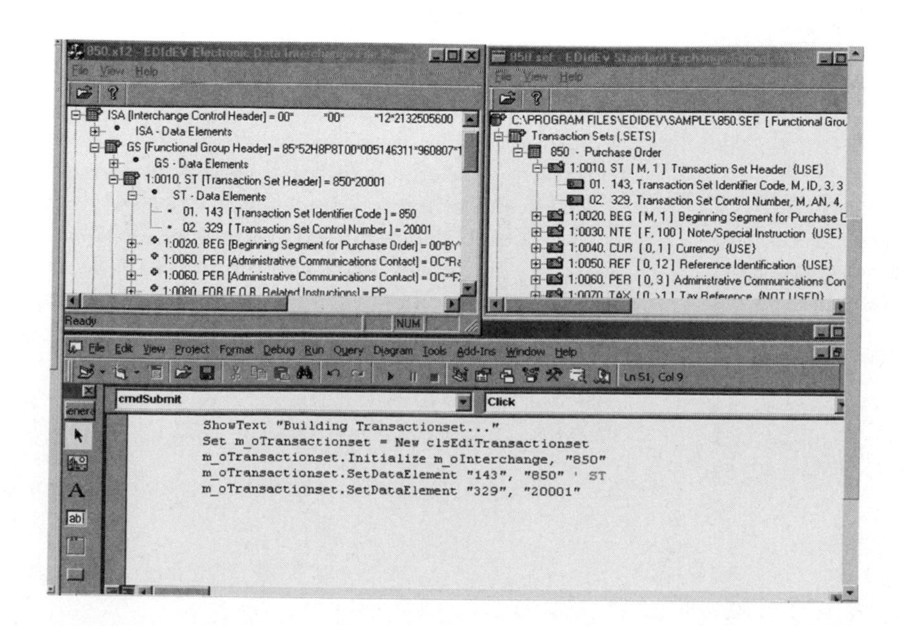

Figure 3.12 Use of EDI compared to traditional transaction handling

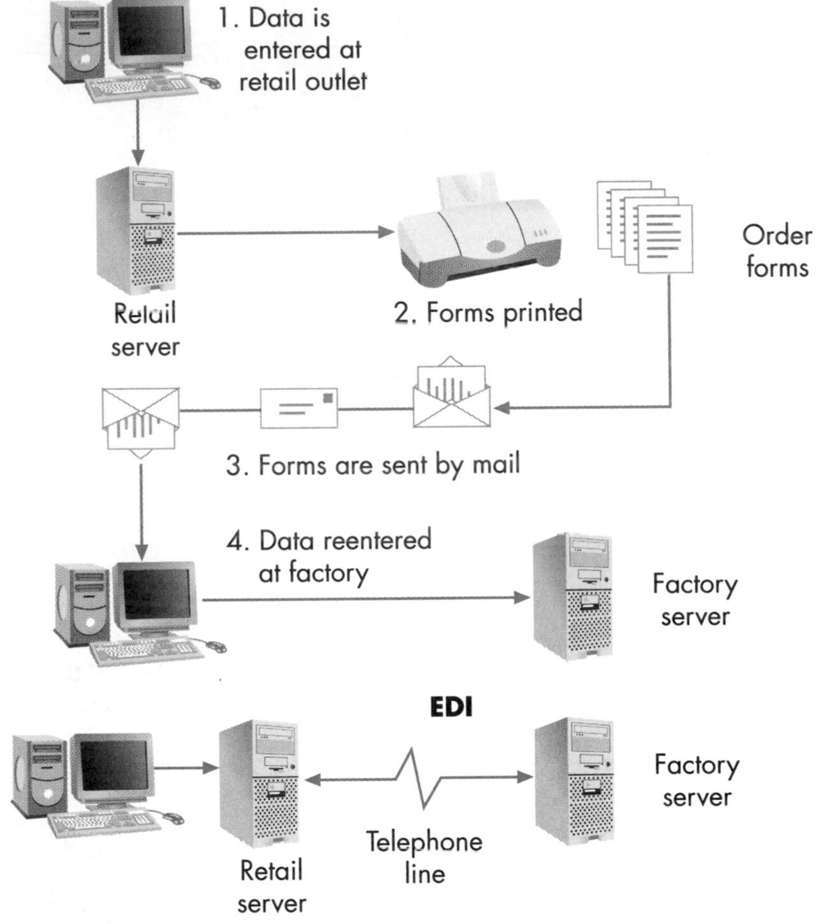

By combining EDI with a point-of-sale inventory system, which automatically subtracts sales from the store's inventory as they are made, a computer at a retail store can automatically order goods, based on units sold, from its supplier. The supplier, in turn, can automatically ship the goods to the retail store and electronically transmit the appropriate documents (invoices and bills of lading or shipping slips). EDI greatly reduces human involvement in the ordering and shipping process, thereby reducing costs and speeding service. Figure 3.12 compares the use of EDI to traditional methods of handling transactions.

value-added network (VAN)
A public network, available by subscription, that provides data communications facilities beyond standard services; often used to support EDI.

EDI and other non-Internet protocols are typically sent over value-added networks. These **value-added networks (VANs)** are available by subscription and provide clients with data communications facilities. The company that runs the VAN assumes complete responsibility for managing the network, including providing any conversions necessary between different systems. In a sense, a VAN adds value to the data by ensuring that it reaches its destination with little effort on the part of the subscriber.

1. What are the three layers of the simplified network layer model?
2. In which layer is the message encrypted? In which layer is the message packetized?

Understanding Local Area Networks

As discussed in Chapter 2, most organizations today use LANs to share information and resources among employees or members and to connect to the Internet. Sharing information enables users to work with the same data or information files and to send messages and files via e-mail. Sharing resources involves the users' ability to share software and hardware. Sharing software means that the organization will not need to purchase a copy of a software package for every computer in the organization. Instead, a number of software licenses sufficient to meet most needs can be purchased and the software then shared. (It is illegal for multiple persons to simultaneously use a software package unless a sufficient number of licenses have been purchased.) Sharing hardware means that users can use the same printers, disk storage, scanners, and so on through the network rather than each user having every possible piece of hardware. Making hardware available through the LAN, especially highly specialized types of hardware, can significantly reduce an organization's hardware costs.

dedicated server network
A network in which at least one of the computers linked to the network acts as a server.

Most LANs are **dedicated server networks** in which at least one of the computers linked to the network acts as a server that is accessed by the *client* computers on the network. Recall from Chapter 2 that a *server* is a computer that carries out one of many specialized tasks enabling sharing of information and hardware. Types of servers include file servers, database servers, Web servers, application servers, communication servers, and so on.

peer-to-peer network
A network configuration in which each computer can function as both a server and a workstation.

LANs can also be implemented through **peer-to-peer networks**, which originally were designed to be used for smaller networks where the emphasis was on sharing files between computers, with each computer functioning as both a server and a workstation. Peer-to-peer networks are significantly less expensive than the dedicated server configuration, but until recently were not thought to be well suited for heavy-duty transaction processing. However, the success in using peer-to-peer configurations for handling large numbers of music file transfers on Napster and other Web sites has led to its being considered for other large-scale transfers of files between users over the Internet. Significant security concerns also arise when using a peer-to peer network for working with sensitive material. One issue that the new Windows XP operating system tries to address is these security concerns. Figure 3.13 shows both the dedicated server and peer-to-peer network configurations.

LAN Components

Local area networks are typically composed of five basic elements: servers, clients, network cabling and hubs, network interface cards, and a network operating system. Because servers and clients were discussed in Chapter 2, the focus here will be on the other three elements.

network cabling
The actual physical wire over which computers communicate.

hub
A device for concentrating connections to multiple network devices.

Network cabling physically connects each computer on the network and connects the hardware peripherals through a hub. A **hub** is a device that allows cables to be connected together. Hubs can also be used as amplifiers or repeaters to extend the range of the network. Twisted pair using copper wire is the most widely used medium for LANs due to its low cost. Coaxial cable and fiber optics are also used in situations requiring high-speed networks.

Figure 3.13 Dedicated server and peer-to-peer network configurations

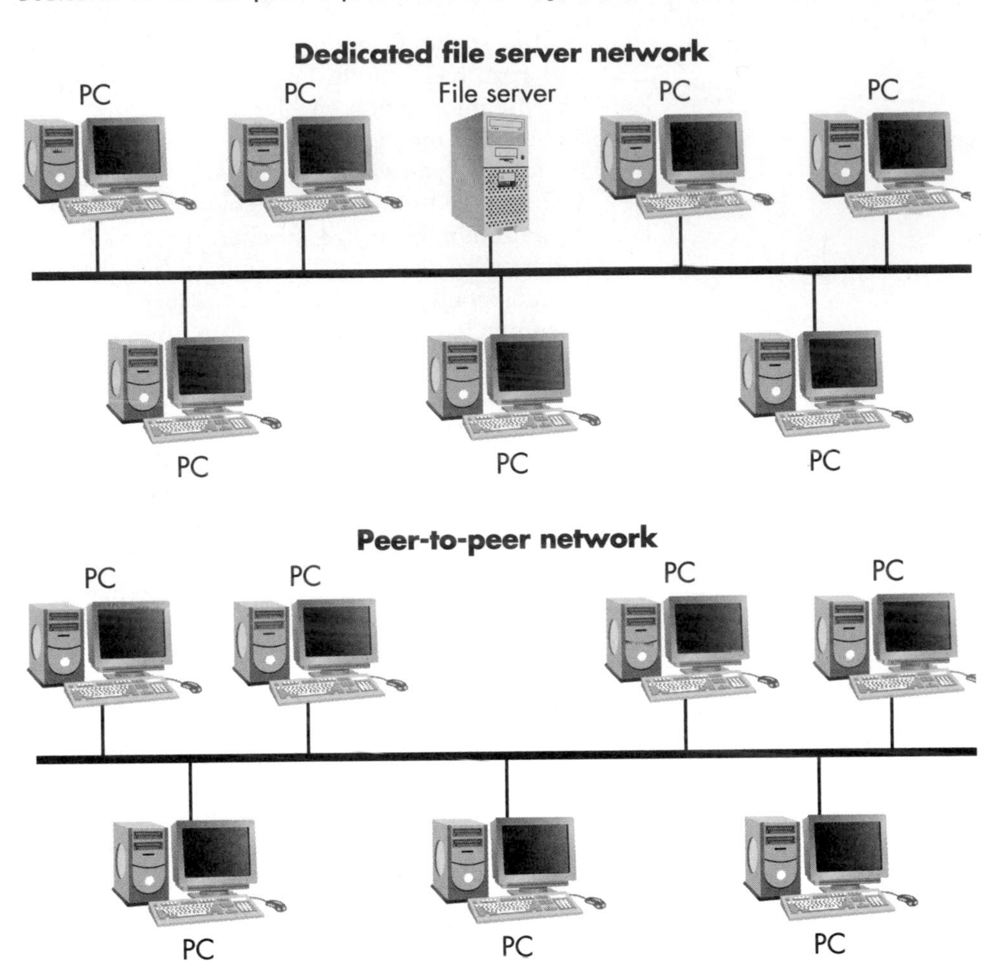

network interface card (NIC)
A card in a PC that connects the PC to the network and that handles all electronic functions of network access.

The network cabling is connected to computers on the network through a network interface card within each computer. The **network interface card (NIC)** handles all the electronic functions entailed in connecting a computer to the network, including receiving information that is intended for a particular computer and sending information from it over the network as needed. Figure 3.14 shows clients connected to the network cabling through NICs and then to a server through a hub.

Network interface cards (NICs) are used to connect computers to the LAN cabling.

Hubs are used to concentrate cabling from multiple computers on a LAN.

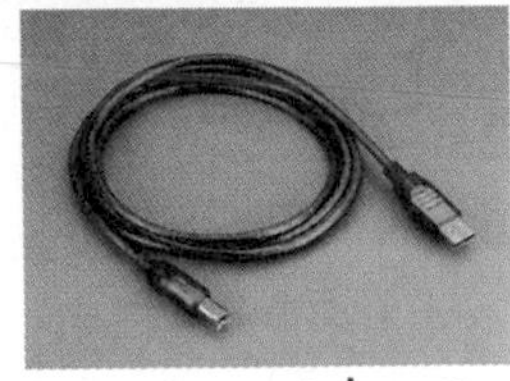

Cabling is an essential part of a LAN.

Figure 3.14 Connecting computers on a network

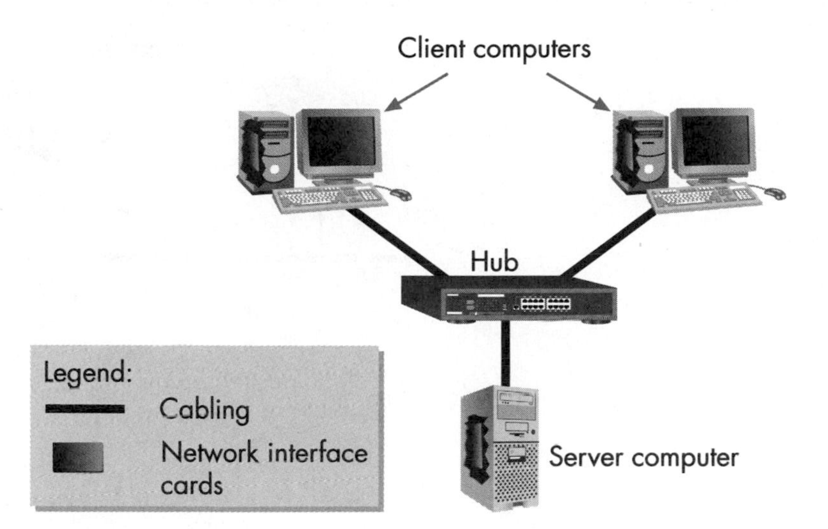

network operating system (NOS)
The software that controls a computer network.

The **network operating system (NOS)** is the software that controls the network. It is primarily located on the server, but a NOS component also resides on each client computer. The server portion of the NOS runs the server and handles operations associated with supervising the network. The client portion of the NOS handles the connection of the client to the network and its communication with the server. By creating different versions of the client NOS software, it is possible to connect dissimilar computers (for example, Apple Macintosh and Windows-based PCs) to the same LAN. The most popular LAN operating systems are those from Novell (Netware) and Microsoft (Windows 2000).

Ethernet LANs

Ethernet protocol
The most popular protocol for controlling LANs; runs on a bus network and uses collision avoidance methodology.

bus network
A computer network in which computers are tied into a main cable, or bus.

bus
A primary cable to which other network devices are connected.

gateway
A combination of hardware and software that connects two dissimilar computer networks.

bridge
A combination of hardware and software that connects two similar networks.

Today, the vast majority of LANs use the **Ethernet protocol** to connect computers and to move information between computers on the network by transmitting packets on a bus network. A **bus network** uses a a main cable, called a **bus**, to connect all clients and servers on the network. With the Ethernet protocol, a computer on the network transmits a message that contains the address of the destination computer. Because all computers are free to transmit at any time, collision-detecting software must be in place to control those cases in which two or more computers try to transmit at the same time. When a collision is detected, each computer is directed to stop transmitting and wait a random length of time before retransmitting its message. This system, which works quite well, is the basis for most LANs in operation today.

If you looked at a bus network, it would appear to form a star because the actual bus is located in a central hub to which all clients are connected. Figure 3.15 shows the logical and physical setup of a bus network.

A client on a LAN can not only share information and software with other PCs on the same LAN, but also communicate through gateways and bridges with other types of computers and with other LANs. A **gateway** is the combination of hardware and software that connects two dissimilar computer networks. For example, it allows a LAN user to access a mainframe network without leaving his or her PC. Similarly, a gateway between a LAN and a WAN enables a LAN user to send e-mail over the WAN. In contrast, a **bridge** connects two similar networks. For example, if two LANs are connected with a bridge, computers on each LAN can access the other network's file server without making any physical changes to the data.

Figure 3.15 Logical and physical setup of a bus network

LANs at FarEast Foods

As mentioned earlier, FarEast Foods has three LANs—one for administration, one for the individuals who create the combinations of different types of Asian foods that are shipped to customers, and one for handling incoming orders. All three LANs are bus LANs that use the Ethernet protocol and are connected via a backbone network that acts as a bridge. The bridge enables users on one LAN to access a server on the other LAN without making special connections. The administration LAN connects the various units of the company so they can communicate about issues facing the company. The order-handling LAN connects the temporary employees who make up and ship the orders with the application server that generates the instructions on the orders to be filled. When the packages are ready to go out, the bar code on the packing slip is scanned and the information is returned via the LAN to the application server, which updates the database server. The customer service LAN connects the employees who fill telephone orders.

1. What are the five basic elements of a LAN?
2. How does the Ethernet protocol work?

Wireless Networks

The fastest-growing trend in networks is wireless networks, both WANs and LANs. As the name implies, no wires are necessary to make the connection. For WANs, no high-speed telephone lines are necessary; for LANs, the internal wiring in buildings is not necessary. This setup gives wireless networks obvious advantages because devices connected to the network can go virtually anywhere—to the top of mountain or to the most remote part of a building.

In the case of wireless WANs, a mobile telephone client is the most popular method of connecting to the Internet and Web. International Data Corporation predicts that wireless access to the Internet will be the most popular method by 2002, and Gartner Group predicts that more than 1 billion wireless handsets will be in use by 2003. Because of differences in the mobile communications protocols, mobile telephones are not yet as widely used in the United States for wireless access to the Internet as they are in Europe and in the rest of the world.

Wireless Application Protocol (WAP)
A protocol designed to enable mobile telephones to access the Internet and the Web.

Short Message Service (SMS)
A system that enables mobile telephones to send and receive text messages up to 160 characters in length.

Global System for Mobile communication (GSM) protocol
The most widely used standard mobile telephone protocol in the world. A digital communication system, it is only now being adopted in the United States.

A number of mobile telephone companies have collaborated to create a special protocol, called **Wireless Application Protocol (WAP)**, just so that their telephones can connect to the Internet. These companies—Ericsson, Nokia, Matsushita (Panasonic), Motorola, and Psion—have also created a company named Symbian (www.symbian.com) to develop and market an operating system named Symbian OS for their wireless devices that will support browsers and other software using WAP. Symbian OS provides contacts information, messaging, browsing, and wireless telephony.

WAP has been transferred to some models of the PDAs running the Palm operating system, enabling them to take advantage of special WAP-based Web sites that have been created. These WAP-based Web sites *cannot* normally be viewed with the most popular browsers, Netscape and Internet Explorer, because they have been specially configured for WAP devices connected to a WAP server.

Another popular reason for connecting mobile telephones to a WAN is to use Short Message Service. **Short Message Service (SMS)** is a service for sending text messages up to 160 characters long to mobile telephones. Although mobile phone systems in the United States can send and receive SMS messages, the widest use of SMS is in systems that use the Global System for Mobile communication protocol. The **Global System for Mobile communication (GSM) protocol** is the most widely used mobile telephone protocol in the world, with the exception of the United States where it is gaining ground. SMS is similar to paging except that the mobile telephone does not need to be active and messages can be held until it does become active. It is also possible to send messages using a Web site to GSM-compatible mobile telephone phones.

This advanced GSM mobile phone is capable of displaying images as well as receiving audio and text.

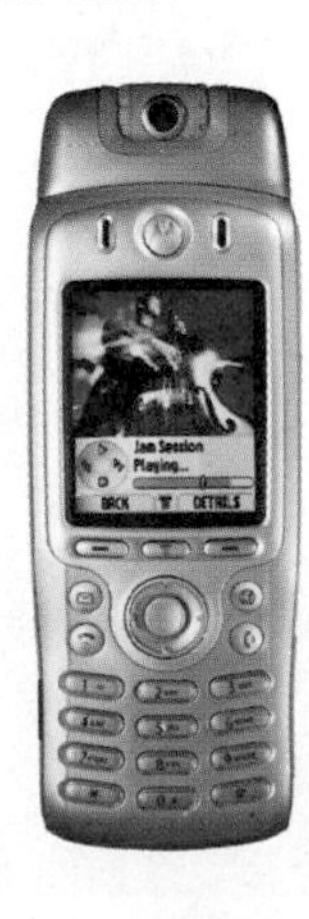

SMS is extremely popular in the parts of the world where GSM is used, with more than 200 *billion* messages being sent in 2001. Beyond the obvious personal applications, numerous business applications are possible:

- Notifying a salesperson about a request for information along with a number to call
- Informing a package delivery person of a pickup that came in after the courier left the office
- Providing a service person with the name and location of the next service call

You can undoubtedly think of many other applications of SMS in the workplace.

Applications such as WAP and SMS are becoming so popular that some mobile telephone companies in the United States either have switched or are in the process of switching to a version of GSM so as to provide these services to their customers.

Wireless LANs

Wireless LANs (WLANs), in which the usual LAN cabling is replaced with wireless transmissions between computers, are becoming increasingly popular as mobile users find that they need to connect to their local network and, possibly, from there to the Internet. WLANs can be very useful in a variety of situations, including eliminating the need for cable in difficult-to-access areas, providing an inexpensive alternative to shared printing, and connecting two networks separated by some obstacle, such as a highway or wall, through which cable cannot be run.

Examples of the use of WLANs in the workplace include the following:

- Using wireless devices in a retail environment to complete activities such as pricing, labeling, handling orders, and taking inventory from anywhere in the store with information being communicated directly to the back office computer
- Using a wireless device connected to a bar code scanner to scan items in a warehouse, thereby producing a list of items and their locations
- Using a wireless device in a medical setting to request medical tests, the results of which are entered into a patient's electronic record; the requesting person can then use the same wireless device to check the results
- For attendees at a conference, checking e-mail on their laptops from anywhere in the conference hotel without having to be connected by wiring

TECHNOLOGY ON THE EDGE

Bluetooth and PANs

One of the newest wireless technologies is being developed by a consortium of companies including Nokia, Ericsson, and Motorola. Named Bluetooth after the tenth-century king who unified Denmark and Norway, it is a form of personal area network (PAN). PAN technology enables wireless devices such as mobile telephones, computers, and PDAs to communicate over a short distance—less than 33 feet (10 meters). The idea behind Bluetooth is to embed a low-cost transceiver chip in each device, making it possible for wireless devices to be totally synchronized without the user having to initiate any operation. The chips would communicate over a previously unused radio frequency at up to 2 megabits per second (Mbs).

As envisioned by its developers, a single device adhering to the Bluetooth protocol would be used as a mobile telephone away from the office and a portable telephone in the office. In addition, it would work as a PDA and quickly synchronize information with a desktop computer (also containing a Bluetooth chip); alternatively, it might act as a remote control to initiate other operations with other devices containing the appropriate chip. For example, a user might initiate sending or receiving a fax or printing or copying a document from anywhere in the office. The overall goal of Bluetooth might be stated as enabling pervasive connectivity between personal technology devices without the use of cabling.

Using Bluetooth, it is possible to communicate over short ranges with other users.

With Wi-Fi, the base station communicates with wireless cards in desktops (bottom) and laptops (top).

Wireless networking hardware uses radio frequencies to transmit information between individual computers, each of which has a wireless network adapter. The individual computers do not communicate directly with each other; instead, they communicate with a wireless network hub or router, which is also used to bridge the wireless network to a traditional Ethernet network or provide a shared Internet connection. The currently popular standard for wireless networking supports a data rate of 11 Mbps, with a typical range through open air of about 220–1100 yards (200–1000 meters). Nicknamed Wi-Fi (for Wireless Fidelity) by the industry group that is supporting it, this approach to wireless LANs uses the IEEE 802.11b standard for short-range radio transmissions.

Its widest use so far is at colleges and universities, where students take their laptop computers to the student center or out on the lawn to work while remaining connected to the campus network.

Eventually, this standard may be surpassed by the 802.11a standard, which allows for data rates of up to 54 Mbps. Figure 3.16 shows how a wireless LAN connects a number of laptops to a hub that, in turn, is connected to the organizational LAN.

1. What is WAP and what does it have to do with wireless networks?
2. What is the function of a wireless network adapter in a WLAN?

With a wireless LAN, you can use laptops virtually anywhere.

Figure 3.16 Laptops connected by a WLAN linked to a LAN and the Internet

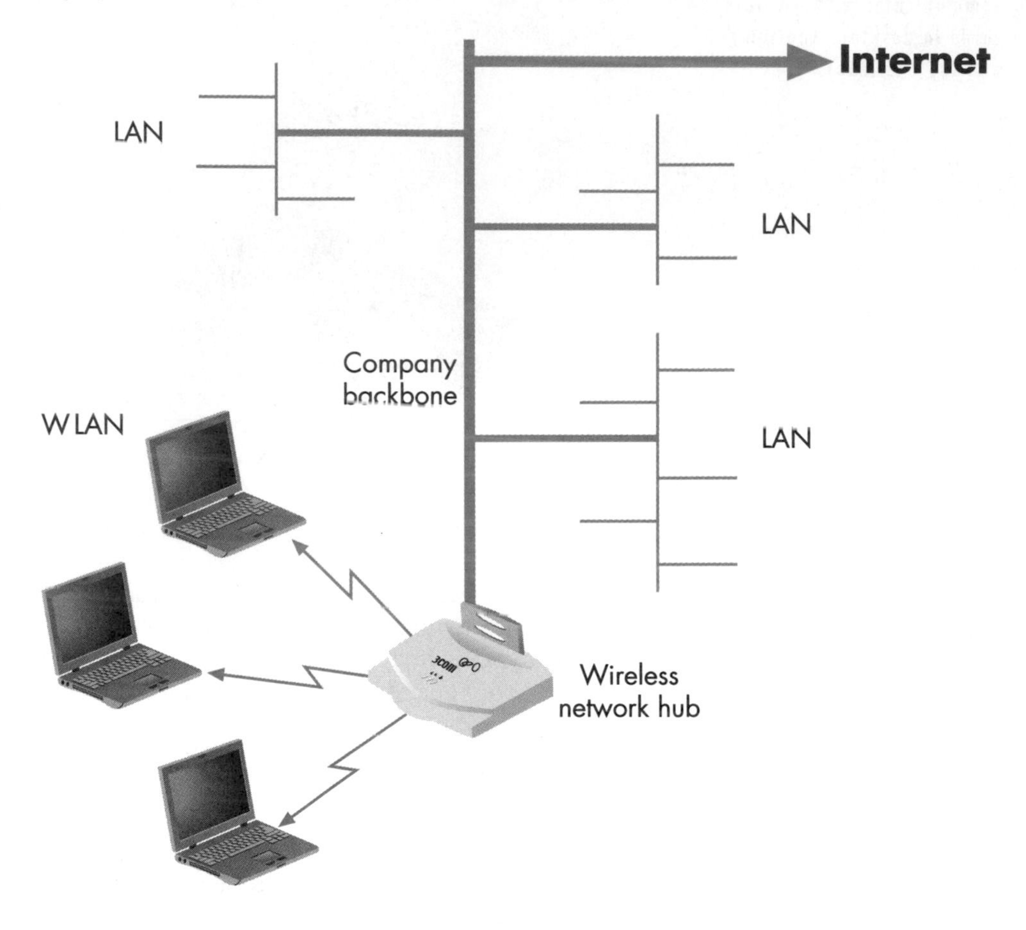

The Internet: A Network of Networks

Thanks to its tremendous growth over the last decade, the Internet has been the subject of much discussion in newspapers, books, magazines, movies, and so on. For many companies, the Internet is the basis for the widespread use of electronic commerce that eventually will lead to an even wider networked economy. Without a doubt, the Internet is the biggest technology innovation to come along since the invention of the computer itself more than 50 years ago.

Originally developed in the 1960s and 1970s as a way of sharing information and resources among universities and research institutions, the Internet began its dramatic growth in 1991, when the U.S. government, which had been subsidizing it, opened the Internet for commercial use. This growth further accelerated when the World Wide Web was introduced in 1994. Today, the Internet is growing so fast that no one can say exactly how many people are using it. Estimates range as high as almost 550 million users in February 2002[1].

A primary reason for the explosive growth of the Internet is the tremendous amount of data, information, and resources that people can access using its applications. These data and information include large numbers of text documents (many of which are indexed by topic or keyword), both images and graphics, and both audio and video recordings that are available over the World Wide Web. Resources

1. http://www.nua.com.

include free or very-low-cost software that can be downloaded, discussion groups on almost any subject imaginable, live chat groups that discuss a variety of issues, e-mail communcation with people all over the world, and access to computers other than your own. Of these applications, the World Wide Web and easy-to-use Web browsers clearly lead hundreds of millions of people to become interested in using the Internet and accessing data and information.

What Is the Internet?

The Internet is not a single network, but rather a network of networks. In fact, the name *Internet* is a shortened version of the term *inter-networking*, meaning that it allows you to work with multiple networks simultaneously. As shown earlier, to connect to the Internet, your computer will usually first connect to a LAN through a NIC or to an ISP through a modem and telephone line. The LAN, mainframe, or ISP is, in turn, connected to a regional network via a high-speed (T-1) telephone line. The regional network, in turn, links into the backbone of the Internet. Figure 3.17 shows a portion of the Internet backbone network in the United States.

host computer
A computer in a network that is connected to the Internet and has a unique Internet address.

With each network, at least one **host computer** is connected to the Internet with full two-way access to other computers on the Internet and with a unique Internet address. In many cases, all computers in the network are host computers because all have Internet addresses, even if for only a short time as with a dial-up system.

Each host computer that connects to the Internet uses TCP/IP for assigning addresses and uses packet switching for exchanging information. By having all networks following the same TCP/IP protocol, users on any network can exchange information with users on other networks with little or no knowledge of their physical location or configuration. An important part of this set of rules is the way in which e-mail and other Internet addresses are assigned. For example, an e-mail address is composed of two parts: the user name and the **server address**. The user name is assigned to a person or organization that is connected to a server, and it is separated from the server address by the at symbol (@). The server address, also known as the **domain name**, consists of groups of letters separated by periods. Moving from right to left, this address goes from

server address
The address of a server computer on the Internet.

domain name
Another name for the server address.

Figure 3.17 Internet backbones in the United States

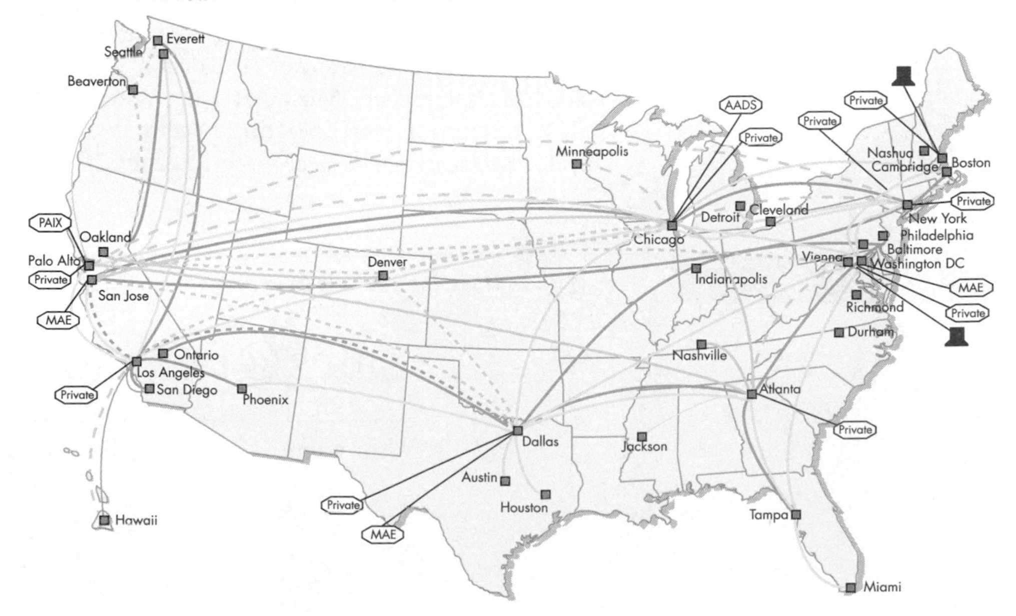

Source:

top-level domain
One of the 14 domain names that define the type of service or area of interest of the server.

the most general (country name or organization type) to the most specific (computer name). The rightmost part of the address is known as the **top-level domain**. It is important to note that these Internet addresses are easy-to-remember versions of the numeric IP addresses that actually identify computers on the Internet. As an example, Figure 3.18 shows the e-mail address of the sales department at FarEast Foods.

In this example, the user name is *sales*, the server name is *fareastfoods.com,* and the top-level domain is *com*. Even though the country (the United States) is not shown, by default it is known to be located in the United States. For servers located in another country, a two-letter suffix is required at the end of the server name as the top-level domain (for example, uminho.pt for the University of Minho in Portugal). Because the Internet originated in the United States, an Internet country suffix is not required for the United States. It is quite common in countries other than the United States for businesses to use a second-level domain name of "co" prior to the top-level domain corresponding to their country. For example, the Tanda Tula game preserve in South Africa has a domain name of tandatula.co.za, where the top-level domain for South Africa is za.

Because the FarEast Foods example is the address of the general e-mail server, it has just the company name (fareastfoods) as the server address. Other servers at FarEast Foods may have additional names to distinguish them from this server—say, returns.fareastfoods.com for the e-mail server for the returns department or www.fareastfoods.com for the Web server.

In 2001, seven new top-level domains were added by the Internet Corporation for Assigned Names and Numbers (ICANN), a not-for-profit company set up specifically for the purpose of administering the domain name system. Table 3.3 shows the top-level domains and the types of organizations that might use them. Of the seven new top level domains, only .biz, .info, and .name were accepting applications at the beginning of 2002. The others are expected to begin accepting applications sometime during 2002.

Using the Internet

A number of software applications run on the Internet, including the World Wide Web, e-mail, chat rooms, and so on. All use the client/server approach, with each server on the network providing data and information to client computers connected to the network. The client computer that you are using must run two types of software to take advantage of the Internet: Internet conversion software and client software. The Internet conversion software enables your computer to work with Internet packets. Most operating systems, including Windows, come with this software built into the operating system; therefore, it is not necessary to run special application software to access the Internet.

Figure 3.18 E-mail address for the sales department at FarEast Foods

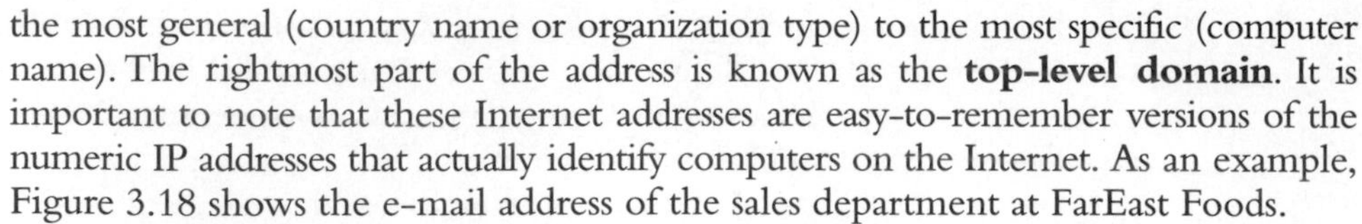

Table 3.3 Top-Level Domain Names

Type of Organization	Designation	Example	Example Organization
Educational institution	.edu	www.uga.edu	University of Georgia
Military	.mil	www.usmc.mil	U.S. Marine Corps
Commercial company	.com	www.fareastfoods.com	FarEast Foods, Inc.
Nonprofit organization	.org	www.icann.org	ICANN
Network provider	.net	www.negia.net	Northeast Georgia Internet Access, Inc.
Government	.gov	www.ustreas.gov	U.S. Treasury Department
Aerospace organizations*	.aero	Not yet active	
Businesses*	.biz	www.webmaster-resource.biz/	Webmaster-Resource.biz (e-commerce company)
Cooperatives*	.coop	Not yet active	
Various*	.info	www.wtcrelief.info	Information on relief for victims of the September 11, 2001, World Trade Center terrorist attack
Museums*	.museum	Not yet active	
Various*	.name	www.pat.mckeown.name	The author of this book
Professionals*	.pro	Not yet active	

*These are the new top level domains approved by ICANN in 2001.

Once you have access to the Internet, you need client software to carry out the desired operation, such as sending e-mail, downloading files from or uploading files to a server, participating in discussion groups, working on someone else's computer, or accessing the World Wide Web. For example, to send e-mail, you would use an e-mail client to generate the message, which then goes to the Internet conversion software that translates it into a form that can be sent over the Internet. Table 3.4 shows the six most widely used Internet operations, which will be discussed in more detail shortly.

At one time, you needed a separate client software package for each of the Internet operations. For example, to send e-mail you needed a separate e-mail client software package. Similarly, to retrieve software or documents using FTP, you needed a separate FTP client software package. Today's client software for the World Wide Web in the form of the Web browser has the capability to carry out *all* of the Internet operations, so you no longer need separate client software packages for each operation unless you especially want them. Let's now take a brief look at each of these applications.

Table 3.4 Internet Operations

Internet Operation	Purpose
E-mail (electronic mail)	Asynchronously exchange electronic messages with other Internet users
FTP (File Transfer Protocol)	Download files (software, documents, or data) from or upload files to a server located on the Internet
Newsgroups	Participate in a wide variety of online discussion groups
Telnet	Work on a computer elsewhere on the Internet
Internet Relay Chat	Synchronously exchange electronic messages with other Internet users
World Wide Web	Transfer text, images, video, and sound to your computer; search for information on the Internet

E-Mail

According to a Gallap poll, e-mail over the Internet is the most popular application, with more than half of all users interviewed saying it is their number one online activity. E-mail has fundamentally changed the way much of personal and business communication takes place. In Table 3.4, e-mail is described as an asynchronous method of exchanging information. **Asynchronous** means that the sender and the receiver are not communicating at the same time, but rather one sends a message that is read and replied to at some other time. In contrast, in **Internet Relay Chat**, which includes chat groups and instant messaging, the communication is **synchronous**—that is, users communicate back and forth at the same time in much the same way as a telephone cconversation.

asynchronous
A type of communication in which only one of the parties can send messages at a time.

Internet Relay Chat
An Internet protocol that enables a user to carry on a conversation by typing messages.

synchronous
A type of communication in which more than one of the parties can send messages at a time.

As mentioned earlier, e-mail typically uses SMTP for sending plain text files. Mail containing graphics, animation, and so on actually uses the Web protocol, which will be discussed shortly. To send an Internet e-mail message, you simply start your e-mail client (often part of a Web browser), enter the Internet address of the recipient, type a message, and then instruct the e-mail client to send it. You can send a file by attaching it to the e-mail message. Reading your e-mail is even easier: Just click or double-click the message line in your Inbox (also known as New Mail in some e-mail systems) window, and the text of the message will appear on the screen. You can also reply to the message, forward it to someone else, or save it for future reference. If the message has a file attached to it and you choose to open the file, the software will either start the software program associated with the file or ask you where to save it.

listserv
A group e-mail function available on the Internet; it enables end users to subscribe to special-interest mailing lists.

A special use of e-mail is a listserv. A **listserv** is server software that can broadcast an e-mail message from one member of a group to all other members. Group members simply *subscribe* to a listserv, and any messages sent to the listserv are broadcast to the group members automatically. Depending on how the listserv is set up, a member may be able to send messages to the listserv or may be able only to receive messages.

E-mail has a number of commercial applications. Many companies are finding that this form of communication has become a very important marketing tool. Incoming e-mail can provide them with queries about their products, information about problems with products, or suggestions of ways to better serve their customers. It can also be combined with other contacts to create mailing lists for either e-mail or postal mail (often referred to as "snail mail") as a means of communicating with customers. At the same time, customers expect companies to respond to e-mail queries, complaints, or comments generated when they click a link on a commercial Web page to send an e-mail to the company. Failure by a company to respond to these messages can result in very unhappy customers. Companies are finding it necessary to develop systems of responding to incoming e-mail in addition to sending an automated response that acknowledges receiving the customer's e-mail message.

In addition, companies are discovering that outbound e-mail can generate revenue in ways never before considered. For example, they may send special offers to customers who have purchased items online or who have sent a question via e-mail. Airlines, for instance, regularly send notices of special fares that can be purchased only online, a practice that has significantly increased ticket sales with almost no additional cost to the airlines. FarEast Foods could very easily use outbound e-mail to alert customers about the availability of special Asian food products or sales on certain combinations of products.

FTP

Companies and individuals frequently need to make software, data, or document files available to a wide audience over the Internet. A software company may want to distribute to current users an upgrade of a software package that it markets. In addition, a company may make a new piece of software available for a trial period (usually 30 days) and then sell it to the user at the end of the trial period. Or a company may make a utility software package or data files freely available. In most cases, the best way to distribute files to a large audience is to place them on a **File Transfer Protocol (FTP)** server and have the users download them over the Internet. A company like FarEast Foods could use FTP to distribute software that enables its customers to convert between various units of measure, such as grams to ounces, or vice versa.

File Transfer Protocol (FTP)
A protocol that supports file transfers over the Internet.

Using an FTP site to download files is straightforward. You start FTP client software or a Web browser, enter the address of the FTP server that you want to use, and enter your user ID and password. At this point, you will see a list of directories or folders on the FTP server from which you can select files to download. By selecting a file and entering or pointing to the download command, the file will be transmitted over the Internet to your local computer. You do not need a user ID and password to access an **anonymous FTP site**; such sites are open to anyone who wants to download files. Some private sites do require a user ID and password as a security measure to prevent protected or personal files from falling into the wrong hands. Figure 3.19 shows a list of files on the client computer and FTP server using popular FTP software. Downloading or uploading files with this client entails a simple

anonymous FTP site
An FTP site that does not require users to have user IDs and passwords.

Figure 3.19 Use of an FTP client

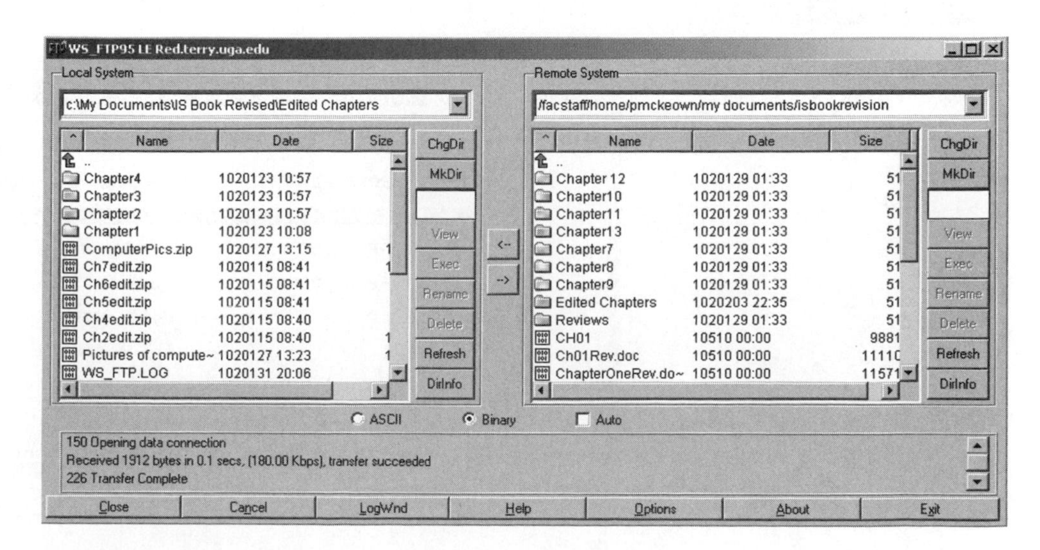

process: highlight the file and click the upload or download arrow. Other systems allow you to drag-and-drop files between client and server windows.

Telnet

Telnet protocol
The capability to use the Internet to log on to a computer other than your local computer.

The original purpose of the Internet was to allow researchers at one university to use a computer at another university. This approach was undertaken to spread the computing load around or to use a special program available only at a distant computer. To make it possible to use a computer at a remote location, the Telnet protocol was made a part of the Internet from the beginning. With the **Telnet protocol**, you actually log on to a computer at a remote location and run the application there, with your computer acting as a terminal. The use of Telnet has diminished greatly over the last few years as organizations have found ways to replace direct access to their computers with Web access. Nevertheless, Telnet remains useful for sending and receiving e-mail from a remote computer by Telneting into your host computer and using a text-based e-mail client there. This ability can be very useful in locations where a slow Internet connection makes a typical e-mail client or even Web-based e-mail difficult to use. As with the other Internet actions, you can use the Web to Telnet into another computer. Figure 3.20 shows the use of Telnet to access a mail system known as Pine on the University of Georgia mail system.

Newsgroups

newsgroup
One of a vast set of discussion lists that can be accessed through the Internet.

Network News Transfer Protocol (NNTP)
The protocol used for newsgroups that makes all messages sent to the specific newsgroup available to all subscribers.

The **newsgroup** Internet application encompasses a vast number of discussion groups on a wide range of topics. A newsgroup is a discussion group about a particular subject and consists of messages written on a series of news servers, each of which transfers messages to each other so that all postings to one newsgroup are replicated on all the other news servers. The groups are organized in a tree structure of discussion topics and use the **Network News Transfer Protocol (NNTP)**, which makes all messages sent to the specific newsgroup available to all subscribers. Typically, you use an e-mail client to access the newsgroups.

Newsgroups are organized into subject hierarchies, with the first few letters of each newsgroup name indicating the major subject category. For example, *rec* refers to recreation and hobbies, *sci* references topics in science, *alt* refers to alternative topics (anything that's not mainstream), and so on. Each major heading includes many subgroups, and subgroup headings are separated from the major heading by a period (for example, soc.culture.australia).

Figure 3.20 Using Telnet to access the Pine e-mail system

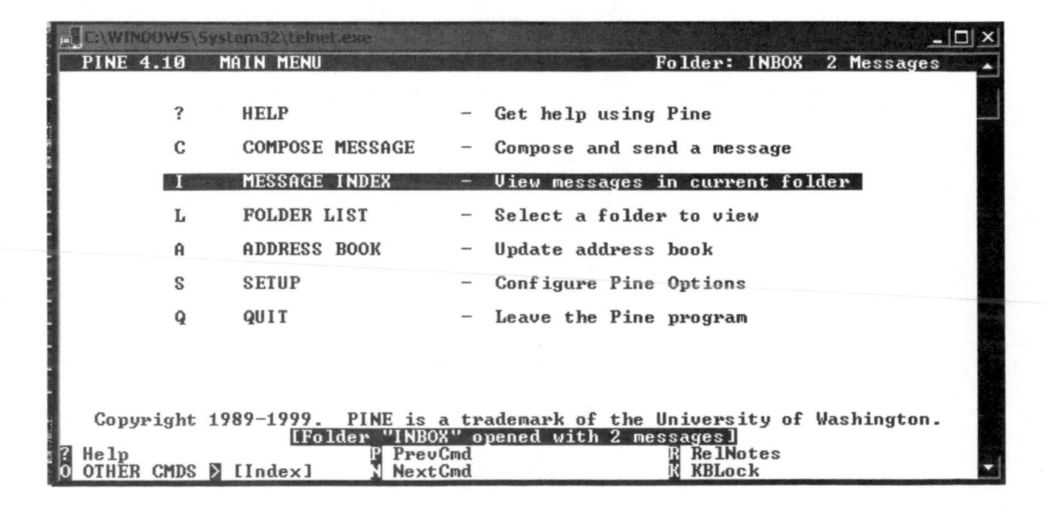

Figure 3.21 Example newsgroup discussion

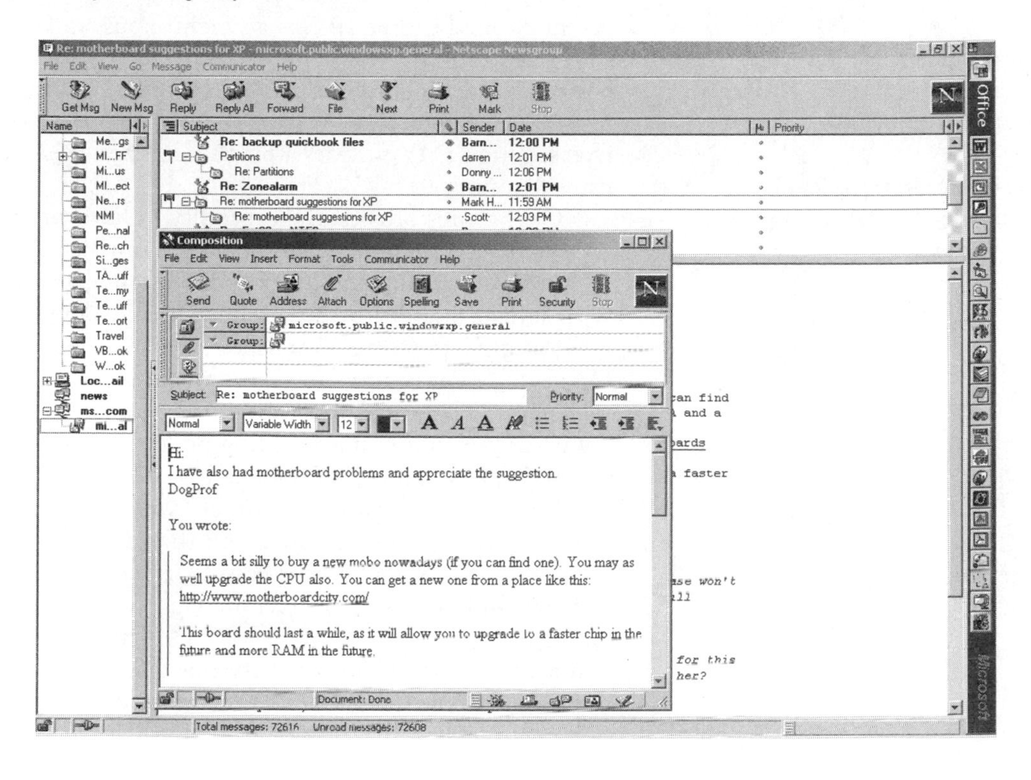

frequently asked questions (FAQ)
A list of popular questions about software or Web topics, along with answers to the questions.

Newsgroup users can post questions or comments to existing newsgroups, respond to previous posts, and create new newsgroups. Messages on newsgroups are *threaded* so that answers to or comments about a newsgroup message appear beneath it in a list of messages, regardless of the data posted, allowing readers to easily follow a discussion. Newcomers to newsgroups should learn basic netiquette and become familiar with a newsgroup before posting to it by reading a newsgroup's **frequently asked questions (FAQ)** list. FAQ lists provide answers to the most frequently asked questions about a topic. Figure 3.21 shows the Microsoft Windows XP newsgroup with a threaded discussion on replacing the main (mother) board on a computer (subscribers' names have been deleted for obvious reasons).

Many companies have begun to monitor newsgroups that are devoted to their products to find out what their customers are saying about them. Because bad news travels fast on the Internet, it is important for companies to quickly detect any emerging problems and respond as needed. For example, you can participate in a variety of newsgroups on food and wine by visiting the Foodcom Web page at www.foodcom.com/ and clicking *Usenet Newsgroups.* FarEast Foods would probably want to monitor newsgroups dedicated to Asian foods to find new ideas for marketing its products as well as to watch for negative or erroneous postings about the company. It might also want to start a newsgroup so customers can ask questions and exchange recipes with other customers.

Internet Relay Chat

chat rooms
Reserved areas that allow users to carry on group IRC conversations on the Internet.

instant messaging (IM)
A form of IRC in which users carry on private conversations.

As mentioned earlier, IRC is a synchronous way to use the Internet to communicate. Chat rooms and instant messaging are two widespread uses of IRC. With **chat rooms**, many individuals can be sending and receiving messages simultaneously regarding a subject of interest to all of them; it is a group conversation. **Instant messaging (IM)** provides a private link between two individuals over which they can communicate.

Chat rooms provide the capability to have an online discussion involving multiple persons, each adding his or her comments; they represent a very popular way of interacting online. A chat room is often a way of meeting other people online with similar interests because each is usually devoted to a single topic. A number of relationships and even marriages have resulted from meetings in a chat room. However, because users in the chat room do not have to provide evidence as to who they actually are, chat rooms have received a bad name as being a way for sexual predators to lure unsuspecting people into dangerous physical meetings, resulting in robbery, injury, or even death.

Instant messaging has become a very popular way to communicate for those not wishing to send an e-mail and then wait for the receiver to read it and reply. With instant messaging, it is easy to determine if another person is linked to the Internet and send him or her a message, which will cause that person's instant messaging software to alert him or her to the message. The individual can then respond, and a dialog can take place between the two parties. Although instant messaging requires a server to create the initial link between the two users, once created, the link becomes automatic, does not require the server any more, and becomes a peer-to-peer network.

The most popular form of instant messaging is available to America Online (AOL) and Netscape users; Microsoft and Yahoo! offer versions that are not compatible with the AOL and Netscape systems. Microsoft would like its system to be compatible, but AOL has refused to open its system to outside users. Figure 3.22 shows a typical instant message exchange.

Figure 3.22 Typical instant message

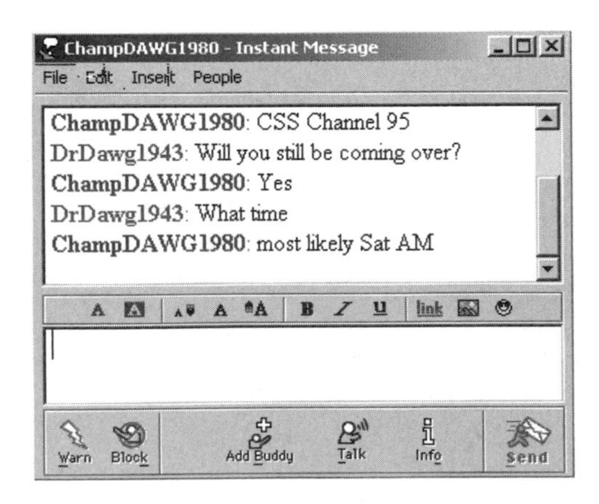

1. What are the six operations used with the Internet?
2. If you need to log on to a distant computer to carry out some type of processing operation, which Internet operations would you use?

Len Kleinrock and Bob Metcalfe

IT INNOVATORS

Both the Internet and Ethernet protocols use packet switching. The theory behind this approach to sending data and information over networks was developed in 1962 by a graduate student at MIT named Len Kleinrock. His theory was put into practice by the Bolt Beranek Newman (BBN) consulting company when it created the Advanced Research Projects Agency Network (ARPANet) in 1969. ARPANet, the predecessor to the Internet, originally linked UCLA (where Kleinrock had become a professor), Stanford University, University of California, Santa Barbara (UCSB), University of Utah, and the BBN offices in Boston. At each location, a minicomputer was used to accept the packets and convert them to a form that the local host mainframe computer could accept. The first application was Telnet, which allowed researchers to use computers at other institutions. In 1972, both e-mail and FTP capabilities were added to the network.

The original network grew slowly. In 1983, when the ARPANet officially became the Internet and the National Science Foundation took over its management, the network included only 583 hosts. After 1983, the Internet began to grow rapidly and that growth rate increased to an explosive rate with the addition of the World Wide Web in 1993. Today, Len Kleinrock continues to work as a professor of computer science at UCLA and is president of a firm called Nomadix, which is dedicated to meeting the needs of mobile users.

As part of his Ph.D. thesis in 1973, Bob Metcalfe studied the only two packet-switching networks in existence at that time: ARPANet and a radio-based network in Hawaii called AlohaNet. When he finished his doctoral research, Xerox offered him a position, where he continued thinking about networks and came up with the idea behind the Ethernet protocol. Xerox was not sure what to make of the Ethernet protocol and allowed Metcalfe to take it with him in 1979 when he founded 3Com Corporation. Today, most LANs in the world use his idea, and the company he founded has annual revenues in excess of $2 billion and is the parent of the company that makes the Palm handheld computer.

Len Kleinrock and Bob Metcalfe were pioneers in network development.

The World Wide Web

hypertext
A method of linking related information in which there is no hierarchy or menu system.

multimedia
An interactive combination of text, graphics, animation, images, audio, and video displayed by and under the control of a personal computer.

Of the six Internet operations listed in Table 3.4, the newest is the World Wide Web (WWW), or simply the Web. The Web is a body of software and a set of protocols and conventions based on hypertext and multimedia that make the Internet easy to use and browse. **Hypertext** is a method of linking related information in which there is no hierarchy or menu system. **Multimedia** is an interactive combination of text, graphics, animation, images, audio, and video displayed by and under the control of a computer. Originally developed to enable scientists to easily exchange information, the Web is now the most popular application on the Internet as individuals and organizations use it to find new and innovative ways to share information with others. It was developed in 1989 at the European Laboratory for Particle Physics (CERN) in Geneva, Switzerland, by Tim Berners-Lee, a computer scientist who saw a need for physicists to be able to communicate with colleagues about their work while it was ongoing rather than waiting until a project was finished. To make this *real-time* communication possible, Berners-Lee wanted to create an interconnected web of documents that would allow a reader to jump between documents at will using hypertext links.

With hypertext, a user can navigate throughout a system of information, with the path followed being limited only by his or her mental connections. For example, if you were reading about information systems on the Web and came upon a hypertext reference, or link, to operating systems, you could click this hypertext link and immediately visit another page that discusses operating systems in more detail. From there, you might jump to another page that discusses Linux as an operating system. Jumps within a document or to a completely different document are possible with hypertext, depending only on the hypertext links created by the author of the documents. The World Wide Web is based on this concept of hypertext, which links documents located on Web servers around the world.

Although used only since the early 1990s on most computers, hypertext actually predates the use of computers. The original idea was proposed by President Franklin D. Roosevelt's science advisor, Vannevar Bush, in a 1945 *Atlantic* magazine article entitled "As We May Think." Twenty years later, computer visionary Ted Nelson coined the term *hypertext*. Nevertheless, hypertext remained a largely hidden concept until Apple released its Macintosh HyperCard software in 1987.

Using Browsers to Access the Web

browser
Client software used on the Web to fetch and read documents on-screen and print them, jump to other documents via hypertext, view images, listen to audio files, and view video files.

point-and-click operations
A method that involves using a mouse or other pointing device to position the pointer over a hypertext link or the menu bar, toolbar, location window, or directory buttons and then click a button to retrieve a Web page or execute a corresponding command.

The Web is a special type of client/server network. (Recall that client/server computing was discussed in Chapter 2.) To access the Web, the client computer uses software called a **browser** that initiates activity by sending a request to a Web server for certain information. The Web server responds by retrieving the information from its disk and then transmitting it to the client. Upon receiving the data, the browser formats the information for display. Web browsers use a *graphical user interface (GUI)* like that available on Microsoft Windows or the Apple Macintosh. With a GUI-based Web browser, you can perform a variety of operations simply by pointing at menu selections or icons representing operations and clicking the mouse button—also known as **point-and-click operations**. For example, you can use a browser to navigate the Web by pointing at a hypertext link in the current document and clicking it. This operation causes the linked document, image file, or audio file to be retrieved from a distant computer and displayed or played on the local computer. You can also type in a specific address to retrieve a desired document or file.

When displaying information, the browser processes formatting instructions that are included in the text file retrieved from the server. For example, assume that the creator of a document stored on a Web server has decided that a certain phrase should appear in boldface when it is displayed. Instead of saving the file with a boldface

Hypertext Markup Language (HTML)
A computer language used to create Web pages consisting of text, hypertext links, and multimedia elements.

Web page
A special type of document that contains hypertext links to other documents or to various multimedia elements.

Web site
An Internet server on which Web pages are stored.

multimedia files
Digitized images, videos, and sound that can be retrieved and converted into appropriate human-recognizable information by a client.

font, the server stores the text with tags of the form <b> and </b> that indicate the beginning and end of the text that will appear in boldface when it is displayed.

The tags in World Wide Web documents are part of a special publishing language called **Hypertext Markup Language (HTML)**, and all documents on the Web have an .html (or .htm) extension. Documents on the Web are referred to as **Web pages**, and their locations are known as **Web sites**. Because HTML is standard for all computers, any Web browser can request an HTML document from any Web server. For instance, a browser running on a PC using Windows XP can readily access files created on a Macintosh that are are stored on a Linux-based server.

Web servers can also store **multimedia files**, which include digitized text, images, animation, video, and audio. The browser retrieves these files and displays them using appropriate software. The transfer of multimedia files from the Web server to the client browser is a key operation that sets the Web apart from the other Internet applications. Because multimedia enables users to view photographs, graphics, and videos and to listen to music, it is a major reason for the phenomenal growth of the Web.

Figure 3.23 shows a fairly simple Web page in which the underlined words are hypertext links. The pointing finger over the underlined words indicates a hypertext link beneath it. Figure 3.23 also shows the HTML source language, or source code, necessary to create the Web page. Tags used in this code are enclosed in angle brackets (< >) and include *h1* and *h3* for two sizes of headings, *center* to center the next, and *b* to make the text appear in boldface. Although knowing HTML is a nice skill to have, you can now create Web pages with word processing-like software such as Microsoft FrontPage and Netscape Composer without knowing the details of HTML.

Building Virtual Tunnels for Data

INTERNET IN ACTION

Although the Internet is the biggest network in use today and continues to grow larger by the day, it still has its problems. The lack of security on the Internet keeps many companies from making greater use of it for transmitting highly sensitive messages. As a result, more than 80 percent of corporate data still runs across dedicated private networks for which companies buy or lease high-speed telephone lines. One way of creating a dedicated private network on the Internet is to use a virtual private network (VPN) that creates *virtual* tunnels in it. VPNs use the Point-to-Point Tunneling Protocol (PPTP) to encrypt data before sending it through the Internet and decrypt it at the receiving end. An additional level of security is provided by encrypting not only the data but also the originating and receiving network addresses.

Until recently, setting up a VPN was a big job requiring hand-coding addresses for each possible pair of offices, encrypting data, and establishing access codes. For example, if a company had 50 offices, setting up a VPN required setting up 1225 Internet addressees, a task that can take 30 minutes per address. For this reason, the VPN market, although filling a definite need, was growing quite slowly. A recent software development by a company called SmartPipes may change that situation. Its software automates the arduous task of hand-coding network hardware and managing thousands of connections with a Web-based program that is as easy to use as buying a book at Amazon.com. For example, Native American Systems determined that it would cost the company at least $125,000 per year to use a private network to connect its corporate offices; the SmartPipes VPN solution will cost the firm only $13,200 for the same level of security. Thanks to the introduction of software like that offered by SmartPipes and other companies, the VPN market is now growing at an annual rate of 68 percent.

Source: Erika Brown, "Network painkiller." *Forbes*, July 9, 2001, pp. 136-137.

Figure 3.23 Simple Web page and HTML source code

Simple Web page

This is a sample page!

With different headings and

different sizes and locations of text

and an underlined text link.

Here is a graphic from the FarEast Foods Web page.

HTML source code for simple Web page

```
<html>
<title>Example of HTML Source Code</title>
<body><center><h1>This is a sample page!</h1></center>
<h3>With different headings and</h3>
<div align=right>different sizes and locations of text</div>
<div align=center><a href="http://www.fareastfoods.com"> and an
underlined text link.</a></center></div><br>
<center><IMG height=84 src="bowl_main.gif" width=85><br>
<b>Here is a graphic from the FarEast Foods Web page.</b></center>
</body>
</html>
```

As you know, people can use a Web browser to fill out forms, such as the form used to order food items from FarEast Foods, and then submit the data over the Internet to a Web server. Before sending data from a form over the Internet, the browser encrypts the data. At the Web server, the data are decrypted and interpreted to determine the necessary information on the customer, his or her credit card data, and the order for items.

Browser Operations

The primary purpose of a Web browser is to retrieve Web pages from Web servers and to display them on a client computer. In addition to being electronic rather than physical, Web pages differ from pages in a book or magazine in other ways. For example, although the amount of information on a physical page is restricted to the size of the paper page, a Web page can extend beyond the area shown on a screen. Table 3.5 describes the differences between Web pages and physical pages.

Many individuals, companies, and organizations have created Web sites that contain information about themselves and their activities, and more pages are added to the Web every day. As mentioned in Chapter 1, an estimated 4 *billion* Web pages existed in early 2001.

Table 3.5 Differences between Web and Physical Pages

Characteristics	Web Page	Physical Page
Form	Electronic	Ink on paper
Amount of information	Can extend beyond a single screen	Restricted to single piece of paper
Types of information	Can include text, images, audio, and video information	Restricted to text and images
Links to other pages	Can be linked to other pages though hypertext	Can be linked only through a separate index
Creation	Can be created with HTML and saved to a server	Can be created using word processor and printer

URL (uniform resource locator)
A standard means of consistently locating Web pages or other resources, no matter where they are stored on the Internet.

service resource
Another name for a protocol on the Web.

A Web page is identified by its address. In Web terminology, the address of a Web page is referred to as its uniform resource locator. It is so named because a **URL (uniform resource locator)** is a standard means of consistently locating Web pages or other resources no matter where they are stored on the Internet. For example, the URL of the instruction page for the FarEast Foods Web site is

http://www.fareastfoods.com/instruction.html

Like every URL, it has three parts: the protocol, the Internet address of the server on which the desired resource is located, and the path of the resource (sometimes hidden). Figure 3.24 shows the three parts of the FarEast Foods instruction page address.

For Web resources, the protocol (also called the **service resource**) defines the type of resource being retrieved. A Web page resource is identified by the letters *http*, which stand for *Hypertext Transfer Protocol*. Some of the other allowable protocols include file, Telnet, FTP, mailto, and news. Table 3.6 lists these protocols (service resources) and describes their purposes.

Figure 3.24 Three parts of a URL

http://www.fareastfoods.com/instruction.html

Protocol — Web server address — Path name

Table 3.6 Internet Protocols

Protocol	Purpose
http	Retrieve Web pages
file	Retrieve files from local hard disk
telnet	Log on to a remote computer connected to the Internet
ftp	Download or upload files from an Internet FTP server
mailto	Send outgoing e-mail
news	Display news group

path
A portion of the URL that includes the name of the home page file plus any directories or folders in which it is located.

The second part of the URL is the name of the Web server—in this case, www.fareastfoods.com. The third part of the URL is the **path** of the Web resource, which includes the name of the Web page file plus any directories or folders in which it is located. In the FarEast Foods example, the path of the Web page document is simply the file name, *instruction.html*. In many cases, the path name will be much longer, because it includes the folder(s) in which the Web page is stored. For example, the URL for the Web site for the syllabus for a course at the University of Georgia is

http://www.terry.uga.edu/~pmckeown/Mist5665/

In this case, the tilde (~) symbol is shorthand for the home directory of the computer on which this file is stored. A tilde is used for security reasons by the person managing the Web server. In this URL, *Mist5665* is a subfolder in the *pmckeown* folder. Note also that this URL ends with a slash (Mist5665/). It indicates that a default HTML file, usually either index.html or default.html, will be accessed. The complete path is interpreted as ~pmckeown/Mist5665/index.html. The use of index.html or default.html files is quite common and saves people unnecessary typing. It also means that you can often guess the URL of an organization's Web page. For example, you would guess correctly if you entered http://www.dell.com/ to access Dell Computer's Web page. In fact, most browsers now have built-in search engines that will search for the *home* Web site if you simply enter the name. For example, entering *Dell* will find and display the preceding Web site.

Once a valid address for a resource has been entered, the next step is automatic: The browser software attempts to connect to the Web server at that address, find the page referenced in the address, and return it to the user's browser. If this operation is successful, then the page appears on the screen. Otherwise, some sort of error message is displayed.

The process of moving from one Web site or page to another Web site or page and then to another, ad infinitum, is known as *surfing the Web*. Web surfing can quickly become a time-consuming process as you follow links looking for information or a product to purchase. You can short-circuit this process to some extent by using one of the numerous Web search engines, into which you can enter a query word or term so as to find pages or sites that match your query. A potential problem with this approach is the large number of Web pages that can be returned, many of which have nothing to do with the query you entered. Chapter 8 will discuss search engines in more detail.

Intranets and Extranets

intranet
An intraorganizational network based on using Internet technology; it enables people within the organization to communicate and cooperate with one another.

Two specialized versions of the Internet have been created to facilitate organizational use of information technology—intranets and extranets. An **intranet** is a LAN that uses Internet protocols but restricts access to employees of an organization. For example, an intranet Web server in the human resources department at FarEast Foods might allow only specified employees within that department plus

extranet
A wide area network using the TCP/IP protocol to connect trading partners.

some executives to view information on compensation for employees in the company. In contrast, an **extranet** is a business-to-business network that uses Internet protocols instead of EDI or other private protocols for transmitting data and information between trading partners. A restricted form of the Internet, it is not open to the general public. For example, a company might want only its wholesale customers to see pricing information on an extranet Web server. The topic of the Internet as well as intranets and extranets will be discussed further as they relate to information systems in Part 2.

The Web and Electronic Commerce

As discussed earlier, the other five Internet operations—e-mail, FTP, Telnet, newsgroups, and IRC—all have commercial uses. Together, however they do not come close to having the economic impact of the Web. In fact, the Web can be said to have created electronic commerce. Virtually all of the volume of sales associated with electronic commerce can be attributed to the Web as companies find ways to make their products available to both consumers and other companies. Similarly, individuals are finding that shopping over the Web is much more convenient than fighting traffic, finding a parking place, and searching from store to store for a particular item. Instead, the search can take place in the comfort of their homes or offices with a browser and a search engine. As this trend continues to grow, so will the networked economy. The strategy of and technology for using the Web for electronic commerce is discussed in detail in Chapters 7 and 8.

1. What is the role of hypertext in the World Wide Web?
2. What is a URL? What are its parts?

Case Study
Using Napster-like Computing on Wall Street

Although mainframe-based and client/server computing remain the predominant approaches to using networks in business and government, the success of Napster for moving large music files between personal computers has increased the interest in using peer-to-peer (P2P) networking. This interest is especially keen on Wall Street, where the tremendous increases in trading volume and required analyses have put pressure on existing systems. For example, the derivatives trading group of First Union Corporation (which recently merged with Wachovia Bank) installed a P2P computing system to help it handle its mounting trading-floor demands.

First Union believes that the system from DataSynapse, which helps customers harness idle and underused computing in their networks, will enable it to boost processing volumes and handle complex portfolio risk analysis throughout the trading day rather than waiting until the end of the day to carry out this processing. The goal was to find ways to generate more revenue out of its trading operations, which required additional computing power, without buying bigger, faster computers.

DataSynapse has developed a system that breaks complex tasks into thousands of smaller tasks that can be run simultaneously on a P2P network of interconnected PCs. This strategy can reduce computation times from hours to minutes, helping financial companies keep up with their increasing processing requirements, such as running simulations that require a fast turnaround time to be useful. At First Union, as many as 100 idle or underutilized PCs in the bank's trading rooms and development areas are being used to carry out needed processing. If more processing capability is required, First Union can simply add more PCs from other departments to handle the stringent trading room demands.

Source: Eric Auchard, "First Union invests in STP for fixed-income derivatives." *Risk News,* April 24, 2001 (http://www.risknews.net/news/story953.htm).

Think About It

1. What factors led First Union to consider using a peer-to-peer approach rather than a mainframe or client/server one?
2. How does this approach to computing compare to the use of parallel computing on supercomputers?
3. Describe two or three other situations in which peer-to-peer computing could be used to improve processing capabilities without adding more computers.

SUMMARY

To summarize this chapter, let's answer the questions posed at the beginning of the chapter.

What are computer networks, and what is their role in the networked economy? A computer network exists whenever two or more computers or other information technology devices are linked through some type of media with the purpose of sharing data, information, and resources. Because the networked economy cannot exist without networks, understanding networks is essential to your understanding of this new economy.

Two types of networks are distinguished based on geographical size: local area networks and wide area networks. A local area network (LAN) is a computer network that is confined to a single geographical area, such as an office, a building, or a group of buildings. Computers connected over a region, country, or the world form a wide area network (WAN). Networks are often connected via a backbone.

What is the network layer model, and how does it work for a wide area network? The network layer model is used to understand the way networks operate; the layers define the standards with which each network must comply. The simplified model presented in this book consists of three layers: application software layer, networking software layer, and physical layer. The application software layer specifies the software on each computer on the network that the user sees and employs to send and receive messages and data between computers, as well as the software needed to encrypt the message or data streams. The networking software layer describes how the message from the application software layer is formatted according to whatever protocol will actually be used to send it over the network. For WANs, the Internet protocols (TCP/IP) and EDI protocols are important within the network layer model. The protocols describe how an address and message are converted into a numeric form and divided into smaller units called packets. They also determine the first stop on the message's route to its destination. The physical layer describes the hardware and media (twisted pair, coaxial cable, and so on) over which a message is sent. Data and information transmitted over networks travel over various media, including twisted-pair wire, coaxial cable, fiber-optic cable, microwave transmission, and satellite transmission. Radio and infrared transmissions are now being used for wireless networks. Packets are transmitted over the Internet via a packet-switching methodology that uses routers. Two other key considerations are the signal type—analog or digital—and the data rate—the rate at which bits are transmitted through the network.

What are the parts of a local area network? The parts of a LAN include the server, client computers, cabling and hubs, the network operating system (NOS), and network interface cards (NICs). The cabling and hubs tie the server and client computers together. The NOS directs the operations of the LAN; it resides on both the server and the clients. Finally, the NIC handles the electronic interface between the servers or clients and the rest of the network. The Ethernet protocol is used in the vast majority of LANs. Ethernet LANs have a bus physical setup and interconnections are usually handled with a hub.

What are wireless networks, and why have they become so popular? As their name implies, wireless networks are computer networks that operate without either the internal wiring associated with LANs or the wired connections associated with WANs. A wireless network has the obvious advantage that devices connected to it can go virtually anywhere. The most popular method of establishing a wireless connection to the Internet and Web (a WAN) uses a mobile telephone as the client. A widely used method of establishing this type of wireless connection involves using a GSM protocol telephone for which the Wireless Access Protocol (WAP) and the Symbian OS operating system have been developed. Wireless LANs (WLANs), in which the usual LAN cabling is replaced with wireless transmissions between computers, are becoming increasingly popular as mobile users find they need to connect to their local network and, possibly, from there to the Internet. Wireless networking hardware uses radio frequencies to transmit information between individual computers, each of which has a wireless network adapter; Wi-Fi is currently the most popular protocol with this setup. The individual computers do not communicate directly with each other; instead, they communicate with a wireless network hub or router, which is also used to bridge the wireless network to a traditional Ethernet network or provide a shared Internet connection.

What is the Internet, and how is it used? The Internet is a network of networks that uses the TCP/IP protocols for addressing computers and sending packets over the network. There is no governing authority or central computer; only the agreement to use TCP/IP ties the networks of the Internet together. The six primary operations on the Internet are the World Wide Web, e-mail, FTP, Telnet, newsgroups, and IRC. E-mail uses the Simple Mail Transfer Protocol (SMTP) to send asynchronous messages over the Internet to individuals and groups. A listserv is a method of easily sending e-mail messages to all members of a group. FTP uses the File Transfer Protocol to transfer files between computers over the Internet. Telnet allows users to log on to a distant computer and use software on that computer. Newsgroups enable users to engage in discussions on a global network of news servers. Finally, IRC (Internet Relay Chat) enables users to communicate in a synchronous fashion in chat rooms or through instant messaging.

What is the World Wide Web, and how does it work? The World Wide Web (WWW), or just the Web, is a body of software and a set of protocols and conventions based on hypertext and multimedia that make the Internet easy to use and browse. In this client/server network the client browser software requests Web pages created in Hypertext Markup Language (HTML) from a Web server. Multimedia files are retrieved separately from text pages. Hypertext allows the user to jump within pages or from page to page. The address of the Web site is called a uniform resource locator (URL) and consists of a protocol, a Web server address, and the path of a Web page.

REVIEW QUESTIONS

1. Give an example of a computer network other than those listed in the text.
2. What are the two primary types of networks as defined by their geographic scope?
3. What are the three layers of the simplified network model described in the text? Which layer contains the software that the user sees and uses?
4. What are two widely used protocols for WANs? What does the IP address consist of?
5. What type of media would you select if you needed extremely fast speed and a very low error rate? If you needed to transmit data to a mobile computer user?
6. What are the two types of signals? Which can be sent between computers without any translation?
7. What is bandwidth, and how might it relate to the transmission of video files?
8. Why is packet switching used to send data between computers over the Internet?
9. What is EDI? What are some of its advantages?
10. What are the two types of LANs? Which has significant security problems?
11. What are the five parts of a LAN? Which controls the network?
12. What is the most popular protocol for LANs? What is the difference between a gateway and a bridge?
13. What is the most popular approach to wireless WANs? What is the most popular protocol for wireless WANs?
14. What protocols have been developed to enable mobile phones to work with the Web?
15. What is a very popular standard for wireless LANs? How fast is it?
16. Why is the Internet described as a "network of networks" ? Which application has been most responsible for its growth?
17. How are addresses defined on the Internet? What organization handles the creation of top-level domains?
18. Of the six Internet operations, which would you use to request information on the Linux operating system? Which operation would you use to download updates to antivirus software?
19. What is the difference between Telnet and FTP? Which would you use to log on to your college's registration system?
20. What is HTML, and how does it relate to the Web? List three protocols that would be used in a URL.

DISCUSSION QUESTIONS

1. Discuss the network layer model using an e-mail message that is not encypted and is sent from a PC linked through an ISP to a PC linked to an Ethernet LAN.
2. Discuss the difference between high-speed Internet connectivity using coaxial cable and digital subscriber line connections.
3. Discuss the difference between a system like that used for long-distance telephone calls, where the circuit must remain open for the entire conversation, and that used for the Internet, which relies on packet switching.
4. Discuss the difference between e-mail and instant messaging. Which can have high levels of security? Which might you use to collaborate with a colleague in another city to make an immediate decision?
5. List several examples of commercial uses for non-Web Internet operations.

RESEARCH QUESTIONS

1. Research and discuss in a two-page paper the setup for the LAN at your college or university. Find out if a backbone connects multiple LANs at your college.
2. Research and discuss in a two-page paper the difference between Wi-Fi and Bluetooth.
3. Research and discuss in a two-page paper the way in which the computers in your college's computer lab are linked to the Internet.
4. Find at least one other Web site for each of the new top-level domain names and summarize their contents in a two-page paper.

CaseStudy

WildOutfitters.com

Claire rubbed her eyes and stared at the vision before her as the light flickered across her face. This time, however, she wasn't gazing at a campfire with Alex. She was staring aghast at the flaming logo that he had put on the home page they were designing for the store.

"What are you trying to say—that the store burnt down?" Claire inquired.

"No," Alex replied. "I just want to have something to get people's attention."

"You're right, the site will need to attract customers, but remember the Web site is an extension of our brick-and-mortar store," Claire explained for the umpteenth time. "It needs to reflect the same style and ambiance that we try to give to our store customers."

"So the flaming logo is out. I guess I can still use it on my personal home page," admitted Alex grumpily. "Given your success with the store, I'll bow to your wisdom regarding the design elements of the store's page. But I do have a few more ideas about jazzing up both the store and the Web page to run by you."

While waiting for their new hardware for the Wild Outfitters store to arrive, the Campagnes have decided to use Alex's old Pentium II computer to start designing their new Web site. Having only Microsoft Word, they have decided to work on mainly the "look and feel" of the site. They will add transaction capabilities to the site later as they obtain more software and achieve greater expertise.

After some initial research into what seems to work and what doesn't seem to work for commercial Web sites, Claire and Alex have decided that successful Web design requires more than just a knowledge of HTML and programming. They have made a list of some things that they should consider when preparing their site.

Navigation: Customers shy away from sites that are difficult to navigate. The Campagnes have decided to keep their layout simple so that customers will always know where they are and where they are going. They hope to make access to any page quick and easy. To accomplish this goal, all links should be clearly labeled on all of their pages. In addition, they hope to use a "three click" rule of thumb, which states that any page on the site can be reached in a maximum of three clicks.

Content: WildOutfitters.com's primary purpose is to sell Wild Outfitter products. Claire and Alex have decided that all content should be aimed at this purpose. Information will be provided to inform customers about the products available and help them make buying decisions. The Campagnes also want content designed to draw people to their brick-and-mortar store. Finally, they would like to include fun content designed to draw customers back to the Web site.

Graphics: Along with product photos, the Campagnes plan to include graphics that reflect a consistent and stylish theme for WildOutfitters.com. The graphics should bring color to the site but not cause too much distraction. Also, the use of graphics should be managed so that the overall size (in megabytes) of each page is not too large, which could cause long download delays for some customers.

Other Elements: The Campagnes have discussed some other design elements that they will need to make decisions about. They would like to choose a text font that looks good in various sizes, for use throughout the site. They prefer a white background but might change their minds as long as they find an overall color scheme for the site that is both attractive and supports legible text. Finally, they have not yet decided on the addition of links to other sites. If this feature is used, they will closely monitor the links to ensure that their site is not associated with anything that would tarnish their image.

Claire couldn't resist needling Alex a little more about his sense of style, saying, "I suppose that next you're going to light up the front of the store with neon and strobes."

"Hey, that's a great idea," Alex replied. "And how about one of those spinning mirror balls over the counter?"

Laughing, Claire exclaimed: "Then we'd have to change our name to Wild Disco Outfitters. Couldn't you just see us in bell-bottomed cargo pants?"

After several minutes of laughter over silly disco jokes, the pair was able to settle down and get back to work.

Think About It

1. Compare and contrast the design elements of a Web page with those of a physical printed page of information. What criteria should you consider for both? What should you consider doing differently when designing a Web page? Are there other criteria for Web design that you can add to the WildOutfitters.com list?

2. One of the main strengths of the Web is the use of hyperlinks that allow a netizen to jump quickly from one Web site to another related site. Many commercial Web sites include hyperlinks to other sites within their pages. What are the advantages and disadvantages of including links to other sites in a commercial Web page? Should the Campagnes include some on their site? If so, what procedures should they put in place to monitor and maintain the links?

3. There is more to designing a Web page than writing HTML. Conduct some research on the Web for various opinions about what constitutes a good design for a commercial site. You can find a few interesting sources to get you started at:

   Yale Style Manual: www.med.yale.edu/caim/manual/

   Authoring and Site Design Pages of C-Net: www.builder.com

   Web Pages That Suck.com: www.webpagesthatsuck.com

   Dr. Jakob Nielsen's Clearinghouse of Usability Information: www.useit.com

   Webmonkey: The Web Developer's Resource: www.hotwired.lycos.com/webmonkey/

Hands On

4. Using a Web editor or a Web-enabled word processor such as Microsoft Word, create a set of Web pages for WildOutfitters.com. Try to incorporate as many of the criteria discussed in the case as possible. At a minimum, your pages should include the following: a home page for Wildoutfitters, a page that provides a sample product review, a sample catalog page with several products listed, a sample product page describing and showing a picture of the product (a link to this page is part of the catalog page), and, if you argued in favor of links, include a page of appropriate links. Feel free to add other pages to your set if you feel they will add to the site and will fit the Campagne's stated criteria.

INFORMATION SYSTEMS IN ORGANIZATIONS

When information technology is used in organizations, it often goes by the name of *information systems*. Information systems help organizations handle the present, remember the past, and prepare for the future. Organizations handle the present through information systems that enable them to sell goods and services as well as to order and pay for raw materials. They remember the past with a variety of types of organizational memory. In preparing for the future, organizations use information systems to help them make good decisions.

Part 2 covers information systems, including transaction processing systems, document management systems, knowledge management systems, and decision support systems. It also focuses on the use of database management systems as support systems for other types of information systems. After reading this part, you should have a good idea of how information systems can ensure that organizations thrive in the networked economy.

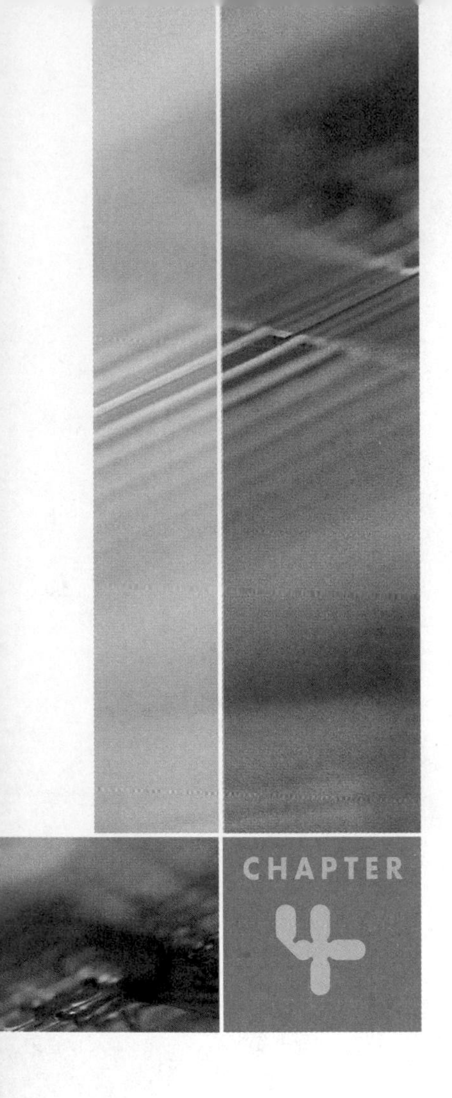

CHAPTER 4

HANDLING THE PRESENT: TRANSACTION PROCESSING SYSTEMS

LEARNING OBJECTIVES

After reading this chapter, you will be able to answer the following questions:

- What types of information systems are present in organizations, and how do they relate to the IS functions?
- What is a transaction processing system, and what functions does it accomplish in an organization?
- What are the various transaction processing methods and activities?
- What are the impacts of the Internet on transaction processing systems?
- How do business-to-business transactions in a networked economy differ from traditional transactions?

Mainstream Companies Turning to Online Auctions

Individuals aren't the only ones who are using online auction companies such as eBay and uBid; companies such as Home Depot, IBM, Coca-Cola, and the *New York Times* are finding that they, too, can generate revenue using these outlets. These large-scale retailers are looking for new channels to sell more goods online, and auctions are a growing part of electronic commerce. They generated $550 million in sales in May 2001 alone, up from $223 million a year earlier. Online auctions account for 10 percent of all electronic commerce spending. Forrester Research estimates that online auction sales will reach $25 billion, or about 25 percent of all online retail income, by 2006. Yahoo Auctions has seen the number of large companies selling through it grow from 20 percent of its listings in 2000 to more than 50 percent in 2001.

As an example of a mainstream company using an online auction, consider Home Depot, which auctions about 300 to 500 different products through eBay. Home Depot and other companies using auctions generally accept bids during a set period of time with a fixed or buy-now price, which enables customers to ensure that they are able to purchase the item. Other companies are using online auctions as a way to move excess and closeout inventory. Instead of working through traditional liquidators, who pay about 10 to 15 cents on the dollar, the companies are selling goods through online auctions where they receive three to five times that amount. Even small companies are finding that online auctions are a much better way to go than depending on a Web site that potential buyers may or may not visit. Many large and small companies are selling goods in online auctions by using the services of third-party partners such as Auctionworks and Andale. For example, Andale claims that its company's clients drive more than 75 percent of eBay's listings each week.

Source: Peralte C. Paul, "Going to Auction." *Atlanta Journal-Constitution*, January 2, 2002, pp. D1, D3.

Information Systems in Organizations

As discussed in Chapter 1, organizations use information systems in many ways to reduce business risks and help managers make decisions. It is safe to say that, today, information systems can mean the difference between success and bankruptcy for many organizations. For example, Wal-Mart's information systems were far superior to those of Kmart, and shortcomings in Kmart's information systems played a large part in its failure to successfully compete with Wal-Mart. Experts blame Kmart's inadequate information systems compared to those of Wal-Mart combined with the downturn in the economy in 2001 for the company's filing bankruptcy in early 2002.

The purpose of information systems in any modern organization is to to carry out the three IS functions discussed in Chapter 1: handling the present, remembering the past, and preparing for the future. These three functions are represented in the IS cycle shown in Figure 4.1. Obviously, no single information system can carry out all of these functions by itself. Because organizations make many different demands on information systems, they have developed several types of specialized information systems to carry out the IS functions. Let's briefly look at the various types of information systems that coexist in an organization to carry out the three IS functions. They include transaction processing systems, document management systems, knowledge management systems, expert systems, management information systems, executive information systems, and decision support systems, all of which are supported by the organizational database.

The transaction is at the heart of the networked economy. Without transactions there would be no economy, so understanding how transactions occur and how information systems process them is essential to understanding what role information technology and information systems play in contemporary corporations. Organizations must process transactions in an accurate and timely manner, or they will quickly experience problems with customers, suppliers, and employees. If an organization does not handle customer transactions, no revenue will come in and the organization will cease to exist. As discussed in the opening case, companies constantly search for new ways to carry out transactions with customers to generate revenue. If an organization does not appropriately handle transactions involving

Figure 4.1 The IS cycle

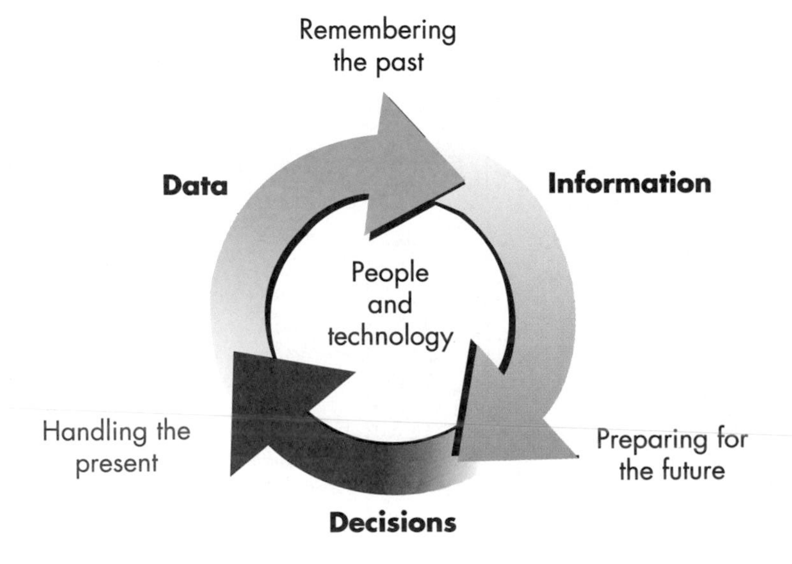

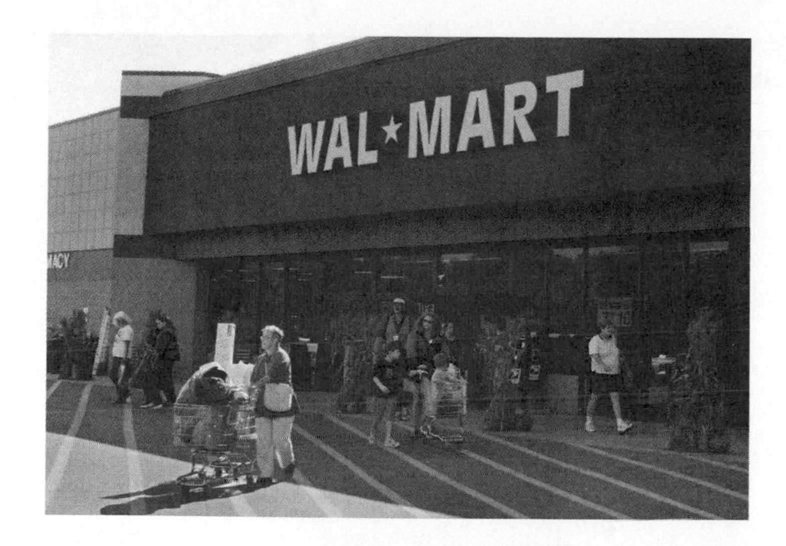

Wal-Mart's industry-leading information systems have enabled it to become the leading retailer in the world, with over 1 million employees.

suppliers, which include ordering and paying for raw materials, the organization will have nothing to offer its customers. Finally, an organization must ensure that it pays its employees in a timely manner or they will cease to work for it.

The organization's transaction processing system typically handles all of these operations. A **transaction processing system (TPS)** converts raw data produced by transactions into a usable, electronic form and forms the basis for handling the present. It generates all internal data that are stored in the organizational memory and then used in various other information systems. You will learn more about transaction processing systems shortly.

transaction processing system (TPS)
A system for converting raw data produced by transactions into a usable, electronic form.

The tasks of understanding and using the data generated by a transaction processing system are essential to the future survival of any organization. For an organization to understand and use data, those data must be stored in a database and made available when needed. In addition to storing transaction processing data, organizations store the large volume of documents created by their operations through some form of document management system. Knowledge management systems and expert systems are used to store the knowledge that an organization uses to run its operations, solve problems, and seek new opportunities. Data from transactions, information from organizational operations, organizational knowledge, and the systems in an organization that store them and then make them available as needed constitute *organizational memory*. Note that databases in an organization are aimed at storage of data for support of other information systems and are not typically considered to be information systems on their own. On the other hand, the systems that are aimed at making information and knowledge available to individuals are clearly information systems. Because components of organizational memory either support other information systems or are such systems themselves, organizational memory is included in the discussion of information systems. Chapter 5 will discuss organizational memory in detail.

Regardless of how well transactions are processed or the resulting data are stored in organizational memory along with information and knowledge, if employees and managers do not make good decisions, the survival of an organization can be in doubt. Information systems that aid in decision making are commonly referred to as *decision support systems (DSS)*. Decision support systems include a variety of types of information systems, which often go by different names such as management information systems (MIS) and executive information systems (EIS). These systems have specific roles in the organization that are often associated with various managerial levels. However, because all of the individual information systems support decision making, they are grouped together under the broad term of decision support systems. Chapter 6 will discuss decision support systems in detail.

Table 4.1 Types of Information Systems

Type	Purpose	Role
Transaction processing system (TPS)	Handles transactions and generates data about them that forms the basis for much of what the organization knows about its operations	Handles the present
Organizational memory	Stores data from transaction processing system as well as information generated by organizational operations and the knowledge necessary to run the organization	Remembers the past
Decision support system (DSS)	Answers questions from management using data, information, and models to facilitate decision making	Prepares for the future

Transaction processing systems, organizational memory systems, and decision support systems are interdependent and ongoing. They serve as the machinery that drives the information systems cycle shown in Figure 4.1, beginning with handling the present through transaction processing and remembering the past by storing transactions in organizational memory systems. Based on stored data, information, and knowledge, the organization uses decision support systems to prepare for the future. The organization then makes decisions about handling new transactions, which are, in turn, processed. The cycle then repeats as shown in Figure 4.1. Table 4.1 summarizes these information system elements along with their purposes and their roles in the IS function cycle.

1. Why is the transaction at the heart of the networked economy?
2. What information systems does organizational memory include?

Transaction Processing Fundamentals

A transaction processing system handles the present by compiling an accurate and current record of the organization's activities and generating data that form the basis for part of organizational memory. To accomplish this task, these systems must provide fast and efficient processing of large numbers of transactions and produce reports that aid in the management of the organization. Because transactions must be correct to have any meaning, a TPS must be able to closely check the accuracy of transactions and correct any erroneous ones. Also, because transactions being processed involve the transfer of funds between consumers and businesses or between businesses, a TPS has a high potential for security-related problems. Even though a TPS repeatedly carries out similar mundane activities, it constitutes the very basis for the ongoing well-being of an organization. If the TPS fails, it can have extremely negative effects on an organization, potentially resulting in a cessation of operations. Table 4.2 summarizes these characteristics of a transaction processing system.

Table 4.2 Characteristics of Transaction Processing Systems

Characteristics of Transaction Processing Systems
Quickly and efficiently handle a large number of transactions
Validate correctness of transactions and highlight or correct invalid data
Have great potential for security problems because of the nature of the transactions
Provide critical operations, so that an organization can quickly cease to operate if the TPS fails

TPS and Business Processes

Before considering transaction processing systems in more detail, you need to understand the business processes that drive the transactions that any TPS must handle. Every organization carries out these business activities to thrive and survive: revenue processes, expenditure processes, conversion processes, and financial processes.

All organizations, whether they are for-profit or not-for-profit, must provide a product or service for which there is a demand. In exchange for this product or service, an organization receives revenue from which it must pay for the direct costs of providing that product or service. The revenue must also pay for overhead costs that are not directly associated with providing the product or service. If revenue consistently falls short of the direct and overhead costs, without additional funds, the organization will eventually cease its operations. This process of receiving funds in exchange for products or services, called the **revenue process**, involves many types of transactions. In general, revenue process transactions include making a sale or accepting an order, shipping goods or providing services through a fulfillment system, and sending invoices (bills) to customers or accepting payment at the time of the sale. In any case, a transaction occurs in which revenue is collected.

revenue process
The receipt of revenue in exchange for goods and services.

Other processes in an organization that involve transactions include the expenditure process and the conversion process. The **expenditure process** handles the transactions associated with the payment of expenses that come from running the organization. The major transactions in the expenditure process include purchasing raw materials and parts, receiving them, and paying for them. The **conversion process** handles the transactions associated with the production of goods and services. Conversion transactions include work order preparation, material requisition, manufacturing and cost allocation, payroll, and inventory.

expenditure process
The process that handles the transactions associated with the payment of expenses associated with running the organization.

conversion process
Transactions associated with the production of goods and services.

Finally, the **financial process** summarizes in accounting terms all of the transactions generated by the revenue, expenditure, and conversion processes. It includes posting all transactions to a ledger and generating an income statement, as well as other accounting documents. Table 4.3 shows the four business processes, their purposes, and the various transactions and activities associated with them. Note that the financial business process is the only one in which no transactions occur; rather, this process records all of the other transactions.

financial process
A summary in accounting terms of all the transactions generated by the revenue, expenditure, and conversion processes.

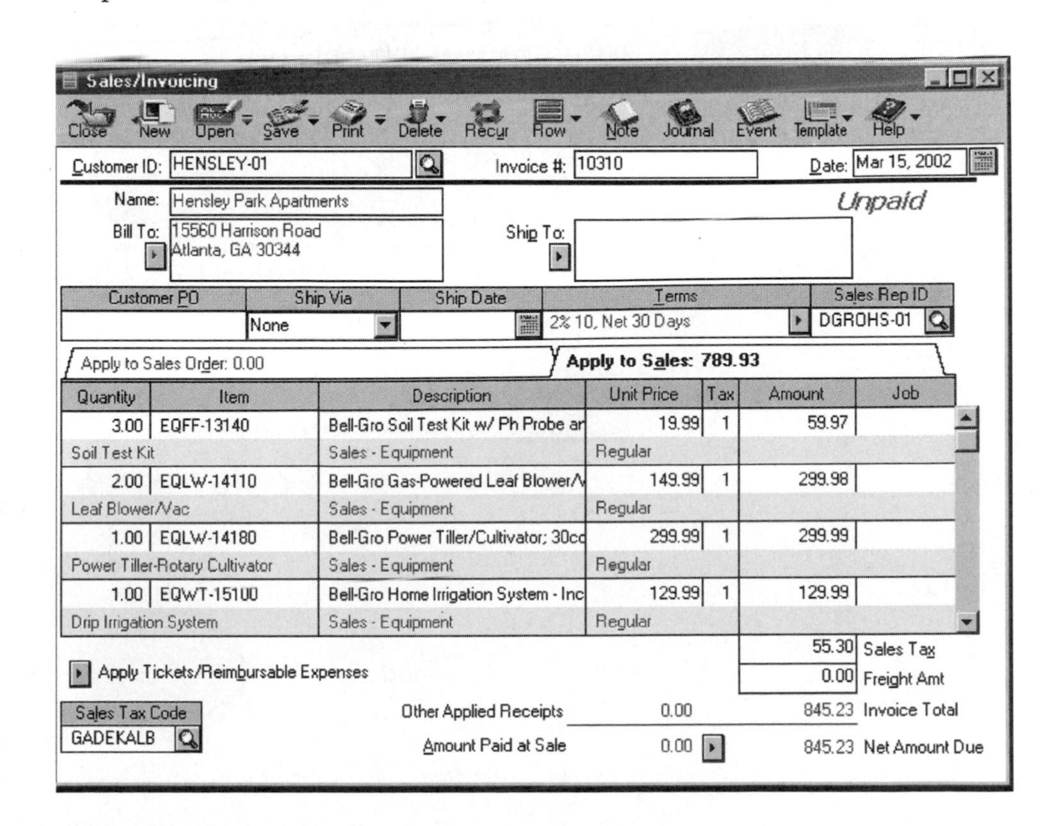

An important element of any accounting system is the capability to generate and track invoices to customers.

Table 4.3 Business Processes and Transactions

Business Process	Purpose	Transactions and Activities
Revenue process	Generates revenue for organization	Transacting a sale or accepting an order, shipping goods or providing services, and sending invoices to customers or accepting payment at the time of the sale
Expenditure process	Handles the transactions associated with the payment of expenses incurred in running the organization	Purchasing raw materials and parts, receiving them, and paying for them
Conversion process	Handles the transactions associated with the production of goods or services	Work order preparation, material requisition, manufacturing and cost allocation, payroll, and inventory
Financial process	Keeps track of all transactions	Post to general ledger and journals

As an example of business processes, consider the interactions between the customer, the company, the food wholesalers, the customer's bank or credit card company, and the package delivery company at FarEast Foods. Figure 4.2 shows the revenue and expenditure processes, labeled according to the order in which they occur (R1, R2, and so on or E1, E2, and so on). All of these processes generate transactions because they involve interactions between customers and businesses or between businesses.

In Figure 4.2, the revenue process begins when a customer uses her Web browser to submit an order to FarEast Foods for various Asian food items (R1). The company then checks the credit card number included with the order. If the credit card number is valid, a package delivery company picks up the order (R2) and delivers it to the customer (R3). The revenue process also involves billing the customer's bank or credit card company; it ends when FarEast Foods receives payment from the bank (R4).

Figure 4.2 Business processes at FarEast Foods

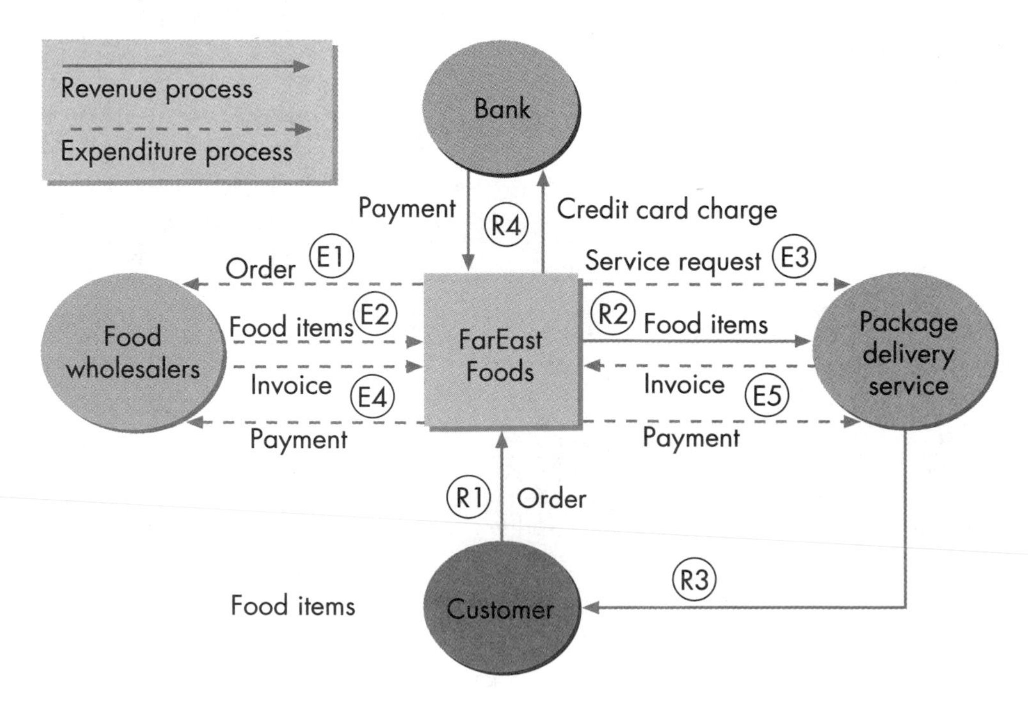

The expenditure process begins when FarEast Foods orders food items from the food wholesalers (E1). It also involves receiving the food items (E2) and requesting a shipment to the customer by the delivery company (E3). The expenditure process ends with receipt and payment of the invoices from the food wholesalers and delivery company (E4 and E5).

TPS and Supply Chains

The transaction processing system does not exist in a vacuum. Every transaction has an effect on the rest of the organization. Just as handling the present leads to the storage of transactional data and remembering the past leads to information being used to handle transactions in the future, transactions have a ripple effect through an organization. In fact, transaction processing is a part of a chain of activities known as the value chain.

value chain model
The chain of business activities in which each activity adds value to the end product or service.

According to Michael Porter's **value chain model**,[1] the organization comprises a chain of activities, each of which adds value to the firm's products or services. The model divides an organization's tasks into primary and support activities.

Primary activities are directly related to the production and distribution of the firm's products and services and create value for the customer. They include inbound logistics (obtaining raw materials, parts, and so forth), operations, outbound logistics (shipping finished goods or providing services), marketing and sales, and service. Combining inbound and outbound logistics with the operations function, in which raw materials are converted into finished goods, yields the supply chain. The **supply chain** consists of the oversight of materials, information, and finances as they move in a process from supplier to manufacturer to wholesaler to retailer to consumer. Although you might think of the supply chain as involving only the inbound side of a firm's operations, its definition has been expanded to include the entire process of bringing in raw materials, converting them into a finished product, and then sending the product to customers. Obviously, supply chain management is an important aspect of any organization.

supply chain
The oversight of materials, information, and finances as they move in a process from supplier to manufacturer to wholesaler to retailer to consumer.

In contrast to primary activities, support activities include administration and management (the firm's infrastructure), human resources, technology, and procurement. Figure 4.3 shows the value chain model.

Figure 4.3 Organizational value chain

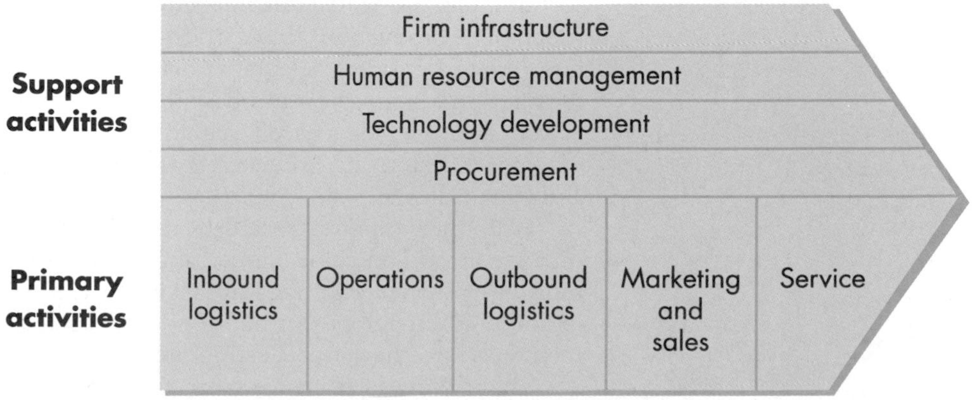

Support activities
Firm infrastructure
Human resource management
Technology development
Procurement
Primary activities
Inbound logistics
Operations
Outbound logistics
Marketing and sales
Service

1. M. E. Porter. How competitive forces shape strategy. *Harvard Business Review*, 1979, pp.137-145.

The raw materials, parts, and sub-assemblies that make up the inbound logistics come in to the loading dock of a company.

value system
The linkage of the value chains of two organizations.

In actuality, each organization's value chain is linked to the value chains of other organizations to create a **value system**. The firm's inbound logistics value chain activity must link to the marketing and sales, service, and outbound logistics activities of its suppliers. For example, an automobile manufacturer's inbound logistics must link to its audio system vendor's value chain systems. On the other side of the organization, its marketing and sales, service, and outbound logistics activities must match up with the inbound logistics activity of its distributors. For example, the automobile manufacturer's outbound logistics must connect with the inbound logistics of automobile dealers. As you can see, a transaction affects not only the organization making it, but also the suppliers of that organization and their suppliers, and so on.

1. Give an example of transaction processing with which you are familiar.
2. What are the primary activities of the value chain?

A Closer Look at Transaction Processing Systems

To meet the needs of an organization, a TPS must carry out a series of transaction processing activities, including input (data collection, data validation, and data correction), processing (data manipulation and database updating), storage, and output (document production). Transaction processing data are collected from cash registers, bar code readers, telephone orders, Web browser orders, and so on. The data captured in a transaction can be internally or externally generated. Internally generated data include payroll statements, shipped orders, and hours worked by employees. Externally generated data include customer orders, customer payments, and bank transfers.

After their capture, these data are *validated* for correctness. For example, a date on a credit card purchase entered as 2014, rather than 2004, is obviously wrong and must be corrected. Although not all errors can be found so easily, the most obvious or inconsistent errors can be found and corrected. Once corrected, the data are reentered into the data collection process.

Transactions identified as correct are used to update the organizational databases and reports are generated about the transactions from the databases. The database updating process entails modifying inventory status, customer data, and supplier data.

Processing transactions can involve a simple count and summation of transactions or a complex analysis of sales by brand or other characteristic. The results of this processing are then stored in organizational databases, from which they can be output in the form of reports. Typical reports include overtime hours worked, payroll by department, and time to ship product. Such an example demonstrates the use of a database management system within organizational memory.

Figure 4.4 shows the activities of a TPS.

batch processing system
A system that combines data from multiple users or time periods and submits them to the computer for processing together.

Transaction Processing Methods

online transaction processing (OLTP)
A process in which each transaction is processed at the time of its entry rather than being held for later processing.

point-of-sale (POS) transaction processing
An input system used to store and process important information that is obtained at the time a sale is made.

Two key operations of any TPS are input and processing. An organization may input and process transactions using two methods: batch and online. In a **batch processing system**, groups of similar transactions are processed periodically. For example, the time that each employee works during a day is input into payroll records at the end of each day as a batch for that day. At the end of the week, payroll checks are written in a batch for all employees for all work they did that week. With batch processing, employees are *not* paid as they work on each work order. In contrast, in **online transaction processing (OLTP)** transactions are processed as they occur. An example of online transaction processing is **point-of-sale (POS) transaction processing**, in which each sale is processed as it occurs, as at a grocery checkout station using a bar code reader linked to a computer.

Figure 4.4 Transaction processing system

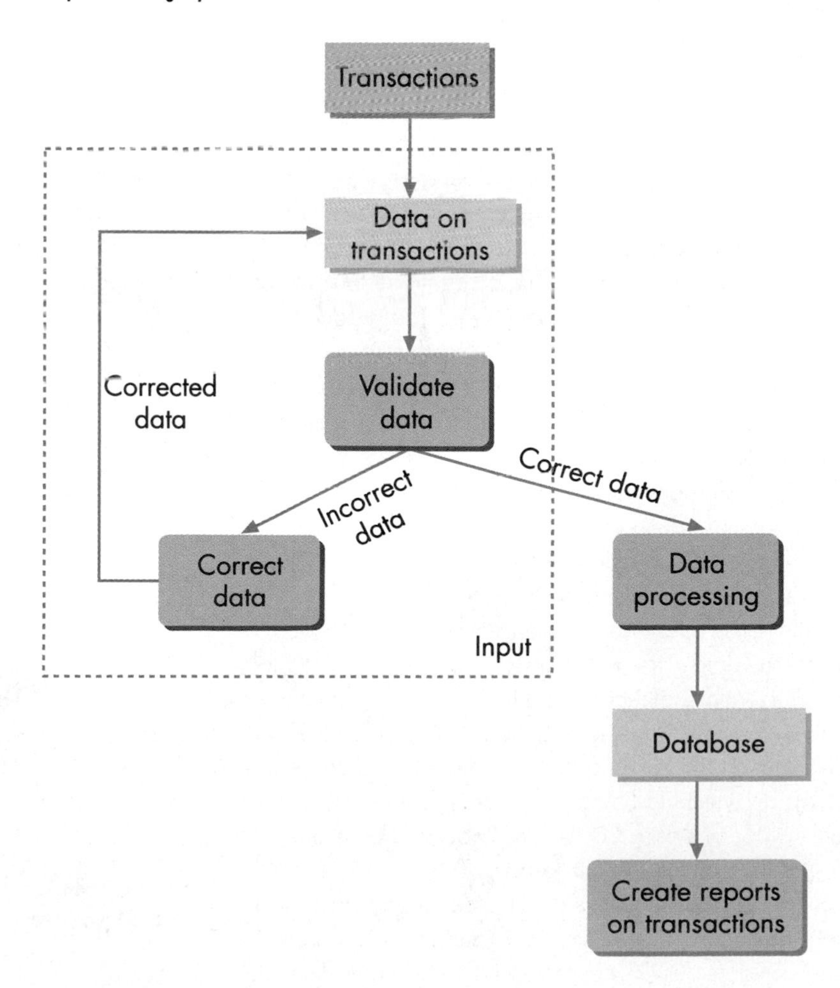

In some cases of OLTP, especially those involving registration or reservations, the processing actually controls availability of the product or service. For example, when customers reserve airline seats using a travel agent or a Web browser, the program must first check the inventory of airline seats to ensure that a seat is available before the transaction can be completed. If a seat is available and the transaction is completed, the program immediately updates the inventory of seats to reflect the reservation. Other examples of processing controlling the availability of products or services include hotel/motel and automobile reservation systems.

In the networked economy, an important part of OLTP entails validating credit cards and placing a hold on, or encumbering, the amount of the sale prior to approving the transaction. Although checking a credit card's validity and encumbering the amount of the sale has been performed for several years in many stores worldwide, the advent of electronic commerce has made this process even more important. It is especially important when a product such as a game or software is being "shipped" electronically. When a customer can immediately download the product over the Internet, once the product has been transmitted, it cannot be recalled.

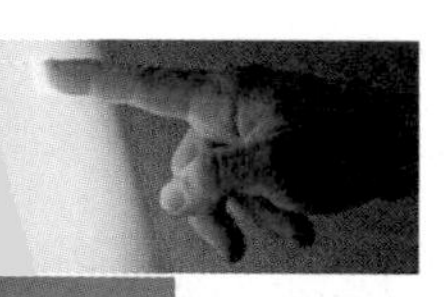

Herman Hollerith

IT INNOVATORS

Transaction processing on computers can be traced back to the end of the nineteenth century when the U.S. Census Bureau held a competition to find a better way to count the population of the fast-growing country. The Constitution requires that a census be undertaken every 10 years, not just for bureaucrats or intellectual curiosity but to apportion seats in the House of Representatives. The 1880 census had not been completed until 1887 and the Census Bureau was very concerned with finishing the 1890 census before work started on the 1900 census. The winner of the competition was a young engineer named Herman Hollerith. Hollerith's approach to tabulating the census became the forerunner of a method of computer input that was used well into the 1970s and led to the creation of one of today's leaders in information technology.

Hollerith's idea was to use punched cards to tabulate the census data. The use of punched cards was not totally new; a French engineer named Jacquard had used them to control the pattern being woven on a loom in the early 1800s. Hollerith realized that he could store data on individuals by punching holes in the cards. They could then be stacked in any desired order and tabulated by completing an electrical circuit through a hole. This completion of the circuit would cause an associated counter to increase by one. The combination of stiff cards and electricity created a very workable system. Early pioneers in computation such as Charles Babbage tried to use mechanical gears to do the same tasks with little success. Based on his work, in 1890 Hollerith founded a company called the Tabulating Machine Company. In 1911, his company merged with two other companies to create a company that would eventually become the International Business Machines Company—IBM.

Herman Hollerith is the father of today's transaction processing systems.

Source: Mark Russo, "Herman Hollerith: the world's first statistical engineer." http:// www.history.rochester.edu/steam/hollerith/.

Travelers have judged the AirTran reservation Web site as one of the easiest to use.

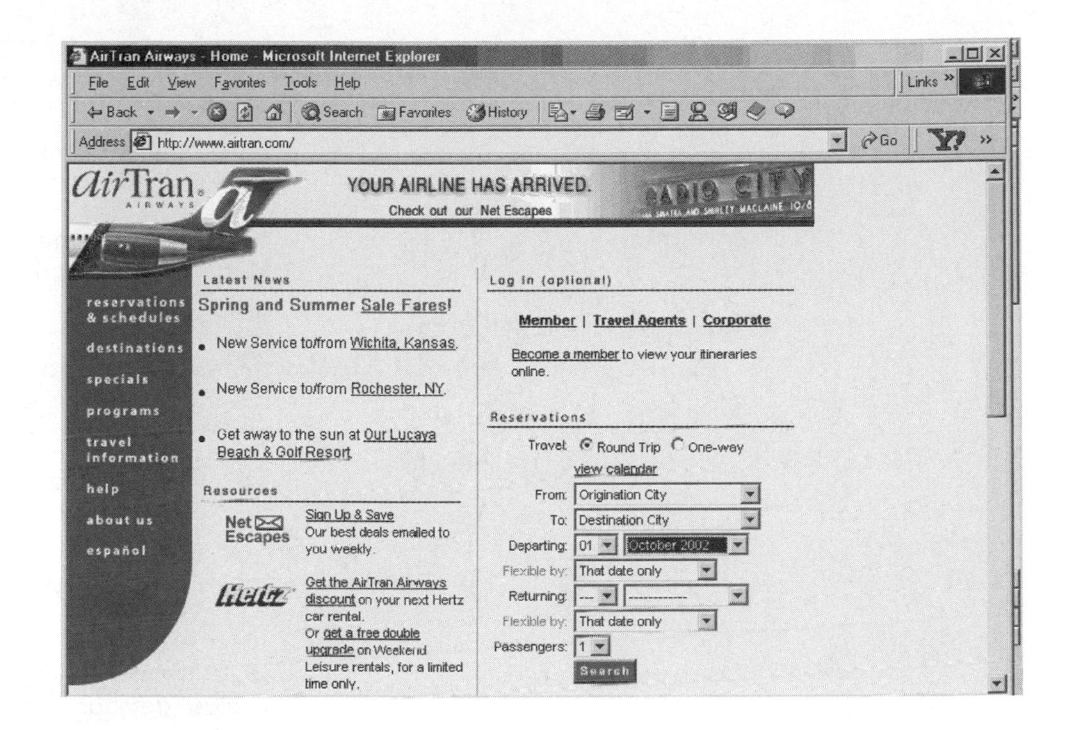

Transaction Processing Activities

As mentioned earlier, you can think of transaction processing as a series of steps proceeding from data gathering and entry, to processing, to storage, to final output of information to users. Let's consider these four steps in more detail:

1. Gathering and entering the data that describe the activities of the organization
2. Processing (arranging and manipulating) the data for use by a large variety of potential users
3. Storing data and reports for easy retrieval when necessary
4. Reporting information to users

back office server
Software that processes data from transactions and uses it to track inventory and send purchase orders when inventory levels fall below a designated level.

To help you understand these four steps, consider the POS transaction processing systems used at grocery and other types of stores. Figure 4.5 shows an example of a POS system. The POS system consists of a bar code reader, a display screen, a device for printing receipts, the TPS, and a **back office server** that processes the data from the transactions. The back office server uses the data collected from the POS system to track inventory and send purchase orders when the available inventory for an item drops below a predesignated level.

Figure 4.5 POS transaction processing system

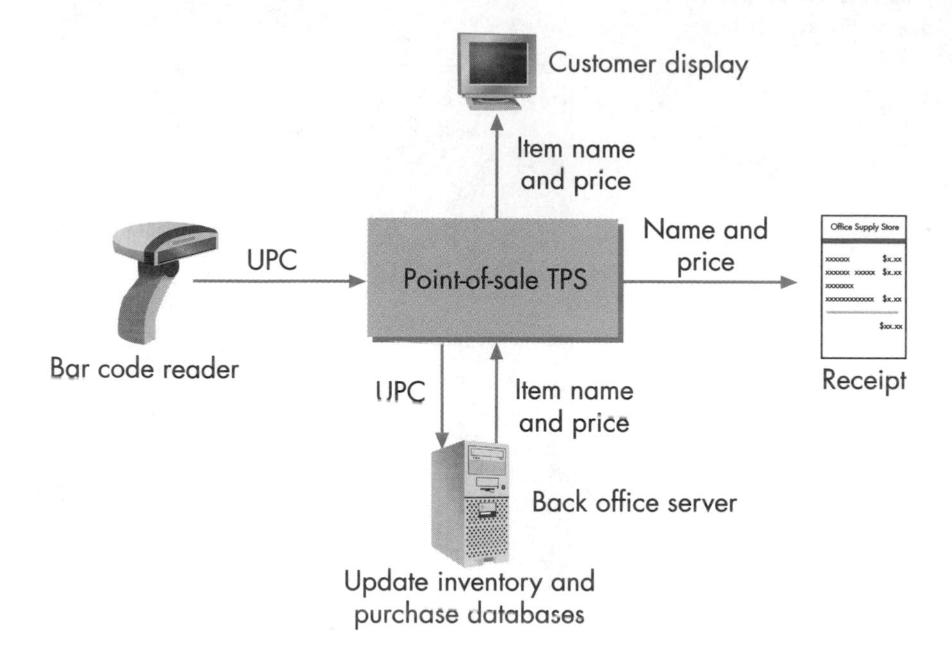

Data Gathering and Entry

Bar codes are used to store information in many situations other than in retail stores.

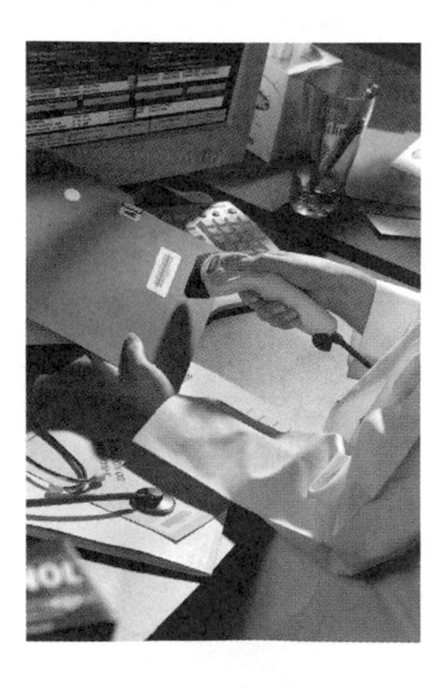

To be used, data must first be gathered on the physical operations of the organization as it engages in its business activities. Gathering of data occurs at all stages of the conversion process—that is, from the input of resources, the transformation of these resources into products, and the distribution of goods and services through sales. Management may then control the operations and make managerial and strategic decisions based on the information it obtains from processing these data.

Data must be gathered from all of the physical input, processing, and output activities in which an organization engages as it markets and delivers its products and services. These activities involve a number of transactions with external parties as well as internal units within the organization, including sales, stock reduction, shipping, billing, cash receipts or collections, purchasing, receiving and stock increasing, and payment transactions. In addition, these transactions may include the conversion of labor and materials into finished goods and services. Data can be gathered and processed in a number of ways. They may be grouped for future processing in a batch mode, or they may be processed as the transaction occurs using OLTP, depending on the situation.

Users may enter data into the computer system using a variety of input devices, such as dumb terminals, personal computers, network computers, bar code readers, smart cards, and the Internet. Of these forms of input, bar code readers are currently the most widely used because of their speed and accuracy of input. Web-based input represents an especially fast-growing form of input, and smart cards have become widely used in many locations outside the United States.

In a bar code system like that used in the POS shown in Figure 4.5, a transaction is accomplished by passing the item over a bar code reader that bounces a laser light beam off the bar code and then measures the reflected light. The most common bar code, the universal product code (UPC), is the standard for most transactions today. Figure 4.6 shows an example of a UPC, with the various elements identified. The measurements of the bars and spaces in the bar code are converted into binary data, which are transmitted to a back office server and used to query the product database for the name and price of the item. The product name and price are transmitted back to the checkout stand, displayed for the customer, and added to the customer's receipt.

Regardless of the form of input, this data entry process requires validation and correction to ensure the accuracy and completeness of the input. In terms of accuracy, bar code and Web-based input are at opposite extremes. Bar codes have an input error rate of less than one in 10 million. On the other hand, because consumers must use a keyboard to input much of the information on a Web-based order, it is subject to frequent errors, especially in the personal information and credit card fields. For this reason, editing and verifying are important parts of the data gathering step in a Web-based transaction processing system.

Processing and Data Manipulation

Processing data into a useful form involves several activities, including calculating counts and sums of the transactions according to some classification, summarizing activities, and updating databases to reflect the transaction. To carry out the calculations, transactions must be classified and stored so that operational and managerial personnel can use them effectively. For example, grocery store transactions can be classified into a number of categories, including groceries, meat, poultry, fish, fresh produce, and nonfood items. By classifying transactions in this way, various managers at the grocery store can quickly look at only the data that concern them. For example, the grocery manager may not care about the sales of nonfood items.

Figure 4.6 Example of UPC bar code

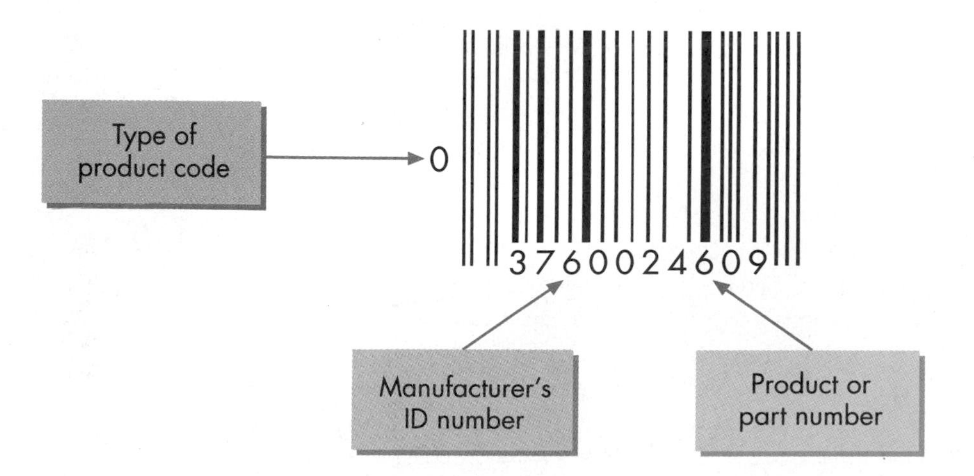

To be useful to management, data must often be manipulated before their storage. This activity may involve combining the data with other data to arrive at values such as the number of units in stock, the total amount of an invoice, or the amount that a customer owes. On the other hand, all data need not be stored; sometimes it is sufficient to summarize data for future use. For example, sales summary data may prove more useful to a sales manager than the details of every single sale for a period of time. Sales patterns that may be very evident when reviewing summary data might never be seen if the manager examines only detailed data.

A retail store might count and sum transactional data by time period, enabling managers to determine when additional employees should be available to handle high-volume periods or when fewer employees are needed. The transaction data could also be counted and summed by other criteria to answer particular questions that management might have. If the store has both in-store and Web-based sales, management would probably want to know the division of sales by each type.

TECHNOLOGY ON THE EDGE

"Super" Bar Codes and Smart Tags Are Coming

The ubiquitous bar code has been around since the 1970s and today appears on virtually every item sold, other than fresh produce. However, bar codes have some shortcomings: They must be visible, and they must be deliberately scanned one at a time. The Auto-ID Center hopes to eventually replace all bar codes with **smart objects** containing microchips and wireless antennas that transmit data to any nearby reader. This system works much like the highway tollbooths that pick up signals from tags on cars as they drive by.

Beyond just computing a price, the smart tags will enable companies to track a product all the way through its supply chain—from the warehouse to the store shelf, and on to the customer's kitchen. Each time a product is removed from a store shelf, information on the sale is sent back to the manufacturer's database, allowing companies to manufacture products in direct proportion to actual customer demand rather than basing their manufacturing decisions on forecasts. This approach will result in better inventory control and, as discussed in Chapter 1, perhaps less inventory buildup that can contribute to economic downturns. The new tags can recognize more than 268 million manufacturers, each with more than 1 million products. The major sticking point with smart tags is the cost—they now cost too much to replace the very inexpensive bar codes. However, the Auto-ID center hopes to get the price down to 5 cents per tag by 2005.

Even if smart tags remain some time away from widespread use, other technologies continue to speed the checkout process. Self-service checkout is already very popular in Europe and is gaining widespread acceptance in the United States at stores such as Best Buy, Office Depot, and Kroger. To help solve the problem of scanning fresh produce, IBM has created a scanning system that is 95 percent accurate in detecting types of produce as they pass in front of a camera. When installed in a few years, this device could have major effects on grocery checkout lines where produce causes a bottleneck.

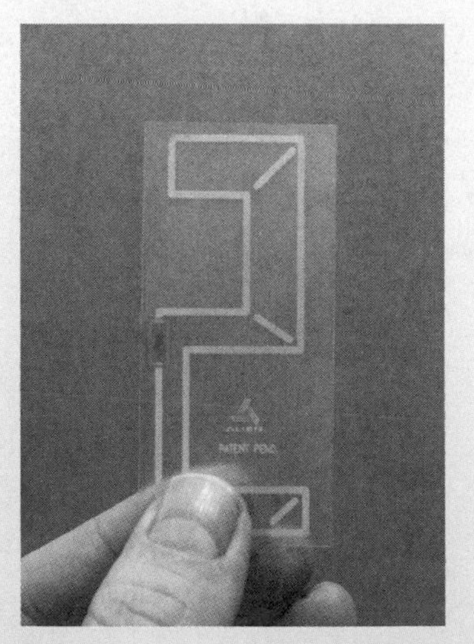

Smart tags such as this one may soon replace bar codes on some items.

Source: Emling, Shelley, "Super Bar Codes." *Atlanta Journal/Constitution*, May 13, 2001, pp. Q1, Q10.

Databases are both queried and updated as a part of the processing operation. You have already seen how the product database is queried in the grocery store example to determine the name and price of the product being scanned. In addition, databases must be updated to reflect the purchase and the corresponding reduction in inventory. Store management might query the purchase database at a later time to gather sales statistics for marketing and buying decisions. Similarly, management might query the inventory database periodically to determine which items need to be ordered. Or, if the store is linked to distributors by EDI or other electronic collaboration technology, the back office server would order the needed items automatically based on rules built into its software. Figure 4.7 illustrates the processing and manipulating of data.

Data Storage

A large number of data elements may need to be stored for some types of transactions. In general, a transaction must be identified by number; the people involved in the transaction, such as the customer number and the sales person number; what was transacted, such as the stock number and amount of merchandise sold; the date when the transaction took place; the department where the transaction took place; and authorization, such as a supervisor's "OK" for overtime hours. In summary, the who, what, when, where, and authorization of each transaction must be gathered, classified, and stored for future use.

Figure 4.7 Processing and manipulating transaction data

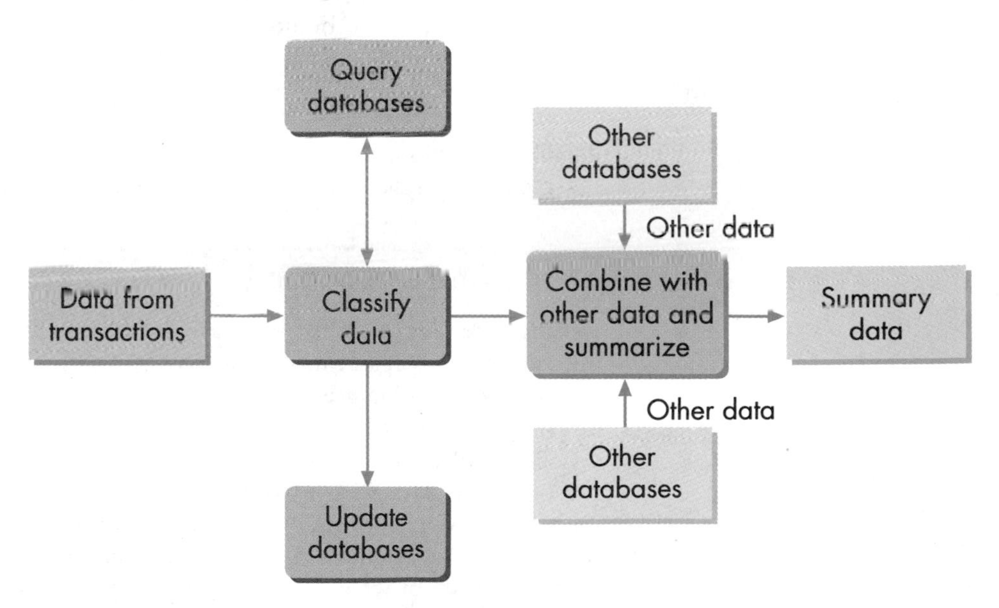

The current and future requirements for information needed to control and manage the organization determine the actual physical and logical structure used to store data. The volume of this transaction data partly explains the growth in the use of extremely large databases, known as data warehouses, to store transaction data for future analysis.

For example, each transaction in the grocery store situation must be temporarily stored locally in a purchase database on the back office server and then transmitted to the corporate database for more permanent storage. Periodically, the data will be added to the data warehouse at corporate headquarters.

Output and Reporting

The transaction processing system ends with the output of information for operational control. This output may take the form of printed periodic reports, such as labor variance reports for plant supervisors, or online queries displayed on a monitor, such as the amount of components on hand displayed by marketing personnel as they transact a sale. These reports may consist of internal documents or messages displayed on computer monitors, or they may comprise formal reports such as financial statements for outside parties. The output may be periodic and follow a set schedule, or it may occur on demand as operational personnel in marketing, production, or financial services seek data to carry on their respective activities.

For example, the grocery store management will look at daily reports on overall sales and sales by type. Management may also request periodic reports on sales by type of item or by brand to determine whether to revise the shelf-space allocation. Managers may find that a certain brand of wine is selling slowly, prompting them to sell the current stock at a reduced price and to cease carrying that brand in favor of another, more popular brand of wine.

1. List the steps that take place each time an item is processed at your local grocery (assuming that a bar code scanner system is used).
2. Why is data editing and verification more important in a Web-based transaction processing system than in a bar code-based system?

The Impact of the Internet on TPS

Because the Internet and Web have become such a prevalent part of life for both individuals and organizations, the way transactions are carried out and the types of interactions possible are changing. In fact, some evidence suggests that consumer use of the Internet has speeded acceptance of self-service gas pumps and the self-service retail checkout systems that are springing up in grocery and other retail stores. Use of the Web to order goods, such as Asian foods at FarEast Foods, is merely one way in which people work through the Internet to transact business. Many other transactions can also take place using the Internet. These transactions lie at the heart of what is commonly referred to as electronic commerce.

eBay—A "Dot Com" That Is Making Money

At the beginning of March 2000, the Amazon.com's and Yahoo!'s of the world were flying high with astronomical stock prices and company valuations. With the Nasdaq crash later that month, most of those stock prices and company valuations dropped significantly. For example, Amazon.com dropped from a high of more than $100 per share to $12.50 per share in early 2002. Similarly, Yahoo! stock dropped from a high of around $225 to approximately $16.50 per share during the same period. Both companies continue to search for profitability, with Amazon losing more than $558 million during 2001 and Yahoo! losing almost $93 million during the same year. On the other hand, another big name in the technology stock world—eBay—realized a net income of more than $90 million for 2001, up from $45.7 million during 2000. Although its stock price is also down from an all-time high of around $125 to around $58.50 per share in early 2002, this decline is a far cry from the huge drops of other technology stocks.

What makes eBay different from the other dot coms that blazed to such heights, only to fall on hard times after the period of "irrational exuberance" that marked the end of the twentieth century? eBay primarily acts as an intermediary to allow companies and individuals to auction off goods to other Internet users. As such, it has very little money tied up in the goods being sold and does not get stuck with unsold inventory. Auction companies such as eBay may actually benefit from a downturn in the economy, when individuals who lose their jobs need to sell possessions to pay bills or, as mentioned in the opening case, when companies need to unload unwanted inventory to meet expenses.

eBay was started in 1995, when one of the founders wanted to trade with other Pez candy dispensor collectors. Since then, the value of goods sold on eBay has increased to more than $5 billion in 2000. By far the biggest such online auction company, eBay has approximately $25 million in sales each day compared to less than $500,000 in sales for its closest competitor. The company has continually looked for ways to service its loyal customers better, including finding ways to guard against fraud and ensure safe transactions. Its latest move is to create a fixed price sales system called Half.com and to go international in countries other than the United States.

Sources: http://cnnfn.com, Miguel Helft , "Ebay's Whitman wows the crowd," *Industry Standard*, July 23, 2001, http://www.thestandard.com.

To understand the wider implications of the Internet for transaction processing, let's first consider the possible participants in transactions: customers/citizens, businesses, employees, government, and not-for-profit organizations. Typically, you think of transactions occurring between a customer and a business, between businesses, and between customers and businesses and government. With the Internet and Web, however it is possible for a much wider variety of transactions to go on between virtually all possible combinations of these stakeholders, including the more common ones: business-to-consumer (B2C), business-to-business (B2B), and business-to-government (B2G). Table 4.4 shows some examples of the possible interactions. For example, customers can interact by sharing information about products or by selling goods to one another using the various auctions. Similarly, customers and employees can interact by sharing solutions to problems or responding to requests for help.

Many new companies and services have sprung up to handle this great variety of possible interactions. For example, because of problems with purchasers not always paying for an item purchased on an auction or sellers not sending the item, a number of electronic payment systems have come into being to help ensure that the transaction is carried out appropriately.

OLTP over the Internet usually involves the use of an application server, like those discussed in Chapter 2, that has the software necessary to handle transactions. This application server actually handles the transaction by interacting with a database server. In large-scale operations such as Amazon.com or e*Trade, many servers in a **server farm** work together to process the transactions. The application server works with a database server to check availability of the item being purchased and to update the customer record or create a new one. Figure 4.8 shows a Web server, application server, and database server handling OLTP over the Internet. Note that this process actually uses a *four-tiered* client/server system created through the addition of the application server between the Web server and the database server. The fourth part of this four-tiered system is the client, which resides on the customer's machine. Although Figure 4.8 shows two separate machines, because a server is actually a piece of software, both the Web server and the application server—although logically different—can physically reside on one computer.

server farm
A group of servers that work together to handle processing chores.

Table 4.4 Stakeholder Interactions Using the Internet

	Customer/Citizen	Business	Employees	Government
Customer/citizen	Share information about products using Web or e-mail	Inquire about products; complain about product or service	Send requests for help on products and services over Web	Inquire about services or fees using Web and e-mail
Business	Sell products or services online; provide after-sale support	Engage in supply chain activities using Internet	Send check stubs electronically via e-mail	Engage in supply chain activities
Employees	Respond to requests for information	Select fringe benefits over the Web and check retirement funds online	Share knowledge about problems	Pay taxes online
Government	Present electronic tax bills; provide information on government services	Provide information on regulations	Provide information on retirement benefits; educate workers about rights	Make financial transfers; share information

Figure 4.8 Use of an application server for OLTP

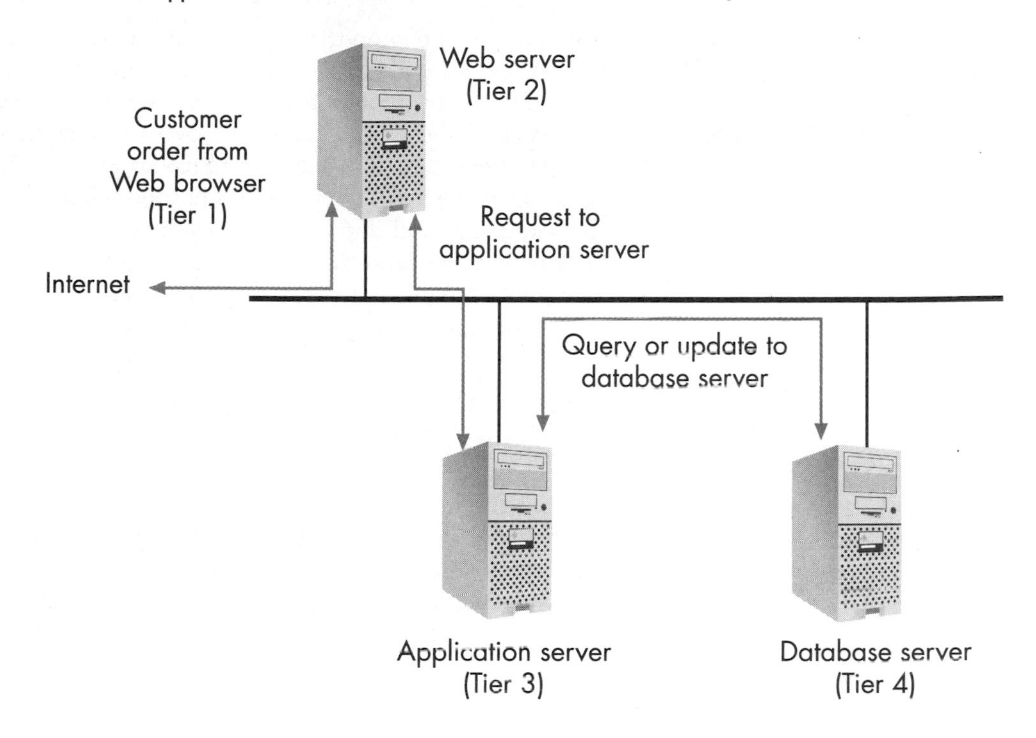

In systems relying on an older mainframe, the application server can be set up to accept requests from the Web server and process them into messages that the mainframe can understand. The mainframe processes the transactions and sends the results to the application server, which, in turn, sends a message to the Web server for transmittal to the customer. This system looks like that shown in Figure 4.8, with the mainframe replacing the database server.

1. List three stakeholder relationships other than the three most common ones (B2C, B2B, and B2G).
2. What is the purpose of an application server in OLTP over the Internet?

Server farms are important in situations where a large number of transactions take place.

Transaction Processing at the FarEast Foods Web Site

This section will demonstrate how an Internet-based transaction takes place by following an order to the FarEast Foods Web site from beginning to end. For a purchase from FarEast Foods, this process begins at the customer's browser, goes over the Internet to the merchant's Web server, and then moves to the application server for checking of the customer credit card. If the credit card has an adequate balance for the purchase, FarEast Foods orders the items from the various food wholesalers, combines the items at its distribution center, and ships the order to the customer. Figure 4.9 provides an overview of the process, starting with a purchaser generating an electronic order from a PC and ending with the purchaser receiving a shipment.

As you can see in Figure 4.9, the food order and fulfillment process includes five steps:

Step 1: The purchaser uses a browser on a PC to connect to the FarEast Foods Web site and generate an order.
Step 2: The order is submitted over the Internet to FarEast Foods.
Step 3: The order is received at FarEast Foods, and the customer's credit card is checked for sufficient funds to pay for the order.
Step 4: The order is processed at FarEast Foods and broken into component food items, which are ordered from food wholesalers.
Step 5: The component food items arrive at the FarEast Foods distribution center, along with food items for many other orders. The specific items for an order are selected and combined into a package, which is then sent to the purchaser via a package delivery service.

Let's consider each of these steps in more detail.

Figure 4.9 Overview of steps in a food order transaction

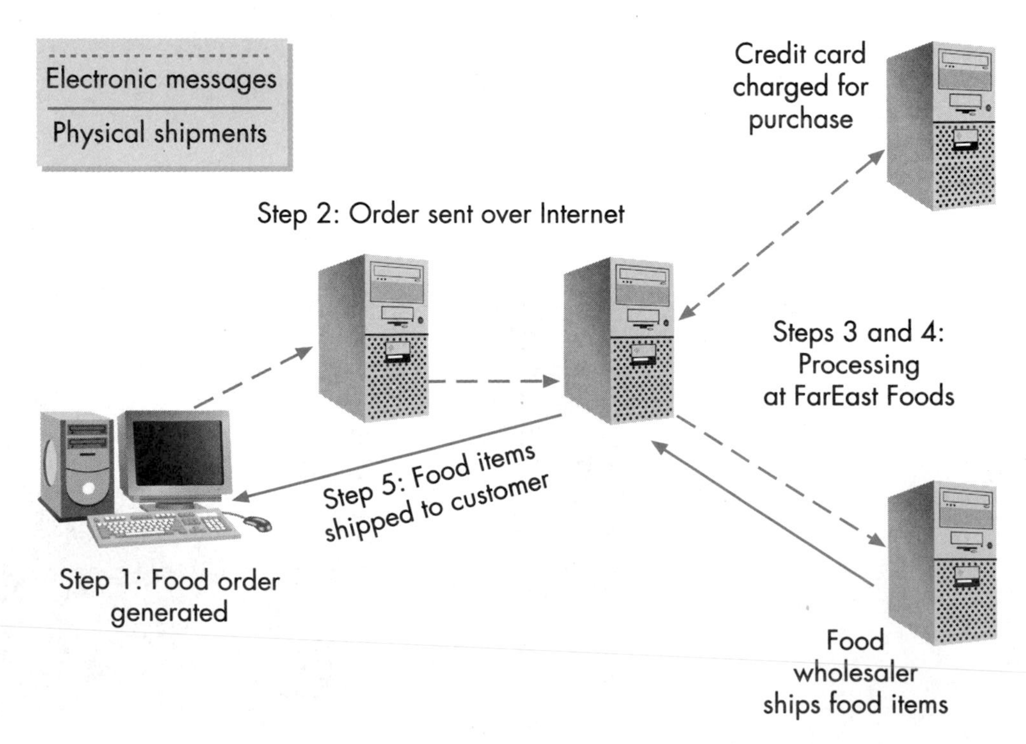

Step 1: Creating an Order

The customer uses the browser on his or her PC to visit the FarEast Foods Web site at www.fareastfoods.com. Once the Web browser connects to the Web site, the purchaser can view an electronic catalog of nonperishable Asian foods. The customer can order desired food items by filling out an online order form and submitting it to the company over the Internet. Customers can pay for their purchases by using a credit card. Figure 4.10 shows a browser screen with a completed order form for FarEast Foods.

The checkout Web page can be programmed to validate some of the information entered by the customer. For example, it can ensure that all fields are completed, determine whether the credit card number is valid by verifying that it has the correct number of digits, and check that the digits sum to an appropriate value.

Step 2: Submitting an Order

encrypt
To convert readable text into characters that disguise the original meaning of the text.

Once a customer has completed the order from by using a browser, the next step is to submit the order over the Internet to FarEast Foods. First, the browser software **encrypts** the order information in such a way that criminals will not be able to intercept it and use the credit card number. Second, the encrypted order information containing the name, the postal and e-mail addresses of the person sending it, and the credit card information is converted into a stream of data bits that is sent from the browser to the PC's modem for actual transmittal over the Internet via telephone line or cable. Figure 4.11 shows this process.

Figure 4.10 Completed food order form in Web browser

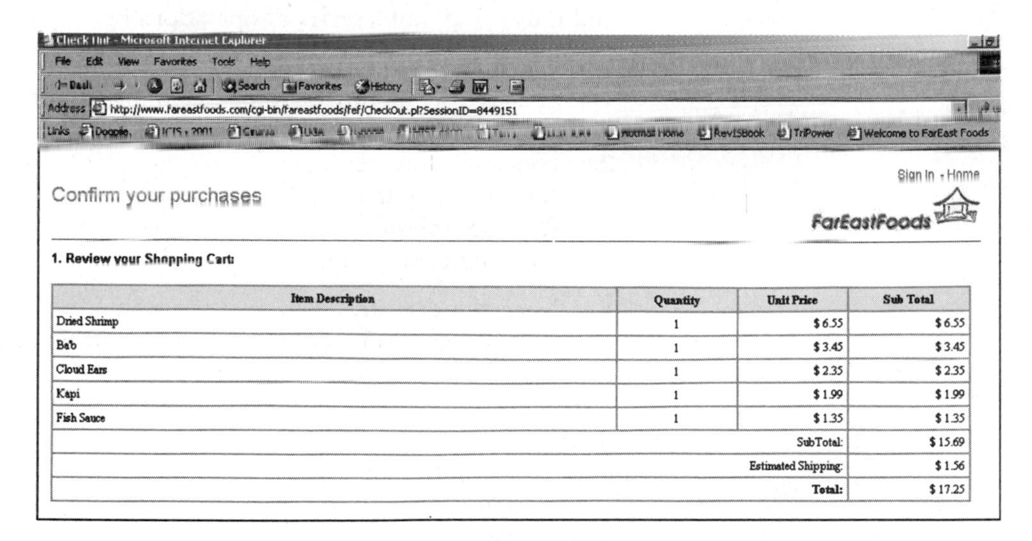

Item Description	Quantity	Unit Price	Sub Total
Dried Shrimp	1	$ 6.55	$ 6.55
Ba'b	1	$ 3.45	$ 3.45
Cloud Ears	1	$ 2.35	$ 2.35
Kapi	1	$ 1.99	$ 1.99
Fish Sauce	1	$ 1.35	$ 1.35
		SubTotal:	$ 15.69
		Estimated Shipping:	$ 1.56
		Total:	$ 17.25

Figure 4.11 Transmittal of order from customer to company

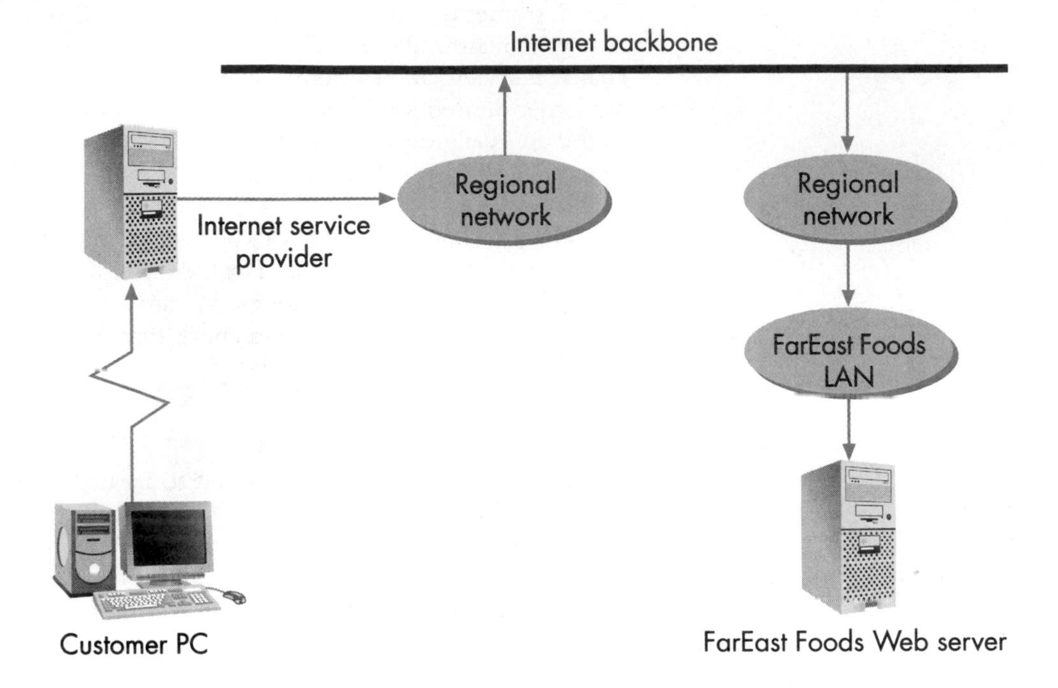

Step 3: Processing the Order at FarEast Foods Web Site

When the order arrives over the Internet, the FarEast Foods Web server receives it and must carry out a series of operations before ordering the food and shipping it to the customer:

1. The Web server processes the incoming data stream and sends it to the application server.
2. The application server computes the total bill and sends a message to the credit card company to check the available funds on the card.
3. The application server updates the customer record or creates a new record for a new customer on the company database.

Once the data stream from the customer browser reaches the Web server, the Web server decrypts it and splits it into the various fields of the order (data on the person ordering the food items, the product numbers of the items being ordered, and the credit card number being used to pay for the order). The Web server then sends the order to the application server for processing, where the total cost of the order is computed.

Next, the application server sends a query to the credit card company to ensure that the card number is valid and that the amount of funds available is larger than the amount of the order. If it is, the card is charged for the amount of the purchase, and FarEast Foods receives a message validating the card. Finally, the application server updates the customer record in the database, or creates a new record if this person is a new customer. The relationships among the Web server, application server, and database server were shown earlier in Figure 4.8.

Step 4: Ordering Food from Wholesalers

If the credit card is valid, the application server generates food orders for the Asian food wholesalers, requesting them to pull the item(s) from their warehouse(s) and ship them to FarEast Foods. Because FarEast Foods does not keep any inventory, depending instead on *just-in-time* ordering from its suppliers to satisfy customer orders, it does not have an inventory database to update. On the other hand, because FarEast Foods typically satisfies customer orders by combining products from multiple suppliers, the application server must generate a *picking form* (also known as a *pick list),* which it sends to the shipping department. This picking form will have a bar code (not the UPC bar code, but one specific to this application) that employees use to select just the right items from a shipment of many items from several food suppliers to satisfy the customer order.

Step 5: Creating and Shipping the Order to the Customer

When the orders arrive from the wholesalers, employees must quickly combine them into a single package using the picking form created in Step 4. The employees scan the package bar code into the application server, and the application server matches it with the original order from the customer. The application server then sends a message to a package delivery company to pick up the combined food items for delivery to the customer. It also sends a message over the e-mail server to the customer indicating that the order has been processed and shipped.

After the delivery company picks up the package, every time the package is handled, the address label bar code is read with a bar code reader to generate data on the package's time, location, and status. These data are transmitted to the package delivery company's application server to be incorporated into its database. Because the package delivery company's database is connected to its Web site, FarEast Foods and the customer can check the history and current status of the package from the package delivery company's Web site. A chip that stores time, location, and status information also may be included on the outside of the package to expedite this process.

When the package is delivered, the delivery driver requests that the purchaser sign an electronic tablet. This tablet digitizes the signature and transmits it to the delivery company's application server and database server as proof of receipt. The digital signature can also be displayed on the package delivery company's Web page.

1. List the steps necessary to create and send an order from a PC to a company such as FarEast Foods.
2. How would these operations differ in a company that manufactures or warehouses the goods its sells rather than ordering them from suppliers as FarEast Foods does?

Picking items from a warehouse is crucial to the order fulfillment process.

Business-to-Business Transactions

Although all of the relationships shown in Table 4.4 are important, the biggest market affected by the Internet will undoubtedly be in the business-to-business area. In 2000, B2B sales totaled $433 billion, up from $244 billion the year before. The Gartner Group forecasts B2B sales for 2002 to reach almost $2 trillion, even with the economic slowdown. Compare 2000 B2B online sales of $433 billion with the $48 billion of B2C sales[2] and the $3.6 billion of B2G sales,[3] and you can see why the B2B market is so important. Figure 4.12 shows actual and forecast B2B sales through 2005.[4]

Whereas consumers will certainly continue to seek out local stores to do much of their shopping so as to deal with people they know in the community, businesses typically do not feel this need. Instead, they are interested in pursuing relationships with companies that supply their needs with high-quality products at a low cost regardless of where their operations are located. For this reason, doing business over the Internet is a natural for many companies around the world. They can reduce all three types of risks—demand, innovation, and inefficiency—through the use of the Internet. For example, as discussed in the Manheim auctions case, automobile dealerships—all small companies—can use the Internet to acquire used cars for resale from Manheim Auctions, which is the world's largest auction source of such cars. Dealerships can acquire stocks of used cars more quickly and easily from the auction company, thereby reducing inefficiency.

Figure 4.12 Actual and forecast B2B sales from 1999 through 2005

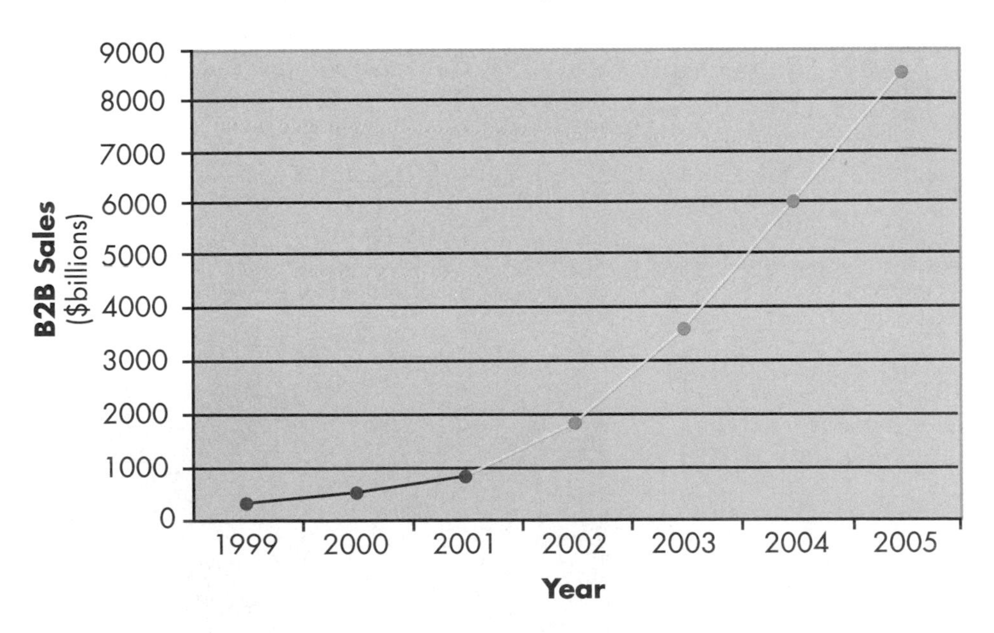

---

2. Forrester Online Retail Index, http://www.forrester.com.
3. "Fed's eclectic e-sales outpace Web heavyweights." *News Press* (Ft. Myers, FL), May 28, 2001, p. 3A.
4. Erich Leuning, "Gartner bearish on B2B Web transactions." *CNET News.com*, March 13, 2001, http://news.cnet.news.com.

Interorganizational Systems and the Internet

interorganizational system (IOS)
A networked information system used by two or more separate organizations to perform a joint business function.

firewall
A device placed between an organization's network and the Internet to control access to data and systems.

Business-to-business transactions are part of an important type of system known as an interorganizational system. An **interorganizational system (IOS)** can be defined as a networked information system used by two or more separate organizations to perform a joint business function.[5] An IOS often involves electronically linking a production company to its suppliers or its customers in such a way that raw materials are ordered, production takes place, and finished goods are sent to the customer to meet demands with little or no paper changing hands. The traditional IOS is based on EDI, which uses value-added networks (VANs) or private networks instead of the regular telephone system, but is too expensive for all but the largest businesses to use. As a result, many businesses have not been able to benefit from IOS. However, the Internet enables smaller companies to take advantage of IOS and carry out business-to-business transactions more efficiently.

Internet technology has created a low-cost platform for linking computers. In addition to using the Internet for business transactions, many organizations are creating internal versions of the Internet called intranets. Recall from Chapter 3 that an *intranet* is a LAN that uses Internet protocols, but restricts access to employees of the organization. It is essentially a fenced-off mini-Internet within an organization. For example, an intranet Web server allows only certain people to have access to the information stored there. A **firewall** is a combination of a computer and software that filters the bits that come into an organization's network, thereby restricting access to the intranet.

In contrast, an *extranet* is a business-to-business network that uses Internet protocols, instead of EDI or other private protocols, for transmitting data and information between trading partners. Similar to an intranet, this restricted form of the Internet is not open to the general public. Figure 4.13 shows these Internet technologies.

Intranets and extranets enable organizations to take advantage of low-cost Internet technology to communicate with and deliver information to different groups of stakeholders. The Internet can enable communication with all stakeholders in the organization, including investors, employees, customers, suppliers, and so on. An intranet, on the other hand, is restricted to sharing of information and computing resources among internal employees. Finally, an extranet aims to share information between trading partners. These three types of Internet-based systems can be classified according to three characteristics: *scope, focus,* and *business processes.* Scope refers to how widely the system is used. Focus refers to the purpose of the system. Business processes refer to the revenue, expenditure, conversion, and financial processes discussed earlier. You can use these characteristics to compare EDI and the various Internet-based systems, as shown in Table 4.5.

EDI and extranets have restricted scope because they focus on business partnerships between two companies. In contrast, the Internet has a global scope because anyone with a computer and Internet access can read information on it. Intranets are also restricted in scope to users within the organization who are allowed access to them.

---

5. J. I. J. Cash, F. W. McFarlan, J. L. McKenney, and L. M. Applegate, *Corporate Information Systems Management: Text and Case*, 4th ed. (Homewood, IL: Irwin, 1994).

Figure 4.13 Internet technologies

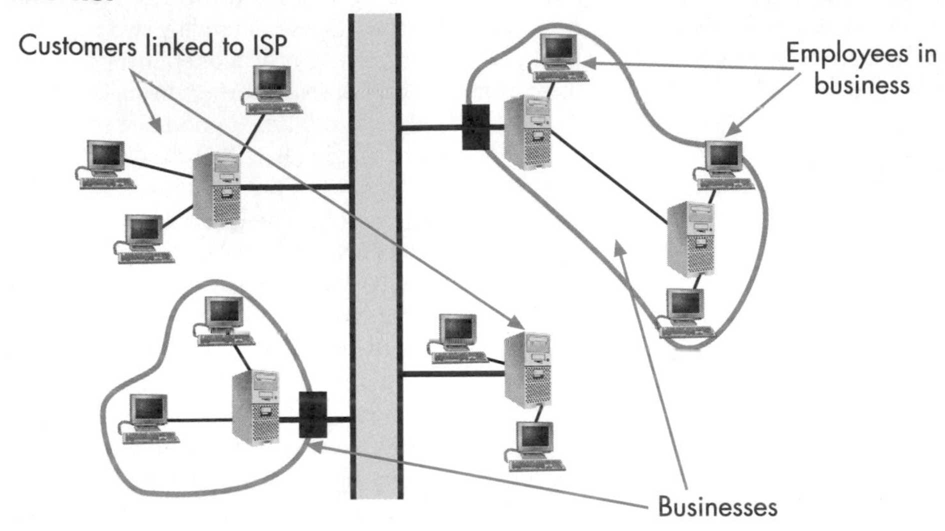

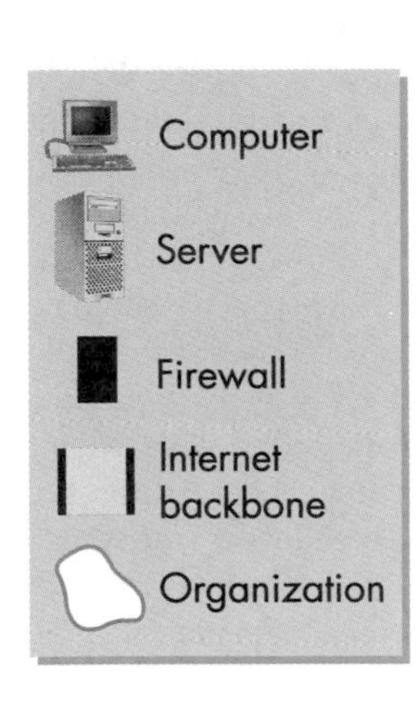

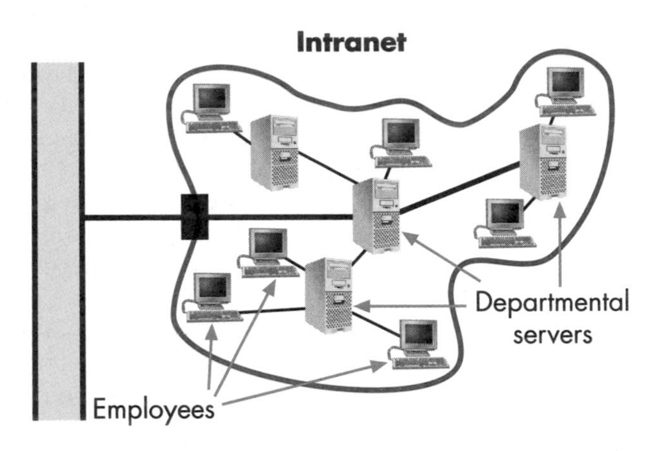

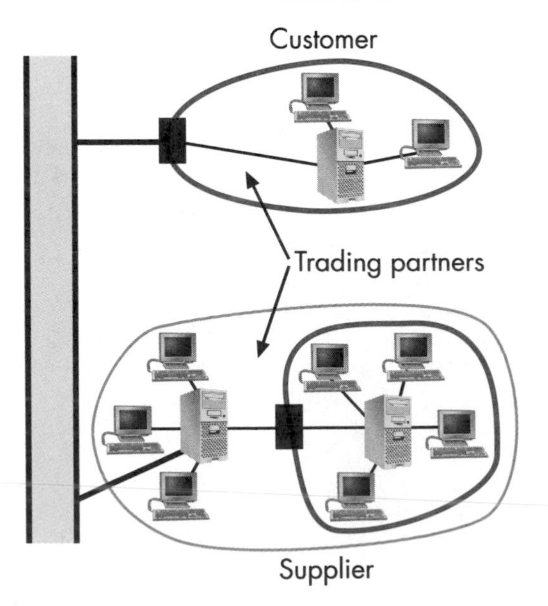

Sources: Adapted from P.G. McKeown and R.T. Watson, *Metamorphosis: A Guide to the World Wide Web and Electronic Commerce*, 2nd ed. (New York: Wiley, 1997).

Table 4.5 Comparing EDI and Internet-Based Systems

Characteristics	EDI	Extranet	Internet	Intranet
Scope	Business partnership	Business partnership	Global	Organizational
Focus	Distribution channel cooperation	Distribution channel cooperation	Stakeholder relationships	Employee communication and cooperation
Business processes	Revenue and expenditure	Revenue and expenditure	Revenue	Expenditure and conversion

Both EDI and extranets aim to foster cooperation with trading partners in distribution channels, whereas the Internet targets stakeholder relationships (anyone who has an interest in the organization). The focus of an intranet is communication and cooperation between groups of employees within the organization.

Finally, in terms of business processes, both EDI and extranets can be used for the revenue and expenditure processes because they involve relationships with business partners—both suppliers and customers. Because of its openness, the Internet is used primarily to bring in revenue to the organization. Finally, intranets are employed for the expenditure and conversion processes because they involve communications and cooperation between employees.

As an example of how extranets affect expenditure processes, consider Manheim Auctions. Automobile dealers use the Manheim Auctions extranet to order used cars for resale, which helps streamline their expenditure processes. Similarly, many companies use the Internet to augment their revenue processes. Intranets are being widely used for the conversion process—employees can easily share design documents or information about production issues.

In Table 4.5, note that extranets and EDI share the same characteristics, indicating that it may be possible to replace an EDI system with an extranet to carry out business-to-business transactions. In general, the costs of business-to-business transactions can be reduced dramatically through the use of extranets to lower costs in the areas of connection, hardware, software, and learning. Extranet connection costs are significantly less than those of traditional EDI because an extranet only requires accessing the Internet though an ISP. A dial-up Internet connection can cost $25 or less per month. Even at the upper end, these costs can be less than half those of using EDI over a VAN. In addition, minimal software costs are associated with an extranet because the most popular browsers are free. Similarly, hardware costs are low (less than $1000 for a high-end computer) and declining. Finally, the learning costs associated with extranets are likely to be lower than those for proprietary EDI software because the Web browser is a commonly used interface. A number of efforts are under way to combine the structure of EDI with the ease of use and low cost of the Internet and Web. For example, Bell Helicopters, Dayton-Hudson department stores, and Do It Best hardware stores have all published plans to use Web-based EDI.

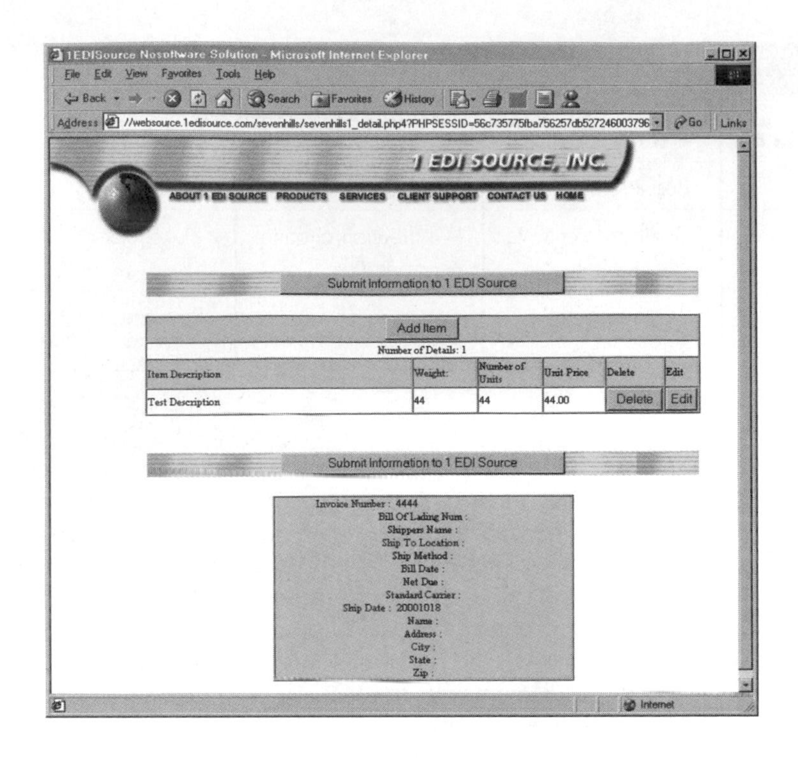

Web-based EDI software combines the best aspects of both protocols—cooperation with trading partners and ease of use.

switching costs
Costs associated with an individual or an organization changing to a new supplier.

A potential inhibitor of companies using electronic means, such as EDI, for business-to-business transactions is the threat of switching costs. **Switching costs** refer to the idea that, once a trading partnership is set up using EDI, the smaller of the two companies may not be able to afford to change to a different trading partner. They are far less likely to serve as a constraint on a business partnership when the total costs of installing hardware and software are an order of magnitude less than the cost of traditional EDI, as they are with extranets. Furthermore, an extranet lowers the cost of learning the supplier's system because it is based on standard Web browsing software. Thus, as in the Manheim Auction case, an extranet model means that the small business invests in highly flexible technology that can be easily adapted to other suppliers and other uses. In this case, switching costs should become a significantly less important criterion for engaging in an extranet compared to a traditional EDI system.

Supporting Business-to-Business Transactions with XML

Extensible Markup Language (XML)
A markup language designed to make information in a document self-describing on the World Wide Web, intranets, and elsewhere.

Another factor that boosts the potential for extranets to replace EDI is the introduction of Extensible Markup Language. **Extensible Markup Language (XML)** is a new markup language designed to make information in a document self-describing on the World Wide Web, intranets, and elsewhere. By comparison, Hypertext Markup Language (HTML) is designed to describe the format in which a document should be displayed in a Web browser. One reason why EDI is so attractive to companies for exchanging purchasing and shipping information is the capability to format this information in a highly structured, standardized fashion. That is, EDI makes it clear that a particular field in the data stream is a part number and that another field is a price for that part. In contrast, the language of the Web, HTML, is a *formatting language* that is meant to display numbers and text in a predefined way on a Web browser. As such, it does not have the structure to impart meaning to items like part numbers or prices. Because it is undesirable for computers to assign meanings to entries in Web pages, HTML is not appropriate for transmitting large amounts of purchasing and shipping information over the Internet.

XML solves this problem by emphasizing the structure and meaning of data. Using XML, companies can define their own tags that their trading partners can understand. For example, the tag <PARTID> would indicate that the field that followed was a part number. An XML file can be processed purely as data by a program, it can be stored with similar data on another computer or, like an HTML file, it can be displayed. For example, depending on how the application in the receiving computer wanted to handle the part number field, it could be stored or displayed, or some other operation could be performed on it, depending on the content of the field. As with the Web and EDI, efforts are under way to combine XML and EDI. You can visit the Web site of a group of IT professionals who are working on this effort at www.xmledi-group.org/.

Figure 4.14 shows an example of XML as applied to a course description. Looking at this example, you can easily see that the course has a code of *Mist 5665,* has a title of *Web Development*, and carries three credits.

Supply Chains and the Internet

Recall that the supply chain involves managing the flow of raw materials into, through, and out of the organization. You can divide the supply chain into three flows:

- Product flow
- Information flow
- Financial flow

The product flow includes movement of physical goods from the supplier, through the organization to the customer. The information flow involves any information associated with ordering of raw materials, status of production, and shipping goods to the customer. The financial flow consists of all aspects of payment to suppliers and receipts from customers. Problems with any aspect of the supply chain can prove disastrous for a company. For example, Nike had significantly reduced earnings for the fourth quarter of 2000 because of problems with its supply chain system that caused it to produce too many unpopular shoes and not enough of the hot sellers.

Because the Internet can be used only for communication, it cannot be used to actually ship physical goods. At the same time, the Internet is a perfect medium to improve the information and financial flows. In these two areas, new Internet-based supply chain software systems are making it possible for companies to have constant and complete knowledge of their products from supplier to final customer. These systems provide this knowledge by showing in real time the sales data, warehouse inventory, production plans, and shipment schedules for everyone and every company in the supply chain. Although only one in five *Fortune* 1000 companies has Internet-based supply chain software in place, many others are in the process of moving in that direction. One estimate suggests that companies spent $7.8 billion buying and installing supply chain software in 2001.[6] In all cases, the goal is to cut inventory, improve forecasting, and improve customer satisfaction.

Figure 4.14 Example of XML

```
<course>
<code>MIST5665</code>
<title>Web Development</title>
<credit>3</credit>
</course>
```

6. Ian Mount and Brian Caulfield, "The missing link: what you need to know about supply chain software." *eCompanyNow*, May 2001, http://www.business2.com.

Business-to-Business at FarEast Foods

FarEast Foods is a fairly small company with yearly revenues of less than $5 million. Because its suppliers are also small companies, it is not among the less than 2 percent of U.S. companies that use EDI. Instead, FarEast Foods has developed XML tags with its more than 50 suppliers that, along with the bar codes assigned to customer orders, make it possible to share information over an extranet. Orders processed by FarEast Foods' application server are automatically divided among the suppliers for specific items, based on bids submitted each day, and sent to the suppliers over the extranet. Each supplier combines a specific item order with other items destined for FarEast Foods and ships them via overnight package delivery. The bar code for each customer order ensures that items are not confused. Figure 4.15 shows the use of XML in the B2B process for FarEast Foods.

FarEast Foods also uses the extranet to post sales information spreadsheets, which suppliers can download and use to forecast future demand for Asian food items that they sell to FarEast Foods. This approach helps the suppliers avoid being out of stock on an item and having FarEast Foods go to another supplier. The suppliers also act as market researchers for FarEast Foods by keeping the company aware of special days or holidays for which it may want to run specials—for example, Chinese New Year.

1. What is the difference between an intranet and an extranet? Which is used for communications within an organization?
2. What are the similarities and differences between HTML and XML? Between EDI and XML?

Figure 4.15 B2B process for FarEast Foods

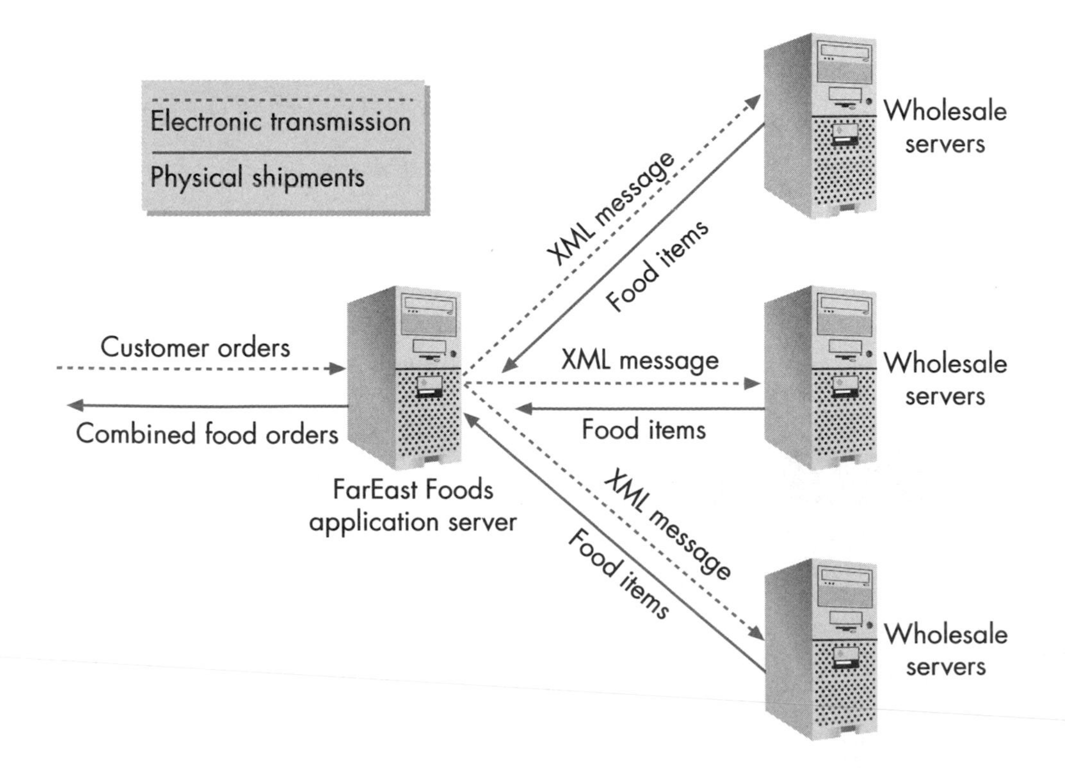

CASE STUDY

Case*Study*

Manheim Auctions

With worldwide sales of more than $1 trillion, the automobile market is a significant part of the global economy. In the United States, 78 percent of all automobiles sold are previously owned. Many of these used cars are supplied to local dealers by Manheim Auctions, the world's largest auto auction company, which whole sales 4 million vehicles annually, worth $38 billion, to dealerships in North America, Europe, and Australia. One particular type of vehicle that Manheim handles is the *program car*; an automobile that has been used by an auto company executive whose lease is complete or that has been returned by an automobile rental company. This segment of the used car market has grown tremendously over the last five years.

Traditionally, dealer representatives attend auctions to purchase vehicles that they believe are in demand in their locale. To reduce the cost to dealers of sending representatives to purchase program cars at auctions, in 1996 Manheim began to use a restricted form of the Internet called an extranet to wholesale program cars to certified dealers. With this system, a dealer with a user name and password can search for program cars by accessing the Manheim Web site at www.manheim.com. The Web page returned to the dealer contains information on vehicles that meet his or her needs, including photographs of the vehicle as well as any problems (dents, scratches, and so on). If the dealer decides to order this vehicle, Manheim will arrange transportation to the dealer's location. The dealers know that Manheim stands behind the vehicles it sells and feel confident that the vehicles delivered will match the photographs and descriptions on the Web page. Since its inception, sales over the Internet by Manheim have grown from zero to more than five percent of its total sales as other types of used cars, beyond program cars, have been included in the online auctions.

To provide end-consumers with access to the cars that are sold to dealerships, Manheim linked their AutoTrader.com operation, the largest business-to-consumer and consumer-to-consumer vehicle Web site, to their auctions. Now, any car that is sold at an auction—physical or online—is automatically added to AutoTrader.com. When a consumer goes to AutoTrader looking for a particular variety of used car located within a specified geographical region, the cars sold to participating dealerships are automatically included on the list provided to the customer along with those from individuals who have paid to include their cars on the site.

Source: Patrick G. McKeown and Richard T. Watson, "Manheim auctions." *Communications of AIS*, June 1999, and updated by the author.

Think About It

1. Why would Manheim choose to initially concentrate on program cars for its online auctions?
2. Was Manheim potentially cannibalizing its core business of auctioning used cars at physical auction sites?
3. Why do you think Manheim chose to use an existing retail used car Web site rather than creating a new one? What synergies did this system add to Manheim's existing system?

The Manheim Online Web site is an important partner to the physical auction process shown here.

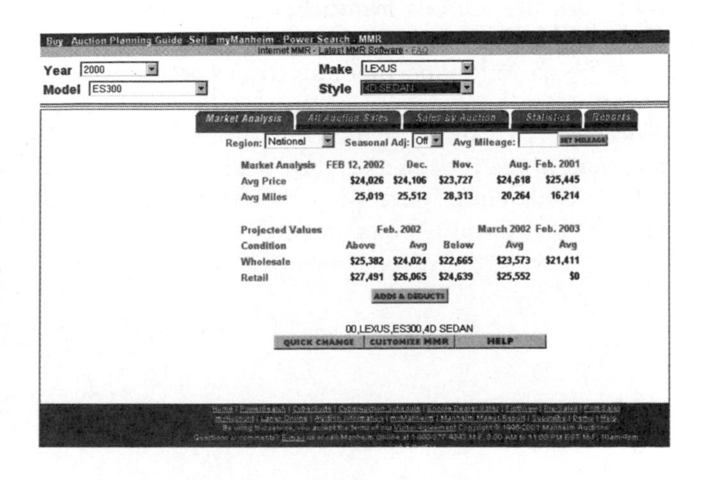

SUMMARY

To summarize this chapter, let's answer the questions posed at the beginning of the chapter.

What types of information systems are present in organizations, and how do they relate to the IS functions? Types of information systems include transaction processing systems (TPS), organizational memory, and decision support systems (DSS). The transaction processing system is essential for handling the present. Its results are stored in organizational memory. DSS are required to prepare for the future. Although databases in organizational memory support other information systems, information and knowledge management systems are information systems on their own. Decision support systems include a variety of types of information systems, which often go by different names; because they all support decision making, they have been discussed collectively in this chapter.

What is a transaction processing system, and what functions does it accomplish in an organization? The transaction processing system is the information system that handles the present by compiling an accurate and current record of the organization's activities. At the same time, it generates data that are stored in organizational memory and used by other information systems to prepare for the future. Characteristics of a transaction processing system include the large number of transactions that it must handle, validation of the correctness of transactions, security problems because of the amount of money involved, and the potential injury to the organization if the system fails. The revenue, expenditure, conversion, and financial business processes every business must carry out are all closely related to transaction processing systems. Transaction processing systems are an important element of every organization's supply chain because they provide the interaction between the organization and its suppliers or customers.

What are the various transaction processing methods and activities? Transactions may be input and processed using batch or online methods. In a batch processing system, groups of similar transactions are periodically processed. In online transaction processing (OLTP), transactions are processed as they occur. A point-of-sale (POS) transaction processing system is an example of OLTP. In some cases of OLTP, especially those involving registration or reservations, the processing actually controls availability of the product or service.

Transaction processing activities include gathering and entering the data that describe the activities of the organization; processing (arranging and manipulating) the data so that they are suitable for use by a large variety of potential users; storing data so that they can easily be retrieved as they are needed; and generating reports of data and information to users.

What are the impacts of the Internet on transaction processing systems? With the Internet, the whole nature of transactions can change and a variety of new transactions can take place involving consumers/citizens, businesses, employees, government, and not-for-profit organizations. In fact, it is possible to have transactions occur among all combinations of these groups or entities, something that was not always possible before the advent of the Internet and its applications, such as the Web and e-mail. Internet-based transaction processing systems usually include the capability to order goods and services using a client/server system in which a customer uses a Web browser as the client and busineses use Web, application, and database servers to process the order and package delivery services to deliver the order to the customer.

How do business-to-business transactions in a networked economy differ from traditional transactions? Business-to-business transactions in the networked economy will continue to use EDI, but will also use all three of the possible Internet technologies: the Internet, intranets, and extranets. All of these technologies are differentiated by their scope and focus, and by the business processes they serve. In many cases, business-to-business transactions are moving to extranets because they serve the same purposes as EDI at a much lower cost. That is, their scope is restricted to business partnerships, and they focus on distribution channel cooperation. The movement to extranets will include the use of XML to provide necessary information about the data being sent between trading partners. Internet-based supply chain systems will provide organizations with knowledge of their product all of the way through the supply chain, allowing them to better predict future demand.

REVIEW QUESTIONS

1. Why are information systems important in an organization?
2. What three functions does every IS system carry out?
3. List the types of information systems used in organizations.
4. List the characteristics of transaction processing systems.
5. List the business processes found in all organizations.
6. What are the purposes of the four business processes?
7. What does Porter's value chain model imply about the way organizations work?
8. What does a value chain have to do with value systems?
9. Which activities are common to most transaction processing systems?
10. What methods are commonly used to input and process data in transaction processing systems? Which method is used in most grocery store checkout systems?
11. What is the most common method for inputting data into a transaction processing system? What two features make it so attractive?
12. What three activities are typically carried out when transaction data are processed?
13. What type of information about a transaction must be captured and stored?
14. How is the Internet changing transaction processing? How is the Internet changing business processes?
15. What four categories of stakeholders are involved in Internet-based transactions?
16. What is an application server, and what is its purpose in Internet-based transaction processing?
17. What are the four elements of a four-tiered client/server system used for Internet-based transaction processing?
18. How do business-to-business transactions differ from consumer-to-business transactions?
19. What are the five steps involved in a typical Internet-based transaction?
20. What is an extranet, and how is it replacing EDI?

DISCUSSION QUESTIONS

1. Discuss a transaction processing system, other than one like a grocery store's TPS, in terms of methods of data gathering and processing, and describe how the TPS activities are carried out. Also, discuss how the Internet might ultimately change the TPS you selected.
2. Discuss the four business processes for a business or organization with which you may be familiar. It could be a for-profit business, a not-for-profit organization, or a club of which you are a member.
3. Discuss your college or university registration system from a transaction processing point of view. What type is it, and what are the input, processing, and output?
4. Discuss why an application server might be needed when using a mainframe system for Internet-based transaction processing.
5. Discuss why HTML is not appropriate for sending large amounts of data between businesses.

RESEARCH QUESTIONS

1. Visit the Web sites of AuctionWorks and Andale, and write a two-page paper on the differences in the services offered by these two companies.
2. Visit the Auto-ID Center at www.autoidcenter.org/main.asp, and research the uses for smart objects discussed there. Write a two-page paper on your findings.
3. One of the leading companies in the area of government use of the Internet is EZGov, Inc. Visit its Web site at ezgov.com, and write a two-page paper on the company's activities.
4. Visit the Manheim Web site at www.manheim.com, and write a two-page paper on new features not mentioned in the book.

Claire pulled the slack out of the rope as Alex hauled himself up over the edge of the rock face. They were taking a break from the store to enjoy a morning climb.

"Man, I'm beat," Alex said, as he found a spot next to Claire to take in the view. "I'm glad you led today. I don't think I would have had enough energy."

"I noticed that you were huffing and puffing," she replied, "even more than usual. Why were you up so late last night?"

Alex fashioned a crude pillow out of his sweater and laid back on the rock to rest his eyes. "It's all of those e-mails we're getting from the Web site. I was up half the night trying to figure out what they want to order," he said with a yawn.

"I noticed we're getting a lot of hits to the site. It looked like several hundred each day."

"Yes, and a lot of them are asking to buy something," he replied wearily.

"Well, nobody said that success is easy," Claire remarked as she watched an eagle drift over the valley below.

Since starting to put their business online, Alex and Claire have come a long way. They have designed, purchased, and set up the hardware infrastructure for their system. They have also developed Web pages to introduce their company to the online world.

The initial pages consist primarily of information about their store and product lines. They also include items designed to draw customers back, such as Alex's outdoor trip reports and Claire's Wild Trail Recipes. In addition to links to more in-depth information about the products and their surrounding area, the site includes an e-mail link so that customers could easily contact the pair with questions or inquiries about purchasing their products. As an added touch, the Campagnes included a counter on their home page to get a feel for how many visitors their site is attracting. The counter has shown that the site has been consistently gaining in popularity since its inception.

Wild Outfitters' current problem stems somewhat from the initial success of the site. Although the pages have allowed customers to view and order products, it turns out that a simple e-mail link is not adequate for processing these orders. For one thing, the messages containing orders are mixed with those simply requesting information. This jumble makes it difficult to organize and prioritize the messages. Also, Alex has found that each e-mail that contains an order is different because no standard format has been provided for the information. As a result, the Campagnes must search through the message for the relevant order information and record it for processing—a slow and tedious operation.

The Campagnes realize that they need to add user interaction and somehow automate the processing of transactions on their Web site. Their first goal is to temporarily provide order forms for each product line. Although this approach may initially limit individual orders to a specific line, the couple feels that it would be the easiest to implement and would begin to standardize the order information. They would continue to use the e-mail link, but only for customer service-related items such as questions and complaints. Eventually, they hope to improve the site by providing a searchable catalog and shopping cart system for order transactions. Also, they would like to expand to accept electronic payments.

After a short while, Claire began to gather the rope and prepare the equipment for their descent. Looking toward a noise from Alex's direction, she saw that he had gotten quite comfortable despite the hardness of the rock. Sitting down next to her husband, she thought to herself, I guess we can rest a little longer—the e-mail can wait. Better check for loose rocks, though, because we wouldn't want an avalanche to start from the snoring.

Think About It

1. Explain why transaction processing would be important to WildOutfitters.com.
2. What types of business processes and transaction activities do you think are needed at WildOutfitters.com? For the transactions in your list, describe the input and output, as well as what should be done to process the data.

3. Search through several commercial sites on the World Wide Web. What types of transactions can you find on these sites? How does each site handle these transactions? Identify the information important in each transaction. What interesting transactions and features did you see?

Hands On

4. A Web page (Wo.html) with a transaction form for one of WildOutfitters.com's product lines is available for downloading from the text's Web page at www.course.com. Obtain the page and include a link to it in the pages that you created for Chapter 3. After you have incorporated the page into your WildOutfitters site, try it out and describe the transaction that takes place. Examine the page's HTML source code. Can you understand the HTML code underlying this page? What changes would you make to improve this page?

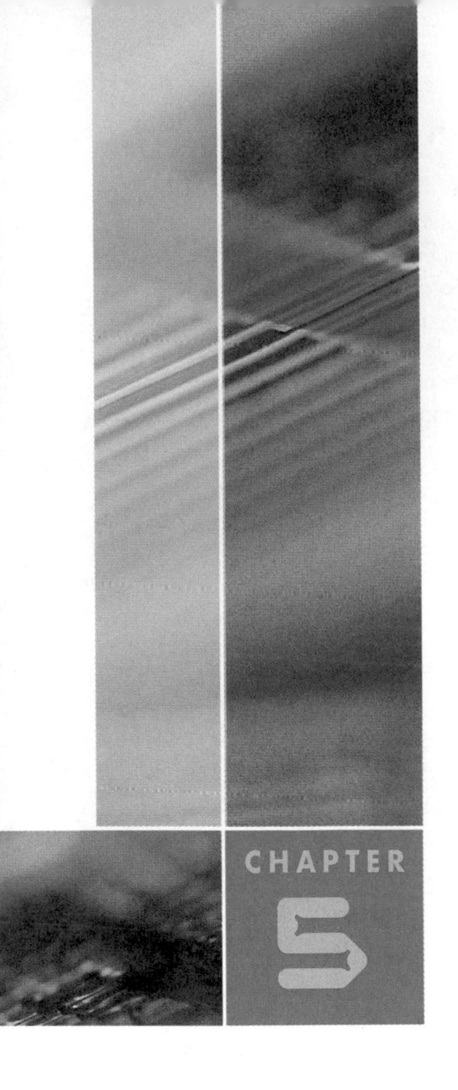

CHAPTER 5

REMEMBERING THE PAST WITH ORGANIZATIONAL MEMORY

LEARNING OBJECTIVES

After reading this chapter, you will be able to answer the following questions:

- What is organizational memory?
- What is a database, and how does it store data in a structured form?
- What are the key elements of a relational database?
- What methods are used to improve the management of information?
- What are the types of knowledge, and how can knowledge be shared in an organization?

Improving Health Information Access

One of the biggest disappointments in the use of information systems has been the failure to implement computerized patient record (CPR) systems in hospitals, clinics, and physician's offices. Even though most people agree that CPR would make dealing with patient records easier, 90 percent of hospitals, clinics, and private practices have no CPR. Many of those that do have CPR remain disappointed in its performance.

Reasons cited for the failure to implement CPR include the huge setup cost in a major hospital, the ongoing cost of supporting such an information system, and privacy issues associated with putting patient records online. However, given that U.S. hospitals generate almost 17 billion pages of medical records per year, solving this problem is an important one. Compounding the problem is the fact that only 5 percent of the typical medical record can be expressed as discrete data. Fully 30 percent of the typical medical record takes the form of transcribed reports, a whopping 40 percent comprises paper documents, and the remaining 25 percent consists of diagnostic images (for example X rays).

The eMRWeb system, developed by Smart Corporation, is a Web-based medical record storage and retrieval system that provides one solution to these problems. It stores patient records on a server located at the company data center and makes the records available over the Web. Unlike with other computerized patient record systems that require the health facility to purchase hardware and software, provide ongoing support, and meet changing privacy requirements, Smart Corporation provides the eMRWeb system on a fee-per-transaction basis, with constant upgrades to ensure that the system meets Health Insurance Privacy Protection Act (HIPPA) regulations regarding privacy of patient records. The medical facility has to do very little other than initially make records available, update them as necessary, and pay a fee for each transaction.

eMRWeb was installed in the Jackson Hospital and Clinic, a 289-bed facility in Montgomery, Alabama, in 2001. This system included on-site digital capture or scanning of documents, indexing of documents for easy retrieval, and enterprise-wide access from an exclusive Web site established for the Jackson facility. Administrators and caregivers can use PCs equipped with Web browsers to access patient records via a highly secure network that makes use of personal passwords and assigned security levels. As the medical records director at Jackson noted, the goal was to improve the hospital and medical staff's access to patients' health information so as to enhance patient care.

Continued

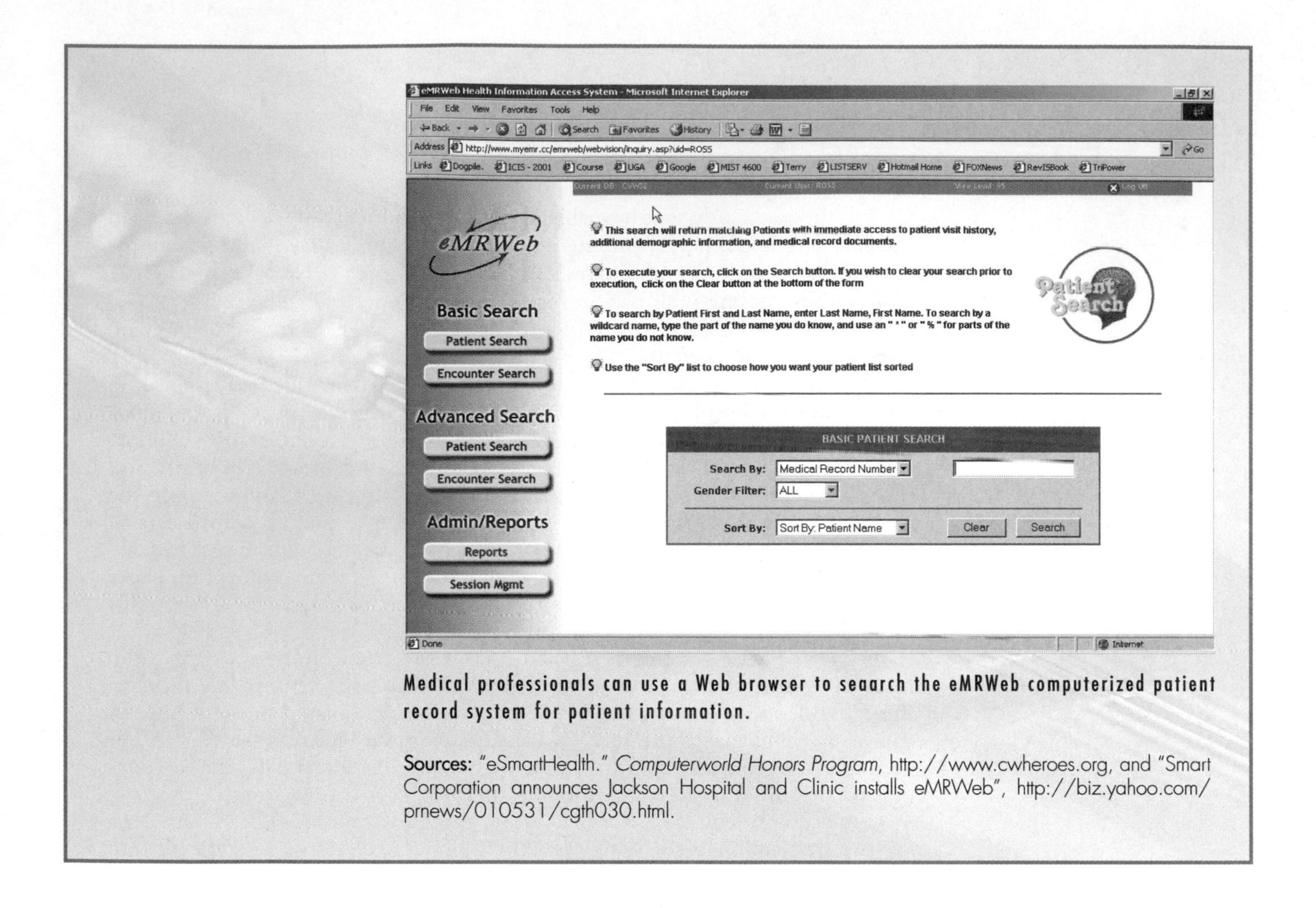

Medical professionals can use a Web browser to search the eMRWeb computerized patient record system for patient information.

Sources: "eSmartHealth." *Computerworld Honors Program*, http://www.cwheroes.org, and "Smart Corporation announces Jackson Hospital and Clinic installs eMRWeb", http://biz.yahoo.com/prnews/010531/cgth030.html.

Organizational Memory

Chapter 4 introduced various types of information systems that are used in organizations. These systems provide management with information and answers to questions that then help management make the decisions that will determine the future well-being of the organization. To review, the purpose of information systems is to handle the present, remember the past, and prepare for the future, as represented by the IS cycle shown in Figure 5.1.

In Chapter 4, you saw that in the process of handling the present, transaction processing systems generate a great deal of data and information that must be stored. By storing transaction data and information, the organization becomes able to remember the past, thereby making it possible to prepare for the future.

The key to the process of remembering the past is organizational memory. *Organizational memory* is an organization's electronic record of data, information, and knowledge that is necessary for transacting business and making decisions. Without a memory, an organization (like humans) will not know how to carry out day-to-day activities or to make the many decisions that are necessary for continued existence. This definition focuses only on *electronic* forms of organizational memory, because, as discussed in the opening case, virtually anything can be digitized into an electronic form and stored in organizational memory for use by other information systems.

Figure 5.1 The IS cycle

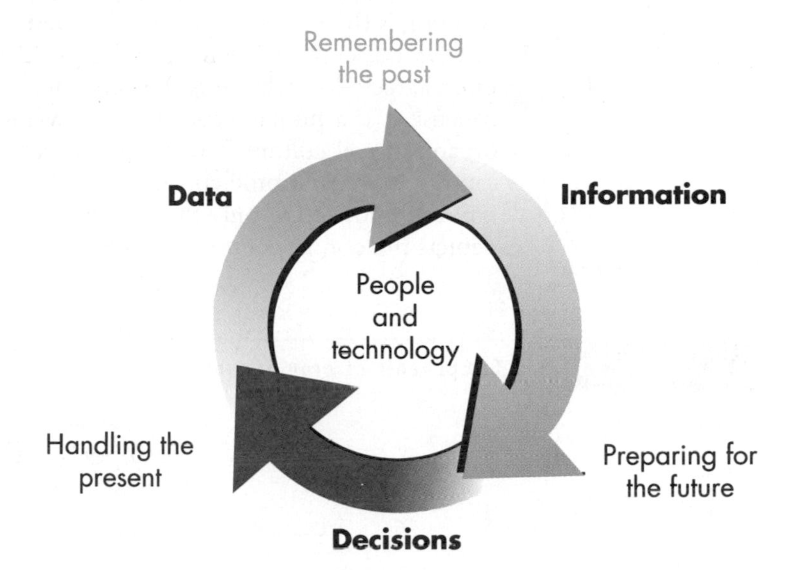

Once a transaction processing system has created data, the next step is to store those data in a database or data warehouse. The stored data are then analyzed in decision support systems (DSS) to create summaries or reports that help the organization prepare for the future by making informed decisions. The data may also be processed into information that is used in a DSS. In addition to data and information, organizational memory is composed of knowledge.

Components of Organizational Memory

Organizational memory consists of three elements: data, information, and knowledge. Recall that data are raw facts, usually in the form of characters and numbers. Information is the result obtained by processing data into a usable form. Information can take the form of text (tables and reports), hypertext, graphics (including charts), images, audio, and video. Text consists of the letters, digits, and symbols used in word processing applications, whereas hypertext is text and graphics that include links to other documents. Graphics differ from images in that they contain embedded data

Many customers use chat rooms to connect with representatives of corporations or other organizations.

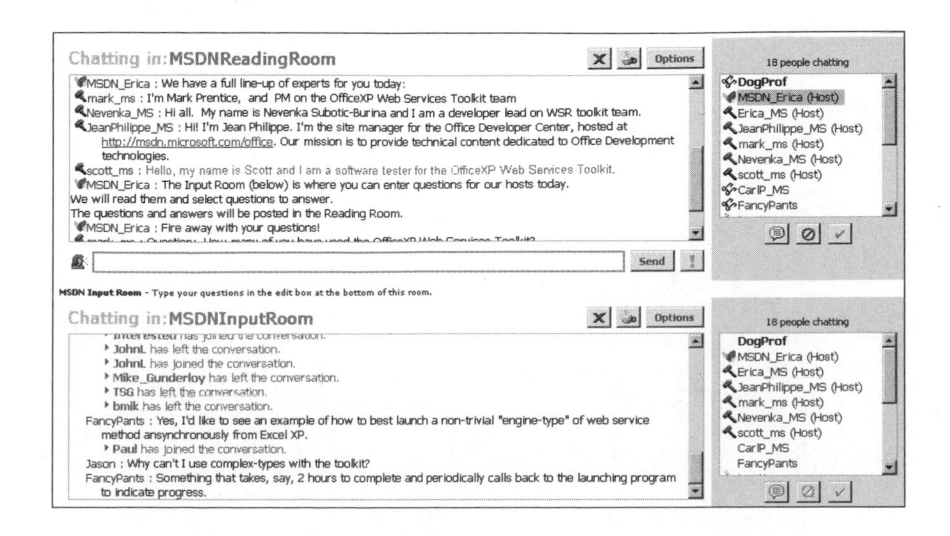

(captions, labels, dimensions, and so on). Audio involves the storage of sounds, whereas video involves the storage of live-action pictures and sound. Knowledge, by contrast, is the human capacity to request, structure, and use information.

Of these three elements of organizational memory—data, information, and knowledge—knowledge is the most difficult to store in an electronic form simply because it is a human capability. Knowledge usually is composed of such things as organizational culture (how things work in the organization), social networks (who can do what), and problem-solving models (how to resolve problems). Of the three types of knowledge, only the models for solving problems can be stored. Figure 5.2 depicts the components of organizational memory.

Figure 5.2 Components of organizational memory

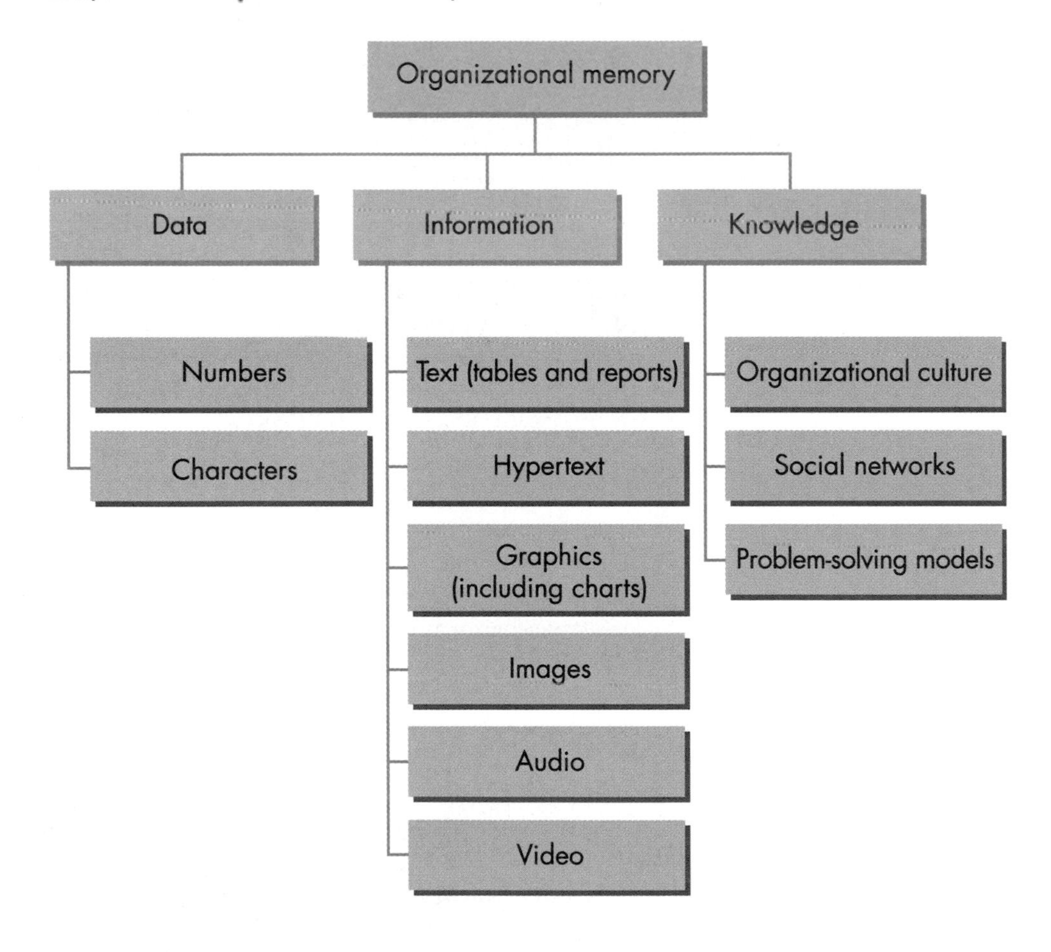

Semistructured Versus Structured Organizational Memory

Organizational memory can be either semistructured or structured. It is not easy to find something of interest in *semistructured* organizational memory, or unorganized data. For example, it is difficult to find a name in an unalphabetized list of names. This type of unorganized data is semistructured. Similarly, information stored in individually created and stored Web pages is semistructured information because finding the information you need is not easy; even the best search engines return many extraneous hits. Also, consider how difficult it is to find the answer to a question by searching the semistructured knowledge stored on many different chat rooms or bulletin boards.

In contrast, consider how easy it is to find a name in an alphabetized list, which is an example of structured data. *Structured* organizational memory makes it easy to find something of interest. Data in a database, for example, are highly structured. It is also relatively easy to find information in a properly managed and indexed Web site because the information is structured. Likewise, it is much easier to find structured knowledge stored in an organized fashion using an expert system.

The key to distinguishing between semistructured versus structured organizational memory is the ease of searching for information. The easier the search, the more structured the organizational memory. Table 5.1 summarizes the concept of structured versus semistructured data, information, and knowledge.

In Table 5.1, you can see that when you move from less structured forms of organizational memory to more structured forms, data, information, and knowledge become easier to find and more useful to management. For example, you can convert unorganized lists into a database and individual Web pages into an indexed Web site. In each of the succeeding sections of this chapter, the structured storage of the three elements of organizational memory will be discussed along with ways each can be used to help management make better decisions.

1. What are the three elements of organizational memory?
2. What is the difference between semistructured and structured organizational memory?

Table 5.1 Structured and Semistructured Organizational Memory

Storage Type	Data	Information	Knowledge
Structured	Database, data warehouse	Indexed Web site, reports, charts, manuals	Virtual teams, document databases, expert systems, frequently asked questions (FAQs), newsgroups
Semistructured	Unorganized list	Web pages, e-mail	Bulletin boards, chat groups

Structured Storage of Data

Data are stored in a structured form in either a database or a data warehouse. A *database* is a collection of different types of data organized so that the data are easy to manipulate and retrieve. Although a database does not have to be stored on a computer, computerized databases are quickly becoming the norm. In this book, the word *database* always refers to a *computerized database*. Databases are essential to all information systems because data must be readily available to the various information systems that aid the organization in preparing for the future.

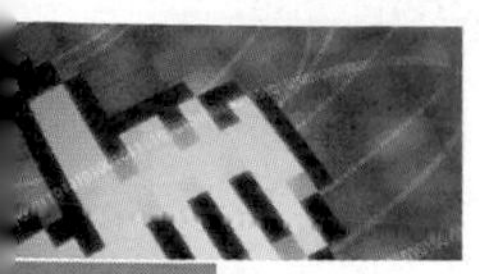

INTERNET IN ACTION

ISWorld Net

The Internet is quickly becoming a place to create organizational memory. One good example involves the ISWorld Net Web site. This site is the single entry point to many resources related to teaching and research in the field of information systems. Approximately 5000 instructors and researchers are thought to make up the global community that takes advantage of this site. Many of these people are listed in a faculty directory, and 2000 or so users subscribe to the ISWorld listserv to send and receive e-mail. ISWorld Net has been described as having the following purpose: "A vision of information systems scholars harnessing the Internet for the creation and dissemination of knowledge." As such, it has become the organizational memory for people interested in information systems.

Started by a group of faculty wishing to share ideas and problems with others in their field, ISWorld Net is composed of many different Web pages stored on servers located around the world. To ensure a common look and feel and to avoid the inclusion of overlapping material, a group of volunteer editors oversee specific areas and ensure that associated Web pages are appropriate. The home page at www.isworld.org is divided into What's New, Top Resources, Research, Teaching, Professional Activities, and Country pages, and it provides a link to a digest of ISWorld listserv messages and an About ISWorld section. Each top-level section, includes several subsections. For example, under the Top Resources section is the IS Faculty Directory, which lists thousands of faculty members interested in information systems. By drilling down in each section, you can find information on almost any information systems topic that you can imagine. As a student of information systems, you should become familiar with ISWorld Net as a source of information.

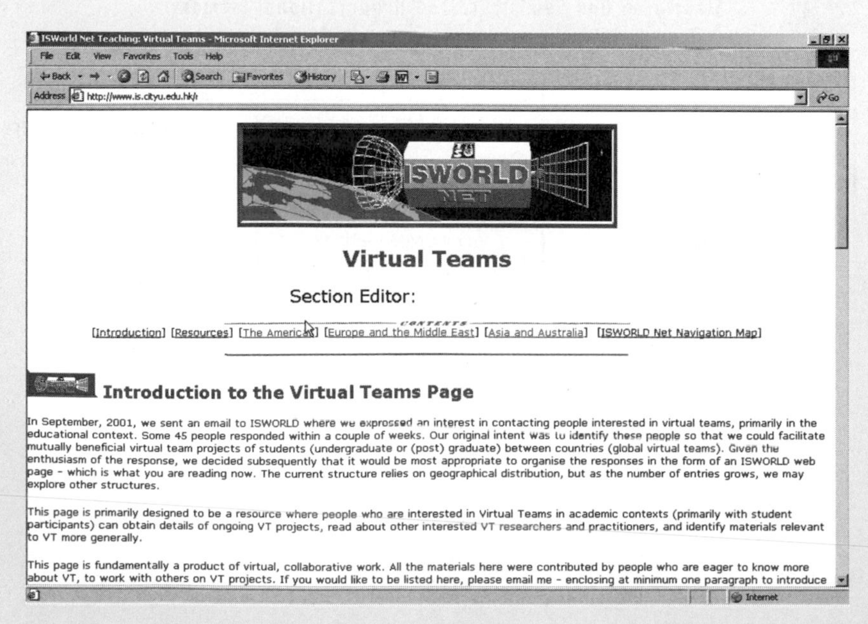

ISWorld—an extensive organizational memory system—is a great source of information and ideas for anyone interested in the field of information systems.

Because computers can manipulate and retrieve information much faster than humans can with manual methods, databases are replacing traditional methods of organizing information, such as books, paper lists, file cabinets, Rolodexes, index cards, and so on. In addition, databases can greatly reduce the space required to store information and can be used to access information from disparate sources. Database management software allows users to create a database, enter information into the database, and then rearrange the database or retrieve information from it as desired. The information that is retrieved can then be output in special report formats. Without a database, many companies that depend on fast access to data would quickly go out of business.

data warehouse
A subject-oriented snapshot of an organization at a particular point in time.

When a transaction processing system generates large amounts of data, they are often stored in a data warehouse. A **data warehouse** is a subject-oriented snapshot of the organization at a particular point in time. Data warehouses enable an organization to detect key facts or relationships within data. Data mining and online analytical processing (OLAP) are typically used to extract and analyze the data stored in a data warehouse. An entire field of study known as business intelligence or customer relationship management (CRM) has arisen to take advantage of the wealth of data now existing in data warehouses. The growth of electronic commerce and the huge amount of data generated by visits to a company's Web page, often referred to as **click stream data**, have increased the sizes of data warehouses into the *terabyte* (trillions of bytes) range. Data mining and OLAP are decision support tools; they will be covered in Chapter 6, along with other decision support tools.

click stream data
Data that are captured about users' activities when they visit a Web site.

FarEast Foods uses its database and data warehouse to good advantage. The company has a group of customers who are interested in special foods that are not normally carried by FarEast Foods' suppliers. For example, a special type of bird pepper from Thailand is not always available. When a supplier offers this product, these customers want to be notified by e-mail. The company's database is essential to finding any solution for this problem. When a special-order product arrives at the supplier, the database can be queried for the names and e-mail addresses of those customers who have requested the product, and an e-mail message can be sent to alert them to the availability of the item.

Data generated by processing thousands of transactions each day as well as non-transaction data gleaned from the log of Web page activity at the FarEast Foods Web site are stored in a data warehouse. FarEast Foods uses data mining or OLAP to learn more about its customers and their product preferences. This information helps the company make decisions about which products to carry and which suppliers to use, thereby preparing it for the future.

Development of Database Management Systems

file management system
Database software that can work with only one file at a time.

When computers were first used to store an organization's data, the data were stored in individual files. To manipulate the data and retrieve records of interest, a special type of software was developed. This software, called a **file management system**, proved very useful for working with lists of records but could work with only one file at a time. The first database software on personal computers was actually a file management system. Over time, as organizations found additional ways to use data, their users created multiple files storing similar data in different locations within each organization. When data are stored in separate files in multiple locations, accessing needed data becomes more complicated, and fundamental questions about data redundancy, data integrity, and data dependence arise.

data redundancy
The repetition of data in multiple files.

Data redundancy is the repetition of data in different files. For example, college offices may create separate files containing much of the same information, such as names, Social Security numbers, addresses, and so on. Such redundancy is costly in terms of the money required to collect and process the data for computer storage and in terms of the computer storage itself.

data integrity
The process of ensuring that data are accurate and reliable.

data dependence
A relationship between data and the software used to store it.

integrated data management
The storage of all data for an organization in a single database.

database management system (DBMS)
Software used to organize, manage, and retrieve data stored in a database.

legacy system
Another name for a mainframe computer system.

relational database management system (RDBMS)
A database system in which elements are represented as being parts of tables, which are then related through common elements.

data hierarchy
The order in which data are organized in a database.

Data integrity is the process of ensuring that data are accurate and reliable. Problems with data integrity occur when the same data are stored in multiple files throughout an organization and the data must subsequently be changed. Obviously, any change in the data must be made in *all* files to maintain data integrity. For example, a student's change of address should be entered in all of the college files. A change missed in only one file can lead to severe data integrity problems for the users of the data and for the person referred to by the incorrect data.

Whenever different departments in an organization collect, process, and store information, they could potentially use different software to perform this operation. This proliferation of software can lead to a problem with **data dependence** between the software and the files. The files of one department may become incompatible with the files of another department because the data storage depends on the programs or hardware used. As a result, it is often very difficult to combine data from the two files stored on different hardware or created with different software.

Taken together, the problems associated with data redundancy, data integrity, and data dependence can turn any effort to combine files from different departments into a painful task. Consider again the example of a college. If a college administrator wished to write a report on the number of students accepted for admission who also requested financial aid and on-campus housing, the process of collecting the necessary data from three different files created by three different departments could be slow and awkward.

The solution to the problems resulting from the use of multiple files in different locations in an organization derives from the use of integrated data management. With **integrated data management**, all data for the organization are stored in a single database. All units of the organization then use a single type of software to access the database. This database software, termed a **database management system (DBMS)**, is much more powerful than file processing software and is capable of handling the data needs of a large, distributed organization.

Database Management Systems

To work with large amounts of data contained in a modern database, a number of approaches to database management systems have been developed for organizing and manipulating a database on large, centrally located mainframes or **legacy systems**. Today, the most popular approach to data management involves the **relational database management system (RDBMS)**, in which two or more tables (files) related through common fields store data elements. Newer client/server systems make wide use of RDBMS, as do older legacy systems and PC-based systems. Given its popularity, RDBMS will be discussed in more detail using the FarEast Foods example; first, however, it is important to understand some of the terminology commonly used in database management.

Database Terminology

Database management, like any field of study, has its own terminology that defines the various elements and operations used in working with databases. The first concept you need to understand is the **data hierarchy**, which is the manner in which data are organized in a database. Figure 5.3 shows a general data hierarchy.

Figure 5.3 Data hierarchy

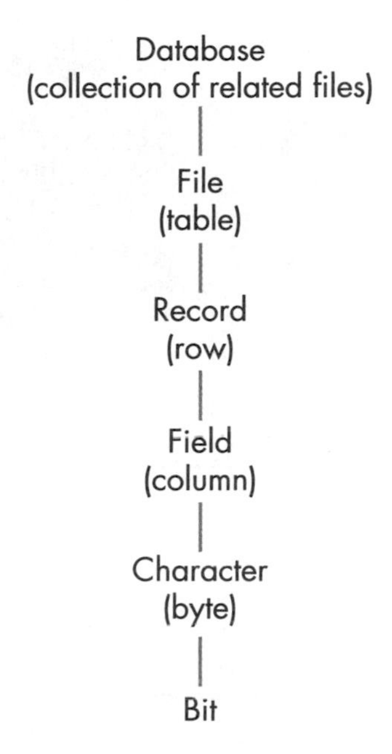

In Figure 5.3, you can see that each item in the hierarchy is composed of the elements *below it.* For example, a database is composed of files, and a file is composed of records. Starting at the top, you have the database, which is a collection of related files that are accessed together to generate needed information. Recall from Chapter 2 that a *file* consists of programs, data, or information to which the user or software assigns a name. Related files share a common purpose—for example, data about FarEast Foods. As another example, consider a database of your DVDs. Each DVD has a number of scenes, so you could create one file with the titles of the DVDs and another, related file with the actual titles of each scene on that particular DVD.

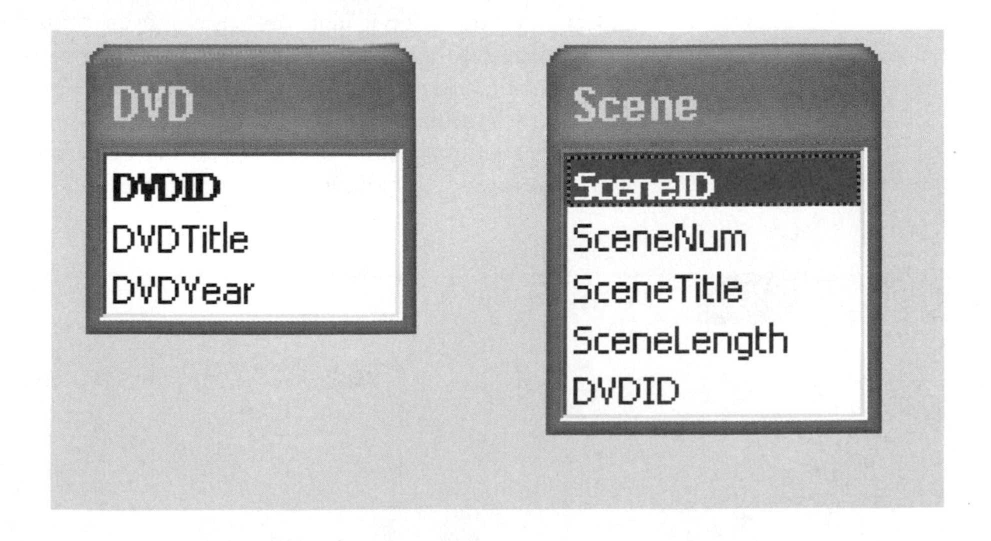

These two tables in a DVD database are related through the CDid field from the CD table.

Figure 5.4 Tables for special orders at FarEast Foods

Customer table	Special-order product table
-Customer ID -Last name -First name -E-mail address (other columns not shown)	-Product ID -Product name (other columns not shown)

table
A database model composed of rows and columns, with rows specifying a particular person, place, or thing, and columns giving the specific details about each person, place, or thing.

record
A collection of fields with information that usually pertains to only one subject (such as a person, place, or event).

field
Part of a database file that stores specific information such as a name, a Social Security number, or a profit value.

field name
An identifier given to a field in a database file.

data types
Specification of the type of data that will be stored in a database field.

character
A byte, or group of eight bits; equivalent to a single character.

The files that make up a database are called **tables.** A table is composed of **records**, or rows, each of which pertains to a single person, place, or thing. For example, if a table contains data on FarEast Foods' customers, then a record will hold data on one of those customers. Each record is composed of multiple **fields**, or columns, which store the specific details about each person, place, or thing described by the record. Examples of fields include a customer's name, address, telephone number, and so on. Fields are given **field names** to differentiate among them and **data types**, which specify the type of data stored in the fields. The contents of each field are stored as **characters** (*bytes*).

As an example of a data hierarchy, consider the relational database at FarEast Foods. This database consists of many tables composed of all the data that the company needs to serve its customers and carry out its day-to-day operations. For example, the firm's database includes a table of data on existing customers; a table of all items sold by the company; a table of data on products not normally available; and so on. The customer data table has many columns, or fields, including those for the customer ID, last name, first name, e-mail address, postal address, credit card information, and so on. This table will have one row for each customer, meaning that each table could have many rows. Both product tables (regular and special items) contain columns to hold product ID, product name, and price values.

Figure 5.4 shows the structures of the customer data (Customer) and special-order data (Special) tables. Only the first four columns of the Customer table and the first two columns of the Special table are shown, even though they include other columns, because these columns are used in the solution of the special-order problem. Figure 5.5 shows the first four columns of the Customer table in a relational database management package, Microsoft Access. (The e-mail addresses are not completely shown for privacy purposes.) Although the FarEast Food database is much larger than just the two tables shown in Figure 5.4, you need to be concerned with only these tables for now.

Figure 5.5 Customer table

Customer : Table

CustID	LastName	FirstName	Email
999-11-0012	Patrick	Chris	devildog@u
999-14-3143	Campbell	Lange	lcampbell@
999-19-1744	Devane	Samuel	samdev@1
999-23-4321	Smith	Joe	JoeGrunt@
999-31-9776	Mullins	Janice	griffingirl@1
999-73-8590	Watson	Betsy	aussie@xy
999-74-3343	Roth	Jerry	BullGuy@a
999-83-6682	Calstoy	Carol	7054cal@u
999-88-1532	Ronson	Suzy	SuzyLegs@
999-89-2269	Hyatt	Ashley	Volunteer@
999-99-1234	Randall	Roy	rrandall@u

In a client/server environment, databases are typically stored on large, very fast database server computers that can handle requests for data. If the database is accessible from the Internet, the Web server must be able to accept questions submitted as data from a Web page and pass them along to an application server. The application server uses a type of software known as **middleware** to formulate these questions into a type of formalized request for data known as a **query**. The middleware then sends the query to the database server, which actually finds the matching values. The database server returns the matching values to the application server, which formats them in the form of a Web page and sends them to the Web server, from which they are sent back to the user. Figure 5.6 illustrates this process. Note that the introduction of the Web server creates a four-tiered client/server model, like that discussed in Chapter 4.

middleware
The software that converts requests for data from a client into queries that are sent to a database server.

query
A formalized request for data sent to a database.

Data Warehouses

Transaction processing systems typically generate a large amount of raw data indicating how the organization deals with its customers, suppliers, and employees—its stakeholders. For example, the more than 600 million credit cards in use globally generate more than 100 billion transactions every year, and a popular Web site can easily have thousands of hits per day. Often, these data are not particularly useful to the organization because they tend to be highly fragmented. The data may be stored on different databases (production and sales) at different locations and on different platforms (UNIX and proprietary mainframe). As a result, analysts find it difficult to access and use the data they need to understand what is happening in the organization. One solution to this problem calls for using a data warehouse to organize these data into a logical collection from which specific data can more easily be found.

Figure 5.6 Querying a Web database

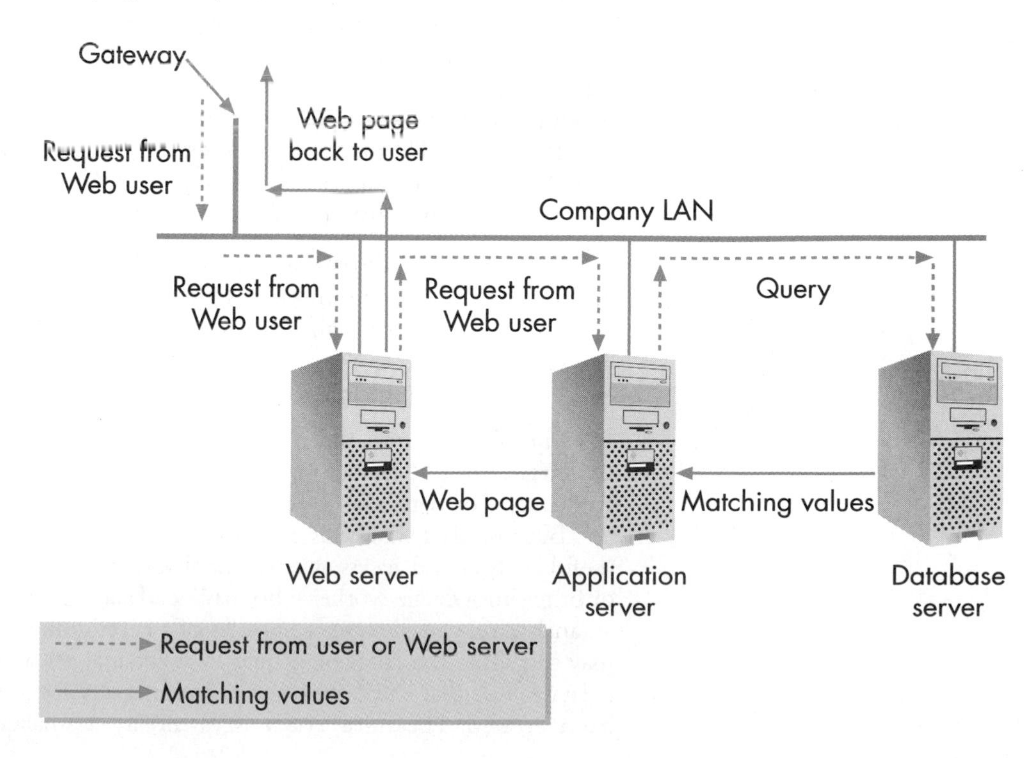

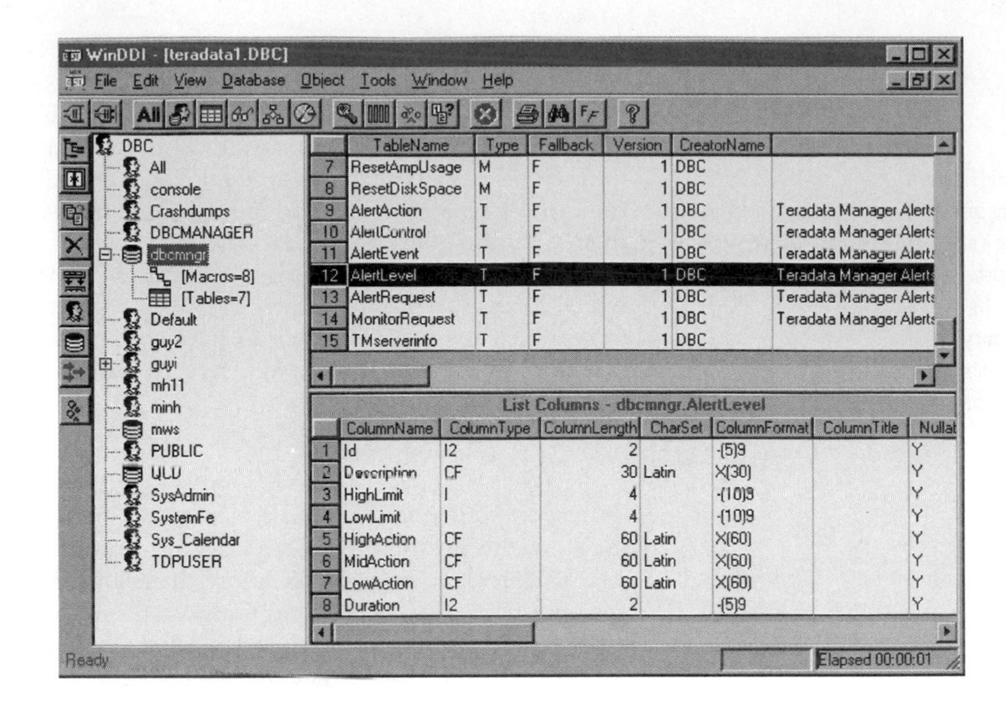

TeraData's WinDDi account management tool, shown here, provides an interface for carrying out database administration tasks.

As defined by its inventor, William H. Inmon, a data warehouse is a subject-oriented, integrated, time-invariant, and nonvolatile set of data that supports decision making.[1] *Subject-oriented* and *integrated* mean that the data in the data warehouse are organized according to subject rather than applications and are consistently named and measured. *Time-invariant* means that the data are accurate as of some point in time (in other words, a data warehouse provides a snapshot of the organizational data). Finally, the data are *non-volatile* because they do not change once loaded into the data warehouse.

The nonvolatility of the data in a data warehouse represents a crucial difference between it and a database associated with a transaction processing system. A database is constantly updated as transactions occur—items are sold, inventory is replenished, prices change, and so on. In a data warehouse, nothing changes once the data are loaded into it. This statement does *not* mean that new data cannot be added to the data warehouse; it *does* mean that existing data are not modified.

Creating a data warehouse is a four-step process:

1. Extract data from databases associated with transaction processing systems.
2. Transform the data into a form acceptable for the data warehouse, with all data using consistent naming conventions and units of measure.
3. Clean the data to remove errors, inconsistencies, and redundancies.
4. Load the data into the data warehouse.

Figure 5.7 shows the process of creating and using a data warehouse. Due to the large amount of data typically stored in a data warehouse, a mainframe-sized application or database server is often used to hold the data.

Because data warehousing emphasizes the capture of data from diverse sources for useful analysis and access, it does not always start from the point of view of the analyst or other knowledge worker who may need access to specialized databases. For example, an analyst may need to see only sales of a certain line of goods, but the data warehouse may store the data chronologically. To meet analysts' and knowledge workers' needs for a data source that emphasizes access and usability for a specific purpose, data marts have been created. The **data mart** is essentially a smaller form of the data warehouse

data mart
A scaled-down version of a data warehouse designed to suit the needs of a specialized group of knowledge workers.

1. W. H. Inmon, *Building the Data Warehouse*, 2nd ed. (New York: Wiley, 1996), p. 33.

Figure 5.7 Creating and using a data warehouse

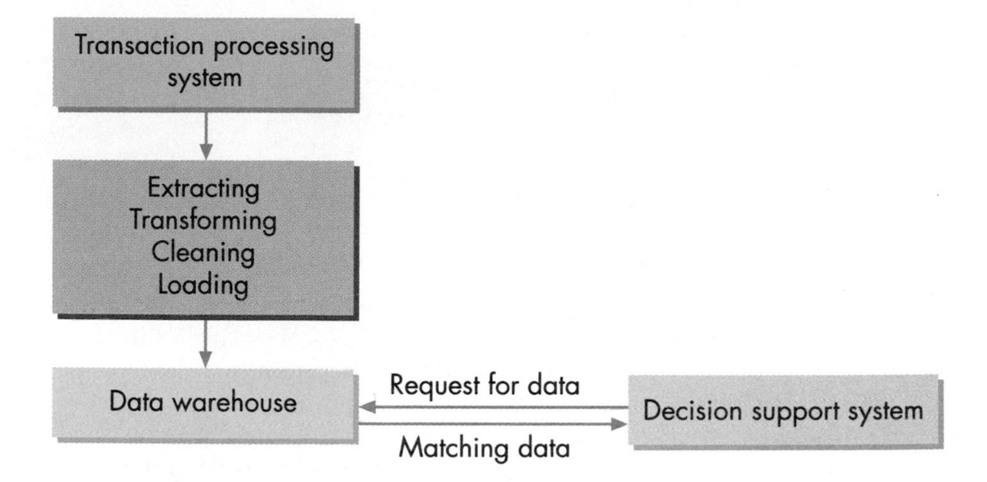

that meets a specific need. For example, a data mart might contain data based on lines of goods for the last six months, rather than data organized chronologically.

1. What problems arise when data are stored in multiple unrelated files in several locations?
2. What are the elements of the data hierarchy?

Relational Database Management Systems

relations
Data organized as a table and used in a relational database.

tuple
A row in a relation used in a relational database.

attribute
A field or column in a relational database.

As noted earlier, a relational database uses a table structure in which the rows correspond to records and the columns correspond to fields. In relational database terminology, the files (tables) are called **relations**, the records (rows) are called **tuples**, and the fields (columns) are called **attributes**. Each row must have the same number of columns, and the same format must be followed throughout. The word *relational* refers to the data model that connects the tables to each other. These relationships are set up during the creation of the model.

The relational data model has become the most popular database model today, and most new database management systems use this model. It was developed in the 1970s by Edgar Codd, who was seeking a way to accommodate an end user's request for data when the database designers had not planned for the request in advance—something that earlier database systems did not handle well. Relational database management systems offer greater flexibility because tables can be added as needed, providing that certain conditions about the relationships are met to avoid redundancy and ensure integrity in the database.

As an example of a relational database, consider the Customer and Special Order Product (Special, for short) tables from the FarEast Foods database shown in Figure 5.4. In addition to these two tables, a Request table contains information on each customer request, including fields for the customer ID number, the product ID number, and a number denoting the customer request. Figure 5.8 shows the relationships among these three tables. The Customer table is related to the Request table through the customer ID number that is common to both tables. The Special table is related to the Request table through the product ID number that is common to both tables. Because both the Customer table and the Special table are related to the Request table, they are also related to each other.

Figure 5.8 Relationships in the FarEast Foods database

primary key
A field or combination of fields that uniquely identifies each record in a table.

foreign key
A primary key for another table placed in the current table.

data model
One of several models specifying how data will be represented in a database management system.

many-to-many relationship
In a data model, the situation in which multiple fields are related to one another.

one-to-many relationship
In a data model, the situation in which one field is related to multiple other fields.

Why are the tables related through these particular columns, rather than through other columns? Two reasons explain this choice. First, each row in a table must be uniquely identified by a **primary key** to distinguish it from all other rows. In the Customer table, the primary key is a customer ID number (either a Social Security number or a company-assigned ID number). Similarly, the product ID acts as the primary key for the Product table, and a request number serves as the primary key for the Request table. As you can see, each transaction is uniquely identified. Second, it must be possible to link or relate a table to another table. The Customer table is linked to the Request table by placing its primary key (the customer ID) in the Request table, where it is known as a **foreign key**. Similarly, the primary key for the Special table (the product ID) is a foreign key in the Request table.

Primary and foreign keys enable you to relate tables and find needed values. For example, you can now determine all customers who have requested a certain product or all products requested by a specific customer. In the first case, FarEast Foods can send e-mail to interested customers when a certain product becomes available. In the second case, knowing what a customer likes means that FarEast Foods can market similar products to that customer by e-mail.

The tables, fields, and relationships used to provide needed data are referred to as a **data model**. Tables can be related in a variety of ways through a data model, including one-to-one, many-to-one, and many-to-many relationships. In each case, the relationship denotes the rows from each table that can be related.

The special-order data model provides an example of a **many-to-many relationship**, in which many rows in one table can be related to many rows in another table. For example, each customer may want many special-order products, and each special-order product may be requested by many customers. However, no direct relationship exists between the Special table and the Customer table. To work with this data model, you need to convert it to two one-to-many relationships through the use of the Request table. In a **one-to-many relationship**, one row in a table is related to many rows in another table. For example, each customer can have many requests, and each product can be mentioned in many requests.

The data model in Figure 5.9 depicts these relationships among the FarEast Foods database tables. In the figure, a three-pronged fork indicates a *many* relationship, and a single line indicates a *one* relationship. The primary key fields in each table are denoted with an asterisk (*).

Figure 5.9 FarEast Food data model

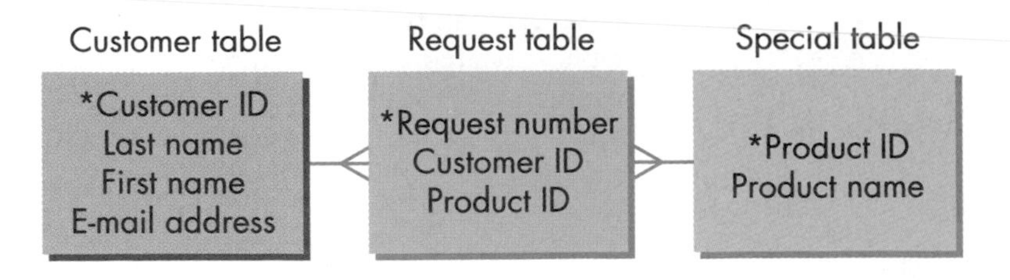

Figure 5.10 FarEast Foods data model in Access

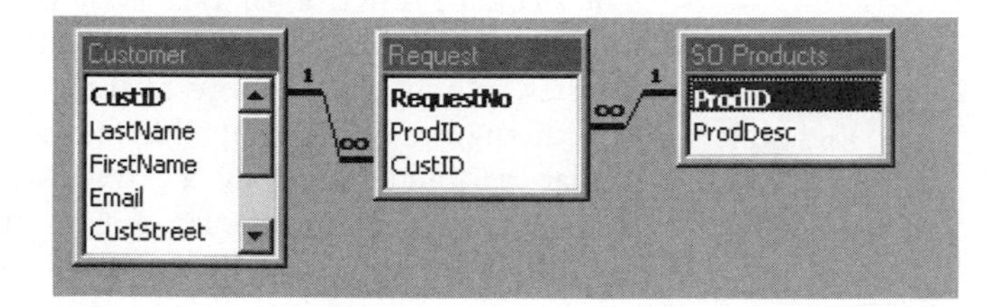

Figure 5.9 uses symbols to provide a succinct way of showing all the necessary information about the data model, in much the same way that various symbols are used to describe a mathematical model. Figure 5.10 shows the same data model in Access. Note that the infinity sign (∞) replaces the three-pronged fork for the *many* side of the model and that the primary keys are now shown in boldface. If values for the Special table are also entered in Access, and requests take place that generate entries in the Request table, the tables shown in Figure 5.11 result. Note that several food items and telephone numbers appear more than once in the Request table, indicating that a customer has made serveral requests and that an item has been requested more than once.

Data modeling and database design are very interesting topics, but going into more detail than is provided here is beyond the scope of this book. For further information on this topic, you may wish to refer to a book on database management.[2]

Figure 5.11 FarEast Foods products and request tables

Products : Table

ProdID	ProdDesc
AM	Aka Miso
BDP	Black Bean Paste
DNM	Dashi-No-Muto
FS	Fish Sauce
HO	Hoisin Sauce
HSS	Hoi Sin Sauce
KLL	Kafir Lime Leaves
KS	Kizami-Shoga
LG	Lemongrass
MP	Miso Paste
MS	Memmi Sauce
PS	Plum Sauce
SBS	Szchuan Bean Sauce
SM	Straw Mushrooms
ST	Shichimi Togarashi
TS	Tonkatsu Sauce
TYS	Tom Yam Soup
WS	Wakame Seaweed

Products table

Request : Table

RequestNo	ProdID	CustID
1	BDP	999-99-1234
2	MP	999-99-1234
3	DNM	999-14-3143
4	LG	999-11-0012
5	SBS	999-11-0012
6	LG	999-89-2269
7	TYS	999-31-9776
8	HSS	999-74-3343
9	KLL	999-74-3343
10	MP	999-88-1532
11	MS	999-88-1532

Request table

2. An excellent discussion of relational databases can be found in Peter Rob and Carlos Coronel, *Database Systems: Design, Implementation, and Management, 5th ed.* (Boston: Course Technology, 2002)

Relational Database Operations with Structured Query Language

update
In a database, to make additions, deletions, or changes to one or more columns for a particular row.

Structured Query Language (SQL)
A computer language for manipulating data in a relational database.

The primary function of a database is to enable users to obtain information from it in a usable form. Users retrieve information from a database by constructing queries, which are questions to the database. For example, a query result might display the names and e-mail addresses of those customers who are interested in a particular product, such as a satay sauce, enabling FarEast Foods to send an announcement to those customers. Once a query has been used to find matching rows, it becomes possible to **update** a row, by making changes to the contents of one or more rows, or to *delete* a row if it is no longer needed. It is also possible to *add* new rows to a table. For example, if a customer changed his or her e-mail address, you could query the Customer table to retrieve this customer's row and update the e-mail address column to reflect the new address. Similarly, if a customer expresses an interest in a product not currently listed in the Special table, you could add a row for that product. Finally, if a product is no longer available, you could find the product name in the Product table and delete it.

For a relational database, queries are written in a computer language known as **Structured Query Language (SQL)**, which is a specific language for manipulating data in a relational database. SQL queries enable database users to find records in a database that meet some stated criterion. An SQL query has the following general form:

SELECT fields **FROM** tables **WHERE** fields match query condition

For example, assume that the marketing department at FarEast Foods wants to know the names and e-mail addresses of customers who have placed special requests so that it can send e-mail to those customers regarding new products that

IT INNOVATORS

Edgar Codd

Without a doubt, the single most important person in the development of organizational databases is Dr. Edgar F. "Ted" Codd, the creator of the first relational model for database management and the father of modern relational database technology. Born in England in 1923, Dr. Codd joined IBM in 1949 as a programming mathematician, participating in the design and development of several important products including IBM's first manufactured and marketed computers (IBM 701 and 702), the IBM Stretch computer (IBM 7030), IBM's PL/1 programming language, and the first multiprogramming control system (STEM). Codd became interested in databases after seeing the enormous effects that Fortran (the first computer language for modern computers) had on the computing market. Because Fortran was designed for engineers and was ill equipped to work with databases, he realized that a language for interacting with databases was needed. After considering a number of approaches, in 1969 Codd published a paper on the mathematical foundation for relational databases. Based on his recommendations, IBM ultimately developed its DB2 relational database systems in 1983.

Codd retired from IBM in 1985 and formed a consulting firm with C.J. Date, a database author and lecturer. He has continued to encourage the development, standardization, and teaching of relational systems. As a result of Codd's work, relational databases are now the norm on all sizes of computers, from mainframes to personal computers. However, because the original relational model he created in 1969 was mathematical in nature, Codd feels that no existing relational database fully meets all of the conditions he created in it. In 1981, Codd received the highest award given by the Association of Computing Machinery (ACM)—the Turing Award.

Source: Sharon Gamble Rae, "ICP interviews: E. F. Codd." *Business Software Review*, October 1985, pp. 57-60.

Figure 5.12 List of FarEast Foods customers who have made at least one special request

LASTNAME	FIRSTNAME	EMAIL
Campbell	Lange	lcampbell@
Hyatt	Ashley	Volunteer@
Mullins	Janice	griffingirl@1
Patrick	Chris	devildog@u
Randall	Roy	rrandall@u(
Ronson	Suzy	SuzyLegs@
Roth	Jerry	BullGuy@a

have recently become available on a regular basis. To find the last name, first name, and e-mail address for all customers who have made at least one request, the SQL command would be

```
SELECT DISTINCT LastName, FirstName, Email
   FROM Customer, Request
        WHERE Customer.CustID = Request.CustID
ORDER BY LastName
```

This query compares the CustID column in the Customer table to the corresponding column in the Request table, with all matching rows being shown. The SELECT DISTINCT command displays each name only once, and the ORDER BY LastName command causes the names to be displayed in alphabetical order by last name. Figure 5.12 shows the result of this query.

access control
Techniques for controlling access to stored data or computer resources.

data sharing
A function of a database query language that coordinates the sharing of database information by multiple end users.

A notable feature of SQL is its capability to provide access control and data sharing on multiuser database systems. **Access control** is a function that restricts access to the data so that only authorized end users can retrieve, update, or delete data. **Data sharing** is necessary to coordinate the sharing of the database by multiple end users. Data sharing must ensure that users do not interfere with one another while working on the same database.

Object-Oriented Databases

object-oriented database
A database that contains a data type called an object, which incorporates both data and the rules for processing that data.

The newest forms of database are object-oriented databases. An **object-oriented database** contains a new data type—an *object*—that contains both data and the rules for processing those data. With a relational database, data are stored separately from the rules or software that actually carries out the processing. For example, when you start a database software package such as Access, you have to tell it which database to retrieve and process. With an object-oriented database, retrieving the database also retrieves the rules along with it. To see how this process works, think about a machine. A relational database is analogous to storing all of the machine parts—for example, gears, chains, covering, information about capabilities, and so on—but none of the rules for putting those parts together to construct the machine. In contrast, an object-oriented database would store the machine parts, associated information, and the rules for assembling the machine as a machine object.

Because you can store any type of data in an object-oriented database, it provides maximum flexibility to store text, numbers, pictures, voice data, and so on, either individually or in myriad combinations. For example, you can store complex data types for computer-aided design applications as objects in an object-oriented database or the multimedia data needed by many video and audio applications. Object-oriented databases are just in their infancy today, and most commercial applications attempt to marry them to the relational model.

1. What is a data model?
2. What is SQL, and what does it have to do with relational database management systems?

Information Management

At the beginning of this chapter, *organizational memory* was defined as an organization's electronic record of data, information, and knowledge that is necessary for transacting business and making decisions. The structured storage of data in databases was discussed in the last section. This section will address the information side of organizational memory. *Information* means the many different ways in which data stored in databases can be processed into a usable form, including text, hypertext, graphics, images, audio, and video formats. Information is typically stored in documents, books, reports, invoices, bills, CD-ROMs, videos, and so on, in both digital and analog or paper form. The newest forms of information comprise the millions of Web pages available on the World Wide Web that store all of the information formats.

Whereas data serve as the basis for all computer processing, information results when this processing is combined with knowledge. Humans use all three to make decisions. The information-based decision support systems mentioned in Chapter 4 (and covered in more detail in Chapter 6) are dedicated to automatically creating information in the form of reports, tables, charts, and so on to keep managers aware of the organization's operations.

The approaches to information management and knowledge management that will be discussed here are used for handling the present and preparing for the future as much as for remembering the past. For example, an employee who wants to know the current status of a plant in a distant location will use information management to find that information and then employ the information to make decisions or to respond to a question from a supervisor. Similarly, knowledge management is often used to help answer questions about how to carry out some activity. Thus, the distinction between organizational memory and handling the present or preparing for the future tends to be more blurred for information and knowledge than it is for data.

Paper-Based Information Storage

Information management has been around ever since the first people decided to organize their written documents—whether—clay tablets, papyrus scrolls, or handwritten or typeset books—into libraries in such a way that individuals could find the documents for which they were searching. Since computers came into widespread use, the need for information management has grown in tandem with the huge increase in the amount of information being generated in many forms—text, graphics, images, audio, and video. Although the concept of the paperless office has generated quite a bit of publicity, and a great deal of information has been stored in electronic form on CD-ROMs or on the World Wide Web, the vast majority of the more than 3 trillion items stored in U.S. organizations still takes the form of paper documents. In fact, the combination of networked personal computer and laser printer has actually increased the rate at which information is being stored on paper in folders in file cabinets. As an example, consider the computerized patient records discussed in the opening case. According to some estimates, the number of paper records is growing by 18 *billion* pages each year.

Organizations of all types face this paper problem because of the many forms, letters, documents, and reports that must be filed away in preparation for the day (which may never come) when they will be needed for reference. A single document may be filed in numerous places in an organization; for example, the originator may keep a copy, his or her office may keep a copy, the recipient may retain a copy made for his or her staff, and so on. It costs an estimated $2 to store a single piece of paper. Even when documents are stored in a different form than paper, that alternative usually consists of microfilm or microfiche, each of which has its own problems.

In addition to the dollar cost of storing paper documents, the paper problem carries other costs, including time to find documents, lack of availability to multiple persons, delays in transmission, problems with manipulating information, and lack of backup. It takes time to find documents. If you need a particular document, you must walk to the filing cabinet to search through hundreds of documents. The same is true even when the documents are stored on microfilm or microfiche—you must still manually search through the documents. Documents also are not easily availabile to multiple persons. If multiple people need the same document, copies must be made, adding to the flood of paper. Even after you find a document, sending it to another person or location can take time, and extracting information can require manual work. Finally, because of the volume of information, there are often no backups in case the originals are lost through fire or natural disaster. The September 2001 World Trade Center disaster resulted in the loss of millions of paper documents, many of which had not been converted into other forms. This loss is now and will be causing problems for many individuals and governmental agencies whenever issues arise that have to do with these documents. Table 5.2 summarizes the problems associated with information stored in paper form.

Document Management Systems

document management system
Technologies used to store and manage information in a digital format.

imaging
The process of converting paper versions of documents to a digital form using some type of scanner and saving them to optical or magnetic secondary storage.

scanner
A device used to translate a page of a document into an electronic form that OCR software can understand.

portable document format (.pdf)
A form of electronic document created with Adobe's Acrobat Exchange that can be easily shared with anyone who has an Acrobat reader.

Today, because of the problems with information stored on paper, most large organizations are moving toward using **document management systems** to store and manage information in a digital format so as to reduce costs and speed access. Organizations engage in an ongoing effort to convert much of their paper to a digital format using imaging. In **imaging**, paper versions of documents are converted to a digital form using some type of scanner and are saved to optical or magnetic secondary storage. A **scanner**, working much like a copier, converts paper documents into a digital form. The results of imaging can be stored as images, in a special digital format, or converted into a textual format. When a document is stored as an image, the resulting digital photographic representation can be stored, printed, or manipulated with various software packages. A special digital file format known as **portable document format (.pdf)** is often used to store the results of imaging. A .pdf image can be viewed and printed from any type of computer, regardless of the type of computer on which it was created.

Table 5.2 Problems with Paper Storage of Information

Problem	Explanation
Storage costs	Paper documents are expensive to store in filing cabinets and storage boxes
Time costs	A large amount of time can be required to find documents
Multiple access (sharing)	Paper documents (or microfiche/microfilm) must be copied so that they will be available to multiple persons
Transmission delay	Paper documents can take a long time to be sent from one place to another
Information manipulation	Information on paper or microfilm/microfiche must be manually extracted; information stored on paper is not easily searchable
Backups	Backups of paper documents are seldom kept due to space problems

optical character recognition (OCR)
The use of a scanner to convert a document to digital form, followed by use of software to determine the letters and symbols present.

fax conversion
The use of optical character recognition to convert incoming fax documents into ASCII format.

fax modem
A hardware component that combines the capabilities of a modem and a facsimile machine.

Converting a scanned image to text requires the use of **optical character recognition (OCR)**. In OCR, a reader device passes over a document and converts it into a digital form. Next, OCR software tries to identify the various letters and symbols in the image by matching them to a predefined set of requirements or patterns. Many OCR packages will apply a spell-checker to find incorrect words or display a special symbol for the user to change to the correct character when it cannot find a match for a character. OCR makes it much easier to convert paper documents to digital form because it relieves the monotony and strain of keyboarding information from a document into the computer.

Another popular use of OCR software involves fax conversion. In **fax conversion**, the OCR software converts incoming fax documents into an ASCII text format. This technique is especially useful when you employ a **fax modem** to route fax documents directly to a computer. Fax conversion allows you to save fax documents electronically rather than as paper. They can also be edited or revised, rather than just being read.

Even when documents are stored in a digital form, problems can occur. First, the system must be capable of identifying and retrieving a given document. Second, once the document is found, the system must have a way to search the contents of the document. Third, it must provide adequate protection against loss of the document due to damage to the storage medium. Finally, there must be security against theft of or tampering with the document. Document management systems must be able to deal with all of these issues to effectively store digital documents and make them available when needed.

Is Personal Information in Your Future?

TECHNOLOGY ON THE EDGE

More than 25 years ago, the first personal computer changed the way people thought about computing. Instead of being seen as huge machines set aside in some distant room, computers became "personal" and people could use them to carry out numerous tasks. One negative aspect of the use of personal computers relates to the location of the resulting data and information—on a single computer located in a home or office. As a consequence, data and information are not always available when needed. For example, consider the chapters of this book, which were written on desktop computers in two locations, plus a laptop at a thrid location It was edited at yet a fourth location. To make this setup work, chapters had to be transferred among all of the computers, always ensuring that the latest version was being used.

Because people often need data and information on computers other than their personal computers, they are starting to think differently about how they use computers. Instead of having a personal computer, people want to have personal information available to them on any kind of machine, no matter where they are working. Just as TV sets are impersonal in that they can all show the same programs, computers will become impersonal by being able to access your data from anywhere. Many companies, including Microsoft, Sun, IBM, and Oracle, are looking for ways to make documents available to users from anywhere. For example, the new Microsoft .Net initiative tackles exactly this objective. Although several companies now make disk storage available over the Internet, the concept of personal information goes beyond this concept to include storing the applications with the data and information. That way, downloading a file will also bring the software to run that file. You could sit down at any computer linked to the Internet and have it be "your" computer regardless of the software stored on it. No wonder some people call this "weightless computing"—you won't be required to carry any sort of computer with you, as any computer will do.

Source: Kevin Maney, "In the future, you'll pluck your info from thin air." *USA Today*, July 20, 2001, pp. B1, B2.

Searching and Indexing Information

Table 5.1, which differentiated between structured and semistructured information, listed Web pages and e-mail messages as examples of semistructured information. Both are difficult to search through to find needed information. Every computer user has experienced the frustration of entering a word or term in a Web search engine and receiving references to literally hundreds of extraneous Web pages that have nothing to do with the desired result. This problem with searching over the Web occurs because of the way search engines interact with Web pages and the efforts that some Web developers make to ensure that their Web pages are included in your search results, regardless of whether they actually match your query. Chapter 8 will discuss this issue in more detail.

Similarly, every day managers often receive tens or hundreds of e-mails that they must process and decide how to handle—write a response, ignore and discard, store for later action, and so on. Even after processing a message, the manager may need to store it and the response for later reference. Searching these stored e-mails at a later time to find one that is needed often proves as difficult as searching the Web because many may match a particular query.

indexing
The process of using data values or descriptors to search through documents.

On the other hand, Table 5.1 included *indexed* Web sites, reports, charts, and manuals under the *structured information* heading. These forms of information are more easily searchable than are semistructured forms of information. **Indexing** is the process of using data values or descriptors to facilitate searching through documents. Many Web search engines go through every Web page they can find and index each word. That explains why you find so many hits with a search engine when you enter a term like *management*—the search engine returns every page containing this word. Many organizational Web sites, or collections of associated Web pages, use a more rational indexing scheme that allows users to search only those pages in the Web site—thereby avoiding unrelated pages or those outside the site. However, this effort remains a hit-and-miss process, with a search engine still returning extraneous Web pages. A search starts with a search engine traveling the Web looking for key words. The search engine builds a list of the words it has found and their locations (their URLs), indexing them based on the importance or *weight* the search engine places on them. It then compresses the data to save space and saves them to disk storage. Then, when you search for a word, the search engine returns to the disk storage, finds the word and the corresponding URLs, and displays them. Figure 5.13 shows how a Web search engine performs indexing, and Chapter 8 will discuss the process of searching the Web in greater detail.

multimedia
An interactive combination of text, graphics, animation, images, audio, and video displayed by and under the control of a personal computer.

The same problems of finding the document you need within the thousands or millions that may be stored on computer or secondary storage at your location or at other locations exist with storage of electronic documents. This is especially true of **multimedia** forms of information, such as images, graphics, audio, and video. Once again, being able to search through this type of information depends on structure, because indexing allows you to use words to find the appropriate multimedia information item.

Several methods of indexing exist, including indexing on just the name of the document, indexing on keywords, and full-text indexing. Although it might seem that full-text indexing would be the best option, it is obviously not appropriate for nontext, multimedia items and can lead to the problems you see with Web-based search engines. For example, you cannot use full-text indexing on an Acrobat document, which is stored as a.pdf image; instead, you must use another form of indexing to access the many documents stored in this format. As with any other tool, indexing must be used with the end goal in mind—providing useful information to management, rather than assuming that one tool fits all situations.

1. What are some of the problems associated with paper forms of information?
2. What is indexing, and what does it have to do with information management?

Figure 5.13 Web search engine indexing

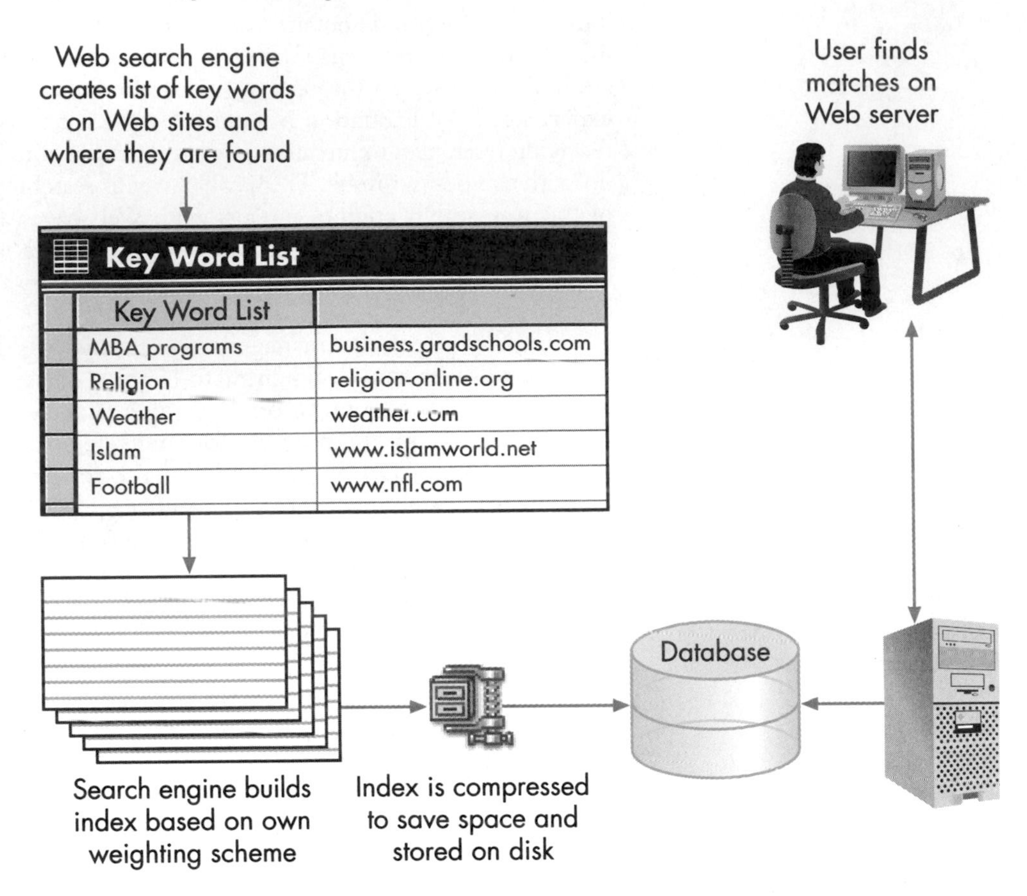
Web search engine creates list of key words on Web sites and where they are found
User finds matches on Web server
Key Word List
Key Word List
MBA programs
business.gradschools.com
Religion
religion-online.org
Weather
weather.com
Islam
www.islamworld.net
Football
www.nfl.com
Database
Search engine builds index based on own weighting scheme
Index is compressed to save space and stored on disk

Knowledge Management

Regardless of the amount of data and information available to an organization, without the knowledge necessary to use them, the data and information will be worthless to the organization. That uniquely human capability called knowledge is of critical importance to any organization's long-term success. Knowledge can be classified in a variety of ways, depending on who has it, whether it is explicit or tacit knowledge, and whether it is semistructured or structured.

Types of Knowledge

Recall that three typical types of knowledge exist: organizational culture (how things work in the organization), social networks (who can do what), and models for handling problems (how to solve problems). Another name for a part of organizational culture is policies and procedures—that is, "This is the way we do things in our business." Other types of organizational culture cannot be codified into policies and procedures because they relate to the organization's ethics and core values. Organizational culture is an example of knowledge that is not generally associated with a single person or team and that continues to exist long after individuals have left the organization. Some of the knowledge of organizational culture can be transferred to new employees through orientation or training sessions, but other elements must be assimilated by working in the organization for a period of time.

In contrast, the other two types of knowledge—social networks and problem solving—are often associated with a person or team; this type of knowledge can walk out the door at anytime. The people who have this type of knowledge are often referred to as **human capital** because they are at least as important to the organization as the building in which they work (and often much more important).

human capital
Those individuals in an organization who have knowledge about the social networks and problem solving.

Another way to classify knowledge is as explicit knowledge or tacit knowledge. **Explicit knowledge** is knowledge that is codified and transferable, whereas **tacit knowledge** is personal knowledge, experience, and judgment that is difficult to codify. For example, knowledge about how to use a word processing package constitutes explicit knowledge because it can be codified and transferred via books and other instructional material. In contrast, the judgment regarding who to talk with in an organization about getting things accomplished is an example of tacit knowledge because it usually cannot be codified. In general, the part of organizational culture that is related to policies and procedures as well as problem-solving knowledge are typically forms of explicit knowledge. The part of organizational culture that is not included in formal policies and procedures as well as social network knowledge are both forms of tacit knowledge.

explicit knowledge
Knowledge that is codified and transferable.

tacit knowledge
Personal knowledge, experience, and judgment that is difficult to codify.

Table 5.1 mentioned bulletin boards and chat rooms as examples of semistructured knowledge, and newsgroups, virtual teams, document databases, expert systems, and frequently asked questions (FAQs) as examples of structured knowledge. As with data and information, you distinguish between semistructured and structured knowledge by noting the ease with which the knowledge can be searched. Bulletin boards and chat rooms represent semistructured forms of knowledge because they are difficult to search for answers to questions. For example, if you have questions about how to do something in Windows XP, you can go to a chat group on the subject and attempt to find the answer to your questions; because of the highly unstructured nature of a chat room, however, your search may prove difficult.

In contrast, structured methods exist for finding answers to questions within newsgroups, virtual teams, expert systems, and FAQs. Newsgroups are **threaded**—answers or comments that relate to a previous question or comment are linked to it. Threading makes it possible to trace a discussion on a particular topic. You can also search newsgroups, although your effort may not be any more successful (or worse) than searching the Web. A structured method known as a document database facilitates collaboration among members of a virtual team, organization, or interest group. Instead of storing tables of data, a **document database** stores related documents. Similarly, expert systems provide a way of storing an expert's knowledge that makes it easy to find answers to questions. Finally, **frequently asked questions (FAQs)** are a form of structured knowledge in which answers are provided for the questions that are most often asked about a subject. Software developers often include FAQs with new software to help users more readily find solutions to their problems. Virtual teams and expert systems will be discussed in more detail shortly.

threaded
An organization of questions and answers in which answers or comments that relate to a previous question or comment are linked to it.

document database
A database that, instead of storing tables, stores related documents.

frequently asked questions (FAQs)
A form of structured knowledge in which answers are provided for the questions that are most often asked about a subject.

Sharing Knowledge

Data and information are usually easy to share because they can be stored as files or documents and provided to other members of a team or organization. In contrast, knowledge is often kept in a person's head and thus is much more difficult to share. The objective of knowledge management often is to simply find a way to share one person's or one team's knowledge with other members of an organization.

Studies have shown that most organizations would like to share knowledge so as to improve their operations or to embed it in their products and services. Often knowledge sharing involves sharing best practices. **Best practices** represent the best ways of carrying out operations as discovered by individuals or groups within an organization. They can include ways to obtain, organize, restructure, and store knowledge. Sharing knowledge can enable companies to reduce the cost and time needed to produce a new product or service (cycle time), increase sales, and, in general, bring about increased customer satisfaction through problem solving.

best practices
A list of the best ways that have been found to carry out operations as discovered by individuals or groups.

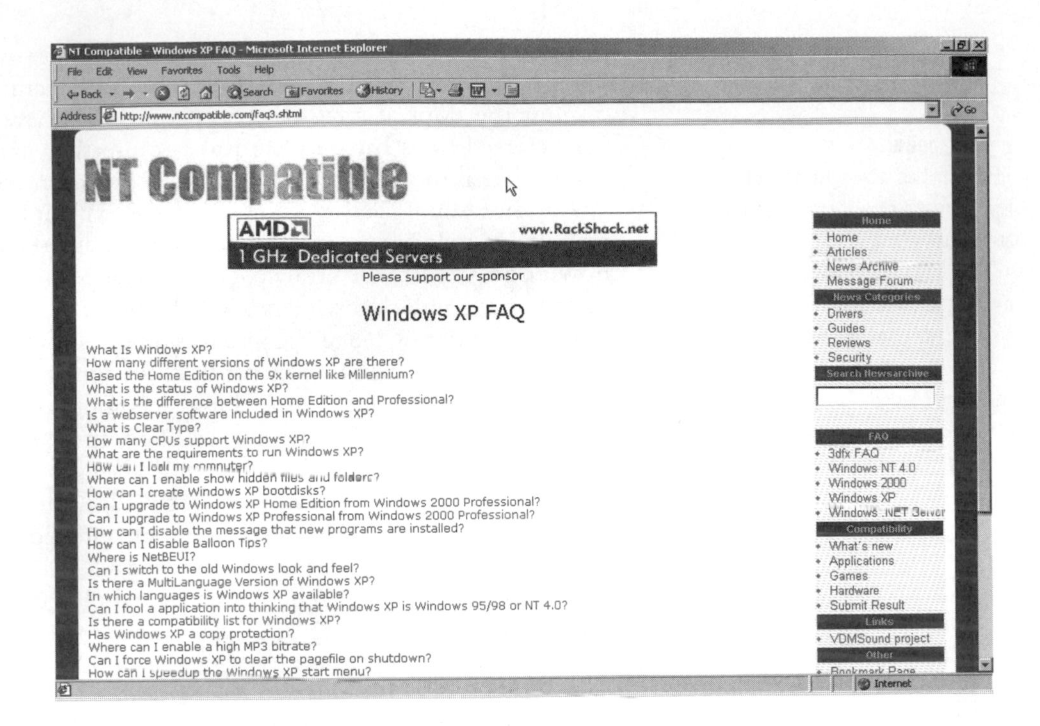

This FAQ provides information on using Windows XP to new and experienced users.

codification
The process of writing knowledge down in some fashion.

personalization
Personal sharing of knowledge.

One way to look at knowledge sharing is to consider how it is done. Knowledge sharing can occur through **codification** of knowledge, which involves writing knowledge down in some fashion. Knowledge can also be shared through personal sharing of knowledge, called **personalization**, in which a personal link exists between those sharing the knowledge. All of the structured approaches to knowledge management discussed earlier—threading, collaboration among virtual team members, and FAQs—are forms of codification; that is, the knowledge is stored in a form that can be retrieved at a later time without the presence of the person or persons contributing the knowledge. In contrast, the personalization approach to knowledge sharing requires the person with the knowledge to be present so as to share it with one or more people. The long-held practice of apprenticeship is a prime example of one person—the expert—sharing his or her knowledge with another person by showing the apprentice how to carry out various activities. In the modern world, personalization means knowing the right person in your organization to go to with a problem.

The type of knowledge being shared, tacit or explicit, reflects the approach followed in knowledge sharing. Although it is often easy to codify or write down explicit knowledge, recording tacit knowledge is far more difficult. For example, although it is fairly easy to codify complex knowledge, such as the rules of grammar and punctuation that lead to acceptable writing styles, it is far more difficult to codify simple things, such as how to ride a bicycle. Learning how to ride a bicycle usually requires personalization, or instruction, from someone who knows how to do it.

expert
An individual who, because of his or her knowledge in a specific area, can provide solutions to problems in that area.

The use of codification or personalization for knowledge sharing can also be a function of the number of people involved in the sharing process—that is, a group versus an individual. In group knowledge sharing, it is assumed that no one person or group has special knowledge that needs to be shared. In other words, everyone in the organization represents a potential source of knowledge that could help others in the organization. Such sharing requires a codification approach to make knowledge available to those who need it. However, almost every organization has one or more individuals, known as **experts**, who because of their (usually explicit) knowledge are essential to the organization's successful operation. These people can cut through superfluous details to get to the heart of a problem and serve as a large source of an organization's human capital. Finding ways to assist these individuals in sharing their knowledge with others in the organization is often essential to the continued success

Table 5.3 Types of Knowledge Sharing

Type of Knowledge Sharing	Type of Knowledge	Number of People Involved	Examples
Codification	Explicit	One-to-many, many-to-many	FAQs, newsgroups, virtual teams
Personalization	Tacit	One-to-one, one-to-few	Apprenticeship, lists of local experts

of the organization, and a personal approach is best, if possible. Table 5.3 compares codification and personalization for knowledge sharing in terms of the type of knowledge and the number of people involved in the knowledge sharing. The table also provides some examples of each type of knowledge sharing.

Virtual Teams

virtual team
A team of people who attempt to use information systems to help structure, focus, and facilitate the transfer of information and knowledge among themselves.

videoconferencing
A way of enabling groups or individuals in different locations to meet at the same time through real-time transmission of audio and video signals between the different locations.

One widely used approach to codification is the virtual team. A **virtual team** attempts to use information systems to help groups or teams of people structure, focus, and facilitate the transfer of information and knowledge among themselves. These working groups are referred to as *virtual* teams because, unlike sports teams, the team members do not always work in the same place or at the same time. Team members may be spread all over the globe working in different time zones on the same projects either synchronously (at the same time) or asynchronously (at different times.)

Videoconferencing facilitates synchronous meetings of virtual teams by enabling groups or individuals in different locations to meet at the same time through real-time transmission of audio and video signals among the different locations. The signals are picked up from cameras and microphones in specialized meeting rooms at each location and broadcast to all other rooms for display on a large screen set up at the front of the room. The fastest-growing type of videoconferencing involves computer-to-computer links, enabling individuals to communicate with one another in real time using both video and audio. Computer videoconferencing is often handled over the Internet using small, relatively inexpensive video cameras, called Webcams, and microphones. In fact, the case on virtual management discusses a new type of videoconferencing that combines inexpensive Web cameras with telephone calls to conduct synchronous virtual team meetings. The use of videoconferencing increased sharply after the events of September 11, 2001, when airlines were either shut down or business people chose to avoid traveling via air.

Videoconferencing has enabled virtual teams to work together over long distances without the need for travel.

Table 5.4 Comparison of Relational and Document Databases

Feature	Relational Database	Document Database
Structure	Collection of related tables	Collection of related documents
Record	Row of a table	Document
Field	Column of a table	Part of a document
Access	SQL query	Tailored report
Replication	Ad hoc through file copying	Scheduled as part of database application

Source: Richard T. Watson, *Data Management*, 2nd ed. (New York: John Wiley and Sons, 1999), p. 443.

Document Databases

Although virtual team meetings are useful for synchronous group sharing of knowledge, making decisions, and solving problems, a document database remains the most popular way to share knowledge within a virtual team on an asynchronous basis. Instead of storing related tables as a relational database does, a document database stores related documents. Table 5.4 compares the features of a document database with those of a relational database.

The most widely used software associated with document databases as a knowledge management tool is IBM's Lotus Notes. Lotus Notes is designed to store and manage large collections of text and graphics. Documents can be organized into a hierarchical structure consisting of sections, folders, and documents. A client/server network application—a Notes server called Domino—manages the document database. The Domino Notes server operates on a variety of platforms and operating systems. Notes clients are installed on users' computers, and can either run independently of the server or be connected to Domino.

To facilitate knowledge management, Notes can operate as a question-and-answer bulletin board system that enables individuals to share their knowledge on a particular subject. For example, a topical area, such as "relational databases," could be created in Notes to which users submit questions. If someone has an answer to the question, he or she submits the answer and Notes will link it to the question. Similarly, users can work with Notes to organize documents containing reference material, such as policies and procedures manuals, organizational charts, telephone and e-mail address directories, and even software documentation. It is also possible to access many Notes functions through a Web browser. Figure 5.14 shows several Notes document databases.

Expert Systems

The best vehicle for sharing tacit knowledge is through some form of personalization—in a personal contact or discussion with the person holding that knowledge. For example, if you are trying to make a sale to a customer in a foreign country and someone else in the firm has already dealt with customers from that particular country, you would want to talk with your colleague about his or her experiences. In so doing, you can avoid mistakes that have been made in the past and take advantage of strategies that worked. A variety of approaches have been suggested for making tacit knowledge available to anyone in the organization who needs it. Employees with several years of experience tend to have their own lists of people they turn to for advice in solving problems. In the preceding example, you might know that Sue is an expert in dealing with customers from Asia, Joe has been successful in making sales in South America, and Alice is a whiz at working with Europeans. In this case, if you are trying to make a sale in Thailand, you would talk with Sue. In addition, searchable lists of individuals who are knowledgeable in various fields, which employees can use to find someone to consult about a particular problem, can prove very helpful to organizations.

Recall that every organization includes individuals whose realm of tacit knowledge in a specific area is so great that they are considered experts in those areas. Whether they are deciding when a commercial soup is ready to be canned, diagnosing diseases, or configuring a computer system, experts use a body of knowledge and rules of thumb to solve problems requiring their expertise. Experts are obviously important to any organization, and when they retire or otherwise leave a company, their expertise and years of experience can be very difficult to replace. Even if experts are available within a company, their area of expertise may not be known to everyone in the company, or they may not have time to personally consult with each person who needs to talk with them.

Because of the importance of experts to all types of organizations, researchers sought ways to store years of expert knowledge in a computer and then make it available to others in the organization. Early attempts at codification of knowledge using databases proved problematic because the tacit knowledge that an expert uses to reach a decision cannot be stored in this fashion. Today, efforts to store tacit knowledge focus on expert systems. An **expert system** is a computer-based system that uses knowledge, facts, and reasoning techniques to solve problems that normally require the abilities of human experts.[3] Expert systems have enjoyed wide use in business, government, and industry as aids to sharing an individual's knowledge. Expert systems differ from the decision support systems mentioned in Chapter 4 and to be discussed in detail in Chapter 6, because they actually suggest courses of action just as an human expert would, rather than simply providing data, information, or solutions. For example, an implementation of an expert system at the Met-Mex Zinc mines in Coahuila, Mexico, led to a 10 percent increase in yields and a 30 percent reduction in labor by controlling the pH balance during the refining process.[4] Although not every user has benefited to this extent, many types of organizations have implemented expert systems.

expert system
A computer-based system that uses knowledge, facts, and reasoning techniques to solve problems that normally require the abilities of human experts.

Figure 5.14 Notes Document Databases

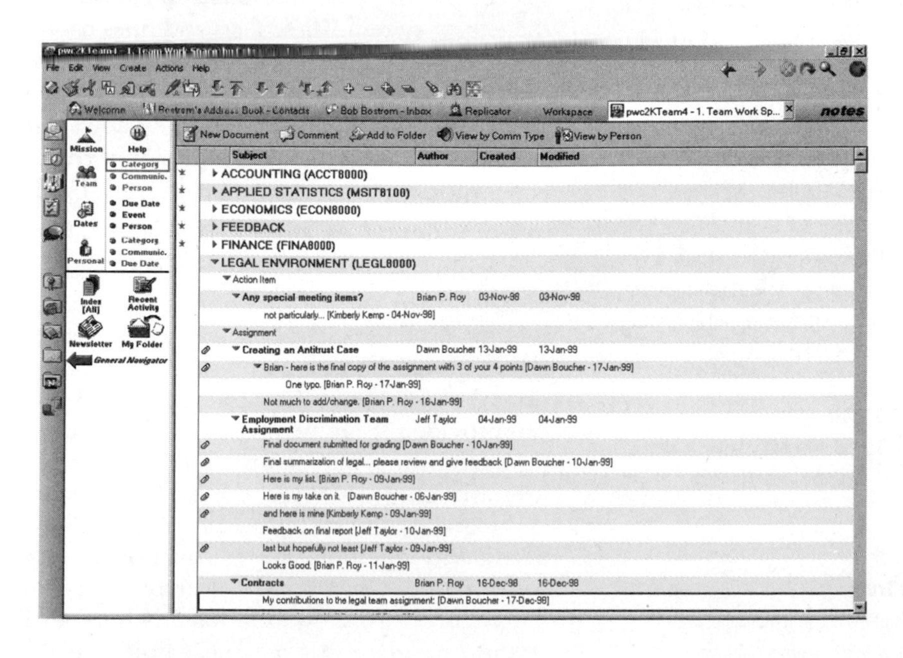

---

3. James E. Martin and Steven Oxman, *Building Expert Systems—A Tutorial.* (Englewood Cliffs, NJ: Prentice-Hall, 1988), p. 14.
4. "Automated operator decision support system and on-line process control significantly improves zinc yields, cuts labor costs", http://www.gensym.com/manufacturing/ss_penoles.shtml.

Figure 5.15 Elements of an expert system

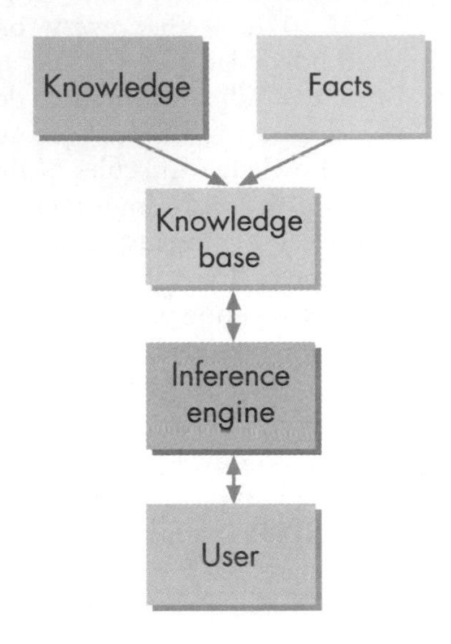

knowledge base
In an expert system, the facts, judgments, rules, intuition, and experience provided by the group of experts.

inference engine
The deductive part of an expert system that uses the information in the knowledge base to make suggestions or ask additional questions.

domain database
The part of an expert system knowledge base that contains the facts about the subject being considered by the expert system.

rule database
The part of an expert system knowledge base that contains the rules to be used by the reasoning element of the expert system.

IF-THEN rule
The rule used in an expert system that, together with facts, create the knowledge base.

logic trace
A trace of the line of reasoning used by an expert system to reach a conclusion.

Note that the definition of an expert system emphasizes problem solving using three elements: knowledge, facts, and reasoning techniques. In an expert system, knowledge and facts are stored in a **knowledge base**.The reasoning capabilities are handled by an **inference engine**, the deductive part of an expert system that uses the information in the knowledge base to make suggestions or ask additional questions. Figure 5.15 shows these elements.

The knowledge base includes all the facts and rules surrounding the problem at hand, and the inference engine works with the facts and rules in the knowledge base to make recommendations. It incorporates two databases: the domain database and the rule database.The **domain database** contains the facts about the problem being solved, and the **rule database** contains the rules used in the reasoning element of the expert system.These facts and rules try to include as much of the expert's experience, intuition, and tacit knowledge of the subject area as possible in the form of IF-THEN rules. An **IF-THEN rule** is one of the rules used in an expert system that, together with facts, create the knowledge base An IF-THEN rule states that, if a condition is true, then a conclusion is true as well. If the condition is not true, the conclusion is not true. For example, an IF-THEN rule might be

IF Water temperature > 100 degrees Centigrade at sea-level, then it is boiling

A unique aspect of expert systems is that they can answer queries from the user as to *why* a particular question is being asked and *how* a specific recommendation was made. This *justification* aspect of expert systems further sets them apart from conventional information systems that cannot provide information as to why and how they carry out their processing.The explanation facility of the expert system can provide the answers to *why* and *how* questions by keeping track of the rules that have been implemented to reach the current state of affairs. For example, when a user asks a *why* question, the explanation facility will specify the rule being tested and explain why the information is needed to test this rule. Similarly, when a user asks a *how* question, the explanation facility will show the line of reasoning, or **logic trace**, that was used to arrive at the current conclusion.

Figure 5.16 shows the logic behind an Internet-based expert system demonstration program that suggests a golf club for a golfer who is on the tee, 225 yards from the hole, with light winds. Note that a 3- or 5-wood is suggested in the bottom window (where CF stands for confidence factor).

Figure 5.16 Expert System for Golf Club Selection

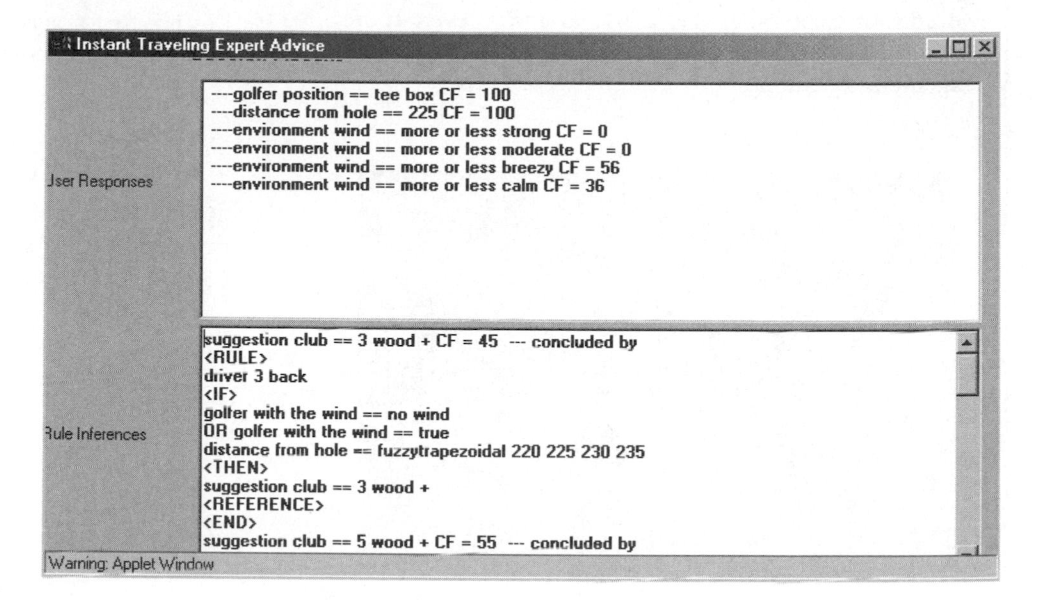

1. Why is a document database considered a form of knowledge management?
2. What are the parts of an expert system?

Case Study: Using the Internet for Virtual Management

A problem facing more and more U.S. companies is their employees' resistance to relocating overseas to manage a branch in a foreign country.The top two reasons are an unwillingness to fracture a spouse's career and a reluctance to disrupt children's lives. In a survey of 82 multinational companies, PriceWaterhouseCoopers determined that 24 percent allowed home-based employees to manage operations in another country instead of forcing them to relocate.

One potential solution to managing distant operations entails allowing employees to *virtually* manage foreign branches using the Internet and other telecommunication technologies. These technologies include e-mail, telephone, fax, videoconferencing, and so on. For example, the Parsippany, New Jersey-based hotel division of Cendant Corporation, which operates more than 6000 hotels in 24 countries, used e-mail and telephone calls to work out the design for a new facade on a hotel in Sharm El Sheikh, Egypt, thereby avoiding some $10,000 to $15,000 in travel costs.

A relatively new technology being used for virtual management is **Web conferencing**, which combines telephone conferencing with visual interaction over the Web. articipants can see each other while discussing issues and simultaneously view charts, graphs, maps and so on. Almost 20 percent of 510 business people surveyed by WorldCom now use Web conferencing. For example, a large chemical company had to change a manufacturing process to meet environmental, health, and safety standards at 132 plant sites where 12 languages were spoken. Using Web conferencing, it was able to complete the project in 10 months instead of the expected 5 years, and it saved $200,000 in travel costs at the same time. Although virtual management will never replace the need to travel to distant branches to mentor employees or to develop company culture, it can help achieve savings on travel and relocation expenses that flow directly to the bottom line. With the events of September 11, 2001, in mind, even more companies may look more favorably upon virtual management techniques.

Web conferencing allows individuals to work together over the Internet without leaving this home or office.

Source: Amanda Ripley, "In control, 10 time zones away." Time.com, April 1, 2001, http://www.time.com/time/global/cover.html.

Think About It

1. Describe the rationale behind the idea of an employee of a multinational company managing a distant operation.
2. What advantages does Web conferencing offer relative to traditional videoconferencing, which uses VANs to transmit signals from cameras and microphones?
3. Discuss some problems that might have to be overcome for Web conferencing to replace traditional teleconferencing methodology.

SUMMARY

To summarize this chapter, let's answer the questions posed at the beginning of the chapter.

What is organizational memory? Organizational memory is an organization's electronic record of data, information, and knowledge that is necessary for transacting business and making decisions. Data comprise raw facts usually in a textual form (letters and numbers). Information results from processing data into a usable form; it can take the form of text (tables and reports), hypertext, graphics, images, audio, and video. Knowledge is the human capacity to request, structure, and use information. Structured organizational memory encompasses forms of memory that can be easily searched, whereas semistructured organizational memory is not easily searched.

What is a database, and how does it store data in a structured form? Data are stored in a structured form in databases or data warehouses. A database is a collection of different types of data organized to make them easy to be manipulated and retrieved. A data warehouse is a subject-oriented snapshot of the organization at a particular point in time. Database management systems can reduce problems associated with data redundancy, data integrity, and data dependence. The data hiersarchy is the way in which data are organized in a database and consists of bits, characters, fields, records, and files.

What are the key elements of a relational database? A relational database uses a table structure in which the rows correspond to records and the columns correspond to fields. Each row must have the same number of columns, and the same specific format must be followed throughout. Each table has a primary key that uniquely identifies each record. When a primary key is included in another table, it becomes a foreign key, which serves to relate the tables. The tables, fields, and relationships that provide needed data collectively make up a data model. Users of a database submit queries to the database so as to obtain information from it in a usable form. Queries are used to retrieve records from one or more tables, and the retrieved records can then be edited or deleted. It is also possible to add new records to the database. Queries to a relational database, which are written in Structured Query Language (SQL), enable database users to find records from a table that meet some criterion or to create composite records from multiple tables that meet the criterion.

What methods are used to improve the management of information? Information is typically stored in documents, books, reports, invoices, bills, and so on, in both paper and digital form. The newest form of information comprises the millions of Web pages available on the World Wide Web that store information in various formats. Much of the information on paper is being converted to digital form through imaging, whereby documents are scanned into a computer. Web pages and e-mail messages represent semistructured forms of digital information, because it is difficult to search through them to find needed information. Indexing entails using data values or descriptors to search through documents. Indexed Web sites, reports, charts, and manuals are considered structured information. Document management systems are used to store and manage information in a digital format.

What are the types of knowledge, and how can knowledge be shared in an organization? The three types of knowledge are organizational culture (how things work in the organization), social networks (who can do what), and models for handling problems (how to solve problems). Social networks and problem solving are often associated with a person or a team that can leave at any time. The people who possess this type of knowledge are often referred to as human capital. Another way of classifying knowledge is as explicit knowledge or tacit knowledge. In general, organizational culture and problem solving are typically explicit knowledge, and social networks knowledge is tacit knowledge. The objective of knowledge management often involves finding a way to share one person's or one team's knowledge with others in the organization. Sharing knowledge can be accomplished through codification and personalization. All of the structured approaches to knowledge management constitute forms of codification; that is, knowledge is stored in a form that can be retrieved at a later time without the presence of the person or persons contributing the knowledge. In contrast, the personalization approach to knowledge sharing requires the presence of the person with the knowledge, who then shares it with with one or more people. Finding ways to assist individuals in sharing their knowledge with others in the organization is often essential to the continued success of the organization. Document databases are typically used for codification, whereas expert systems are used as a substitute for working with an expert on a personal basis. Expert systems use rules and facts to suggest solutions to problems.

REVIEW QUESTIONS

1. How does organizational memory fit into the IS cycle? What are the components of organizational memory?
2. How are data stored in a structured form?
3. What is click stream data and where is it stored?
4. What is the relationship between a database and files? Between a record and fields?
5. What is the difference between a field name and a data type?
6. Assume you are designing a database for a school club and you need information about its members, including their local addresses, telephone numbers, and e-mail addresses. What fields would you need, and what field names might you assign to them?
7. Why does an application server query a Web database? Do you always need separate machines for the Web, application, and database servers? Explain.
8. What is a data warehouse? How can it help the organization?
9. What are the various types of database management systems? Which is most widely used today?
10. What is a data model? A primary key? A foreign key?
11. What is SQL? How is it used in database management?
12. Describe three problems with paper storage.
13. What is imaging? What can be the output of imaging?
14. After a search engine has found Web sites that contain interesting words, how does it make that information readily available to you?
15. What are the various types of knowledge? Give an example of each.
16. What is an FAQ? Give a situation with which you are familiar in which a FAQ would be helpful.
17. What are ways of sharing knowledge? Why has one method become more popular since the events of September 11, 2001?
18. Give three differences between a relational database and a document database.
19. Why are expert systems considered a type of organizational memory?
20. Write an IF-THEN rule for the following situation: When hourly employees work more than 40 hours in a week, they receive overtime pay.

DISCUSSION QUESTIONS

1. Discuss the difference between structured and semistructured organizational memory, giving examples of each type other than those mentioned in the text.
2. Assume that a database of DVDs is available on the Web. Discuss the process that would occur if you requested to see information about a specific DVD. Make whatever assumptions you feel are necessary for your discussion.
3. Discuss the difference between the index of this book and an index of Web sites.
4. Document management systems are aimed at storing and managing information in a digital form. Discuss the problems that might be associated with this approach to storing information that might not be found with storing information on paper.
5. Discuss a topic for which you are an expert and the ways your knowledge could be incorporated into an expert system.

RESEARCH QUESTIONS

1. Go to the Web site for the demonstration version of the eMRWeb product at www.smartcorp.net/contents/prods/emrweb-demo.html. Test the product; being sure to fully investigate the information on the sample patient. What problems did you find with the demo? Write a two-page paper on the results of your test.
2. Go to the Web site for ISWorld at www.isworld.org and search for items of interest to you. Specifically, pick a country other than that in which you reside and investigate its ISWorld Web page and the editor of that page. Write a two-page report on your findings.
3. Visit the Web site for Microsoft that discusses the changes that .Net will bring to computing at microsoft.com/net/defined/netchange.asp. Write a two-page paper on what changes Microsoft expects to see from .Net.
4. Read the article "In control, 10 time zones away" (*Time.com,* April 1, 2001, www.time.com/time/global/cover.html), from which the end-of-chapter case was excerpted. Select one of the other examples of virtual management mentioned in the article. Research that company, and write a two-page paper on its use of Web conferencing.

WildOutfitters.com

"...three, four, five..." It was inventory day at Wild Outfitters. Claire and several friends were counting the items in the store while Alex tried to make sense of the sales transaction printouts that the couple had filed away for safekeeping. Twice each year, the Campagnes invited friends and neighbors to help them audit the inventory and then treated them to a barbecue as thanks for their help. The event was fast becoming a tradition.

Alex walked from the backroom office with an armload of papers and said, "Going through these records never seemed to take this long before. All of the new orders from the Web site have really made the pile bigger."

"It doesn't look like you'll be done with those anytime soon. We're almost done in here. Maybe I should get the food started on the grill," Claire offered.

"Good idea, otherwise we could have a hungry worker revolt," replied Alex, still keeping a sense of humor despite the work that would delay his fun.

Wild Outfitters' filing system worked well enough in the past, primarily due to the store's low sales volume. Now that both their store and online sales have increased, however, the Campagnes are becoming overwhelmed by the amount of data generated. Their current practice is to simply print out the transaction and file a copy of each sales transaction in chronological order. When it's time to do the books, one of them compiles the information and performs calculations for all sales since the date of the last inventory. The couple has been able to determine when to restock an item through simple observation of the number on hand. Now, with the increased sales volume, their current file system is rapidly becoming unmanageable. In addition, they are running out of items unexpectedly at times, which forces them to put valuable customers on hold.

In reality, their transaction records contain a lot of valuable information—if only it were more readily accessible. A number of data items are printed on the invoice for each sales transaction. Part of this information identifies a customer: the customer's name, street address, city and state, ZIP code, daytime telephone number, and e-mail address, if available. The invoice also includes information about the products purchased: the name, price, and quantity of each item purchased along with product details, such as size and color. In addition, each invoice is assigned an order number and marked with the type of sale: in-store or online.

The Campagnes realize that they need to do something to improve this situation. They would like to take advantage of their new network infrastructure to automate these recordkeeping tasks. They are ready to incorporate database technology to provide organizational memory for Wild Outfitters.

When Alex finally joined the festivities, Claire was taking the last of the ribs off the grill.

"It's a good thing that I kept a close eye on the barbecue rib inventory," Claire teased. "I would fear for my life, if I had to make you wait for a back order."

"Fear can be a good motivator. You should keep that in mind and get a database set up soon," Alex said, as he gnawed hungrily at his food.

Think About It

1. What benefits would organizational memory provide for Wild Outfitters? Explain.
2. How would you categorize the sales records of Wild Outfitters? Are they composed of data, information, or knowledge? Are the contents structured or semistructured organizational memory?
3. What steps would you take to make these records become a part of organizational memory? What other kinds of organizational memory do the owners of Wild Outfitters have?
4. Would an information management system or knowledge management system benefit the Campagnes? If so, explain why.

Hands On

5. You may download a database file (WildOutfit.mdb) for this case from the textbook's Web page at www.course.com. Use Microsoft Access or a comparable database management system to answer the following questions:
   a. What are the names of the tables in the database?
   b. For each table, what are the primary keys and foreign keys?
   c. What are the relationships between the tables (many-to-many, one-to-many, one-to-one)?
   d. Create a query to determine the number of sales transactions that occurred from the Web site.
   e. Create a query to find the total number of camping equipment items sold.
   f. Create a query to find the names and ZIP codes of all customers who purchased the Litewalkers Hiking Boots.
   g. Create a standard invoice using your DBMS reporting function. The invoice should include the following items: customer information; a listing of the transactions that make up the invoice, including each product, quantity, and price; a calculation of the subtotal, tax (use a 5 percent sales tax rate); and total.

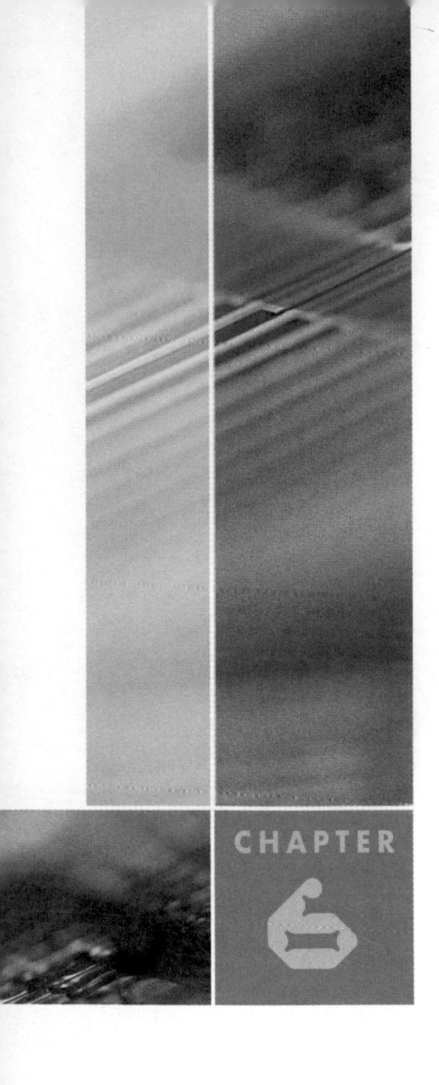

CHAPTER 6

PREPARING FOR THE FUTURE WITH DECISION SUPPORT SYSTEMS

LEARNING OBJECTIVES

After reading this chapter, you will be able to answer the following questions:

- What is a decision support system?
- How are decisions made?
- What types of decision support systems are in use today, and who uses them?
- How does an information-based decision support system support decision making?
- In what ways does a data-based decision support system provide answers to questions?
- How does a model-based decision support system answer questions from decision makers?

Using Fresh Data to Save Money at Red Robin

Like most restaurants, the Red Robin chain of more than 170 "gourmet burgers and spirits" restaurants is always looking for ways to save money. Many food items are good only for a short period of time, and preparing food that is not sold can result in significant loss through spoilage. Similarly, failure to have enough popular menu items loses sales and produces unhappy customers when they are not available. Getting good information about what is selling and what is not can be difficult in the restaurant business. At Red Robin, the problem became so bad that regional managers resorted to calling restaurant managers directly to try to determine what was selling so they could make decisions on menus for the different geographical regions. Purchases of supplies were then based on the resulting menus. Red Robin's problems resulted from the storage of sales, inventory, marketing, and other operational data in separate, aging databases that made it difficult to obtain up-to-the-minute reports.

To attack this problem, Red Robin's management decided to start over, by discarding old reporting systems and creating a new Web-based system that uses multidimensional data "cubes" that would provide the necessary data to managers to help them make decisions. Under the old system, managers would have to call the IS staff and ask them to generate a report. This process often took 24 to 36 hours, by which time the data was out of date. Now, managers can access the data instantaneously by going to the Web.

In addition to providing up-to-date data, the system can help managers watch for mistakes. For example, Red Robin ran a special promotion on some of its drinks that involved a give-away costing $1 for each item. The company thought that it was running out of the give-away item and prepared to order 250,000 more of them. After consulting the new system, however, managers realized that the stores had started the promotion one week early. They were able to cancel the order, saving $250,000.

In the three years since it became operational, Red Robin's Web-based system has resulted in annual savings of $1 million on food and other supplies. At the same time, it has increased sales of higher-margin items.

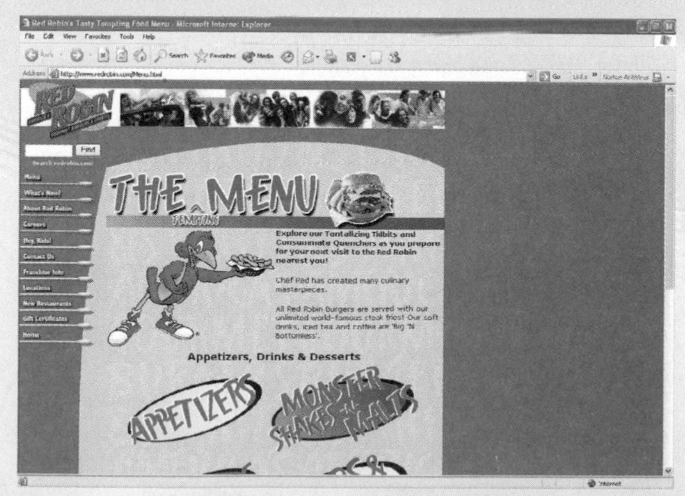

Red Robin's corporate managers require a decision support system to help ensure that restaurants have all the food items needed for their wide variety of menu choices.

Source: Valarie Rice, "House specialty: Fresh data." *eWEEK*, July 10, 2001, http://www.zdnet.com/eweek/stories/general/0,11011,2783719,00.html.

Preparing for the Future

When an organization uses transaction processing systems to handle the present (see the IS cycle model in Figure 6.1), data on the transactions are generated and stored in a database as a part of organizational memory. Decision support systems (DSS), like the one discussed in the opening case, use these data, along with the information and knowledge stored in organizational memory, to help the organization prepare for the future. This chapter discusses the various ways that decision support systems help organizations prepare for the future. Decision support systems aid in the decision-making process, and the resulting decisions will, in turn, determine how day-to-day operations (the present) are handled in the future. Each DSS uses data or information from organizational memory or from external sources, as well as mathematical models, to support decision making.

Decision making has always been key to the long-term survival of organizations. In the networked economy, simply making good decisions isn't sufficient; the organization must make them quickly as well. Failure to react to the dynamic environment of the networked economy and to adequately handle the increased level of innovation risk can result in a CEO losing his or her job or a company going out of business. For example, the well-known and once-successful company Polaroid filed for bankruptcy in the fourth quarter of 2001 due to problems dealing with the changing photography market.

Making good decisions quickly often requires a DSS that can provide the decision maker with data, information, or answers to questions in a hurry. Without such support, decisions may be based on hunches, rumors, or bad information. In DSS, as in many other areas of information technology, the movement is toward Web-based systems, like that implemented by Red Robin, that managers and decision makers can access from their Web browsers. Managers and decisions makers can no longer wait for paper-based reports, as they did in the past; they need the information or the answers to questions yesterday!

To understand the types of systems that decision makers need to help them make decisions in the fast-paced networked economy, let's start by defining *decision support systems* as information systems that provide decision makers with information, data, or answers to questions. Included in this broad definition are three basic types of DSS: information-based DSS, data-based DSS, and model-based DSS. Traditionally, some of

Figure 6.1 Information systems cycle

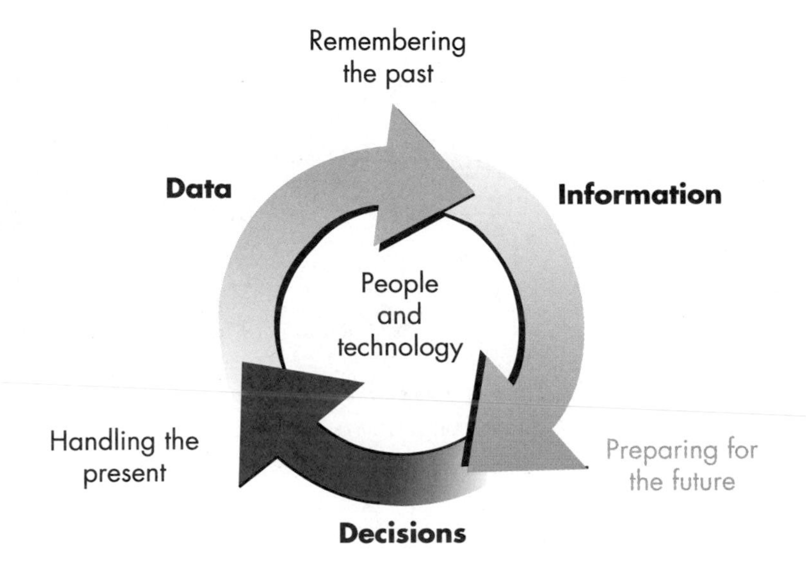

these systems have gone by other names, including management information systems, executive information systems, and database management systems. Because all of them are aimed at supporting decisions and differ only in the way in which they do so, they are grouped under the broad category of decision support systems.

business intelligence
Information systems aimed at helping an organization prepare for the future by making good decisions. Also called *decision support systems.*

In 1996, the Gartner Group coined the term **business intelligence** to refer to DSS that provide decision makers with data, information, and answers to questions. This term was defined to encompass the types of information systems that are commonly included in DSS. To avoid confusion, the older term *decision support system* will be used in this book, but you may see *business intelligence* used interchangeably with it.

Without the data, information, and knowledge stored in organizational memory, it would not be possible to do much of anything with a DSS. The link between organizational memory and the task of preparing for the future with a DSS as shown in Figure 6.1 cannot be overstated. You *must* have data, information, and knowledge to use a DSS to make decisions to prepare for the future. Now, let's look at how people make decisions.

Decision-Making Concepts

It is universally agreed that decision making is a key managerial activity—maybe *the* key activity—that often decides the fate of organizations. That is, good decision making usually ensures the long-term survival of an organization, whereas poor decision making can quickly lead to the demise of an organization. Decision making means recognizing problems, generating alternative solutions to the problems, choosing among alternatives, and implementing the chosen alternative. Figure 6.2 shows one model of the decision-making process.

Figure 6.2 Model of decision-making process

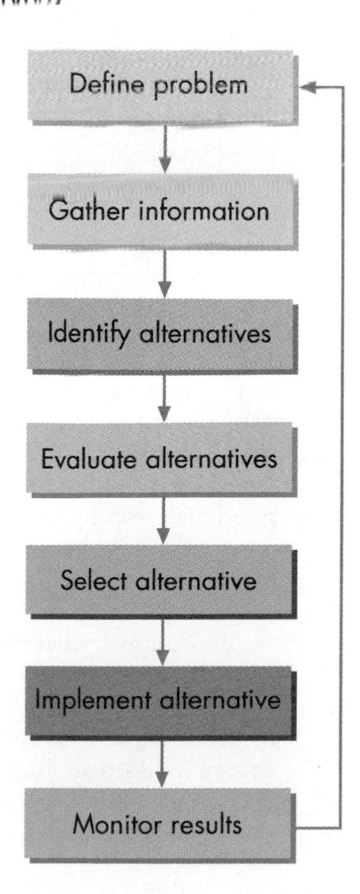

As illustrated in Figure 6.2, the decision-making process begins with the definition of the problem requiring a solution or decision. You must be careful at this stage to find the real heart of the problem and not just its symptoms. Problems are not inherently bad; they may actually represent opportunities that can be exploited to gain a competitive advantage. Defining a problem requires effectively monitoring internal activities and the business environment to determine the existence and nature of the problem or opportunity. Information systems constitute an important part of this monitoring process because they provide decision makers with the information necessary to define the problem. Of course, decision making does not occur in a vacuum—environmental constraints as well as strategic resources and organizational constraints apply to the process. These constraints ensure that the final decision can be implemented to solve the problem.

After defining the problem, you gather information on the problem or competitive opportunity. The information systems discussed in this chapter have been developed for just this purpose—providing business intelligence to decision makers.

In the next step, you identify alternatives to resolve a problem or to exercise a competitive initiative. Again, an effective information system must support the process of generating an array of feasible alternatives. The decision makers then evaluate these alternatives in light of criteria established by the organization. Sometimes evaluating alternatives is not so simple a task, given various organizational and political constraints as well as conflicting objectives within the organization. Nevertheless, you must select the most appropriate alternative.

Next, you implement the selected alternative course of action. Implementing the alternative is often the most difficult part of decision making because it requires convincing others in the organization to buy into the decision and enthusiastically work to make it a success. Many good decisions have failed to solve the original problem because they were not successfully implemented.

Finally, as shown in Figure 6.2, you must monitor the results of the implementation to provide feedback to management for review of the selection criteria, the alternatives, and the decision. An effective information system is also necessary to carry on these monitoring activities.

Throughout this entire process, the decision maker needs to access the necessary information to size up the problem or the competitive opportunity, to assess the alternatives under consideration, to decide on selection criteria, to evaluate the alternatives, to implement a course of action, and to monitor compliance with the decision. This task can prove very difficult, and information systems play an important role in supplying the needed information.

Types of Decisions

structured decisions
Decisions made by following a set of rules and usually made on a repetitive basis; decisions that can be programmed in advance. Also called *programmed decisions*.

Decisions can be structured or unstructured. **Structured decisions** (also called **programmed decisions**) are made by following a set of rules, usually on a repetitive basis. The type of problem addressed by structured decisions is especially amenable to solution by computerized mathematical models. Examples of structured decisions include ordering raw materials or parts based on the current level of inventory, assigning checkers in a grocery store based on the time of day and day of the week, or deciding what action to take when a machine varies from acceptable tolerances.

unstructured decisions
Decisions that involve complex situations and often must be made on a once-only or ad hoc basis using whatever information is available. Also called *unprogrammed* or *ad hoc decisions*.

In contrast, **unstructured decisions** (also known as **unprogrammed** or **ad hoc decisions**) involve complex situations and often must be made on a once-only or ad hoc basis using any available information. No clear-cut solution methodologies exist for these tough decisions, and their resolution requires a high-degree of human intuition and judgment. Examples of such decisions include deciding whether to move into a new product line, selecting an applicant to hire for a critical job, or choosing which merger or buyout offer to accept.

Table 6.1 Comparison of Types of Decisions

Type of Decision	Information Required	Identification of Alternatives	Selection of Alternative
Structured	Well defined	Limited	Use rules
Unstructured	Not well defined	Ambiguous	Use intuition and judgment

Table 6.1 compares structured and unstructured decisions in terms of the information required, identification of alternatives, and selection of alternatives.

At one time, management experts thought that various types of decisions could be associated with managerial levels of the organization. That is, structured decisions were associated with those employees who managed the day-to-day operations of the organization, whereas unstructured decisions were made solely by managers who dealt with the tactical and strategic issues facing the organization. With the movement toward empowering employees at all levels to make decisions about their jobs without waiting for approval from their supervisors, this separation is not as distinct.

1. List the steps in the decision-making process.
2. Describe the differences between structured and unstructured decisions.

Types of Decision Support Systems

As mentioned earlier, three types of decision support systems exist: information-based, data-based, and model-based. This section discusses all three, and then an entire section is devoted to each type later in the chapter.

information-based DSS
A generic name for many types of information systems that have the same goal: to provide decision makers with the information they need in an appropriate form.

With an **information-based DSS**, individuals at all levels of the organization can view information needed to make decisions. For example, an employee can determine the current status of a production process for which he or she is responsible, a salesperson can check the availability of a product or the status of an order, and the company CEO can request a comparison between the performance of the company's stock and that of its competitors. With this type of DSS, software processes specified internal or external data into information that is then displayed to users in a format that makes it possible for people to make needed decisions.

In more and more cases, and especially for managers and executives, this information is being reported in the form of Web pages. Usually, information system professionals in the organization have spent a great deal of time determining the information needs of employees, managers, and executives to ensure that the right type of information reaches them. Once they receive it, the employees, managers, and executives must spend even more time studying the information to understand its implications for the decisions they must make and for the organization as a whole.

management information system (MIS)
A system for providing information to support operations, management, and decision-making functions in an organization.

Information-based DSS have traditionally been known as management information systems or executive information systems, depending on the format of the report and the audience targeted. **Management information systems (MIS)** were originally developed as paper-based reports on internal operations. Although these reports were once considered adequate for mid-level managers, the people using them soon found that they did not always provide the type of information required by top-level executives. Instead of the reports provided by an MIS, top-level executives needed specialized information from a wider variety of sources in a

executive information system (EIS)
A personalized, easy-to-use system for executives, providing both internal and external data, often in a graphical format.

specific format. To meet this need, **executive information systems (EIS)** were created and used graphical on-screen displays to provide information to executives. Today in the networked economy, workers at all levels need information in an easy-to-use format—a format once reserved for executives. As a result, the differences between MIS and EIS have, to a large extent, simply disappeared. For this reason, this book refers to any system that provides information to an employee or manager as simply an *information-based DSS.*

data-based DSS
A decision support system that is aimed at exploring databases and data warehouses to analyze data found there so as to answer questions from decision makers.

A **data-based DSS** enables analysts and others within the organization to use data in databases and data warehouses to answer questions about the organization's operations. Quite often these questions take an *ad hoc* form—that is, they are created for the purpose at hand rather than being planned carefully in advance. The questions are answered through a variety of approaches, including SQL, online analytic processing (OLAP), and data mining. As discussed in Chapter 5, SQL is a special language for querying databases. OLAP enables a user to selectively extract data from a database using different points of view. In contrast to SQL and OLAP, data mining analyzes data to identify the previously undiscovered relationships. The opening case describes a data-based DSS that enables Red Robin's managers to understand the differing patterns of food and supply demands at the company's restaurants around the United States.

model-based DSS
A decision support system that combines data from the database or data warehouse with mathematical models to answer questions asked by management.

When questions require mathematical or statistical analyses, organizations can turn to a **model-based DSS**. As mentioned earlier, historically the term *decision support system* applied only to this type of system, but the definition has since been expanded to include information-based and data-based systems as well. Model-based DSS combine data from a database or data warehouse with mathematical models to answer questions asked by management. This effort often involves generating forecasts, carrying out statistical analyses, simulating situations, or searching for the best allocation of resources.

DSS Users

A wide variety of people in an organization may use some type of DSS, ranging from executives to operational personnel and customers or suppliers. In some cases, however, the person using a decision support system to answer a question is not the same as the person making the final decision. Individuals with titles such as *analyst* or *DSS chauffeur* actually interact with the decision support system to find answers to questions posed by the actual decision maker. In particular, these personnel may work with data-based and model-based decision support systems, because a great deal of specialized knowledge is often necessary to manipulate these systems effectively.

Table 6.2 shows a taxonomy of DSS users, in which users are classified according to their job descriptions, DSS skills, time availability, and type of software client used. *Job description* refers to users' responsibilities in the organization or ways in which they interact with the organization. *DSS skills* refers to users' capability to work with a complex DSS product to find needed information, data, or relationships. *Time availability* refers to the time they have to work with a DSS product. *Software used to interact with DSS* refers to the way in which users interact with the DSS product—using a Web browser or using software installed on the user's PC.

Let's look more closely at the type of software used to interact with a DSS. Locally installed software is used primarily by analysts and others who wish to *drill down* into the information and underlying databases by digging deeper into them to gain a greater understanding of root causes of the problem. These users can also extract data and use their local software to analyze it for later presentation to managers or executives.

In contrast, use of Web browsers appeals to many managers and executives who wish to review a report or look at an external news source, but not take the time to analyze the underlying data. These users may see things in the report that cause them to ask

Table 6.2 Taxonomy of DSS Users

Type of User	Job Description	DSS Skills	Time Availability	Software Used to Interact with DSS
Executive	Provides overall direction for entire organization	Low	Minimal	Browser
Manager	Provides direction for subunit of organization	Low to moderate	Low	Browser
Analyst	Works independently with data, models, and information on a DSS to discover facts and relationships	High	High	Software on PC
DSS chauffeur	Works directly with executive or manager to use DSS to answer questions	Moderate to high	High	Software on PC or browser
Operational personnel	Works under direction of manager with some independent decision making	Moderate	Moderate	Browser
Customers/suppliers	Interact with organization supply chain to provide materials or acquire finished goods	Low to moderate	Low to moderate	Browser

questions that an analyst using special software would attempt to answer. Operational personnel also use Web browsers to check the status of their current task or job. Likewise, customers and suppliers employ this type of software to check the status of an order or shipment. Although some DSS users will employ specialized software for their work, they and virtually all others in the organization will use a Web browser to access general-purpose information. Web browsers have become the application of choice for all employees of the organization because they are familiar to virtually everyone today, are easy to use, and are easy to maintain by information systems personnel. They also enable all employees, regardless of their position in the organization, to use the same software; the only difference involves the contents of the Web pages viewed.

In most cases of DSS use, work takes place over the organizational intranet. Recall that an intranet is a network that uses Internet protocols, but restricts access to employees of the organization. For example, an intranet Web server at FarEast Foods might allow only full-time employees to access the information stored on it, but lock out temporary employees. Similarly, customers could access data they need through an extranet. Recall that an extranet is a business-to-business network that uses Internet protocols for transmitting data and information between trading partners.

Note that the differentiation among the three types of decision support systems is somewhat artificial. It is possible to carry out activities associated with one type of DSS with software originally intended for use with one of the other two types. For example, tools developed primarily for extracting data can also support some types of information reporting. In fact, all three types of DSS can typically be accessed from the same Web site over an intranet, called an **enterprise information portal (EIP)**, that allows the individual seeking help with a decision to use any type of information system. An enterprise information portal acts as a Web-based doorway to the decision support systems through which the user enters to find the information, data, or solutions needed to make a decision. Figure 6.3 depicts this process graphically.

enterprise information portal (EIP)
A Web site on an intranet that allows the individual seeking help with a decision to use all three types of information systems without having to worry about which one is being used.

Figure 6.3 Use of enterprise information portal

Browsers

Portal interface

Information

Data

Models

Although this chapter discusses each type of DSS separately to help you understand each system's role in supporting decision making, you should be aware that they may be integrated into a single decision support system that is available over the Web. Before you learn about the three types of decision support systems, you need to develop a basic understanding of how decisions are made and what role information plays in decision making.

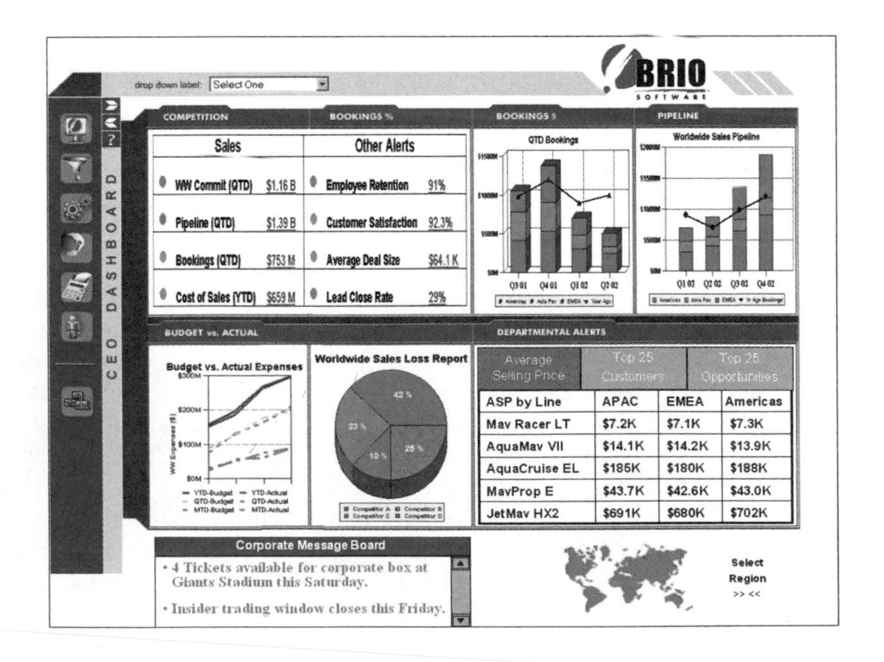

This report from an information portal shows a great deal of information and data about sales.

The Corrosion Portal

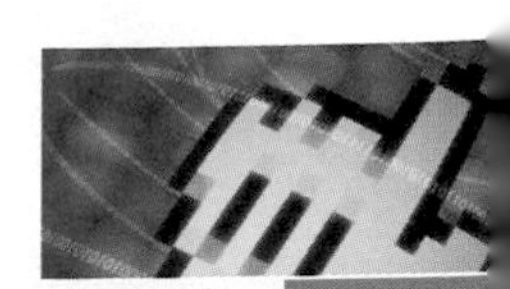

To many people, the topic of corrosion is something that they either ignore or know little about. However, corrosion—the oxidation of iron, copper, or aluminum resulting in rusted iron, corroded pipes, or tarnished doorknobs—is a $350 billion per year problem. Although not all corrosion is bad (wine is the by-product of grape corrosion), it can have nasty effects when corroded bolts fail or corroded pipes give way.

To help engineers and technicians combat this problem, Houston-based Integriti Solutions has created a portal dedicated to issues associated with corrosion. Designated as a "Best of the Web" site by *Forbes* magazine in 2001, it offers a wide variety of resources related to corrosion, categorized into three areas: a corrosion knowledgebase, the corrosion community, and the corrosion marketplace. Each area boasts myriad resources. For example, the knowledgebase section includes subheadings such as "hot topics", "handbook", "technical library", and so on. Many of these links connect to very helpful tables, tutorials, and how-to discussions. The community section provides sections on news, online conferences, discussions, an events calendar, and so on. An interesting link connects this portal to the *Corrosioning* journal that contains articles on corrosion. Finally, the marketplace section has links to an online directory, a request for quotes, the Corrosion Network, and a series of storefronts.

An especially interesting feature of this portal is the capability to run the Predict software from it. Predict addresses one of the most significant problems in corrosion evaluation—the assessment of corrosion rates in steel exposed to harsh environments. This software allows the user to specify data for the input parameters and watch how that parameter affects the corrosion rate for the entire system. For example, the user can specify pH-related data, temperature/gas-water ratios, the velocity and type of flow, and so on. The Predict software can be a very useful tool for modeling the rate of corrosion.

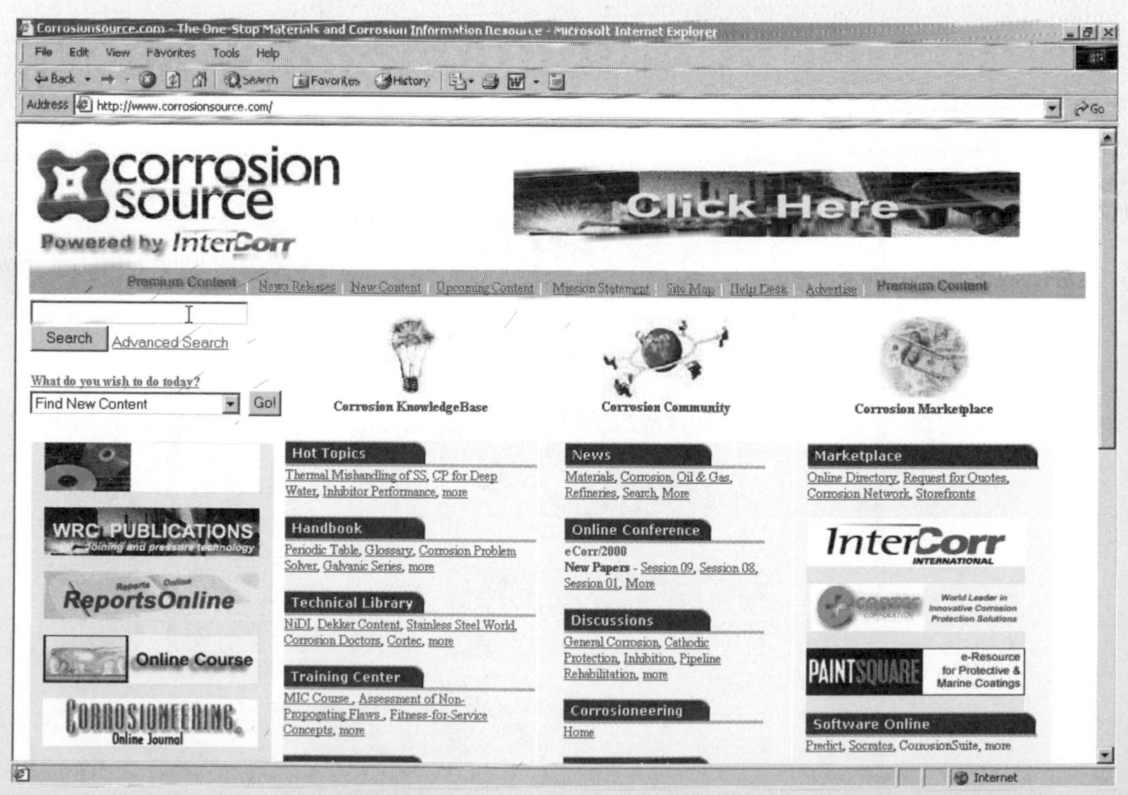

The corrosion portal provides a wealth of information on corrosion and related topics, including software to help make calculations about corrosion.

Source: Josh McHugh, "A question of rust." *Forbes.com*, http://www.forbes.com/best/2001/0521/040.html.

Using a DSS to Support Decision Making

As its name implies, a decision support system aims to support decisions at all levels of the organization. It does so differently at each level, however. For example, if an employee has responsibility for only a single machine, operational information on the current status of the machine can be sent to a Web server that formats it as a Web page, which the employee can retrieve on a regular basis. Based on this information, the employee can decide how to correct any problems that arise. Such **operational decisions**, which control the day-to-day operations of the organization, are usually considered to be structured because those decisions can rely on rules.

operational decisions
Operations that control the day-to-day operation of the organization.

Similarly, a salesperson can dial into a company intranet to find needed information on product availability, price, or specifications. This information can help the salesperson decide what terms to quote to a potential customer and help avoid demand risk due to poor customer service.

On the other hand, a manager who is responsible for one or more units of the organization will want to see summary information regarding those units so as to make tactical decisions. **Tactical decisions** implement policies that result from the **strategic decisions** of policy makers, which determine the organization's long-term direction. For example, assume that FarEast Foods has a policy to send out shipments within 24 hours of receiving an order. The manager in charge of the shipping department determines that orders are actually taking up to 30 hours to be shipped. In this case, the manager must discover why the problem exists and make one or more decisions on ways to reduce the time to ship an order. Just discovering the problem requires an information-based DSS that gathers data on shipments and summarizes them on a Web page that the manager can access. Determining the cause of the delay will require the manager to investigate the shipping process, possibly by looking at the Web pages created for the individual workers on their jobs. In a sense, the manager must drill down to more-detailed information to find the causes of the problem.

tactical decisions
Decisions made to implement the policies created by strategic decisions.

strategic decisions
Decisions that determine the long-term direction of the organization by creating policies.

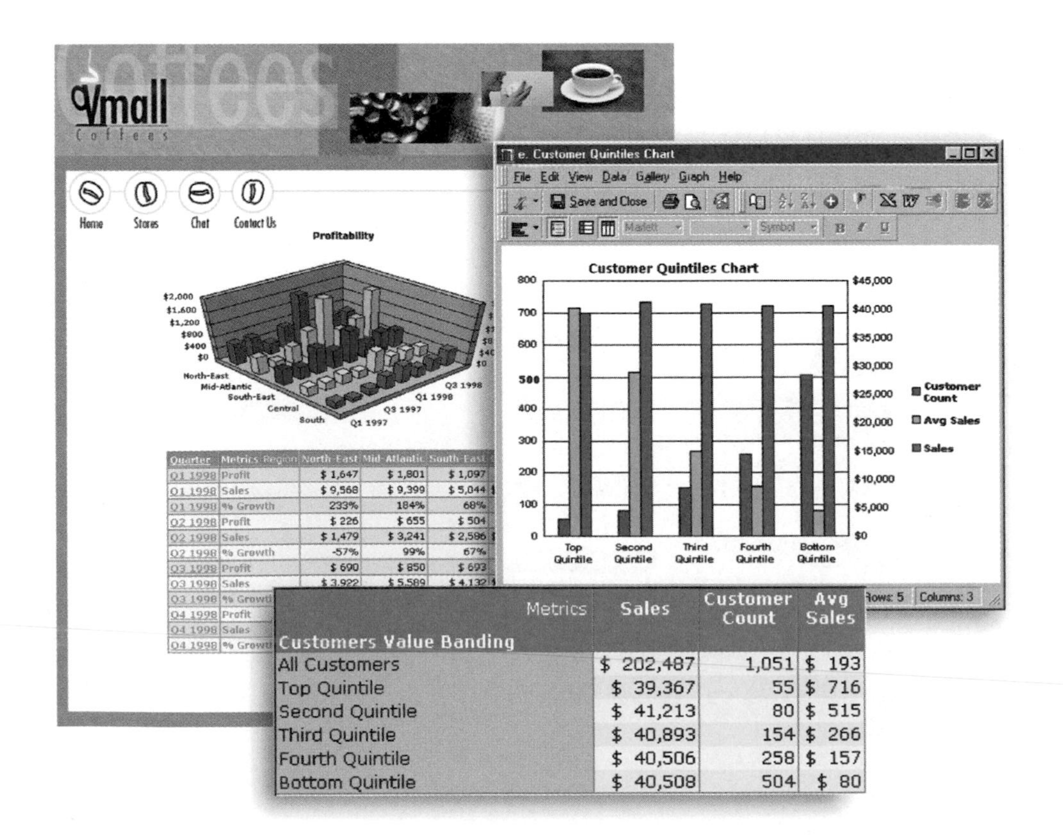

DSS software such as this product provides the user with a large number of options for displaying and working with data.

Similarly, executives responsible for policy making will want to see internal information on the entire organization. They will also seek out external information on the organization's suppliers, customers, and competitors so as to stay aware of developments in the environment. Strategic decisions tend to be even more unstructured than tactical decisions, often requiring a different type of knowledge and judgment on the part of the managers. If an executive sees a problem in the summary information, he or she may want to drill down to the tactical level or even to the operational level to determine the source of the problem. For example, if the CEO at FarEast Foods determines that the company is losing market share, he or she will seek information both internally and externally to shed light on the cause of this problem. In the process shown in Figure 6.4, the flow of summary information moves from the operational level to the tactical level, and from there to the strategic level. In contrast, the drill-down operation to obtain more detailed information goes in the opposite direction. The figure also shows the flow of external information into the strategic level.

Figure 6.4 Decision-making levels in the organization

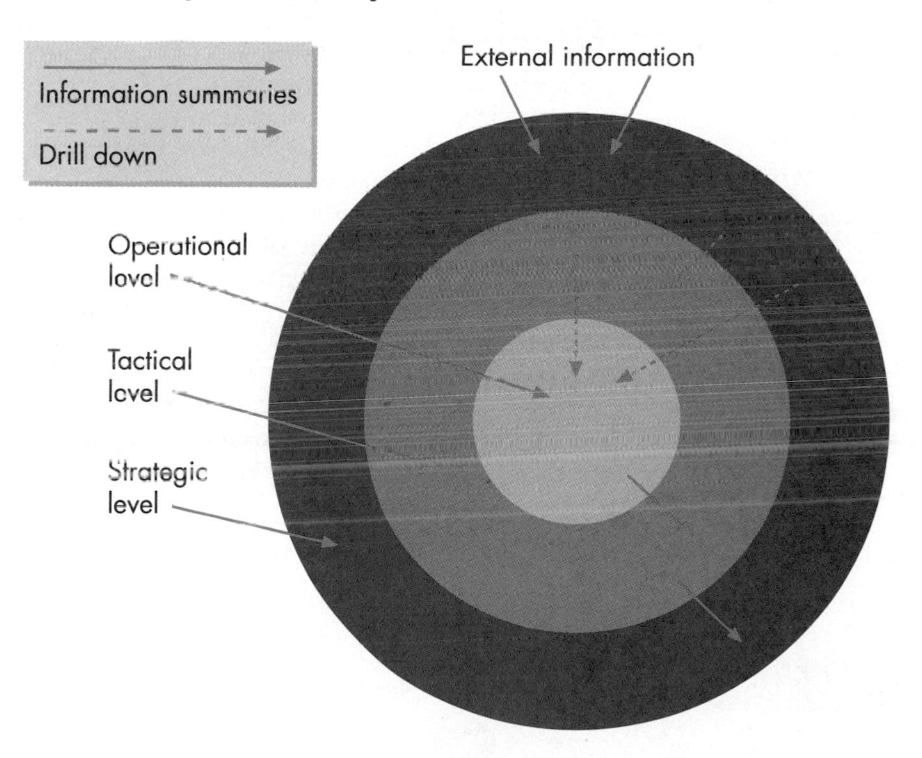

Decisions at FarEast Foods

Recall that all organizations face three types of risks: demand risk, innovation risk, and inefficiency risk. In Chapter 4, you learned about the need for FarEast Foods to successfully deal with inefficiency risk by carrying out transaction processing more efficiently. In that case, the company was trying to find faster and less costly ways to take orders, accumulate food items into an order, and ship those orders to its customers. In Chapter 5, you discovered how FarEast Foods was using organizational memory. In this chapter, you will examine decisions that FarEast Foods must make to avoid demand risk. The company is considering several ways to increase demand for its products:

- Increasing market penetration
- Developing new markets
- Developing new products or services
- Diversifying

The first method for avoiding demand risk, increasing market penetration, involves finding ways to increase the amount of an existing product or service sold into an existing market. For example, FarEast Foods managers might decide to use an e-mail marketing campaign to make current customers aware of specials that are offered each week. Developing new markets involves finding ways to sell existing products or services into new markets. For example, FarEast Foods might decide to try to sell its Asian food products in Brazil to take advantage of the very large population of Japanese people who have immigrated to that country. Developing new products involves selling new products or services in existing markets. For example, FarEast Foods might want to add Caribbean food products and spices to its existing catalog of Asian foods for sale in the U.S. markets that it currently serves. Finally, diversifying involves selling new products and services in new markets. For example, FarEast Foods might decide to sell its new Caribbean foods and spices in the Bahamas and Jamaica.

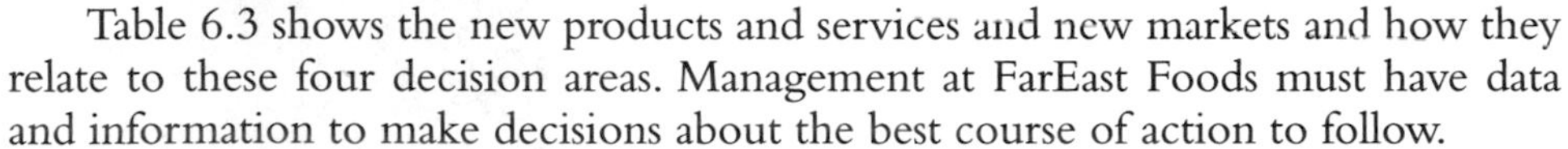

Table 6.3 shows the new products and services and new markets and how they relate to these four decision areas. Management at FarEast Foods must have data and information to make decisions about the best course of action to follow.

1. List the three types of decision support systems.
2. Which type of decision support system is described in the opening case on Red Robin restaurants?

Table 6.3 Ways of Avoiding Demand Risk

		Markets	
		Existing markets	**New markets**
Products	**Existing products**	Market penetration	Develop markets
	New products	Product development	Diversification

Information-Based Decision Support Systems

Information systems play an important role by providing information to the organization in a form that can help it prepare for the future by aiding in decision making. As shown in Table 6.4, all seven steps of the decision-making model need data, information, and answers to questions. For example, the first step of the decision-making process is to define a problem requiring a decision. Typical data or information required to complete this task include data and information on problems in current operations, new opportunities available to the organization, or threats to the continued existence of the organization.

An organization uses a DSS to obtain the data, information, or answers to questions for the various steps of the decision-making process shown in Table 6.4. This section concentrates on showing how information is provided to decision makers; later sections will focus on providing data and answers to questions.

The primary method used to provide information to decision makers is a report—that is, a series of pages that impart information about a specific topic. At one time, most reports generated by an information-based DSS were printed on paper. Today, however, most such reports are available online on the Web for employees to either read on screen and print only what they need for future reference or to print the entire report. Entire sets of documentation now reside on the Web. For example, you can find the contents of many reference books developed by Microsoft on its programming products by going to msdn.microsoft.com/library. Although the Web has also affected data-based and model-based DSS, they still depend largely on software installed on the employee's PC for analyses.

Beyond providing reports to employees in a standard, easily accessible format, the Web provides the capability to link one report to other related reports or Web pages, to related Web sites outside the organization, or even to the underlying data used to generate the report. This capability dramatically expands the reach of the executive, manager, or operational employee to find the information he or she needs to make the best decisions for the organization. At FarEast Foods, for example, the shipping manager might look at a Web page report containing information on the on-time delivery of shipments.

Table 6.4 Information Required for Decision-Making Steps

Decision-Making Step	Information Required
1. Define the problem	Problems with operations; new opportunities; threats from competition
2. Gather information	Information about the problem defined in the first step
3. Identify alternatives	Alternative solutions to problem; ways to take advantage of opportunities or deflect threats
4. Evaluate alternatives	Information on consequences of selecting each alternative identified in previous step
5. Select an alternative	Rules or shared knowledge from organizational memory
6. Implement the alternative	Individuals or organizational units that must be involved in implementation; problems with implementation process
7. Monitor the results	Information on whether the decision was implemented and on the results of implementing the decision

Types of Reports

Regardless of the managerial level—operational, tactical, or strategic—and the format of distribution—paper or Web-based—virtually every type of information-based DSS generates three types of reports: scheduled reports, exception reports, and demand reports. Each serves a different purpose for the managers and employees of the organization, but all are important.

scheduled report
A report generated by an information-based DSS on a regular basis, containing summary reports of the results of the data processing operation.

Scheduled reports offer periodic and historic information on the organization's operations and are automatically generated by the information system on a schedule (for example, daily, weekly, monthly, or quarterly). They often result from the process of categorizing and summarizing the information produced by a transaction processing system. Managers use reports containing periodic information to make decisions or to decide to request more information. For example, the FarEast Foods shipping manager might receive a daily report on the time required to ship items. This report might include average shipping time, worst case, best case, and other statistics. Based on this report, if problems arise with late product shipments (more than 24 hours after an order is received), the shipping manager might decide to drill down to the operations level to seek information that reveals whether a particular product or group of products is causing problems or whether a particular employee is to blame. Figure 6.5 shows a typical scheduled report for the shipping manager at FarEast Foods.

exception report
A report generated by an information-based DSS when some condition falls outside a previously defined acceptable range.

Exception reports are generated when some condition falls outside of a previously defined acceptable range. The definition of what will cause the generation of an exception report, called a threshold value, must be determined in advance by management and must be programmed into the information system software that generates the reports. In many cases, a manager using a scheduled report will detect a problem and then request an exception report so that whenever this problem arises again, he or she is immediately alerted. In this way, an exception report helps the manager detect problems at an early stage, an essential part of good management. At the same time, an exception report does not overwhelm the manager with unnecessary information.

For example, an exception report for the FarEast Foods shipping manager might be generated whenever some fraction, say 5 percent or 10 percent, of the shipments goes out late. The shipping manager must determine the threshold value that will prompt the creation of an exception report. Such a report might prompt the manager to investigate the possible reasons for excessive late shipments. In this case, the exception report provides immediate information on a known problem area so the manager does not have to wait for a scheduled report or search through the report for the needed information. Figure 6.6 shows an exception report for the shipping manager at FarEast Foods that will help her watch for problems in on-time delivery of orders.

Figure 6.5 Scheduled shipping report at FarEast Foods

FarEast Foods

Scheduled Report

Time to ship (hours)

September 29, 2004 Average 14	Worst 48	Best 4	Total Orders Shipped 354
Previous Day Average 8	Worst 26	Best 4	Total Orders Shipped 227
Previous Week Average 11	Worst 52	Best 2	Total Orders Shipped 1358

Figure 6.6 Exception report for FarEast Food's shipping department

FarEast Foods

Exception Report

Late shipments September 29, 2004

Hours Late	Product	Destination	Via	Number of Shipments
24	Wakame Seaweed	AZ	UPS	3
24	Wakeme Seaweed	AK	FedEx	5
12	Nori Sampler	NY	UPS	1
8	Aka Miso	AZ	UPS	2
4	Wakamo Seaweed	TX	FedEx	6
4	Miso Paste	IL	UPS	2
2	Aka Miso	FL	UPS	1

Today, managers can receive exception reports in real time (in other words, as problems occur) through a Web browser at their desks or over wireless devices—either mobile phones or wireless PDAs—rather than wait for paper-based reports to be written, printed, copied, and delivered. With wireless technology, exception reports can be delivered immediately regardless of the recipient's location. For example, a sales person going to meet a client can receive immediate information over his or her wireless device on late-breaking problems faced by the client and possible solutions to those problems that the sales person can offer. Given the speed at which decisions must be made in the networked economy, real-time exception reports are essential for avoiding or solving problems.

demand report
A report generated by information-based DSS upon a request by a manager.

Finally, **demand reports** comprise specialized reports that a manager may request when needing specific information on a particular subject. Often such a request results from an unexpected outcome in one of the other reports or from the receipt of outside information. For example, the FarEast Foods shipping manager might request a demand report on the effects of adding a second shift as part of an effort to improve the department's on-time performance. Once again, a variety of technologies, including the Web, e-mail, or wireless devices, can be of great use to the manager. At her desk, she can click a link on a Web page to request more information on something in a scheduled or exception report. She can also send e-mail requesting more information on a report that has no links. If she is out of the office, she can use her wireless device to request information on a customer's problem from the customer's office or her taxi.

As the example of the shipping department at FarEast Foods suggests, scheduled, exception, and demand reports are typically customized to satisfy the needs of various departments. The reports in the finance department or the personnel department will differ from those in the shipping department and from those in other departments or areas in any company. Although all employees use the same Web browser, and although the Web pages used to deliver the information may have the same overall design, the information being delivered will be different. One strength of a Web-based information-based DSS is the capability to deliver the right information to the right person in the organization.

An Information-Based DSS in Action: The Balanced Scorecard System

Robert Kaplan and David Norton developed the balanced scorecard system for measuring organizational health.

As an example of the use of an information-based DSS, consider a new approach to measuring organizational health. Developed by Drs. Robert Kaplan and David Norton in the early 1990s, it requires an information-based DSS: the balanced scorecard. Before studying the inner workings of the balanced scorecard approach, you first need to take a brief look at traditional financial approaches to measuring organizational health.

Since shortly after World War II, organizations have depended on financial measures provided by their accounting departments to determine their overall health and to make decisions for the future. For example, management might look at earnings per share(EPS) as an indicator and make decisions that would ideally lead to higher EPS values. Unfortunately, although financial data are precise and objective, they often fail to tell the whole story of an organization's health. Financial data provide *lagging indicators*; that is, they show what happened in the past. To keep an organization healthy, *leading indicators*, which give management some idea what may lie ahead, are needed. Furthermore, the use of financial indicators is strongly associated with industrial economy companies. They often work poorly as indicators of organizational health in the networked economy.

The balanced scorecard approach aims to help a networked economy organization clarify its vision and strategy and translate them into action. It does so by providing feedback about the internal business processes and external outcomes so as to continuously improve strategic performance and results. When employed correctly, the balanced scorecard transforms strategic planning into a reality that helps the enterprise succeed in the networked economy. The balanced scorecard approach views the organization from four perspectives:

- Learning and growth
- Business process
- Customers
- Finances

When using a learning and growth perspective, the balanced scorecard approach helps the organization answer the following question: To achieve our vision, how will we sustain our ability to change and improve? When using a business process perspective, it emphasizes how the organization can satisfy stakeholders by excelling at specific business processes. The balanced scorecard approach also uses a perspective that examines how the organization should appear to its customers. Finally, this technique employs the traditional financial perspective. These four perspectives are linked through a double-feedback loop that provides and measures internal business process outputs and the outcomes of business strategies.

Figure 6.7 illustrates the balanced scorecard approach. In the figure, the grids hold the firm's high-level objectives, the measures that will be used to determine how well the objectives are being met, targets for the objectives, and the particular programs implemented to reach these targets. For example, a firm's objective might be to make itself more financially viable. The measures to determine how well this objective is being met might focus on improvements in return on equity or return on assets. The targets might comprise specific numeric values based on company established goals, industry-leading firm numbers, or industry benchmarks, or they might simply consist of improvement over past performance. Finally, initiatives might involve having a review committee consider formal proposals costing more than $250,000 that will increase revenues or decrease costs.

Figure 6.7 Balanced scorecard approach

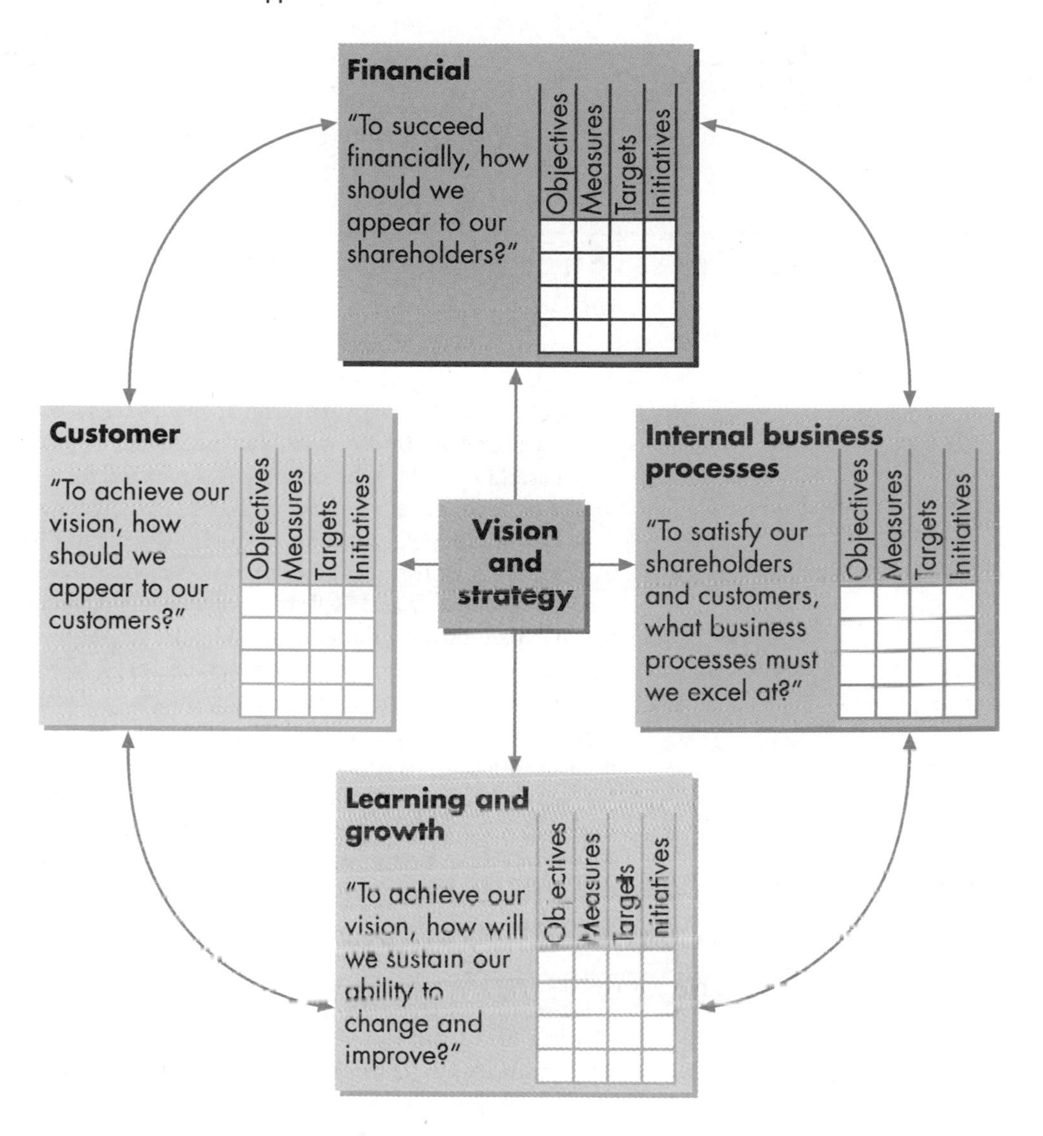

Source: "What is the balanced scorecard?" The Balanced Scorecard Institute. http://www.balancedscorecard.org/basics/bsc1.html.

metric
The measurement of a particular characteristic of a program's performance or efficiency.

A key element of the balanced scorecard approach relates to outcome metrics. A **metric** is the measurement of a particular characteristic of a program's performance or efficiency. In the balanced scorecard approach, metrics measure how the organization is doing compared to its defined objectives in all four areas. To carry out comparisons to outcome metrics, an information system must provide the required information to the management team; the most appropriate type of information system for this task is an information-based DSS. An information-based DSS tries to do exactly what is required by the balanced scorecard approach—provide information on operations that can be used to rate the organization's performance in the areas it has defined as important. You can find more information on the balance scorecard approach at www.bscnews.com/ or www.balancedscorecard.org.

1. Why have information-based DSS changed more than other types of DSS with the introduction of the Web?
2. What three types of reports are commonly used in an information-based DSS? How has wireless technology changed them?

Data-Based Decision Support Systems

When making decisions, employees and managers depend a great deal on Web-based information that has been developed by others. In many other situations, however, the decision maker has questions for which there is no available information. In these cases, the decision maker may turn to an analyst for help in finding answers to his or her questions in a database or data warehouse. Finding answers in this way requires some type of data-based DSS. This type of analysis is referred to as *ad hoc analysis* because the decision maker generates the questions based on a specific need or immediate problem.

The use of SQL queries to request information from a relational database, discussed in Chapter 5, constitutes one type of data-based DSS. In fact, Edgar Codd developed relational databases and SQL to satisfy the need for a system that could support queries to a database beyond those planned by the database designers—something that was not possible with the existing databases of the 1970s. In addition to SQL, two other methodologies are widely used for querying databases; online analytical processing (OLAP) and data mining. Both have become popular over the last few years for answering questions about data or finding previously unknown relationships in data.

Both OLAP and data mining are based on the data warehouse concept. To review, a data warehouse (or its smaller cousin, the data mart) is a subject-oriented snapshot of the organization at a particular point in time. It is typically used for analyzing past data rather than for handling current transactions and is designed to help decision makers understand more about their organization and its environment.

Online Analytical Processing

online analytical processing (OLAP) A software tool that enables an analyst to extract and view data from a variety of points of view.

Online analytical processing (OLAP) is a software tool that enables an analyst to extract and view data from a variety of points of view. It is used for planning and decision making using an analysis of existing data in a data warehouse. Typical applications include market analysis and financial forecasting. For example, in analyzing a market, an analyst at FarEast Foods might ask questions about units sold by market, time period, and product, such as "What were unit sales in the Seattle market for Fish Sauce during the third quarter?" Trying to answer this type of question with a relational database table like that shown in Figure 6.8 would be difficult, especially if the analyst wants to consider many combinations of time, product, and market. Note that in the figure only one of the three dimensions (time) varies, enabling the analyst to see the difference in units sold.

Figure 6.8 Table in a FarEast Foods relational database

Product	Market	Time	Units
Fish Sauce	Seattle	Q1	1185
Fish Sauce	Seattle	Q2	1303
Fish Sauce	Seattle	Q2	1521
Fish Sauce	Seattle	Q3	1779

multidimensional databases
A database with two or more dimensions in which each dimension represents one parameter that can be varied to determine an effect on a variable of interest.

Instead of using a tabular representation of the data, OLAP relies on **multidimensional databases**, in which each dimension represents one parameter that can be varied to determine an effect on a variable of interest. Varying multiple parameters produces a *n*-dimensional cube, in which each edge of the cube represents one dimension. For the current example, the variable is the number of units sold, and the dimensions consist of products, markets, and time. You can represent this scenario as a three-dimensional cube, as shown in Figure 6.9. In this example, the markets might be Seattle, Los Angeles, San Francisco, and so on; the time period might be First Quarter, Second Quarter, and so on; and the products might be Cloud Ears, Prawn Cracker, and so on. One value, therefore, might be 850 units of Cloud Ears sold in the second quarter in the Los Angeles market. The use of a multidimensional cube to represent data more closely matches, in many ways, the way that employees and managers actually think about their data. For example, it is more natural to view data by product and market over time than as a series of flat tables.

Figure 6.9 Multidimensional database for FarEast Foods

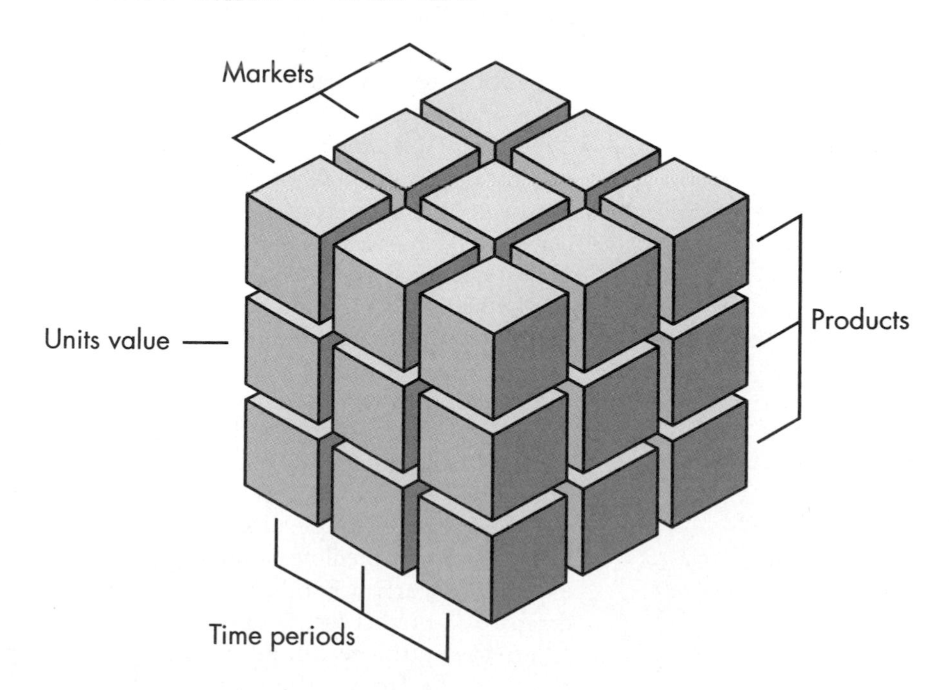

Table 6.5 Comparing OLAP and OLTP

Characteristics	OLAP	OLTP
Operation	Analyze	Update
Screen format	User-defined	Unchanging
Data tranformation	Considerable	Little
Level of detail	Aggegrate	Detail
Time	Historical, current, projected	Current
Orientation	Multidimensional cells	Records

Source: Paul Gray and Hugh J. Watson, *Decision Support in the Data Warehouse* (Upper Saddle River, NJ: Prentice-Hall, 1998), p. 125.

Multidimensional databases for use with OLAP are typically stored in data warehouses because they are used for analyzing historical data. In contrast, online transaction processing (OLTP) is used for updating databases. Because of the similarity in the names, the two are often confused. In reality, they are very different, as shown in Table 6.5.

In Table 6.5, you can see that the two systems have quite different purposes. The purpose of OLAP is to analyze data. The purpose of OLTP is to update a database as each transaction is processed. OLAP uses a screen format customized to the user's needs, whereas the OLTP screen is standardized. OLAP performs numerous calculations on the data to provide information to the user, whereas OLTP performs few, if any, calculations. OLAP works with aggregate (summarized) data that have been pulled from the detailed database maintained by OLTP. The time frame for OLAP is typically historical, with projections being based on this data. By comparison, OLTP works in a current time frame as it handles transactions in the present. Finally, like most database management systems, OLTP is record-oriented, whereas OLAP focuses on the contents of a multidimensional cell of the database.

slicing and dicing
A technique that enables the user to extract portions of the aggregated data and study them in detail.

Data in a multidimensional database can be analyzed in many ways. An analyst can choose to view data from different perspectives (for example, that of the product manager or of a financial analyst) by slicing through the cube from different directions. Commonly known as **slicing and dicing**, this technique enables the user to extract portions of the aggregated data and study them in detail. For example, product managers at FarEast Foods can study one product across many time periods and markets by slicing along the product dimension. Similarly, a financial analyst can study unit sales for all products and markets over one or more time periods by slicing across the time dimension. An analyst at FarEast Foods could use OLAP to determine the unit sales of Satay Sauce for Boston for a single quarter, for the entire year, or for as many quarters as data exist. Figure 6.10 shows market and time views of data in a multidimensional database.

In addition to slicing and dicing data in a multidimensional database to answer *who* and *what* questions about the data, OLAP gives the analyst the capability to drill down into the database to answer *what if* and *why* questions. These calculations are usually more complex than just summing data; they often involve projections of future results or predicted values that would result from a particular alternative. For example, at FarEast Foods, a manager might use OLAP with the multidimensional database to predict future unit sales of a particular product or in a specific market. Alternatively, this technique might be used to predict the effect on unit sales for next year if a product (for example, Plum Sauce) were discontinued.

Figure 6.10 Views of a mulitidimensional database

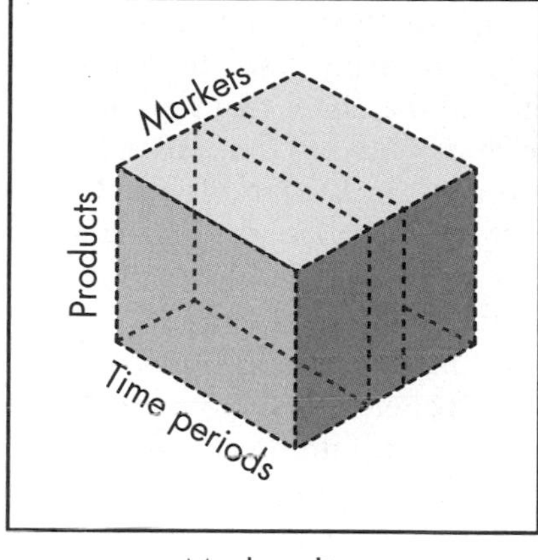

Market slice

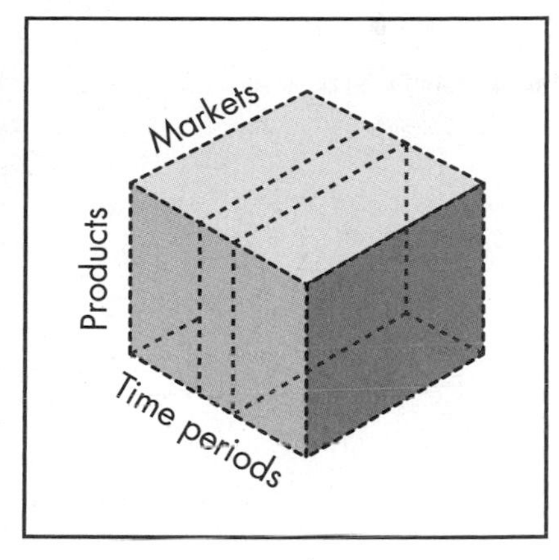

Time slice

William Inmon

IT INNOVATORS

Accurate and consistent databases are key to both data-based DSS and model-based DSS. In the late 1980s, however, it became obvious that the databases used in transaction processing would not meet these criteria for working with a DSS because they are constantly being updated. To solve this problem, various companies began to explore specialized databases, which were essentially snapshots of the TPS databases. They gave a variety of names to these databases, including the one that has stuck: data warehouse. In 1992, William H. "Bill" Inmon, a database technology veteran of many years, wrote *Building the Data Warehouse,* a book that codified the ideas behind the field and laid down the four basic rules for defining a data warehouse. Thanks to this early work, Bill Inmon has been dubbed "the father of data warehousing." As both OLAP and data mining depend on data warehouses, Inmon's work also bolstered their use.

Inmon has written 39 books on data management, data warehouses, design review, and management of data processing that have been translated into nine languages. A regular contributor to trade journals, he has more than 500 articles to his credit. Inmon founded and is currently the chief technology officer of Pine Cone Systems, a Colorado–based company whose mission is to create innovative software tools that address the complex and critical issues of managing and controlling data warehouse and data mart environments. Inmon conducts seminars and conferences in the United States, Europe, Canada, Asia, and Australia and holds two software patents for meta-data management.

William Inmon is commonly referred to as the "father of data warehousing."

Sources: http://www.pine-cone.com and private communication with William Inmon.

Data Mining

data mining
A search for relationships within the data in a database or data warehouse.

Another way to analyze data in a data warehouse, in which a specific database form is not required, is to use data mining. **Data mining** entails a search for relationships within the data. In many cases, the relationships discovered in this way are unexpected or are not the direct result of questions from decision makers. For example, an analysis of unit sales data for FarEast Foods might result in the discovery of a seasonal variation in the sales of wasabi. This recognition might lead the marketing manager to create e-mail marketing campaigns with special prices on this product to bolster its sales in typically slow months.

Data mining generally includes searching for associations, sequences, classification, clustering, and time-based relationship patterns or tendencies in the data. Table 6.6 shows these five patterns, the objective of searching for each, and relevant examples.

For example, data mining has proved very helpful to both customers and credit card companies by stopping the use of stolen credit cards. It does so by first finding patterns of typical credit use for the customer. Then, if use of a credit card in an atypical manner is detected (for example, for a very large purchase) the company can call the card holder and ask if he or she used the card for that purchase. If not, the sale is stopped and the card is canceled on the spot.

You can think of the process of data mining as a seven-step process, as shown in Figure 6.11. The process starts with the analyst understanding the goals of the use of data mining and selecting the target data for discovery. These two steps involve the analyst knowing what relevant knowledge about the application exists, what the goals of the data mining project are, and what data exist in the data warehouse. As a part of this process, the analyst lists all variables deemed relevant. In the third step, the analyst uses his or her knowledge to simplify the problem by reducing the number of variables being considered. The analyst then applies one of the data mining approaches—association, sequences, classification, clustering, and time-based relationships—based on his or her knowledge of the objectives of data mining.

Once the analyst selects the data mining approach, the actual data mining can begin by using a variety of software products developed specifically for this purpose. Data mining software is called **siftware** because it *sifts* through the data. Many of these products rely on statistical methods or clustering analysis, whereas others employ a variety of other methods.

siftware
Software used in data mining to sift through the data looking for elements that match the search criteria.

Table 6.6 Data Mining Approaches

Pattern	Objective	Example
Associations	Finding an event that is correlated to another event	Finding that a purchase of beer is often associated with a purchase of chips and nacho sauce
Sequences	Finding that the occurrence of one event leads to a second event	Finding that some percentage of all persons who become certified in scuba take a trip to an ocean resort within six months
Classification	Finding patterns that lead to recognizing rules about the data	Analyzing credit card utilization, which results in recognizing patterns of normal usage
Clustering	Finding new ways to organize data into groups	Grouping customers by the amount and frequency of purchases from Web sites
Time-based relationships	Finding relationships in the data that are time-based and can be used to predict future values	Finding that the sale of items associated with newborns increases nine months after a power outage

Figure 6.11 Data mining process

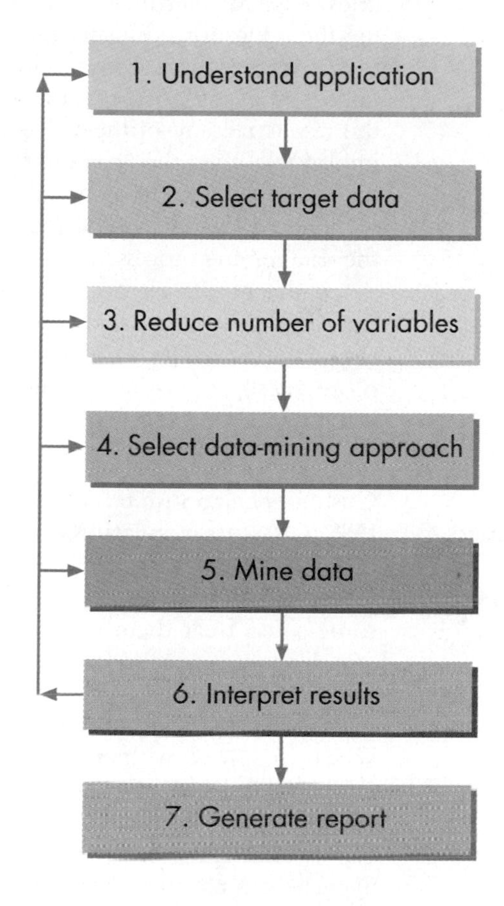
1. Understand application
2. Select target data
3. Reduce number of variables
4. Select data-mining approach
5. Mine data
6. Interpret results
7. Generate report

neural networks
Computer hardware, using multiple processors, and software systems that seek to operate in a manner modeled on the human brain.

artificial intelligence (AI)
Hardware and software systems that exhibit the same type of intelligence-related activities as humans, including listening, reading, speaking, solving problems, and making inferences.

One of the more useful data mining methods is a neural network. Popularized as the brains behind the cyborg in *Terminator 2,* **neural networks** are hardware and software systems that seek to operate in a manner modeled on the human brain. Like the expert systems discussed in Chapter 5 and applications such as game-playing computers, robotics, and computers that can understand commands in everyday language, neural networks represent a type of **artificial intelligence (AI)** in that they seek to replicate human intelligence and decision-making capabilities.

Neural networks differ from other software approaches to data mining in that the user does not have to decide in advance the approach or data to be used in the analysis. Instead, the neural network is *trained* to look for patterns in the data. As with training animals or educating humans, the learning process depends on repetition to train the software component of the neural network.

Neural networks are most useful for detecting patterns in data, and they have been applied in finance, speech and image processing, and data mining. In the financial area, neural networks have applications in risk analysis, fraud detection, and predictive modeling. In the data mining area, companies are using them to search through their databases for useful information. For example, a neural network could search through FarEast Foods' data warehouse for inactive customers who are most likely to purchase again if contacted by e-mail with a special "getting reacquainted" offer.

Once the data mining has yielded its results, the analyst must again use his or her knowledge to determine whether the results are meaningful. This effort involves checking the results for possibility, internal consistency, and plausibility. *Possibility* means that the results are physically possible. *Internal consistency* means that the results are not contradictory. *Plausibility* means that the association or relationships found are believable. If the results fail any of these tests (for example, if they are impossible, inconsistent, or implausible) then the users can rerun the analysis with a new target data set, a different set of simplified variables, or a different data mining approach (for instance, looking for sequences rather than associations). Figure 6.11 shows this capability to *iterate* through the data mining process in the form of the return arrows to the various steps.

If the results are possible, consistent, and plausible, the analyst then interprets them and generates a report to the decision maker who asked the original question about the data. The decision maker can then take actions based on the results, ask more questions, or do nothing, depending on the situation.

Data-Based DSS in Action: Customer Relationship Management

customer relationship management (CRM)
Techniques applied to organizational data with the goal of segmenting customers in order to serve them better.

Customer relationship management provides an excellent example of a data-based DSS. **Customer relationship management (CRM)** applies a variety of techniques to organizational data with the goal of segmenting customers in order to serve them better. Although the idea behind CRM is fairly simple—know your customers and treat them uniquely—the implementation is much more complicated and depends on the use of a data-based DSS working with a data warehouse. [1]

CRM attempts to create a 360-degree view of customers by looking at more than just sales transactions. Companies should collect data and information on every interaction they have with customers, whether or not it generates a sale. Obviously, this effort involves collecting, storing, managing, and analyzing massive amounts of data on each and every customer. Armed with the 360-degree view of customers, companies can learn to optimize future interactions based on past interactions. For example, credit agencies can use their data warehouses to find individuals who tend to make only the minimum payment on a credit card and sell their names to credit card companies. The credit card companies can then send preapproved credit card applications to this group of consumers, knowing in advance that, if the individuals use their card, the customers will end up paying a large amount in interest and finance fees. Similarly, CRM allows the jeweler at your local mall to advertise more efficiently by mailing flyers only to individuals who live within 10 miles of the mall and have an income of at least $100,000. Also, if a company has access to multiple data warehouses, either by purchasing access or by entering into reciprocal agreements with other companies, it becomes possible to combine data from them to produce needed pictures of consumers.

The process of implementing CRM involves four key phases:

1. Capture
2. Analyze
3. Plan
4. Interact

touchpoint system
A contact point through which a company interacts with the customer, including the Web, e-mail, personal sales, direct mail, call centers, and so on.

In the *capture* phase, the CRM system extracts customer data from operational systems and goes through the necessary steps to store the data in a data warehouse. This phase is carried out through any of multiple touchpoint systems. A **touchpoint system** is a contact point through which a company interacts with the customer, including the Web, e-mail, personal sales, direct mail, call centers, and so on.

---

1. Much of this discussion is based on *Harnessing Customer Information for Strategic Advantage: Technical Challenges and Business Solutions*, edited by Wayne Eckerson and Hugh Watson and published by The Data Warehousing Institute, 2001.

In the *analysis phase*, the data are used to develop reports on customer behavior, segment the customers into groups, and create predictive models upon which future marketing plans are based.

In the *planning phase*, the results of the two previous phases are used to generate rules for optimizing customer interactions. The goal is to deliver the right offer to the right customer at the right time through the right contact points.

Finally, in the *interaction phase*, the organization implements the rules developed in the planning phase in the same touchpoint systems from which data were captured in the capture phase. During the interaction phase, the company strives to optimize customer interactions without annoying the customers. Data are then captured at these touchpoints, and the process starts again, as shown in Figure 6.12.

Note that the four phases are wrapped around business applications, which are in turn wrapped around the customer database. For example, consider a casino that first mines its warehouse data to identify market segments (such as young, male, high-velocity players who visit a casino for the first time and have high lifetime potential value to the casino). For a particular marketing campaign, the company scores the customers in the data warehouse to see whether they fit the profile for the campaign. If so, they join a list of people slated to receive the campaign offer (such as $50 in free chips and a complimentary night's stay if they visit the casino in the next two weeks). The list of IDs is then passed to the operational data store to pick up contact information and the offer is sent out to them.

Note the similarity between this diagram and the IS cycle model shown in Figure 6.1. The capture and interaction phases match up with handling the present and remembering the past, and the analysis and planning phases match up with preparing for the future.

1. How do neural networks differ from other computer systems?
2. How does the database used by OLAP differ from that used in an SQL query?

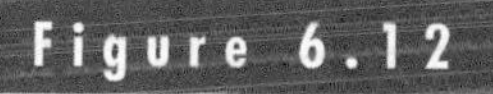

CRM phases

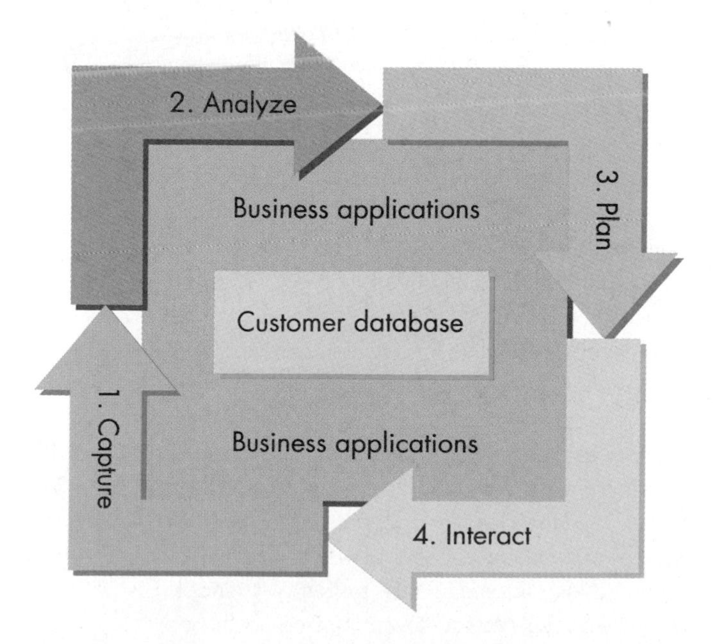

Source: *Harnessing Customer Information for Strategic Advantage: Technical Challenges and Business Solutions*, edited by Wayne Eckerson and Hugh Watson (Seattle: The Data Warehousing Institue, 2001).

Model-Based Decision Support Systems

Information-based DSS aim to provide decision makers with information from internal and external sources. Data-based DSS analyze data from data warehouses so as to answer questions from decision makers. In neither case do the systems provide alternative solutions to the problems they might reveal. A third type of analysis system, known as model-based DSS, is specifically designed to help managers find such solutions. A model-based DSS explores alternative actions through the use of quantitative and graphical models.

Model-based DSS have been used for diverse activities, such as ski resort development, financial planning, and school bus routing. A model-based DSS can be as simple as a spreadsheet on a PC or as complex as a full-scale financial planning and forecasting package on an application server or workstation.

Parts of a Model-Based Decision Support System

model base
A collection of models that can be used to analyze the data from a database.

A model-based DSS has three major parts: the database, the model base, and the user. The database for the model-based DSS contains the data necessary to carry out the needed analyses. The **model base** contains a variety of types of models that

TECHNOLOGY ON THE EDGE

Retaining Customers at Mazda with CRM

An important measure of customer satisfaction for automobile manufacturers is customer loyalty—that is, customers who purchase the same brand of new car as they currently own. For Mazda, getting its CRM program under control is critical to improving customer loyalty. Prior to mid-2000, Mazda had more than 60 programs that touched the customer in one form or another. As a result of this fragmented approach, Mazda had not developed a cohesive understanding of its customers. To remedy this problem, the company licensed CRM software from E.piphany Inc. Installing the software involved cleaning and integrating Mazda's four major databases: marketing, vehicle, service, and warranty. For example, the marketing database was sorted by customer name, whereas the vehicle database was sorted by vehicle ID number. Just carrying out this preliminary step of the CRM process is a huge job, but has the potential to save the company more than $1 million by reducing duplication of effort.

Once its database duplication problems are resolved and the databases are integrated, Mazda hopes to build a cohesive picture of its customers, segmenting them by their value and loyalty to the company. The overall objective of the project is to provide call-center representatives with a complete view of each caller's history with Mazda on their computer screen. Combined with the power to make decisions to retain loyal customers, this system will enable Mazda to identify a loyal Mazda owner (having purchased multiple cars) who is having warranty service problems. This person would be tagged as a retention risk, and the company could take proactive action to keep him or her as a loyal owner. This effort might involve extending goodwill dollars and granting work under warranty, even if the vehicle is 1000 miles beyond the warranty period.

Mazda hopes that the new system will make all necessary information available to customer service representatives on one screen rather than several.

Source: Mitch Betts, "Mazda wants 360-degree view of customers." *Computerworld*, February 12, 2002, http://www. computerworld.com.

user
In a model-based DSS, the person working with the models and data to generate alternative solutions to a problem.

can be used to analyze the data from the database. The **user** is the decision maker or analyst who seeks the solution to a problem.

data management system
The part of a decision support system that retrieves information from a database as needed.

model management system
The part of a decision support system used to select a model that can help find a solution to a problem.

user interface system
The part of a decision support system that handles the interactions between the analyst and the computer.

To make a model-based DSS work, the user must select a model and data to use in analyzing the problem. Three software systems make this possible: a **data management system**, which manages the retrieval of data from a database as needed; a **model management system**, which helps to select a model that can be used to arrive at a solution; and a **user interface system**, which handles the interactions between the analyst and the computer.

The data management function must be available to the analyst so that he or she can retrieve needed information from the company's database on demand. The analyst also selects a model using the model manager and combines the data and the model to develop alternative solutions. The models comprise a group of computer programs that will perform specific operations on the data from a database. The various alternative solutions can then be portrayed in tabular or graphical form. To make a model-based DSS easy to use, the user interface allows the decision maker to change data and models as needed. For example, if the user requests a graph of data on the screen, the user interface allows the user to select the data to be graphed and the model to be used to create the graph. After selection of a model, the user interface combines the model with the data to arrive at the desired graphical display. To a large extent, the user interface should make the data management and model management functions invisible to the user by handling all of the interactions. The user interface turns the model-based DSS into a truly useful tool by allowing the analyst to move about freely within it.

macro language
A computer language built into personal productivity software, such as spreadsheets, that extends their capabilities.

Note that modern spreadsheets have all of these capabilities. A user can combine a spreadsheet's database capabilities with easy-to-create models. The use of a **macro language**, which is a built-in computer language that enables the building of custom-designed spreadsheet applications, makes it possible to turn a spreadsheet into a powerful model-based decision support tool. Figure 6.13 shows the interaction between the user and the parts of the model-based DSS using the various software systems.

Using Models

model
A simplified version of a system that allows an analyst to understand the system's important parts.

In addition to obtaining needed information from an information-based DSS or data from a database or data warehouse using a data-based DSS, the decision maker or analyst needs to use software to determine the effects of changes in the information or data. To observe these changes, a model must accurately represent the physical, economic, or financial situation under study. A **model** is a simplified version of reality that captures the interrelationships among important variables in the situation. Once the model has been created, the analyst may then use it to answer questions.

Figure 6.13 Interaction between user and parts of a model-based DSS

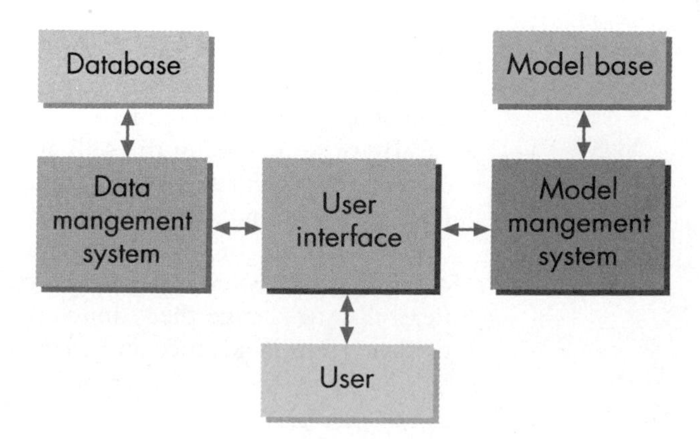

forecasting model
A process that uses currently available information to predict future occurrences.

statistical model
A model in which the objective is to learn about tendencies within the data set or to prove that differences exist between parts of the data.

optimization
The use of a mathematical technique to find the best solution to a model.

simulation
The use of probability-based models to imitate a real phenomenon.

Four primary types of models exist: forecasting, statistical, optimization, and simulation. A **forecasting model** combines historical data, assumptions, and a forecasting formula to predict what the future may hold. For example, companies constantly use forecasting models to predict sales based on various independent variables, such as previous sales, current orders, and economic conditions.

In a **statistical model**, the objective is to learn about tendencies within the data set or to prove that differences exist between parts of the data. For example, companies such as Manheim Auctions, which was discussed in Chapter 4, use statistical models to determine the value of a used car based on age, miles driven, and condition.

In **optimization**, the user implements a mathematical equation that will determine the best possible solution to the model. For example, you may wish to find the maximum profit associated with a particular product when production is constrained by the availability of raw materials, labor hours, and so on.

Finally, **simulation** uses probability-based models to imitate a real phenomenon. Usually, simulation is performed when the model developed proves too complex to respond well to optimization techniques or when the problem calls for more than one acceptable solution. With a computer, you can simulate several years of conditions and results in just a few seconds. After compiling and analyzing the results, you can use them to determine the effect of a change in one or more of the independent variables in the model. For example, suppose an airline was thinking about a new pricing plan that would allow travelers to fly anywhere in the country during a one-month period for a flat price of $1000. Simulation would allow the company to investigate the ramifications of this plan on its operations. As a result of the simulation, the airline might decide to increase the price, lower the price, or drop the idea altogether.

Model-Based DSS in Action: Revenue Management

revenue management
The application of disciplined tactics that predict customer demand at the micro-market level and optimize product price and availability so as to maximize revenue growth.

As an example of a model-based DSS, consider the use of revenue management to increase revenue for a wide variety of companies, including airlines, rental cars, and hotels to name just a few. If you have flown commercially, rented a car, or stayed in a hotel or motel room, you have been affected by the application of this model-based DSS because all of these industries use revenue management to set their prices. As defined by the father of revenue management, Robert G. Cross, **revenue management** involves the application of disciplined tactics that predict customer demand at the micro-market level and optimize product price and availability to maximize revenue growth.[2] When practicing revenue management, companies work to sell the right product to the right customer at the right time through the right sales channel at the right price. (Note the similarity to the objectives of CRM.) Proponents of revenue management note that whereas more than 90 percent of most executives' time is spent on cost-related or capital expenditure issues, revenues are largely determined by price. You can see the importance of price in Table 6.7, which is based on the results of a study by McKinsey and Company. The table compares the effects on profit of 5 percent improvements in sales expense (cost), sales volume, and selling price, assuming that all other factors remain constant. Note that a 5 percent decrease in sales-related expenses (cost of goods sold, travel, and so on) increases profits by only 3 percent, whereas a 5 percent increase in selling price yields a 50 percent increase in profits.

Because increasing the selling price has such a dramatic effect on profits, revenue management seeks ways to increase price rather than seeking to reduce costs. It does so by using price discrimination to distinguish among different types of customer demand. In the airline industry, revenue management takes the form of charging business customers who buy their tickets at the last minute, but want the flexibility to change them, much more than leisure customers who must buy their tickets weeks in advance and have no flexibility to change the tickets.

2. Robert G. Cross, *Revenue Management Hardcore Tactics for Market Domination*, (New York: Broadway Books 1997), p.51.

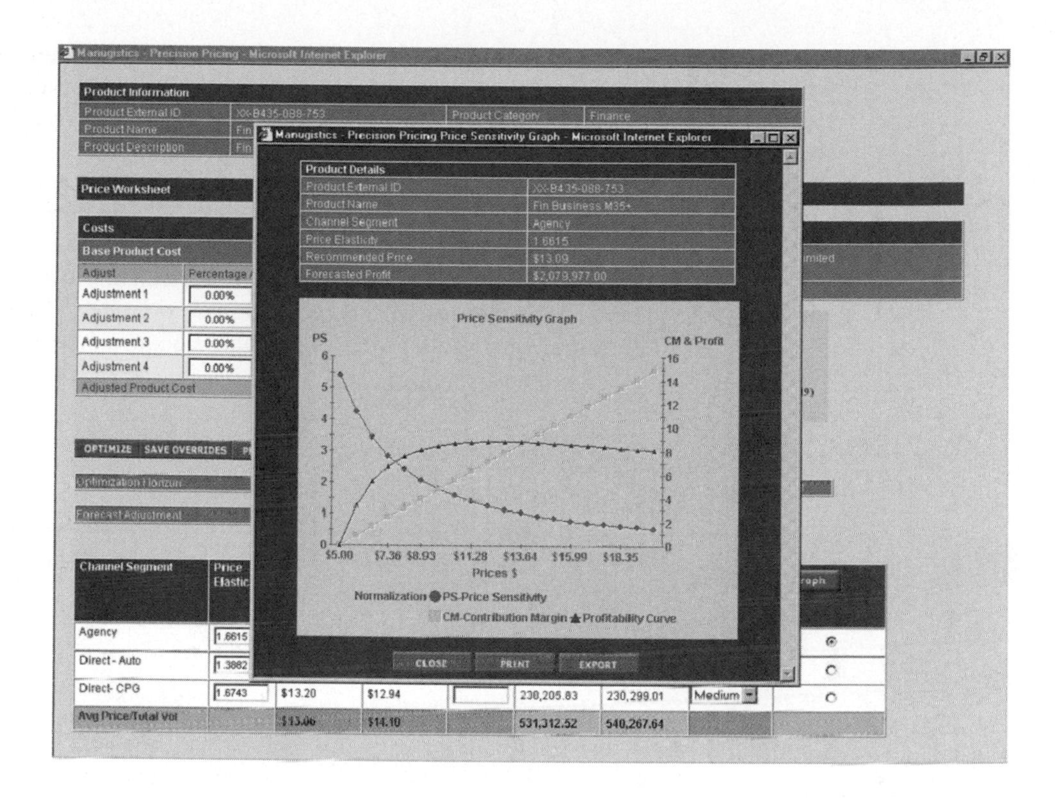

Manugistics uses sophisticated software to help its clients determine prices that will increase their revenue.

Revenue management consists of a four-step process:

1. Segment the market
2. Measure the response by market segment
3. Optimize price to maximize profit
4. Dynamically recalibrate to yield long-term profit maximization

In the first step, customers are segmented according to price-response characteristics. Segmenting the market requires a central repository of corporate market knowledge, like a data warehouse, to determine how customers will respond to differing prices. In the airline example, passengers are broadly segmented into business-oriented travelers and leisure travelers because each group has different responses to price.

In the second step, the company uses its data warehouse to determine how each customer segment will respond to different prices. Essentially, this step predicts customer demand based on historical data from the data warehouse. For example, will leisure travelers increase their airline trips to Florida if the company reduces ticket prices to that destination?

Table 6.7 Effect of a 5 Percent Change in Profit Elements on profits

Item	Increase in Profit
5 percent decrease in sales expense	3 percent
5 percent increase in sales volume	20 percent
5 percent increase in selling price	50 percent

Measuring the market response by market segment requires a thorough analysis of price and demand data to determine how each and every segment will respond to price changes. Once the market response of each segment is known, the firm can use optimization models to determine how much of each product should be sold at each price. In the airline example, the optimization model determines how many seats should be priced for leisure travelers and how many should be priced for business travelers.

Finally, when the prices generated through the optimization model are applied to the company's product, data are collected and used to revise the market segments and the estimates of market response to prices. Prices are recomputed using the optimization model and applied again. Figure 6.14 graphically depicts this process.

1. List the three models most often used in a model-based DSS.
2. What type of model is used in revenue management?

Figure 6.14 Revenue management process

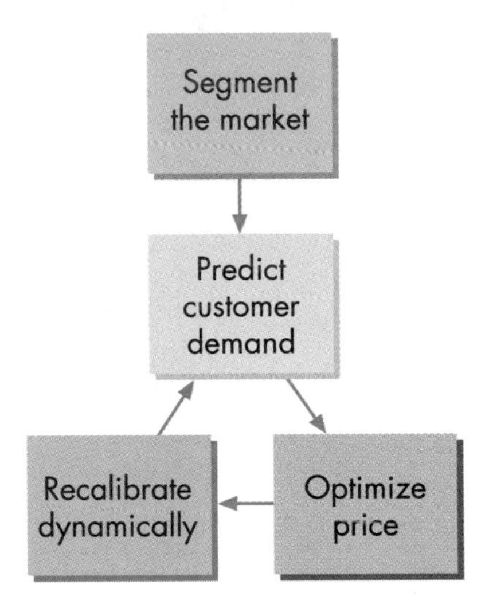

CaseStudy

Flight Scheduling at United

In addition to using revenue management as a tool for increasing revenues, airlines are always looking for other ways to increase revenue or decrease costs. One continuing point of interest is scheduling aircraft. Because several hundred thousand passengers are traveling each day between many different airports on a variety of types of aircraft with potential stopovers in intermediate cities, scheduling the right aircraft to the appropriate route is a huge problem for airline companies. These problems are so complicated that they can be solved only through computer modeling. To understand this scheduling problem, consider that each route between cities may be composed of one or more legs—a leg is a flight from one airport to another. For example, you might fly from Houston to New York on legs from Houston to Atlanta and from Atlanta to New York, flying on the same plane or different planes on each leg. The airline company wants to match up particular aircraft with demand on each leg in such a way as to maximize revenue and minimize costs at the same time.

Airline companies have traditionally used two models—the flight assignment model and the through assignment model—to solve this problem. Airlines use the flight assignment model to assign aircraft to legs based on seat demand. They use the through assignment model to assign the same flight number to two individual legs so that passengers have no need to change planes. Because travelers are willing to pay more to avoid changing planes, the airline company wants to create as many of these "through connections" as possible. Until recently, it was not possible to combine these two models into a single model because of the extreme complexity of the resulting problem. However, researchers at the University of Florida and Massachusetts Institute of Technology have created a solution technique that can solve such problems on a personal computer within a few seconds.

Called the Very Large Neighborhood Search technique, this method starts with a workable solution and then searches for better solutions. United Airlines has provided funding for the researchers to study its scheduling problems, and they have been able to increase the number of through connections, which could produce an increase in revenues of as much as $25 million a year. United is very excited about this solution methodology and is planning on implementing it in their scheduling system.

Source: R.K. Ahuja, J. Goodstein, A. Mukherjee, J. B. Orlin, and D. Sharma. "A very large-scale neighborhood search algorithm for the combined through-fleet assignment model." Technical Report. Industrial & Systems Engineering, University of Florida, Gainesville, 2001.

Think About It

1. How is this problem like or unlike the revenue management problem discussed in the text?
2. Would you apply revenue management techniques before or after the combined through-fleet assignment problem has been solved? Explain your answer.
3. Assume that Nothing But the Best Airlines (NBBA) has a hub in Denver (all flights go through Denver) and flies routes between Atlanta and Los Angeles, Atlanta and San Francisco, Houston and Los Angeles, and Houston and San Francisco. Determine the number of possible legs and number of possible through routes for this problem. What happens to the problem if NBBA has only two aircraft that can each fly two legs a day?

SUMMARY

To summarize this chapter, let's answer the questions posed at the beginning of the chapter.

What is a decision support system (DSS)? A decision support system is an information system that provides a decision maker with information, data, or answers to questions. It helps the organization prepare for the future by using data from organizational memory as well as other types of data and information.

How are decisions made? The decision-making process includes seven steps:

1. Define the problem
2. Gather information
3. Identify alternatives
4. Evaluate alternatives
5. Select an alternative
6. Implement the alternative
7. Monitor the results

During this process, the decision maker needs to access the necessary information to size up the problem or the competitive opportunity, assess the alternatives under consideration, decide on selection criteria, evaluate the alternatives, implement a course of action, and monitor compliance with the decision. Decisions can be either structured (programmed) or unstructured (unprogrammed), depending on whether predefined rules are applied. Decisions can also be defined as operational, tactical, and strategic depending on who is making them and how they affect the organization.

What types of decision support systems are in use today, and who uses them? Three types of decision support systems exist: information-based DSS, data-based DSS, and model-based DSS. An information-based DSS enables individuals at all levels of the organization to view information needed to make decisions, usually through intranet-based Web pages. A data-based DSS enables analysts within the organization to use data in databases and data warehouses to answer questions about the organization's operations. The questions are answered through a variety of approaches, including SQL, online analytic processing (OLAP), and data mining. A model-based DSS combines data from the database or data warehouse with mathematical models to allow analysts to answer questions asked by management. This effort often involves generating forecasts, carrying out statistical analyses, searching for the best allocation of resources, or simulating situations. Intranet-based enterprise information portals (EIP) allow the individual seeking help with a decision to use all three types of information systems without having to worry about which one is being used.

How does an information-based DSS support decision making? Information-based DSS include many types of information systems that have the same goal: to provide decision makers with the information they need in an appropriate form. The primary method for providing this information is via a Web browser over an intranet. It supports decision making by gathering the types of information required at all steps of the decision-making process—namely, information on problems in current operations, information on new opportunities available to the organization, and information on threats to the continued existence of the organization. Virtually every type of information-based DSS generates three types of reports: scheduled reports, exception reports, and demand reports. Each serves a different purpose to the managers and employees of the organization. Scheduled reports reflect periodic and historic information on the organization's operations. Exception reports are generated only if some condition—usually an exceptional event—activates the generation of a report. A manager may request specialized demand reports when he or she needs specific information on a particular subject.

The balanced scorecard is a good example of how management can use an information-based DSS to provide overall direction for the organization.

In what ways does a data-based DSS provide answers to questions? When a decision maker has questions for which there is no available information, he or she can use some type of data-based DSS to find the answers. Data-based DSS include SQL, OLAP, and data mining. Both OLAP and data mining are used to access data in a data warehouse. OLAP is a software tool that enables an analyst to extract and view data from a variety of points of view using a multidimensional database. It can involve a slice and dice approach to answer *who* and *what* questions or a drill-down approach to answer *what if* and *why* questions about the data. In contrast, data mining provides a way to search for relationships within the data in a data warehouse. Typical data mining approaches include searching for associations, sequences, classifications, groups, or time-based relationships in the data. Neural networks represent one method of implementing data mining.

Customer relationship management (CRM) is a good example of the use of a data-based DSS to learn about customer attributes and needs and to develop a marketing plan that more precisely targets the customer.

How does a model-based DSS answer questions from decision makers? A model-based DSS uses quantitative and graphical models to find solutions to problems facing decision makers. A model-based DSS has three major parts: the database, the model base, and the user. The database contains the data necessary to carry out the needed analyses.

The model base contains a variety of models that can be used to analyze the data from the database. The user is the decision maker or analyst who seeks the solution to a problem. Three software systems—a data management system, a model management system, and a user interface system—support a model-based DSS. The data management system manages the retrieval of data from a database as needed. The model management system selects a model that can be used to help arrive at a solution. The user interface system handles the interactions between analyst and computer. Models, which are simplified versions of reality that capture the interrelationships between important variables in the situation, are a group of computer programs that perform specific operations on the data from a database. Four primary types of models exist: forecasting, statistical, optimization, and simulation.

Revenue management offers a good example of a model-based DSS. It aims to increase revenues by determining prices that match the demands of differing segments of a customer base.

REVIEW QUESTIONS

1. How do decision support systems fit into the information systems cycle?
2. What is another name for a decision support system?
3. List the three types of decision support systems in use today. Which system was once the only one called a decision support system?
4. What should you avoid during the problem definition step of decision making? Why?
5. What is another name for a structured decision? An unstructured decision?
6. What types of decisions are normally associated with day-to-day operations of the organization? What types of decisions are used to implement policy?
7. How do the users of decision support systems differ? What was once believed about the relationship between managerial levels and types of DSS used at those levels?
8. What is the difference between a Web-based DSS and an installed-software DSS?
9. What is an EIP, and what is its purpose?
10. What two types of information systems are included under the name information-based DSS?
11. Name the three types of reports generated by an information-based DSS. Explain the purpose of each.
12. How has the way in which reports are provided to users changed over the last few years?
13. What two methodologies are typically included in a data-based DSS? Where do the data originate?
14. What is the difference between slicing and dicing and drilling down in an OLAP?
15. What five patterns are commonly searched for in data mining? How are neural networks used in a data-based DSS?
16. How does CRM fit the concept of a data-based DSS?
17. What are the three key components in a model-based DSS? How are they implemented?
18. What role does a model play in a model-based DSS?
19. Name and discuss the four types of models commonly used in DSS. Which one tries to find the best solution to a problem?
20. What are the four steps in the revenue management process?

DISCUSSION QUESTIONS

1. Discuss the following statement: The only way to avoid making bad decisions is to have the right data, information, or answers to questions.
2. Discuss the concept of an enterprise information portal as it relates to decision making.
3. Discuss the roles of analyst and chauffeur in the use of decision support models.
4. Because today's information-based DSS focus on providing reports in Web format, they have become a less visible part of an organization's information system. Does this trend indicate that DSS is losing its importance to the organization?
5. Discuss one of the four types of models used in a model-based DSS.

RESEARCH QUESTIONS

1. Visit the Web site for the EIP portal for Brio at www.brio.com/products/brio_portal/index.html and prepare a short presentation on it.
2. Visit the Web site for the Corrosion Source at corrosionsource.com and write a two-page paper on what you find there.
3. Visit the Web site for Re:cognition, a CRM product from Lumio at www.lumio.com/. Determine what the product does, and write a two-page paper on it.
4. Visit the Web site for Manugistics, a leader in the field of revenue management (or, as the firm calls it, Pricing and Revenue Optimization), and visit the page dedicated to this topic. Write a two-page paper on the services that Manugistics offers clients in this area.

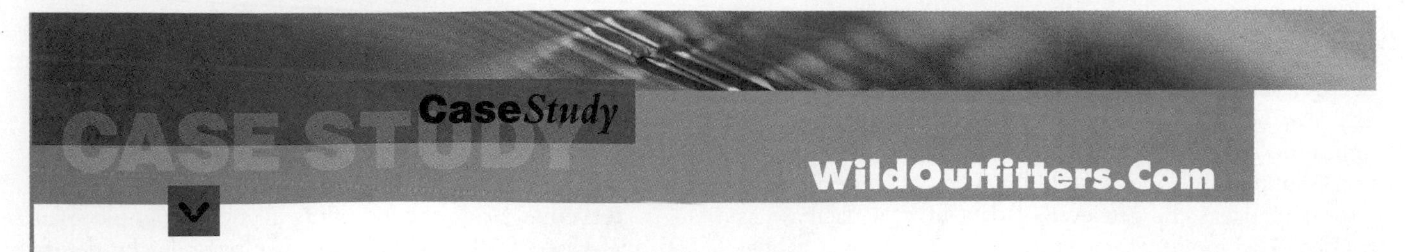

WildOutfitters.Com

"I'm telling you, it was the funniest thing that I ever saw! I don't know how that outfit stays in business with service like that..."

Alex was leaning over the counter of the Wild Outfitters booth at the Gauley Riverfest, as one of his raft guides described what he had seen that day on the river. The Gauley Riverfest is an annual festival put on by rafters and paddlers timed to correspond with the release of water into the Gauley, a Class IV–VI river in West Virginia. Water is released during four weekends in the fall of each year. The festival has developed into a trade show for river-running equipment and services, as well as a chance to party after a day of paddling and rafting on a world-class whitewater river. The Campagnes had set up a booth with the goals of advertising their presence in the area and selling a few products.

Alex was only half-listening to the guide. His mind had begun to wander to the subject of the Campagne's own river-guiding services. For some time, Alex and Claire had been offering guided raft trips. They were now thinking of expanding into adventure travel packages. Typical adventure travel packages revolve around an outdoor activity, such as whitewater rafting or mountaineering. Based on their own experience, they believed that they were qualified to add these adventure travel packages to the Wild Outfitters product line. The problem was that the revenue from the raft trips barely covered their costs. The Campagnes wanted to find a way to make the rafting service more profitable before expanding into other types of trips.

The raft-guiding service had grown quite a bit in the last months. Currently, Wild Oufitters was running as many as 15 trips per day on three different rivers during peak season. Each trip included five or six 8-person rafts carrying customers and guides—one guide per raft. The customers, rafts, and equipment were transported from the Wild Outfitters store to the river on buses customized for that purpose. Until now, they had been charging a flat rate of $30 per person.

Alex smiled as he thought of the diverse clientelle that he and Claire had the pleasure to serve. Just this morning Claire had led a group of businesspeople who were on a corporate retreat. They both agreed that the corporate groups were fun but they usually required more supervision than scout troops. Most of their clients tended to be vacationers. Some made reservations ahead of time, whereas others were in the area and found a way to fit in a raft trip for fun. No matter what the customer's background, Claire and Alex had been able to provide a safe and enjoyable experience.

"How they could have the rafts, the guides, and the clients at the river and ready to go without the rest of the equipment I'll never know," the man went on. Laughing, he gave Alex a good-natured smack on the back, breaking him out of his reverie. "I've heard of being up the creek without a paddle, but never thought I'd see it with my own eyes." Alex decided to leave the business thoughts behind and began to pick up the conversation again. After all, Gauley Riverfest only happens once a year.

Think About It

1. Could Wild Outfitters benefit from the use of revenue management? Explain why or why not.
2. List the possible market segments discussed in the case. Can you think of others?
3. What types of information systems would the Campagnes need to obtain to implement revenue management for their raft-guiding services? For each system that you describe, discuss the data required and its source, the purpose of the system, and the types of models used, if any.

Hands On

4. Conduct an Internet search for the key term "revenue management." Based on your search, is this topic popular? What types of businesses use revenue management? For what types of products?

ELECTRONIC COMMERCE: STRATEGY AND TECHNOLOGY

Electronic commerce, which involves carrying out transactions over networks, has become a growing part of the world economy. It is not restricted to the Internet, however—many transactions occur within organizations or between trading partners. A study of electronic commerce should look at both strategy and technology. Electronic commerce strategy deals with what companies should be doing to use electronic commerce to reach their goals, whereas electronic commerce technology deals with how companies go about this endeavor.

Chapter 7 covers electronic commerce strategy, and Chapter 8 focuses on electronic commerce technology. After reading these two chapters, you should have a good idea of what electronic commerce can do and how to do it.

CHAPTER 7

ELECTRONIC COMMERCE STRATEGIES

LEARNING OBJECTIVES

After reading this chapter, you will be able to answer the following questions:

- What is electronic commerce, and how does it relate to the networked economy?
- What types of Web sites exist, and how do they relate to electronic commerce?
- How is the nature of electronic commerce changing, and what are the Internet effects?
- What electronic commerce strategies should be used to deal with price transparency and customer demand for perfect choice?
- How can electronic commerce strategies be used to take advantage of the communications flip-flop?

Electronic Commerce at Dell Computers

If there were a poster child for electronic commerce, it would be Dell Computers. Started in 1984 by 19-year-old college student Michael Dell in his Texas dorm room, Dell has achieved great success in the personal computer market by selling directly to its customers and eliminating the middleman. Dell initially sold its products through telephone orders, but more recently has aggressively moved into electronic commerce as a way of connecting with its customers. Total direct sales in 2001 topped $80 million *per day*, with 50 percent of all sales conducted via the Web. While overall shipments of PCs were declining by 5 percent worldwide, with 125 million units shipped, and declining by almost 13 percent in the United States, with 42 million units shipped, Dell's shipments actually increased to nearly 17 million PCs. This increase enabled Dell to supplant Compaq as the number one manufacturer of PCs. Also, while many companies showed losses for the year because of the global economic slowdown, Dell managed to remain profitable despite reduced revenues due to its lower prices. In the home market, Dell has grown from being eighth in U.S. market share in 1999, to being first in market share, with over 20 percent of the home market at the end of 2001, beating out the previous leader in home computers, Hewlett-Packard.

The Dell Web site (www.dell.com) has made the direct sales model the heart of its approach to electronic commerce. By providing greater convenience and efficiency to its customers, the Web site simplifies the process of selecting and purchasing a computer. Dell custom-builds every computer so that customers get exactly what they want and, by having no middlemen with old inventory to push, the customer always receives the latest in technology. Although Dell serves all markets, it has proved extremely successful in the corporate market by creating "Premier" Web pages for each of its large customers to facilitate firms' purchasing processes. Dell has also improved its support services by making the Web work as a front end to its internal databases so that its customers and suppliers can see the same support information it uses internally.

Michael Dell has used electronic commerce as a part of his company's strategy to become the top-ranked computer maker in the world.

Sources: Matt Hamblen, "Update: PC market declines in 2001; slow turnaround expected." *Computerworld*, January 18, 2002, http://wwws.computerworld.com; and Mikako Kitagawa and Charles Smulders, "Battle rages in US home PC market." *Gartner Research*, February 15, 2002, http://www.gartner.com.

Electronic Commerce and the Networked Economy

In Chapters 4, 5, and 6, you considered how organizations use various types of information systems, including transaction processing systems, database management systems, data warehouses, and decision support systems, to handle the present, remember the past, and prepare for the future. Chapters 7 and 8 will complete that discussion by considering how businesses and organizations around the world are taking advantage of an exciting application of information systems—electronic commerce. This chapter discusses strategies for using electronic commerce, whereas Chapter 8 covers the technologies that underlie this application. In a sense, this chapter focuses on what companies should be doing with electronic commerce, and Chapter 8 explains how they are doing it.

To understand the need for an electronic commerce strategy, you must first understand business strategy. In general, **business strategy** consists of the long-term plans that describe how a firm will achieve its desired goals. Usually, the overriding goal of any business or organization is to survive. Subgoals that help a business achieve this goal include increased profitability, increased revenue, lower costs, and so on. For example, as discussed in the opening case, Dell's business strategy involves selling computers directly to customers rather than going through retailers. Based on this concept of strategy, **electronic commerce strategy** is the manner in which electronic commerce is used to further the goals and aims of the business or organization.

business strategy
The long-term plans that describe how a firm will achieve its desired goals.

electronic commerce strategy
The manner in which electronic commerce is used to further the goals and aims of the business or organization.

Defining Electronic Commerce

Electronic commerce is the process of carrying out business transactions over computer networks. Notice that the definition is not restricted to the Internet and Web, because a firm may use other types of networks or network protocols to carry out transactions as well. Increasing profitability, gaining market share, improving customer service, and delivering products faster are some of the organizational performance gains made possible by electronic commerce. However, electronic commerce entails more than simply ordering goods from an online catalog. It involves all aspects of an organization's electronic interactions with its **stakeholders**, the people who determine the future of the organization. Thus, electronic commerce includes many activities, such as establishing a Web site to support investor relations and communicating electronically with college students who represent potential employees. For example, Dell Computers has links on its Web pages for all stakeholder groups, including customers, suppliers, investors, potential partners, and so on, and the company has found ways to link customers and suppliers directly into its customer support databases.

stakeholders
The people who determine the future of the organization, such as stockholders, employees, and customers.

Electronic commerce should not be an altogether new concept to you because each of the first six chapters of this book mentioned it. Many of the cases at the beginning or end of the chapters have described the use of electronic commerce to provide customers with online access to goods or services. In addition, the book's example company, FarEast Foods, uses electronic commerce to market, sell, and distribute a variety of Asian food items. Whether it is called e-commerce, e-business, e-tailing, or some other name, electronic commerce is changing the business world by providing a new outlet for goods and services.

Whereas at one time the debate in most businesses focused on *whether* to engage in electronic commerce, today the debate concerns exactly *how* to go about engaging in electronic commerce. Companies such as Dell Computers, which do the best job of using the Internet and the World Wide Web, are the most likely to be successful, whereas some other companies that falter in this arena may find themselves out of business. Even companies that do not actively market their services over the Web are affected by electronic commerce through the wide variety of Web sites that enable a user to make an informed decision by comparing companies offering similar products. Although the subject of electronic commerce could take up an entire book (and does),[1] this chapter will try to summarize most elements of this important topic as they relate to your understanding of information technology and the networked economy.

Although its rate of growth has slowed somewhat over the last couple of years, electronic commerce continues to expand both in the United States and globally in terms of its volume of retail sales, percentage of total retail sales, and number of individuals purchasing online. For example, the research firm IDC found that global electronic commerce sales in 2001 were 68 percent higher than sales in 2000, with total volume of $600 billion. The United States accounts for 40 percent of all electronic commerce sales, and business-to-business sales account for 83 percent of the total. IDC has predicted that, by the end of 2002, more than 600 million people worldwide will have Web access and that electronic commerce sales will top $1 trillion. It also predicted that in 2006, 80 percent of the U.S. population will use the Web at least once per month, but that the fraction of total electronic sales attributable to the United States will drop as other regions of the world begin to catch up. Electronic commerce sales in Asia and Canada are growing at a rate of more than 80 percent per year, while sales in Europe are growing nearly 70 percent annually. Finally, B2B sales are predicted to account for 88 percent of all electronic commerce sales in 2006.[2]

Although electronic commerce certainly existed prior to the beginning of widespread use of the Web in 1994, you can attribute the explosion in electronic commerce sales volume directly to the growing popularity of the Web. The Web made finding and communicating with companies to purchase goods and services much easier, enabling many more people to become comfortable with the process. And, as discussed in Chapter 3, you can use the Web to carry out all of the activities over the Internet that once required separate software packages. Today, the vast majority of all electronic commerce applications occur through the Web, or the Web could be used for them even if a separate software application exists. For that reason, the discussion in this chapter will concentrate on use of the Web.

Types of Web Sites

Given that the Web is the primary source of electronic commerce activities, to develop and implement an electronic commerce strategy, you must understand the types of business-related Web sites in wide use. You can classify business-related Web sites using the same technology classification system introduced in Chapters 3 and 4: Internet, intranets, and extranets. Recall that the Internet is global in scope, whereas an intranet is limited to the organization, and an extranet is based on business partnerships. Figure 7.1 illustrates these three Internet technologies.

---

1. Gary P. Schneider and James T. Perry, *Electronic Commerce, 2nd ed.*, Cambridge: Course Technology, 2001.
2. Elaine X. Grant, "Study: E-commerce to top $1 trillion in 2002." *E-Commerce Times*, February 13, 2002, http://www.ecommercetimes.com.

Figure 7.1 Internet technologies

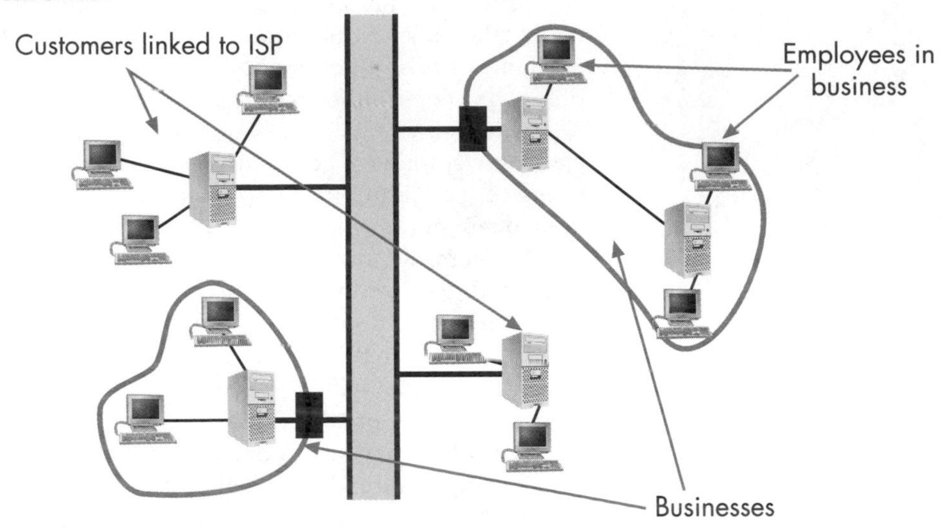

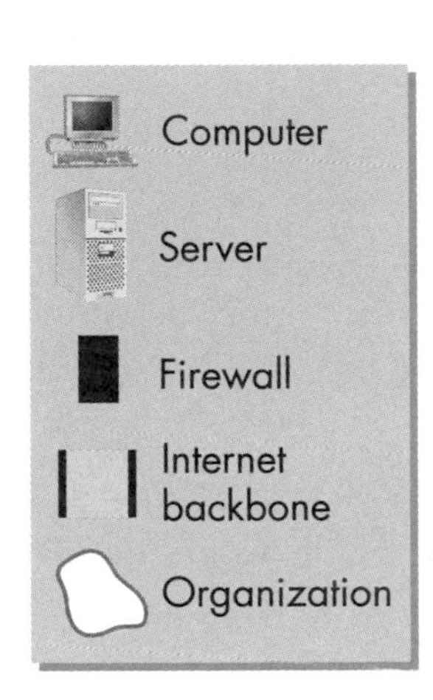

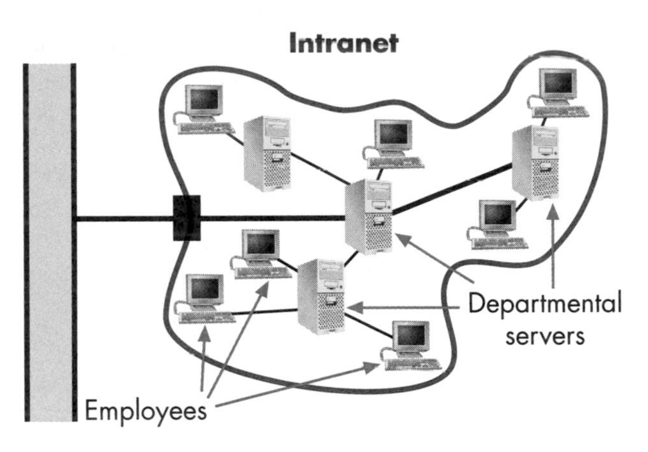

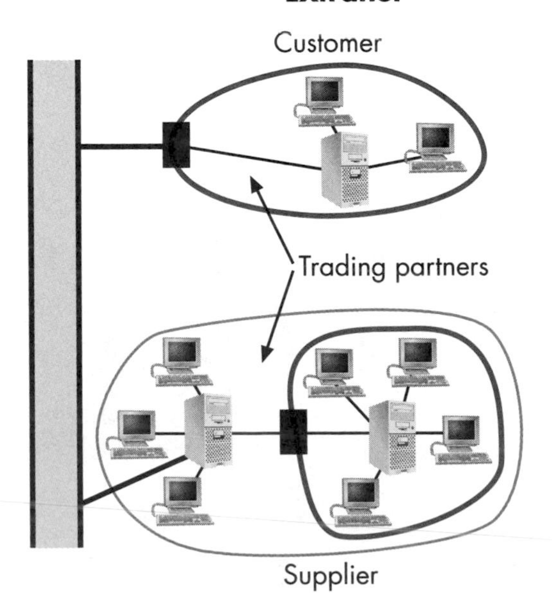

You can categorize electronic commerce Web sites according to the three Web topologies and the audiences served, as shown in Table 7.1. Intranets are included in this classification because their ability to facilitate organizational communication improves organizational performance, which is an important part of the definition of electronic commerce strategy. The use of extranets for handling business-to-business transaction processing, which is an essential part of electronic commerce, was discussed in Chapter 4. This classification system will serve to guide the discussion of other aspects of electronic commerce.

Although a company will usually recognize the audience to which it markets its products and services—consumers or other businesses—it may not be very sure about the appropriate business model to use in developing its Web presence. Eight business models for Web sites exist, each of which has the same goals—to find a customer need, build demand, fulfill the demand, and then repeat the process. Table 7.2 lists the eight types of business models for Web sites, along with a description, the Web site technology used, and an example. The types of Web sites are arranged from those with which you are probably most familiar to those that may be new to you.

In Table 7.2, you can see that the first four classifications involve business-to-consumer models that attempt to generate revenue by selling goods and services over the Internet (e-tailers/e-malls and communities of interest), by advertising (portals), or by charging a fee for a service (auctions). The fifth business model, informediaries, generates revenue by charging customers for information that they need. The last three business models target the business-to-business value chain using intranets and extranets.

Table 7.1 Electronic Commerce Classifications

	Web Technologies		
	Internet	**Intranet**	**Extranet**
Scope	Global	Organizational	Business partnerships
Type of electronic commerce	Business-to-consumer	Business-to-employees	Business-to-business

Table 7.2 Business Models on the Web

Business Model	Description	Technology	Electronic Commerce Application	Example
E-tailers/e-mall	Retail sites aimed at the Web consumer	Internet	Business-to-consumer	Lands' End (www.landsend.com)
Portals	Sites that are major starting points for users when they connect to the Web	Internet	Business-to-consumer	Yahoo! (www.yahoo.com)
Auctions	Sites that enable an online shopping population to bid on items and offer items for sale	Internet	Business-to-consumer	eBay (www.ebay.com)
Communities of interest (COI)	Sites that create niche content and context for their members	Internet	Business-to-consumer	Wine.com
Informediary	Sites that gather, organize, and link to new information and services on the Web	Internet, extranets	Business-to-consumer or business-to-business	WebMD (www.webmd.com)
Process/services improvement	Sites aimed at improving supply chain management processes internally and externally	Intranets, extranets	Business-to-business	Manheim Online (www.manheim.com)
Value chain service provider	Sites for companies that focus on dominating a specific function of the value chain	Extranets	Business-to-business	UPS (www.ups.com)
Value chain integration	Sites for companies that integrate all steps in a market's value chain	Extranets	Business-to-business	Manugistics (www.manu.com)

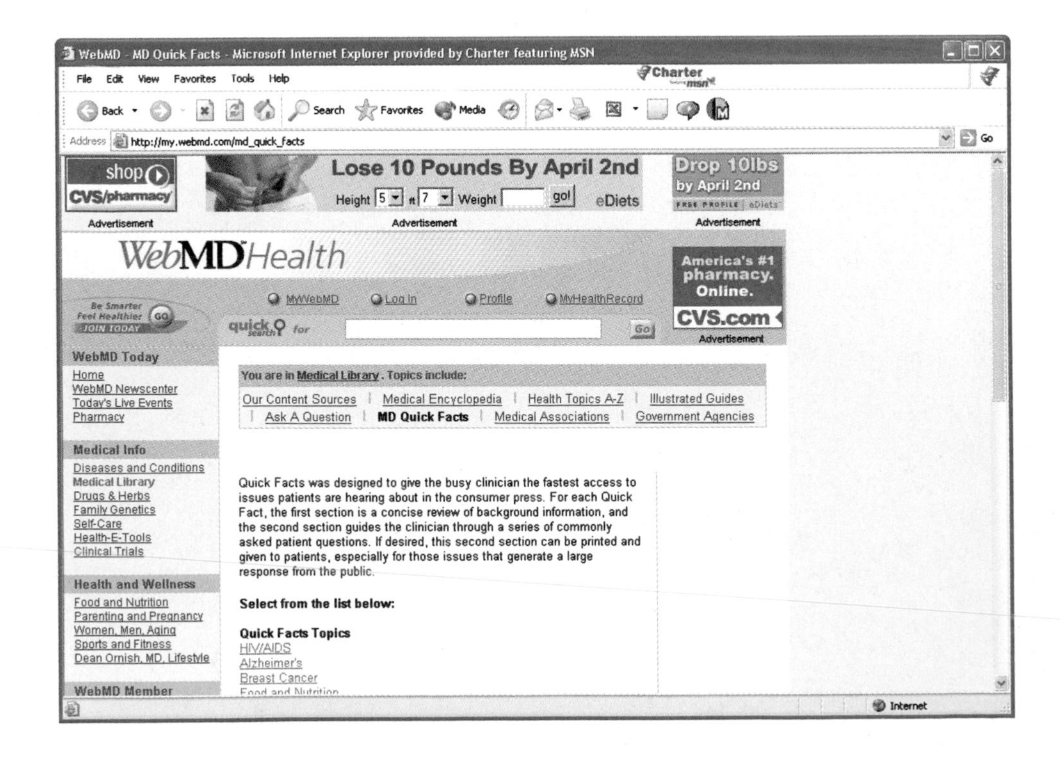

The WebMD site is an example of an informediary Web site.

The company whose progress you have been following in this book, FarEast Foods would be classified as an e-tailer that uses the Internet to sell food items directly to consumers through its Web site, fareastfoods.com. It uses a value chain service provider such as UPS to handle the distribution of the food items to its customers, and it may hire a process/services improvement company to handle its supply chain management to move food items from the suppliers to the company. From this example, you can conclude that business-to-consumer companies use business-to-business Web sites to help them better manage the value chain. This chapter will concentrate on electronic commerce strategies for the first five business models and their associated Web sites. Business-to-business electronic commerce was discussed in Chapter 4.

Limitations of Electronic Commerce

In thinking about electronic commerce, you should realize first and foremost that computer networks, including the Internet, can do only one thing: transmit electronic messages between computers. Although these messages can include data, information, and software, you cannot move anything physical over a computer network. Because electronic commerce relies on computer networks, the shipment of any physical good purchased using electronic commerce must still take place through traditional means. That is, when you buy a physical good over the Web, it must be picked from a shelf in a warehouse, packed for shipment, and physically moved from the warehouse to the customer via a package delivery service. Although many electronic commerce systems rely on a Web page and ordering system, it is still important to have the so-called back-office elements in place to handle order fulfillment. Beyond order fulfillment, any company in the business of accepting orders and shipping goods to customers must be prepared to handle returns—another physical process, and one that can prove more complicated than the actual order fulfillment.

The fact that order fulfillment and returns continue to represent a part of the business process even when electronic commerce is used explains why the more successful implementations of electronic commerce have involved companies that were already engaged in accepting orders by telephone or postal mail and fulfilling those orders. Firms such as Dell (discussed in the opening case) or Lands' End (discussed in Chapter 1) are good examples of this type of company. Telephone- and mail-order companies also have vast experience in handling returns, so returns from Web orders pose no special problems for them. In essence, electronic commerce extends their existing, profitable business model by providing another channel for interacting with customers. As noted in Chapter 1, Lands' End has found some

FedEx, like UPS and other package delivery companies, is classified as a supply chain service provider.

novel ways to use the power of the Web to improve its customer service, which should positively affect the company's bottom line. FarEast Foods offers another example of a mail-order company that has successfully incorporated the Web as another type of channel without substantially changing its back-end operations.

The limitations of computer networks pose little or no problem when the good being purchased is itself electronic or can be converted to an electronic form. Electronic goods include computer software and games, audio and video products, news sources, and electronic books (e-books). Because these electronic goods can be transmitted directly over the Internet or a local network as a series of bits, no picking, packing, and shipping problems occur. Likewise, returns are a simple matter; if a problem occurs with the electronic product, the company can send out a new one to replace the defective product. A common strategy used by companies dealing in electronic products is to make reduced-functionality versions of their products available for no charge or full-functionality versions available for a test period, typically 30 or 90 days. This strategy assumes that the user will either pay to upgrade to the full version from the reduced-functionality version or buy the full version when the test period ends.

Figure 7.2 compares electronic commerce with physical products to electronic commerce with electronic products. The dotted lines in the figure indicate an electronic message, and the solid lines indicate a shipment of physical goods. In the latter case, a return would reverse all of these shipments.

1. What is electronic commerce?
2. What are the limitations of electronic commerce?

The Changing Nature of Electronic Commerce

By the middle of 2000, much had been written about the ways in which electronic commerce would change the very nature of business. Some IT professionals theorized that electronic commerce would enable companies to dramatically increase profits by cutting costs while keeping prices up. At the time, common business wisdom said that a company need only create a Web site and the customers would just show up. Investors of all types rushed to put money into any company whose name carried a dot-com extension (the "dot coms"). As a consequence, dot-com companies with little or no revenue or real plans on generating revenue had higher valuations than many of the biggest *real* companies. Unfortunately, many of the dot coms proved to be badly mismanaged, both in terms of spending habits and in developing products that would eventually generate revenue.

Figure 7.2 Ordering and processing of physical and electronic items

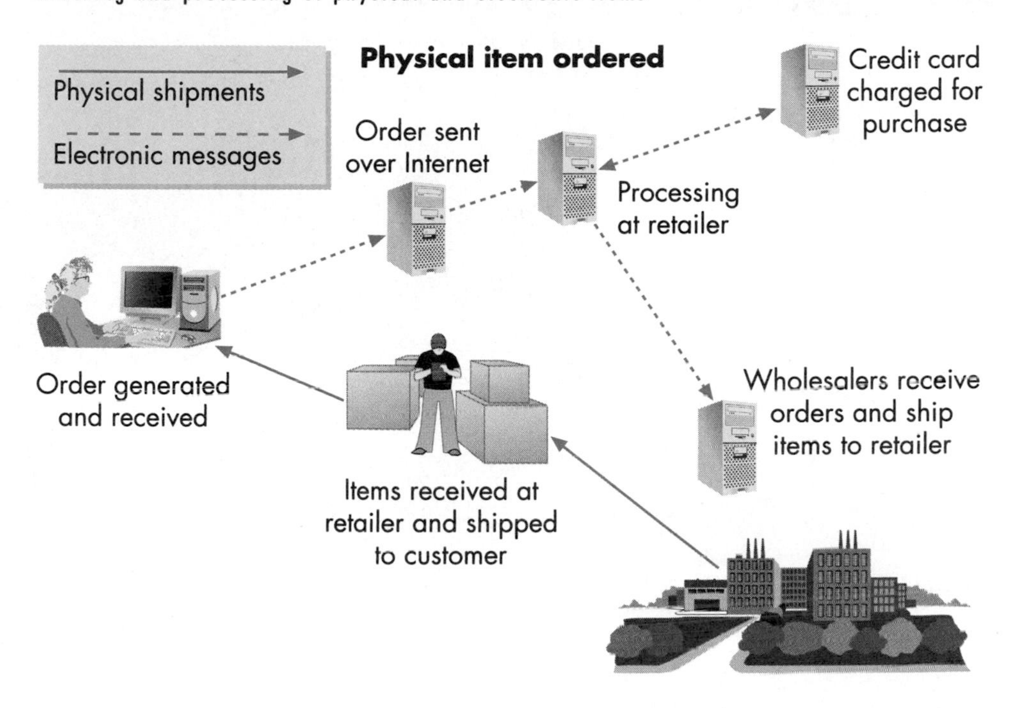
Physical shipments
Electronic messages
Physical item ordered
Order sent over Internet
Processing at retailer
Credit card charged for purchase
Order generated and received
Items received at retailer and shipped to customer
Wholesalers receive orders and ship items to retailer

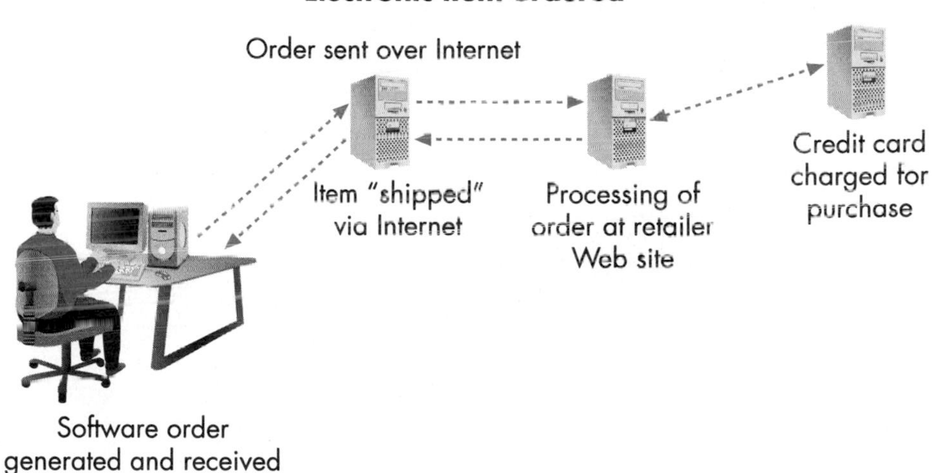
Electronic item ordered
Order sent over Internet
Item "shipped" via Internet
Processing of order at retailer Web site
Credit card charged for purchase
Software order generated and received

In March 2000, the promises of the dot coms started to unravel when the Nasdaq stock market dropped dramatically. Many investors hesitated to put money into companies with no workable business plans, and many of the dot coms went bankrupt. Only a few of the original dot coms, such as Monster.com discussed in Chapter 1, remain in existence today, and even fewer have managed to reach anything close to the revenues originally projected.

Figure 7.3 Stock Prices for Yahoo!

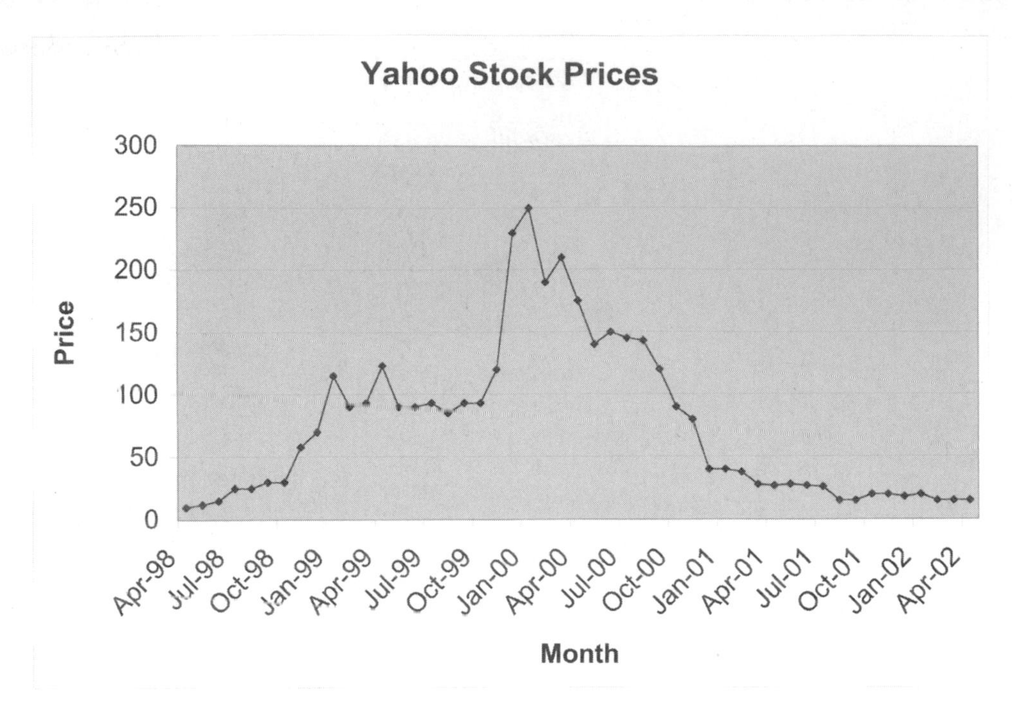

Source: http://bigcharts.marketwatch.com.

As an example of the wild values for Internet stocks, consider Figure 7.3, which shows stock prices for Yahoo! for the three-year period from March 1999 through February 2002. In early January 2000, its stock price reached a high of almost $250 per share, only to sink to as low as $8 per share in late 2001.

Companies have found out the hard way that electronic commerce does not automatically mean increased profits. In fact, it can actually mean increased downward pressure on prices as consumers learn more about manufacturer and retail price structures through Web searches. Electronic commerce can also result in a need for dramatic changes in advertising due to a change in the nature of communication brought on by the Internet. Finally, it can lead to more demanding consumers who want greater choice in their purchases. These changes will require companies to rethink their electronic commerce strategies, as discussed later in this chapter.

In a sense, the business world has gone from an exciting but unreal period to a more mundane but sustainable period of electronic commerce. With a few exceptions, such as Monster.com and eBay, dot coms with no connection to physical companies have, to a large extent, been overtaken by clicks-and-mortar companies, which combine physical locations with a Web presence. Even the poster child for electronic commerce, Amazon.com, took several years to reach profitability at the end of 2001, and some writers predict that it will eventually fail or be taken over by a traditional company such as Wal-Mart.[3]

Effects of the Internet and Web on Businesses

When electronic commerce was first introduced in the mid-1990s, it was hailed as an almost automatic way for companies to increase their customer base, increase revenue, reduce costs, and generate higher profits for their stockholders. It has since become clear that the Internet and its multimedia application, the World Wide Web, can

3. http://www.mcleanreport.com.

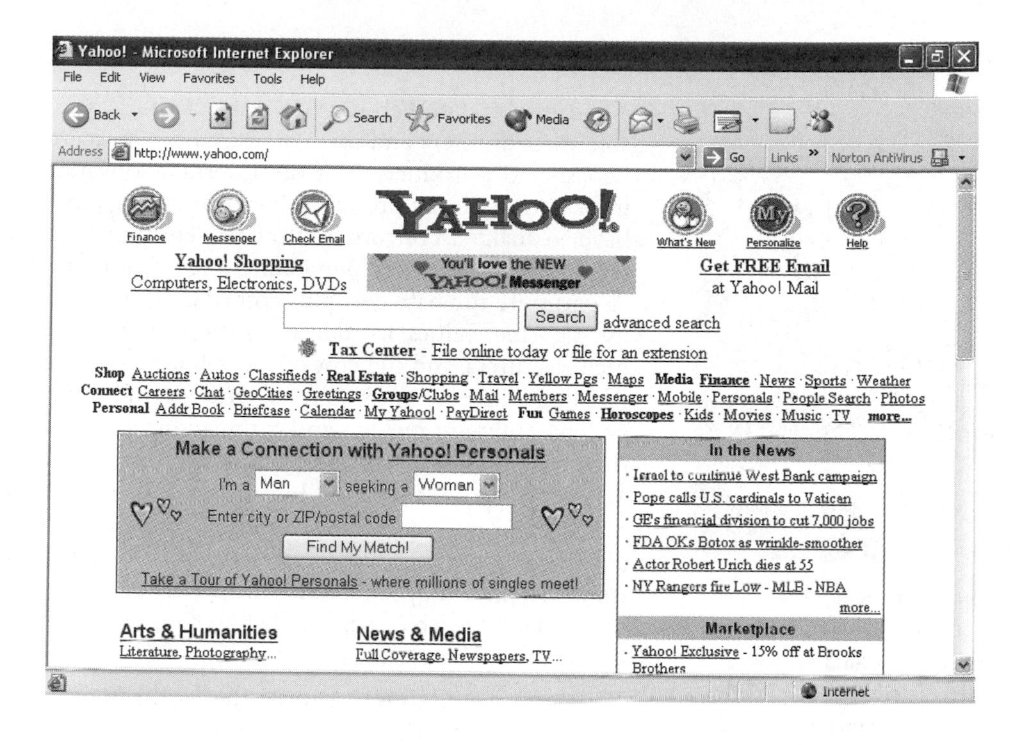

Yahoo!'s stock prices have dropped dramatically since March 2000.

Internet effects
Somewhat unexpected effects that the Internet can have on manufacturers and retailers, including price transparency, communications flip-flop, and customer demand for perfect choice. Also called *Internet threats*.

price transparency
The capability of customers to use the Web to learn about the prices of products from a variety of sellers.

communications flip-flop
The reversal of the seller controlling the flow of information through print and media advertising to the customer controlling the flow of information by first finding Web sites and then choosing which ones to visit.

perfect choice
The demands by customers for a wide range of products and a choice in how to buy them at a variable price.

have decidedly different effects on manufacturers and retailers. Three **Internet effects**, (or threats) you should consider are price transparency, communications flip-flop, and customer demand for perfect choice. In **price transparency**, customers use the Web to learn about the prices of products from a variety of sellers. In **communications flip-flop**, instead of sellers controlling the flow of information through print and media advertising, the customer controls the flow of information by first finding Web sites and then choosing which ones to visit. Finally, instead of sellers offering a minimum of choices at a fixed price, customers demand **perfect choice** in which they have a wide range of products available and a choice in how to buy them at a variable price.

In each Internet effect, one of the three risks faced by every business—efficiency risk, demand risk, and innovation risk—is exacerbated. Recall that efficiency risk is the risk associated with not providing the product or service efficiently, demand risk is the risk arising from a lack of demand for the good or service, and innovation risk is the risk of not keeping up with the competition through constant innovation. Table 7.3 describes each Internet threat and indicates the risk that each exacerbates. The following sections examine each Internet effect in more detail, noting why it increases a particular risk.

Table 7.3 Internet and Web Threats

Threat	Description	Risk
Price transparency	Consumers can quickly and easily determine prices of products and services from a variety of vendors using the Web	Efficiency
Communications flip-flop	Instead of retailers and manufacturers directing communications at customer, the customer controls communications by deciding which Web sites to visit	Demand
Perfect choice	Instead of customers being restricted to a few options at a fixed price, the Web and Internet enable them to seek a a wide range of products and purchase options at variable prices	Demand, innovation

Price Transparency[4]

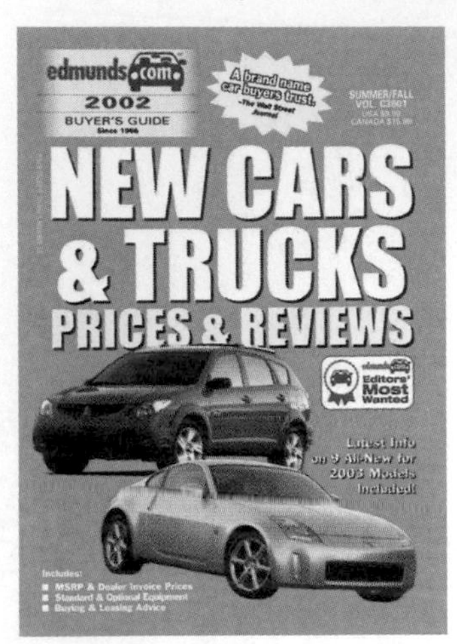

Edmunds has been very successful in moving from a magazine format to a Web format for making information on car prices more accessible to consumers.

The Internet and Web have dramatically increased the amount of information available to consumers and their industrial counterparts, purchasing agents. Prior to the widespread introduction of the Internet, consumers found it difficult to search for information about products. For example, a prospective automobile buyer would have to make a concerted effort to determine the cost of a car to the dealer. This search usually involved purchasing one or more books or magazines and then trying to compute the base cost plus cost of options. Often, the books or magazines would list incorrect values. In addition, dealers often discounted cars based on rebates from manufacturers, and the cost of the vehicle to the dealer remained a well-kept secret. Although invoice costs were advertised, they often were not what they seemed to be. This ability of retailers and manufacturers to hide their costs gave them the power to extract higher prices and profits from the goods and services they sold.

With the advent of the Internet and Web, consumers can search for information quickly and easily with a few mouse clicks and keystrokes. Today prices are no longer a deep, dark secret. In fact, prices are often *transparent* to consumers, who can determine whether they are getting the lowest price available on a purchase. Information on prices is available over the Web in a number of ways, including the following:

- More efficient searches
- Reverse auctions that make the "price floor" more evident
- Web sites that specialize in providing information on a product or service

Thanks to more efficient searching, you no longer have to go from seller to seller asking for information that often turns out to be wrong; with the Web, you can enter search terms and find needed information in minutes. In fact, an intelligent agent can carry out the search for you, sending back cost information as it is

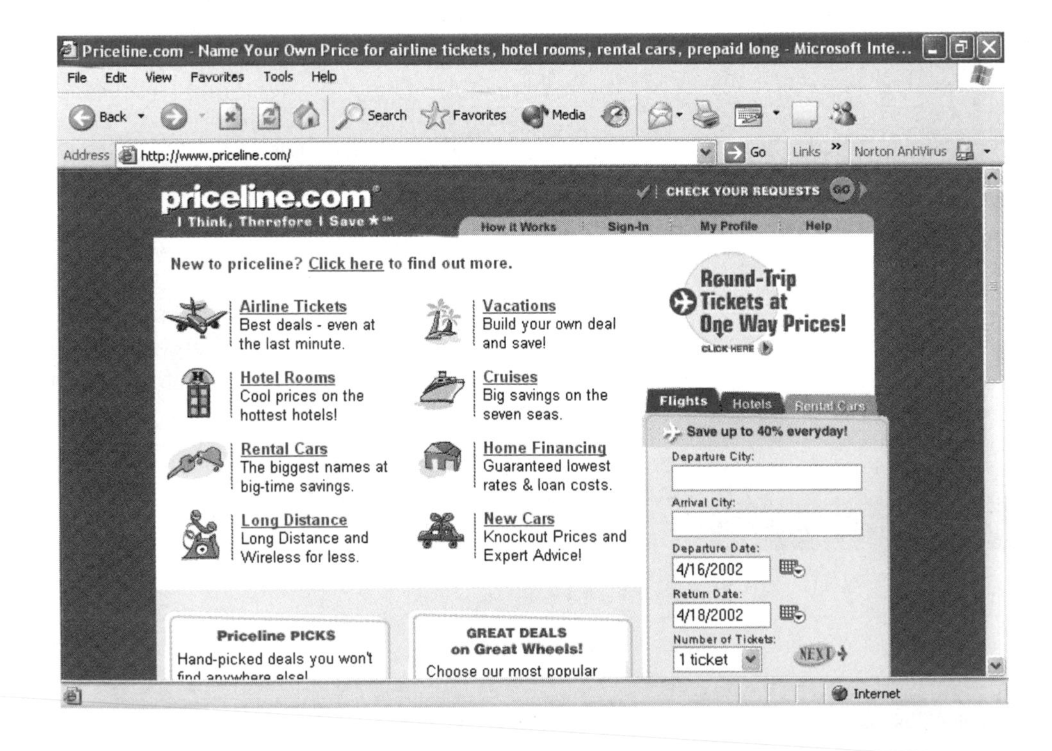

Priceline.com changed the way transactions take place by enabling buyers to "name their own price".

4. This section is based on Indrajit Sinha, "Cost transparency: The Net's real threat to prices and brands." *Harvard Business Review*, March-April 2000, pp. 43-50.

intelligent agents
Computer programs that can be trained to carry out a search over the Internet for needed information. Also called *bots*.

found. An **intelligent agent** is a software program that gathers information or performs some other service on some regular schedule without your immediate presence. Typically, the agent searches all or some part of the Internet, gathers information based on parameters you have set, and presents it to you on a periodic basis. Also known as **bots**, intelligent agents come in many flavors, including those used to shop the Web for the best price on a particular item. (For a complete listing of available bots, go to www.botspot.com.) Intelligent agents have also found a home in the business-to-business context, where Web sites are set up to enable buyers to find sellers, and vice versa, thereby widening the number of suppliers from which a company can choose.

Through auctions and reverse auctions, products and services long assumed to have a set price have become available for less on Web sites such as eBay and Priceline.com. Priceline.com is a reverse auction that allows buyers to "set their own price" and see if any sellers will accept that price for items such as airline tickets, hotel rooms, or home financing. Although you may ultimately choose not to use this particular Web site, visiting it can give you an idea of just how low a price can go on a product that will have no value after a certain date.

Finally, instead of having to resort to a friend or relative in the industry to obtain a good deal, you can go to a Web site that makes such information freely available. A growing number of such sites provide reliable and independent information on products and services as well as testimonials or complaints about the products. For example, in the automobile industry, Edmund's provides a wealth of information on almost anything to do with a purchase. On a more general note, ClarkHoward.com provides information on the best deals as well as scams to avoid on a wide variety of products and services. Figure 7.4 shows mySimon.com, a bot that provides a listing of prices, including shipping costs, for the first edition of this book from a number of book sellers.

Figure 7.4 Web site providing information on products and services

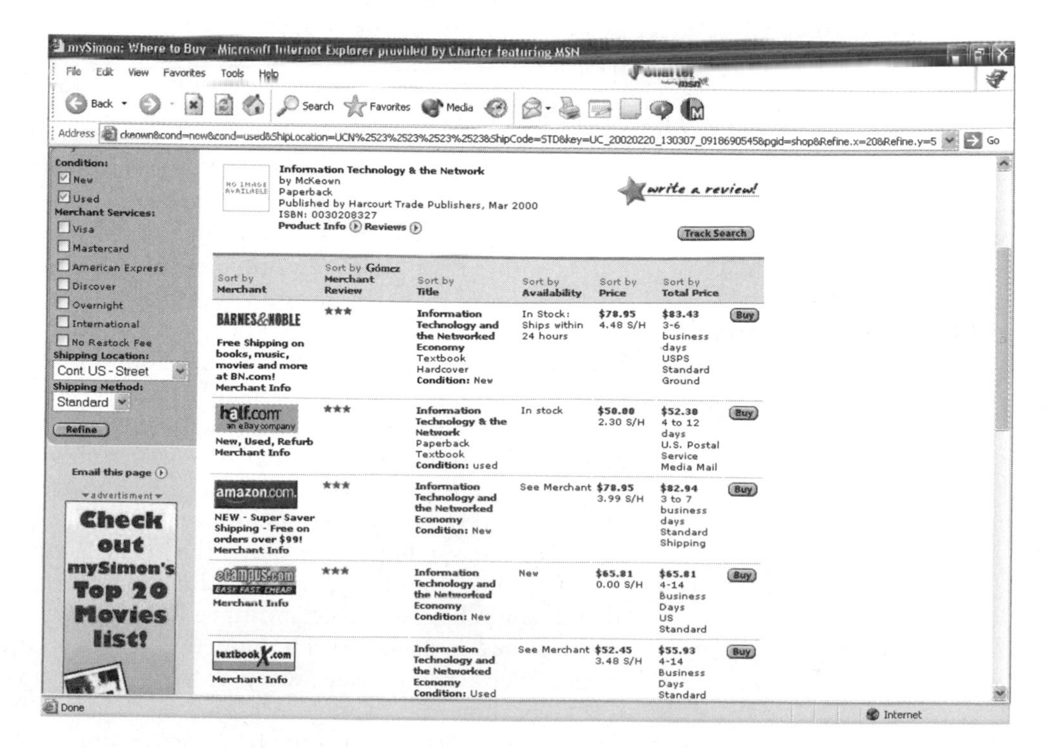

Price transparency can exert a decidedly negative effect on prices in a number of ways. First, sellers may have difficulty in obtaining high margins. If customers know the competive prices, they are less likely to pay an unusually high price for the product or service. Most consumers are willing to pay a fair price but balk at going beyond that. This phenomenon is not new: In 1984, AT&T was forced to lower its long distance prices when low-cost competitors such as MCI and Sprint entered the market. Once a monopoly in the long distance market, AT&T had a difficult time dealing with the low-cost competitors, but finally had to drop its prices to compete.

Second, when customers are aware of lower prices for the same product or service, they often make decisions based on price alone. This trend turns the product or service into a commodity that is purchased on price alone. Such price competition drives down prices, cutting into a firm's profits.

Third, knowledge of prices reduces brand loyalty, especially when no discernible difference exists between store brands and well-known but more expensive brands. Customers can now easily determine information about the quality of products from consumer-based Web sites or from other customers through newsgroups. The exchange of this type of information about quality can reduce brand loyalty keeping manufacturers from charging higher prices based on brands alone.

Finally, if consumers can determine prices and quality from the Web, they may quickly become unhappy with a company that is deemed to charge unfair prices. Given the speed with which bad news can spread over the Web, once one consumer perceives a price as being unfair, millions of consumers will probably know this information within a short time.

Price transparency exacerbates efficiency risk for companies by making it more difficult for them to be inefficient. With customers having a good idea about the prices, inefficient companies will be threatened by loss of profit as customers move to competitors that can provide similar goods and services at a lower price. No longer can a company hide its inefficiencies behind high prices; customers simply won't stand for it. If two companies provide a similar good or service with similar cost structures, the more efficient company can deliver it at a lower price, eventually driving the less efficient competitor out of business. Firms must find ways to become more efficient or discover novel ways to price their goods and services if they wish to stay in business.

Communications Flip-flop

customer convergence
The Web marketing concept stating that firms must describe their products and services in such a way that potential customers converge on the relevant Web pages.

Electronic commerce has reversed the flow of communication. Instead of sending messages to customers through typical print, radio, or television advertisements, a firm now wants customers to converge on its Web site. **Customer convergence**—the idea that firms must describe their products and services so that potential customers converge on the relevant Web pages—is the key to Web marketing. Unless customers find a firm's Web site, the entire effort is wasted. This fact means a firm must ensure that its Web site is readily found by a variety of search engines. Even more importantly, a firm with many products must ensure that potential customers converge on the page that describes the product or service of greatest potential interest. For example, a camera manufacturer must first make certain that photography enthusiasts find its Web site and then help each potential customer, possibly with the aid of an expert system, navigate to the pages describing the cameras or lenses of interest. Figure 7.5 shows the effect of customer convergence.

The shift from traditional broadcast advertising to customer convergence fundamentally changes the relationship between the advertiser and the customer, as summarized in Table 7.4. The initiative moves to the customer, who now decides which Web pages to view, when to view them, what part of the message to read, and even how the message is presented. For example, a customer may decide to ignore an advertiser's message by never visiting its Web site or disregard an important element of a message by not playing audio clips.

Table 7.4 Communication Flip-flop

Traditional Advertiser	Web Customer
Decides audience	Decides advertiser
Decides schedule	Decides schedule
Decides message content	Selects and customizes content
Decides media for distribution	Decides how message is presented

The communications flip-flop phenomenon represents a fundamental change in the nature of the relationship between buyers and sellers. With electronic commerce, the customer is royalty, and customer service is the primary element that will keep customers coming back to a Web site to make purchases. As buyers increasingly use the Web to learn about new products and services, sellers must find ways of attracting visitors to their Web sites, enticing them to read their messages, buy their products, and, most importantly, return in the future. In many cases, sellers get customers to revisit their Web sites by gathering information about each customer that they can use to better serve him or her on future visits. For example, if you visit an online bookstore or CD mart, it will often alert you to new books or

Figure 7.5 Customer convergence

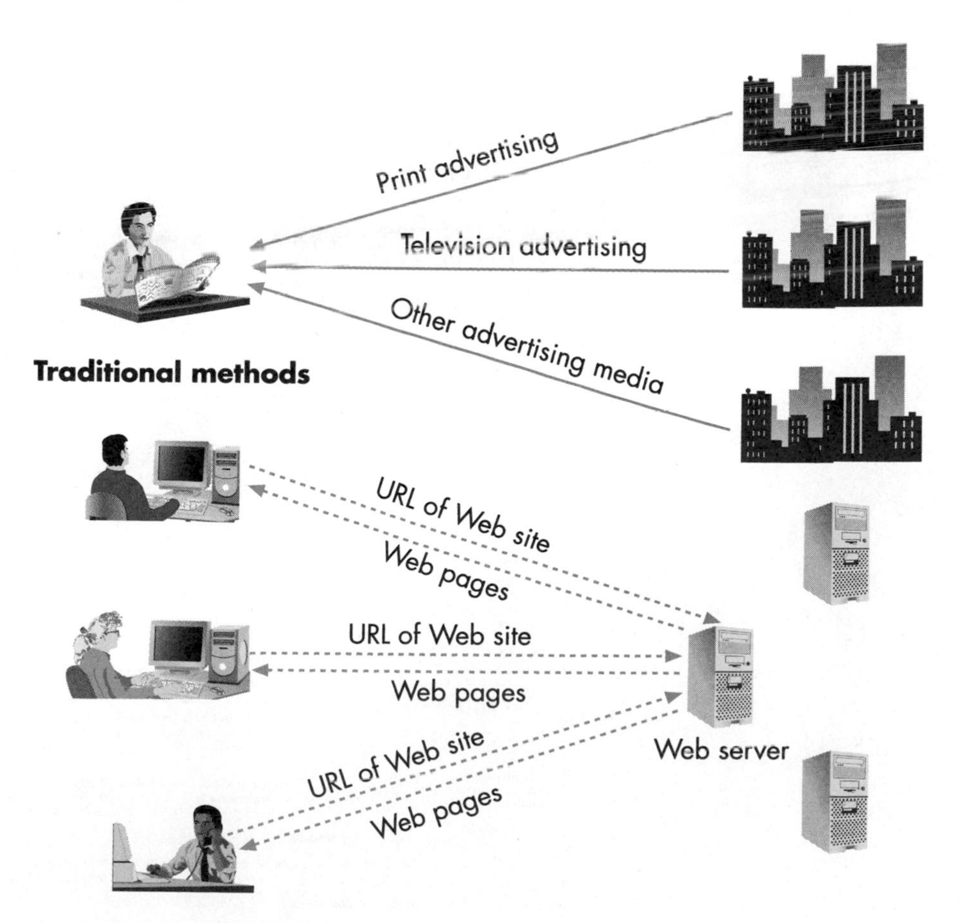

mass customization
The process of making each visitor to a Web site feel that the site has been customized to his or her particular needs.

CDs based on what the Web site has learned from your previous visits. Referred to as **mass customization**, this personal touch for every buyer offers a way to build customer loyalty because visiting a competing Web site will require that you teach it your interests and buying patterns—a time-consuming task.

As another example of the dramatic changes in communication brought on by the Internet and Web, consider the effect on companies that have specialized in acting as middlemen or brokers among retailers and consumers or among manufacturers and their customers. In traditional commerce, people are used to the role of such intermediaries. For example, few people want to try to deal directly with the banana grower, so they are happy to buy just the bananas they want from a market. With the reversal in communication facilitated by the Web, however, many of these intermediaries become unnecessary. For example, as discussed in the opening case, electronic commerce can remove the need for intermediaries through programs such as Dell Direct.

disintermediation
The process of eliminating intermediaries.

The process of eliminating intermediaries is termed **disintermediation**, and it casts a large shadow over many segments of business whose primary function was once gathering information from suppliers and passing it along to consumers. This group includes travel agents, insurance agents, and stockbrokers. In the case of travel agents, Web sites such as Travelocity and Orbitz, as well as those of the individual airlines, often offer lower prices than those available from travel agents. At the same time, the airlines are cutting or eliminating the agent commissions, putting further pressure on them to charge fees for their services. Other brokers are feeling similar pressures from Web-based systems. For example, why go to an insurance broker when you can just as easily find the prices and terms on the Web? Similarly, why pay the commission on a stock trade when you can essentially do it yourself for much less? You can probably think of many more cases where the Web could eliminate intermediaries. Figure 7.6 shows one of the more popular travel Web sites that has contributed to the disintermediation of travel agents.

Figure 7.6 Popular travel Web site

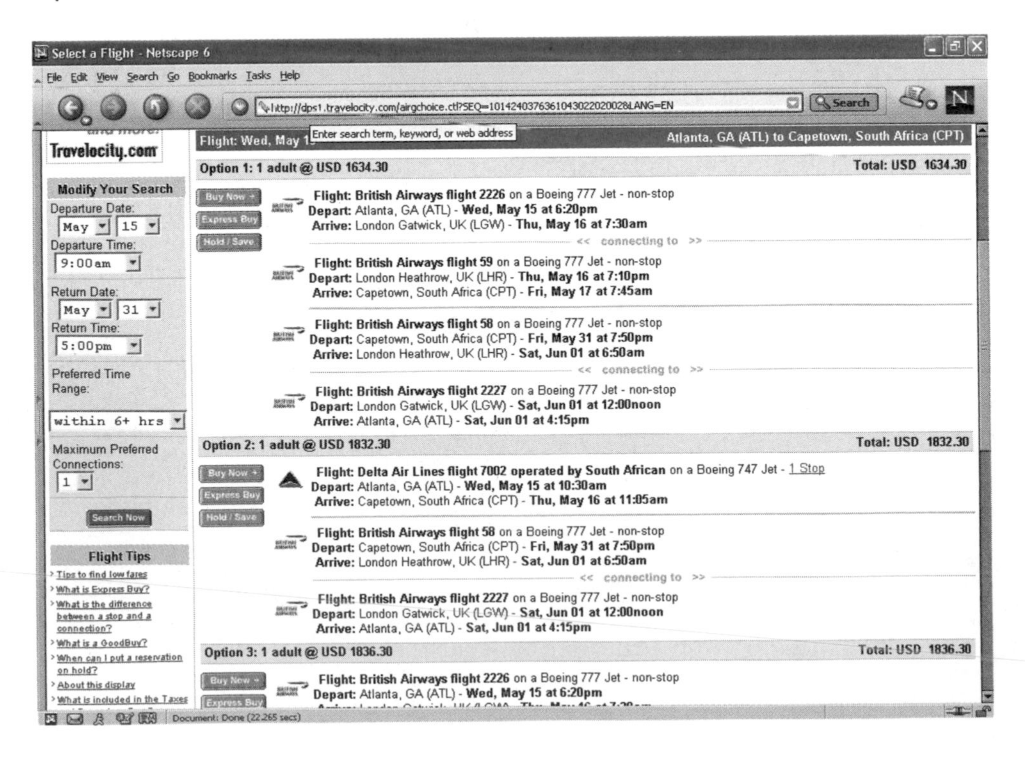

Travel agencies, insurance agencies, and stockbrokers have suffered from disintermediation.

reintermediation
The process of creating new intermediaries in electronic commerce.

Although many intermediaries will cease to exist, others have found ways to recast themselves to meet new needs in electronic commerce by creating auctions or other markets not available prior to the introduction of electronic commerce. This process is known as **reintermediation**. Examples of reintermediators include companies such as Manheim Auctions, eBay, and e*Trade. For example, as discussed in Chapter 4, Manheim Auctions has moved from being strictly an intermediary at physical auctions to serving as an electronic intermediary through Manheim Online. It has also added AutoTrader.com to make the same capabilities available to the consumer used car market.

Communications flip-flop exacerbates demand risk by potentially reducing demand for a company's goods or services. With convergence on a Web site becoming a more important way of advertising a product, failure to attract visitors to the site can reduce demand for the product. For example, if your company chooses to concentrate on print advertising at the same time as your customer base is moving toward the Web, you could very well see a reduction in demand. Similarly, intermediaries that serve only to provide information to customers about goods and services may suffer a drop in demand for their services if customers then communicate directly with companies providing the goods and services.

Perfect Choice[5]

In the industrial economy, economists assumed that the perfect model was perfect competition, in which a large number of buyers and sellers competed to trade for standardized products. In the networked economy, consumers are more interested in perfect choice than in perfect competition. That is, they want a wide selection of products and a choice in how they buy them. They also want their products and services customized to fit their preferences.

Instead of the perfect competition model, in which the first automobiles could be had in "any color as long as it was black," in today's perfect choice model, consumers want the widest possible choice. For example, online book and music stores make it possible to purchase almost any book or music CD rather than being restricted to those items available at a local store. AutoTrader.com enables you to choose how wide a search for a used automobile to conduct—from less than 15 miles to any point in the United States.

5. Richard T. Watson, "Perfect choice." *Ubiquity* 32 (2), 2001, http://www.acm.org/ubiquity/views/r_watson_2.html.

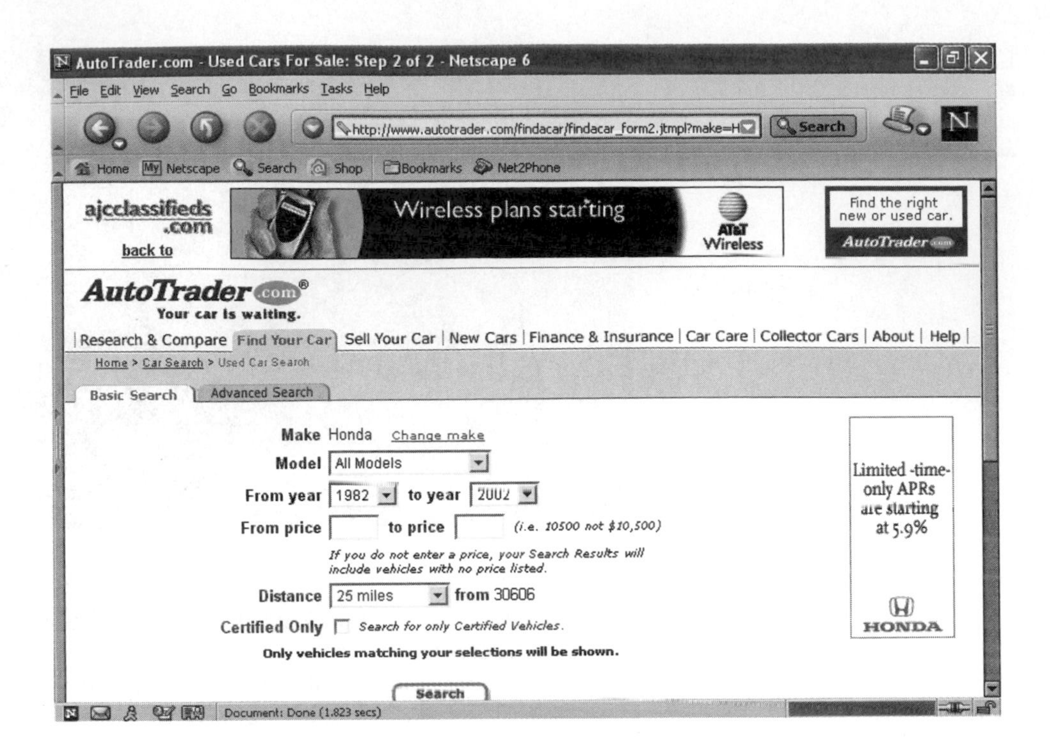

The AutoTrader.com Web site is an example of reintermediation.

Consumers also want to choose the channel through which to make a purchase. For the books and music business, channels might potentially include the Web, paper catalogs, television, local new or used book stores, or auction sites such as eBay. Many traditional retailers such as Home Depot, JCPenney, and Best Buy have recently adopted the clicks-and-mortar approach by offering goods over the Web. On the other hand, previously telephone-or Web-based retailers, such as Gateway, are going the other way: They are setting up physical locations.

Consumers also want component choice, like that popularized by Dell, which allows them to build a computer by selecting the components that will go into it. Some experts predict that the day of ordering an automobile, motorcycle, bike, or high-end sound system online with exactly the options you want is only a few years away. Figure 7.7 shows the Dell Web site, where you can build a computer to suit your specific needs.

Perfect choice exacerbates demand risk by causing companies that fail to provide choice in products, components, or channels to lose demand. It also increases innovation risk by forcing companies to find more and better ways to provide customers with the choices they want. Failure to provide the choices sought by informed customers—choice of products, choice of channels, and choice of components—could result in a more innovative competitor increasing its customer base or its demand by providing additional choices at its rival's expense. For example, Dell has found ways to offer customers more choices, while other computer companies look for ways to merge with other companies to survive in this market.

Figure 7.7 Dell Web site

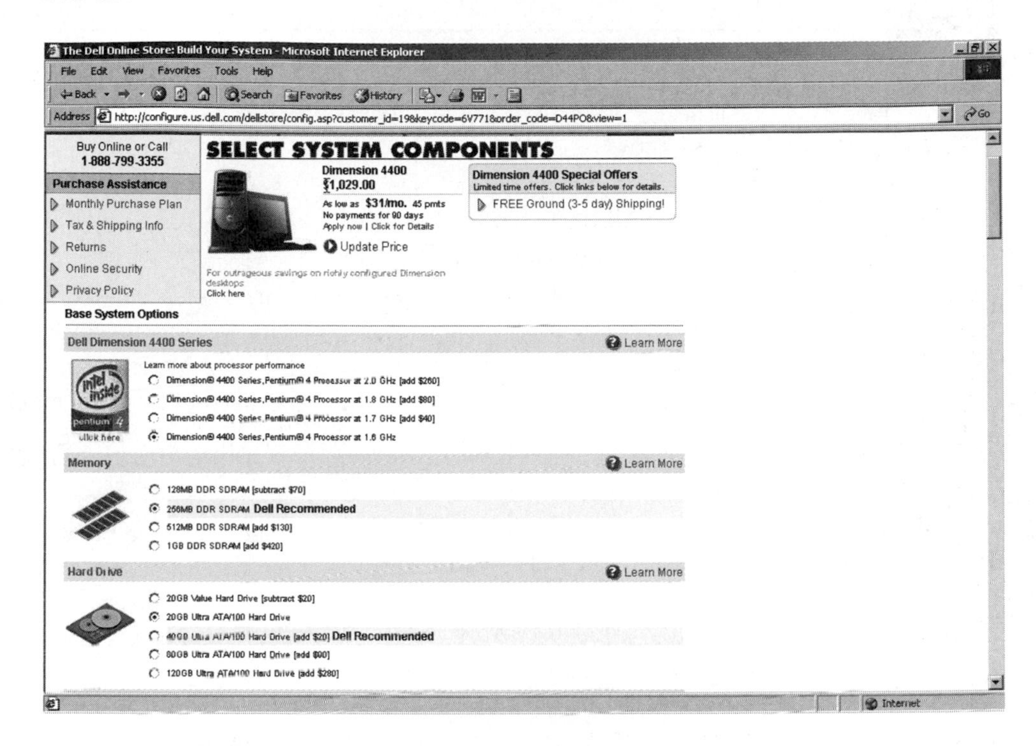

Internet and Web Effects on FarEast Foods

FarEast Foods has been affected by the Internet and Web in several ways. Price transparency is affecting FarEast Foods because customers can get a good idea of the prices of its food products by searching the Web for prices of competing products and for prices of foreign competitors for similar products. Communications flip-flop is also hitting FarEast Foods hard. Although the company has traditionally mailed catalogs to prior customers, it has seen a decrease in response to them. Even though the firm has created a Web site, fareastfoods.com, it is not happy with the number of prospective customers visiting it and the number of purchases from those users who do visit the Web site. Finally, FarEast Foods is being affected by customer choice. Customers are requesting items that the company does not normally carry, and demand for access to their products through local distributors is increasing.

1. List the three Internet effects.
2. Which Internet effect most strongly influences the price a company can charge for its good or service?

INTERNET IN ACTION

Tesco.com

One of the biggest failures on electronic commerce in the United States has involved online grocery delivery companies such as Webvan. In the United Kingdom, however, the largest grocer has proved much more successful with its electronic commerce venture. It is worthwhile to compare the approaches taken in the two countries to learn something about what works and what doesn't in the online grocery business and other similar business models.

Getting Started

Tesco started small by picking and packing from existing stores and leveraged its brand, suppliers, and database of affinity card users to launch its online grocery service for just $56 million. The venture started with one store in 1996, then gradually rolled out its service until about one-third of the company's stores were involved. This put it within reach of 91 percent of the U.K. population. In contrast, Webvan spent *$1.2 billion* to create its service but had no customers initially. Interestingly, none of its board members came from the grocery industry. Webvan tried to enter 24 markets in the first three years, and it opened warehouses in three market areas in its first 15 months. It kept building warehouses even when none of the first three had broken even.

Delivery Costs

Bucking the trend of the late 1990s toward free delivery of goods, Tesco.com charged $7.25 per order for delivery. It now receives more than 70,000 online orders weekly and collects $27 million annually for deliveries alone, making the difference between profit and loss. In contrast, Webvan tried to woo customers with free delivery for orders less than $50. As a result, it *lost* between $5 to $30 on every order it delivered, adding millions in unrecovered costs.

Online Versus Bricks-and-Mortar

Tesco.com does not have to make it on its own, because the grocer views its online grocery service as a way to extend its brand. In fact, half of Tesco.com's customers come from rival stores, and the parent company hopes they will shop at its supermarkets after purchasing from its Web site. In contrast, Webvan operated as an online company with no bricks-and-mortar component. Given its failure, the Webvan approach might not be the best way to go for online grocers.

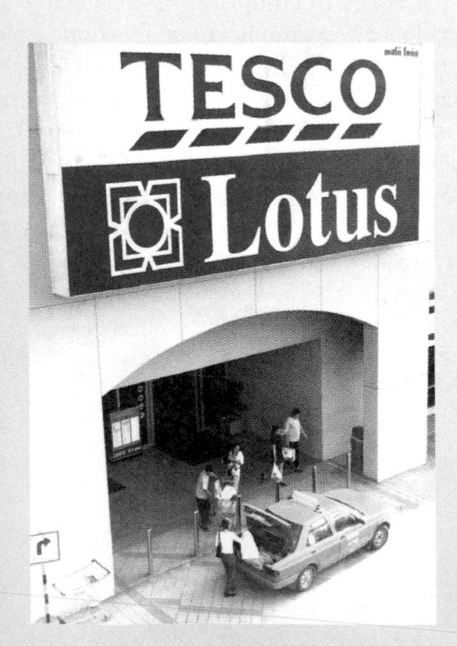

Tesco used a very different approach to Web-based grocery shopping from that used by Webvan and has been very successful.

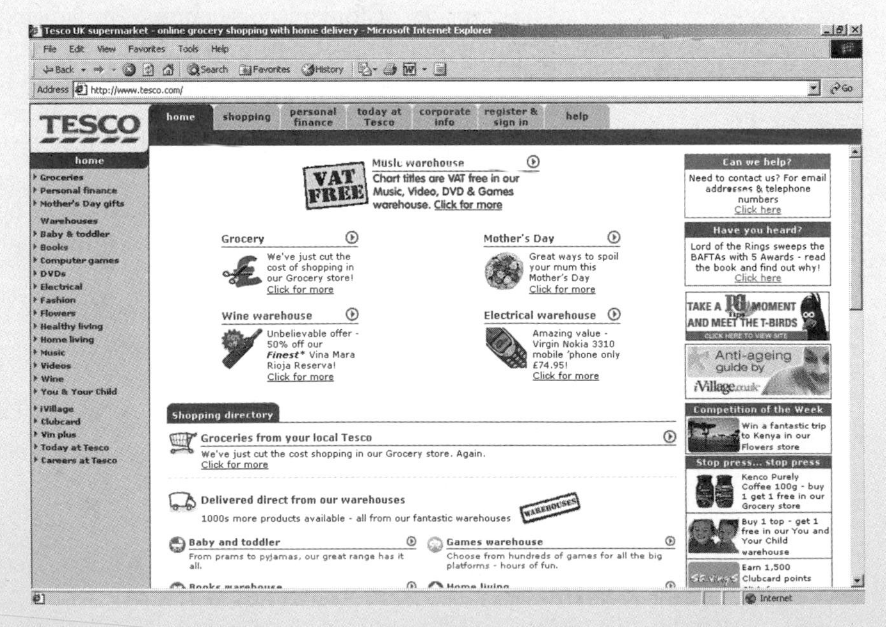

Tesco has been successful with its online grocery whereas several other companies have not.

Source: Andy Reinhart, "Easy does it." *BusinessWeek e.biz*, October 1, 2001, pp. EB 28-29.

Electronic Commerce Strategies

Given the effects of the Internet on businesses, companies need to implement electronic commerce strategies that counter these threats. In *all* cases, they should emphasize innovation, because innovation is the one thing that is most difficult for other companies to imitate. To counter the price transparency effect, companies need innovative pricing strategies that enable them to increase revenues without encountering customer resistance. In response to the communications flip-flop effect, they need innovative Web strategies that attract visitors to their Web sites, encourage them to purchase either there or at a physical location, and keep them coming back. Finally, to counter the perfect choice effect, companies need innovative strategies that will provide customers with the choices they demand in the networked economy. Table 7.5 summarizes the strategies employed to counter the Internet threats.

This section discusses strategies for dealing with price transparency and perfect choice, and the next section focuses on strategies for dealing with communications flip-flop. Responding to communications flip-flop by using innovative Web site design requires a more in-depth study.

Electronic Commerce Strategies for Handling Price Transparency

Recall that price transparency arises when customers use the Web to search for information on retail companies as well as industrial suppliers. Companies then have a much more difficult time hiding their costs and charging high prices, which results in a potential for lower profits. In the past, companies have dealt with intense price competition and the associated efficiency risk by increasing their efficiency so as to reduce their cost structures. In the networked economy, that approach may not suffice because other companies can quickly work to match this strategy. Instead, to combat problems with price transparency, companies must undertake innovative steps. These approaches involve new ways of marketing products that move the firm away from price competition and toward competition based on quality, differentiation of products, one-to-one marketing, or the experience provided by the Web site. These strategies using electronic commerce include the following:

- Improving the benefits that goods and services offer
- Using creative pricing strategies
- Bundling goods and services
- Using customer relationship management (CRM) to create one-to-one marketing
- Creating a Web-based experience for the customer that transcends price

Table 7.5 Strategies to Counter Internet threats

Internet Threat	Strategy
Price transparency	Innovative pricing strategies
Communications flip-flop	Innovative Web site design
Perfect choice	Innovative ways of providing customers with goods and services

Improving the benefits that goods and services offer counters price transparency because customers almost always value quality very highly. Even in a period of price transparency, higher-quality goods and services will command high prices. Companies need to continually enhance the quality of their products and find ways to communicate that quality to buyers. This strategy includes understanding that customers might pay more than expected. For example, a study of prices paid for used vehicles showed that dealers would pay more online for the same vehicle than the amount paid at an associated physical auction—even when they knew the prices being paid at the auction. The justification for these higher prices related to the services being offered and the fact that no employee time was lost traveling to and from the auction when the vehicle was purchased online.

When customers have a good understanding of prices on comparable products, it becomes necessary to use creative online pricing strategies. Several approaches are possible, including the revenue management techniques discussed in Chapter 6. Companies can also implement **tiered pricing**, in which goods and services are offered at different price points to meet different customers' needs. For example, in the United States, America Online (AOL) offers five different pricing plans depending on the services provided and the payment conditions. AOL offers even more plans for customers in different parts of the world where connections are different. The same is true of long distance plans and mobile telephone contracts.

tiered pricing
Goods and services being offered at different price points to meet different customers' needs.

Tiered pricing differs from revenue management in that customers in a tiered pricing system segment themselves either by their needs or their geography. The differences in pricing are openly shown to the customers, and they segment themselves based on the price tiers. In contrast, a revenue management system segments customers according to their demand and price-response characteristics using data on past purchases and predictions of how each segment will respond to different prices. Essentially, this approach forecasts customer demand based on historical data, and a price is set that will optimize revenue across all customers. A problem with both tiered pricing and revenue management is that customers can use the Web to search out the best deal among all companies, thereby pushing the companies back toward price competition.

To find ways other than pricing to compete, companies should consider bundling goods and services, thereby providing a higher perceived value for the price of the bundle. This tactic has proved especially profitable in electronic commerce, where part of the bundle is an electronic service for which the variable costs are essentially zero. For example, Charles Schwab has avoided direct price competition with online brokers by providing online services such as research tools. These additional services enable Schwab to charge somewhat higher prices without additional variable costs for the services. In another example, until recently, Internet service providers offered customers a discount on the purchase of a PC in return for signing up for three years of Internet service. This service costs the Internet service provider virtually nothing, while ensuring it of a steady stream of income for the three-year period. Obviously, from a consumer's point of view, this deal is less attractive because the contractual period of three years is probably at least 50 percent longer than the usable life of the computer.

Another strategy that moves firms farther away from price competition is the use of CRM to create a one-to-one marketing experience for the customer. The Web generates a huge amount of data on customer buying habits, preferences, amount typically spent, and so on. Armed with this data, companies can use CRM to tailor their products and prices on their Web sites to the individual buyer. This customization creates a closer match to the customer's real needs, which, in turn, increases the probability that the visitor will return to the Web site for another purchase.

The use of CRM can also create switching costs for customers because they must retrain a new Web site to understand their preferences. Although you can provide your data to a single site that shares this information with many electronic commerce sites, thereby reducing the switching price associated with data entry, many Web sites learn about you from your actions, and such data are much more difficult to share over multiple Web sites. In these cases, switching costs remain significant.

The ultimate way to avoid price competition is to use the Web to stage experiences for customers. Instead of using sensory cues such as smell, feel, taste, or sound to encourage customers to purchase on the basis of emotion rather than price, as is often done with print, radio, and television advertising or personal selling in a shop, electronic commerce merchants must find other ways to sell their products based on other factors. One approach employs the Web as a theater to stage unique personal experiences for which customers are willing to pay. These personal experiences can be aesthetic, entertaining, educational, or escapist. In every case, however, the intent is to provide customers with something that has nothing to do with price and that encourages other purchases.

Electronic Commerce Strategies for Dealing with Perfect Choice

With perfect choice, networked economy customers are much more demanding about the goods and services they purchase. They are no longer willing to take what is available on the shelf or car lot; instead, they seek to fulfill specific wishes or needs with their purchases. This demanding nature of consumers actually has a positive side for the firm because it moves customers away from focusing on price and toward focusing on finding a product that meets their needs. Because customized products are seldom standardized, comparisons become more difficult. In addition, the customer is often more than willing to pay a higher price for a customized product.

There are several ways to provide a wider degree of choice to the customer, including virtual showrooms, increased channel choices, wider component choice, and

Tim Berners-Lee

IT INNOVATORS

When you think of electronic commerce, you probably immediately think about the World Wide Web because it is the primary technology that has facilitated so much of the commercial activity on the Internet. Interestingly, whereas creation of the Internet involved the work of a diverse group of people, the Web was invented by one individual—Tim Berners-Lee. While working at CERN (the European Particle Physics Library in Geneva) for six months in 1980, Berners-Lee began thinking about ways to store random associations of disparate things on computers. As a result of this thought, he wrote a program that actually saved information with random links. This program, named Enquire, although never published, was the predecessor of the Web.

In 1984, Berners-Lee went back to work for CERN. In 1989, he proposed a project that would use hypertext to link random pages of information. This project, known as the World Wide Web, enabled users to work together by sharing information in a *web* of hypertext documents. Berners-Lee began work on the first Web server software and the first browser in 1990, making both available within CERN in December 1990 and on the Internet at large in summer 1991. Between 1991 and 1993, Berners-Lee worked on the design of the Web based on feedback from Internet users. The first Windows-based browser, Mosaic, was released in 1994—and the rest is history.

In 1994, Berners-Lee joined the Laboratory for Computer Science (LCS) at the Massachusetts Institute of Technology (MIT), where in 1999, he became the first holder of the 3Com Founders chair. He is also a director of the World Wide Web Consortium, which coordinates Web development worldwide. *Time* magazine named Berners-Lee one of the 100 greatest minds of the twentieth century. He is also the co-author of *Weaving the Web*, in which he discusses the invention of the Web and his ideas about its development in the future.

Tim Berners-Lee invented the World Wide Web and has remained at the forefront of its development.

Source: http://www.w3.org/People/Berners-Lee/Longer.html.

use of mobile technology. For example, Borders decided that it could offer its bookstore customers a much wider variety of items by teaming with Amazon.com to create a virtual store that offered many more books than were available at any bricks-and-mortar store. Barnes and Noble has a similar setup with its own Web site, but goes one step further by allowing returns to either the Web site or to any store. This returns policy combines the best of the Web with the best of the physical store.

The creation of a Web site by L.L. Bean provides another good example offering additional channel choices. After selling through indirect channels for years, Apple moved to a Web ordering system and is now adding physical stores as well. The use of online auctions by large retailers and manufacturers such as Home Depot is another example of creating channel choices.

mobile commerce
The use of laptops, mobile telephones, and personal digital assistants to connect to the Internet and Web to conduct many of the activities normally associated with electronic commerce.

Possibly the largest new channel offered by electronic commerce involves mobile devices using wireless networks to create mobile commerce possibilities. **Mobile commerce** entails the use of laptops, mobile telephones, and personal digital assistants (PDAs) to connect to the Internet and Web so as to carry out many of the activities normally associated with electronic commerce. Instead of being restricted to customers who visit their place of business or use a computer to visit their Web sites, firms can now engage in two-way interaction with customers through these mobile devices. Companies can use the special form of the Web developed for mobile telephones to make information available to customers and to handle orders from them. They can also use outgoing e-mail and short message system (SMS) messages to alert customers to product opportunities. For example, in South Africa, a mall sent messsages to customer mobile telephones as they entered the building, alerting them to sales and special deals at stores in the mall. Finally, customers can use the same e-mail and SMS capabilities as well as voice to request information on products or to actually purchase them.

multimedia messaging system (MMS)
The transmission of richer content types, including photographs, images, voice clips, and eventually video clips, over mobile devices.

These capabilities will be further enhanced by the introduction of **multimedia messaging systems** (MMS) that involve the transmission of richer content types, including photographs, images, voice clips, and eventually video clips. An advertisement for MMS shows one person on a sailboat and another person at a store using a Web camera to send pictures of items to the first person. MMS assumes that the system works with the digital GSM mobile telephone protocol—it is widespread in the rest of the world and now being introduced in the United States. A study by the consulting firm Accenture predicts that the worldwide market for Internet-ready wireless devices will grow by an astounding 630 percent by the year 2005.

U-commerce
An extension of mobile commerce in which the *U* stands for a number of things, including ubiquitous, universal, or unison.

Surpassing mobile commerce in terms of its potential as a channel is **U-commerce**, in which the *U* stands for a number of things, including ubiquitous, universal, or unison. *Ubiquitous* means that computer-based devices will be everywhere in consumer durable devices, and they will all be connected to the Internet with their own IP addresses. They will be able to communicate with one another and with other devices to make life easier. For example, imagine a car that can detect when its battery is weak, search the Internet for the nearest dealer, and set up an appointment for a replacement that suits your calendar. *Universal* means that one universal mobile device will replace all of the devices you carry now and will work anywhere in the world with no roaming or long distance charges. This device might uniquely identify you in a way that no plastic ID card ever could. Finally, *unison* means that there will be complete agreement between all computerized devices you use in terms of your calendar, address book, to-do list, data files, applications, and so. You will no longer have to worry about which device you stored a letter on—it will be available no matter where you are. U-commerce is not too far away thanks to the increasing prevalence of mobile devices, so be on the lookout for examples of U-commerce being used by companies to market and sell their products and services in new and different ways as they seek ways to deal with the Internet effects.[6]

6. Richard T. Watson, "U-commerce: The ultimate." *Ubiquity*, http://www.acm.org/ubiquity/views/r_watson_1.html.

Another strategy for dealing with perfect choice gives consumers increased component choice. It involves providing customers with more information—something at which electronic commerce excels. Using this information, customers can build their own systems, whether they be computers, cars, or high-end sound systems. In essence, the economy will move from capacity-based "push" systems to consumer-based "pull" systems in many industries. In fact, General Motors already allows customers in Europe and Brazil to configure, pay for, and track new car deliveries online. A similar system will likely appear in the United States within the next few years.

Electronic Commerce Strategies for FarEast Foods

FarEast Foods has decided to implement some changes in an effort to deal with the effects of the Internet and Web. First, its management has decided to add higher-quality products to its line for which it can charge higher prices. The company is working on a strategy to sell its products through auction sites such as eBay to provide for variable rather than fixed pricing. It is also studying several ideas to reward frequent purchasers with discounts. Finally, the marketing division plans to start using CRM to learn more about its customers, with hopes of installing a one-to-one marketing system.

3G Mobile Devices

TECHNOLOGY ON THE EDGE

The basis for much of mobile commerce is a new type of mobile device termed a **3G** (third-generation) **device**. To understand this terminology, you need to know that the first-generation mobile telephones were analog devices, similar to land-line telephones. The second-generation (2G) telephones were digital. All of the world except the United States already uses 2G telephones with the GSM protocol. The United States is in a transition era, as it moves from analog mobile phones to digital ones. In fact, mobile communications providers now commonly offer dual-band phones that can deal with either protocol.

In 1999, the migration to packet-switched networks at high-speed data rates was introduced in the United States with the 2.5 generation. The same approach is currently used on the Internet and will offer a highly efficient means for transmitting both voice and data. The 3G standard will provide a speed increase and one consistent digital standard. Because 3G has been developed in concert by all of the mobile telephone manufacturers, it will be a global standard for communications—it became available in Japan in October 2001, with Europe hoping to follow suit in 2002—if the firms can work out some compabililty problems. Implementation in the United States will follow at a later time after the developers resolve some problems with the Defense Department's use of radio frequencies accessed by the 3G standard. Motorola unveiled its first 3G phone in early 2002 for delivery to Hong Kong several months later.

What will 3G mean to mobile telephone users? The primary goal is faster network speeds (eventually, up to 2 Mbps), global interoperability among different operators' networks and mobile phones, and more sophisticated application services. Such advanced application services would allow people to use their phones to listen to high-quality music that they could download from the Internet, send video postcards, engage in videoconferencing in almost any location, and access a range of services not possible with 2G mobile phones. Mobile commerce will thrive in a 3G environment that makes it possible to send text, voice, graphics, audio, and video over a wireless network. The use of MMS will also depend on the implementation of 3G to make high-speed mobile access to the Internet and Web available. For example, in Japan the first 3G phones had bright screens capable of showing 4096 colors, could access the Web at 384 Kbps, and could send and receive e-mail consisting of 5000 characters, far more than available with SMS.

Source: http://www.group3g.com.

For customers demanding greater choices, FarEast Foods' management wants to brand and sell its products through a series of established Asian food distributors. This branding will be linked into the fareastfoods.com Web site, and customers will be directed to their closest distributor if they choose not to purchase over the Web. The company is also looking at ways to move into mobile commerce via SMS so as to reach customers with digital mobile phones.

1. What strategy should be used to deal with price transparency?
2. What strategy should be used to deal with perfect choice demands?

Electronic Commerce Strategies to Deal with Communications Flip-flop

Because communications flip-flop has changed the direction of communication between seller and buyer such that customers now converge on an organization's Web site, any electronic commerce strategy needs to be developed around a company's Web site. Early on, many companies took the approach, "If we build it, they will come." This reasoning has proved faulty, however. With more than 6 billion Web sites and millions being added every day, the buyer has many options available. To counter this problem, companies must know how to develop Web sites that attract potential customers, offer them goods and services they need, and cause them to want to return.

To understand how to build this type of Web site, you must first recognize the opportunities provided by Web advertising and the various types of Web sites possible. Only after you understand Web advertising and the types of Web sites can you move toward creating an electronic commerce strategy that will attract customers who will purchase goods and services via the site. As with other electronic commerce strategies, the emphasis will be on innovation because it represents the only way to develop a long-term competitive advantage.

Web Advertising

The Internet enables firms to develop Web sites where their products and services are described and promoted in considerable detail. A particular advantage of Web advertising is that the firm can change it very quickly. Advertisements for traditional media (print, radio, and TV) are less flexible. For example, a glossy brochure may take weeks to prepare and distribute, and the danger exists that many customers will continue to refer to an older version. By comparison, a firm can update its Web site easily and quickly, and customers always see the latest version. Web advertising means that firms can react quickly to changing demand and adjust customer communication with great speed.

A company can utilize both traditional media and Internet communication (e-mail, lists, and news) to create open communication links with a wide range of customers. Today, common practice is for companies to use print, radio, and television advertising to provide URLs for their Web sites as a way to push customers toward them. In terms of Internet communication, both outbound and inbound e-mail can facilitate frequent communication with Internet-active customers. Outbound e-mail can alert subscribers to sales and special deals available only on the Web or at stores in close proximity to the customer. Inbound e-mail provides the company with access to customer ideas, complaints, and suggestions.

Companies can create e-mail lists to enable any customer to request product changes or new features. The advantage of a list is that another customer reading an idea may contribute to its development and elaboration. Also, a firm can monitor relevant newsgroups to discern what customers are saying about its products or services as well as those of competitors.

A firm can use the Web to pilot new ways of interacting with customers and other stakeholders. For example, it might experiment with different ways of marketing and delivering products and services, or it might develop new communication channels with employees. Above all, firms need to be innovative in using the Web and in finding ways to access creative innovations.

pop-up windows
Browser windows that are generated automatically when you visit certain Web sites and that must be closed separately from the browser window you originally opened.

A fairly new, but somewhat annoying type of Web advertising is the **pop-up window**, which appears with a link to a different Web site when you visit a given URL. These browser windows must be closed separately from the browser window that you originally opened. For example, at the CNN site at cnn.com, a separate window pops up to offer visitors a variety of versions of cnn.com depending on their geographic location. In some cases, these windows can pop up at such a rapid rate that you can't exit the site, forcing you either to close the pop-up windows more rapidly than they are opening or to reboot your computer.

Attracting Customers

attractor
A Web site that continually attracts a high number of visitors.

Once a company has decided on a business model to use in developing its Web presence, it must then find ways to attract customers to its site—not just on a one-time basis, but on a repeat basis. A Web site that attracts very few visitors or the wrong type of visitors represents a very poor investment. Thus, organizations need to consider whom they want to attract to their Web site and how they might attract those users. Because a Web site is essentially located on a one-way street with no other stores on the street, it must be an **attractor** to bring in customers on a repeat basis. As a rule, an organization should concentrate on attracting the most influential stakeholders—that is, these groups that can determine an enterprise's future. Usually, it will want to attract prospective customers, but other groups can influence the future of the organization and thus be the target of a Web site. For example, a firm might use its intranet Web site to communicate with employees, or it might attempt to attract and inform investors and potential suppliers. After selecting the targeted stakeholder group, the organization needs to decide on the degree of personalization of its interaction with this group.

The use of pop-up windows is an approach to marketing that can be annoying.

There are many ways to attract people to your Web site for an initial visit, including advertising on other popular Web sites with links to yours or in other media (television, newspapers, and so on), having your site prominently displayed on a portal site such as Yahoo!, or offering games or giveaways. To attract people on a repeat basis, however, your site must provide customers with something they want—a line of products, a service, or information. Once you have attracted them for the first time, the real trick is to turn visitors into repeat customers. All Web sites that are good attractors, especially for repeat visits, share the feature of **interactivity**, which is the capability to interact with the user in some way. In most cases, the greater the interactivity, the greater the site's perceived attractiveness. The only caveat to this statement is that a high degree of interactivity should not make the site slow and unresponsive. In all cases, companies must avoid creating Web sites that reproduce their print advertising and publicity—so-called **brochure-ware**—that usually bores visitors.

interactivity
The capability of the Web site to interact with the user in some way.

brochure-ware
Web sites that reproduce print publicity and advertising documents.

Although most Web sites allow users to click to Web pages on the same site, or to Web pages on other sites, this level of interactivity often does little to increase the site's appeal. To achieve attractiveness, a developer often uses a programming language to make the page respond immediately to a user's requests or data entry. Another approach is to link the Web server to a database so that it can query a product database to determine whether a product is in stock and then query a credit card database to verify that a credit card number is acceptable. Both approaches are discussed in Chapter 8.

A Two-Stage Model to Attractiveness

Beyond interactivity, it is often necessary to identify the strategic properties of a Web site that will make it attractive to selected stakeholders. Figure 7.8 shows a two-stage model for identifying which features will make a Web site attractive to stakeholders. The first stage involves identifying the target stakeholder groups and using some sort of **influence filter** that makes the site more attractive to the selected stakeholder group. Once this group has been attracted, the site's developer should develop a method for customizing the Web site to meet the needs of the stakeholders, called the **target refractor**.

influence filter
A Web site feature that makes the site more attractive to a specific stakeholder group.

target refractor
A method for customizing a Web site to meet the needs of stakeholders.

Influence filters determine the group that will be attracted to the site. For example, the Kellogg's Web site attracts children by including many features that make it fun for them to visit. One such feature lets young visitors create their own electronic greeting card (e-card), and another allows them to drive a race car. Although adults may visit this site because of the brand name, they probably would not be attracted back to the site. Figure 7.9 shows the Kellogg's racing game on the company's Web site.

Figure 7.8 Two-stage model for attracting visitors to Web site

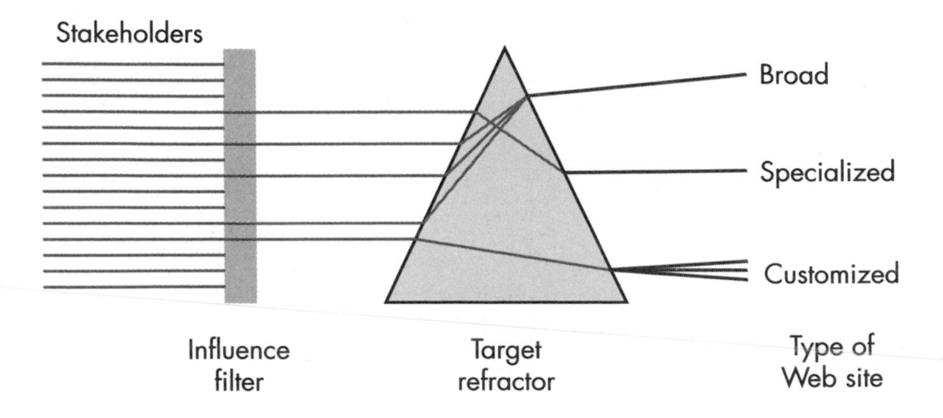

Source: Richard T. Watson, Sigmund Akeisen, and Leyland F. Pitt, "Building Mountains in the Flat Landscape of the World Wide Web." *California Management Review,* Winter 1998, pp. 36–56.

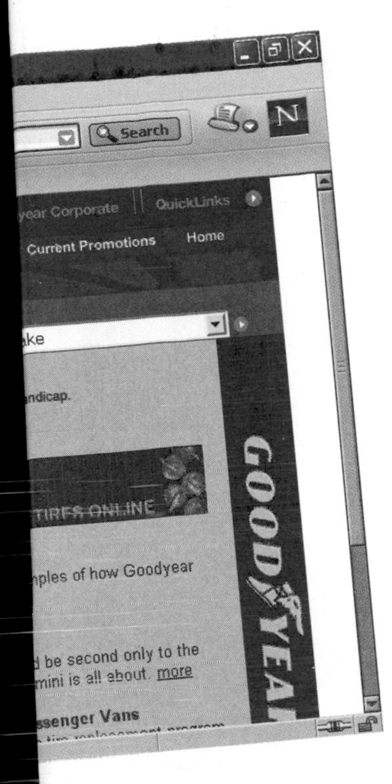

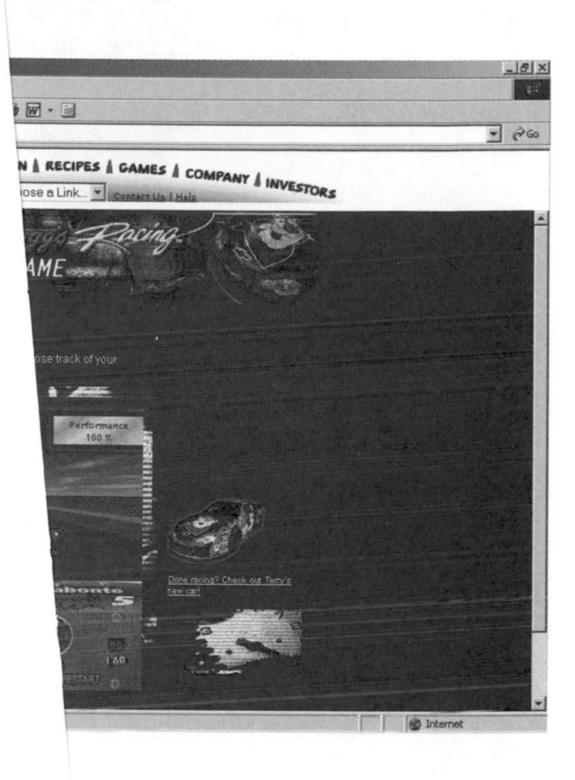

eals to a narrower audience.
el tracking system, has decided
a receipt number to determine
are for preparing transportation
fewer visitors, but nearly all of
The Premier Web sites that Dell
specialized customization.
nteractive relationship with indi-
s just that. Database technology
to be personalized to meet the
b pages exist, such as MyYahoo!,
s, after completing a registration
future visits. For instance, with
MyCNN content), you can deter-
from Horoscopes to News to
your page. On future visits to the
verything of interest to you. This
to save appointments and events.
intments and events, the user can
em appear whenever the user vis-
"My" sites is controlled by a user
s.

attracts repeat visitors through a number
ake it slightly different on each visit. For
as well as the prices on existing items.
cial offers to repeat buyers as an attractor.
gned to appeal to specific groups (young
ood), filter visitors but do not attempt to
xample, all children see the same set of
o make a site worthy of repeat visits, the
way to each visitor. That is, different vis-
different set of pages or have access to
might think about allowing a child to
f countries and languages. It could go
it makes this selection automatically
which can be detected from the cus-

volves the use of a target refractor to
a Web site: broad, specialized, or per-
b site attempts to communicate with
ople in one stakeholder category. For
's Web site,[7] with its information on
Web site provides content with min-
ny visitors may not linger too long
r attention.

The Goodyear Web site is an example of a broad attractor.

With specialized customization, a Web site ap
United Parcel Service (UPS), for instance, with its par
to focus on current customers. A customer can enter
the current location of a package and download softw
documentation. A specialized Web site may attract
those who follow the link find the visit worthwhile.
creates for its large customers are an example of very

personalized Web site
A Web site that creates an interactive relationship with customers.

The marketer's ultimate dream is to develop an i
vidual customers, and a **personalized Web site** do
and back-end application software permit a Web site
needs of the individual. For example, many "My" We
MyScudder, MyNetscape, and so on. With these site
form, the visitor can then select what to see on
MyNetscape (which includes the now-terminated N
mine the content included in your page, ranging
Sports. You can also select the layout and settings for
MyNetscape site, you can obtain a quick view of e
site also includes calendar software that you can use
Because no one else can change or view these app
schedule appointments from anywhere and have th
its this Web site. All of the information on these
name and password for security and privacy reason

My Netscape is an example of an adaptable site.

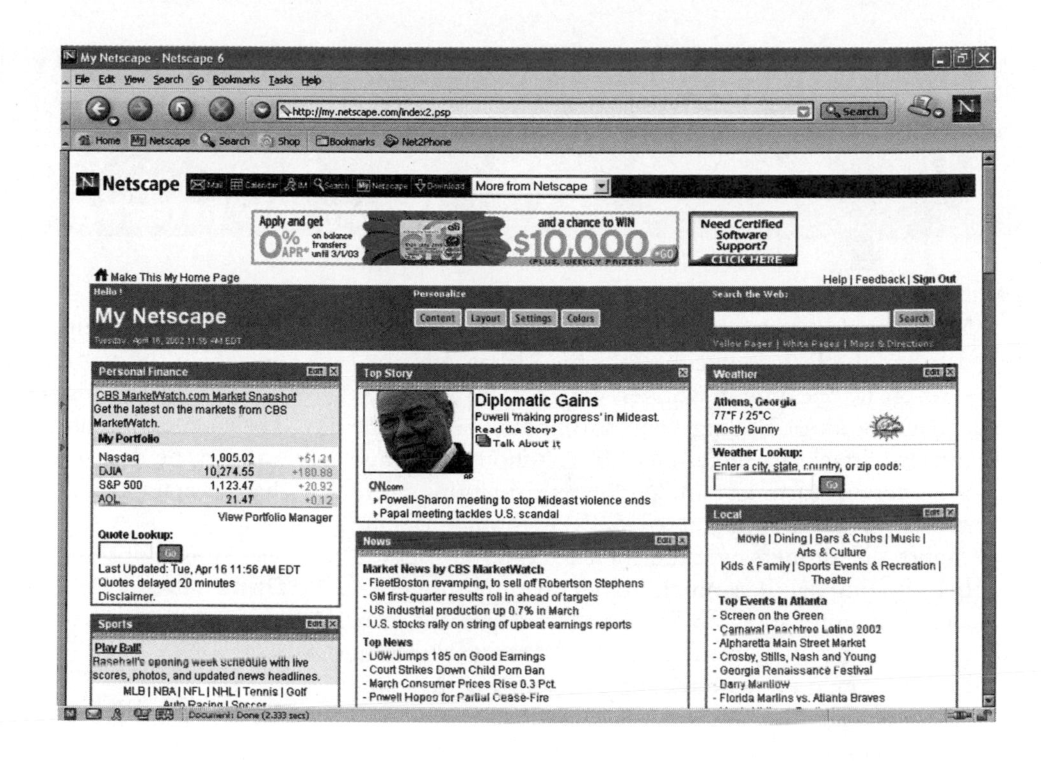

adaptable site
A Web site that can be customized by the visitor.

adaptive site
A site that learns from the visitor's behavior and determines what should be presented.

cookie
A small data file containing data about the user that is kept on the user's computer and read by the browser when the user visits a particular Web site.

Two types of personalized attractors exist. **Adaptable sites** can be customized by the visitor, as in the case of the "My" Web pages. For example, with the My Netscape Web site, the visitor establishes the content, layout, and settings that fit his or her needs by answering questions or selecting options.

In contrast, an **adaptive site** learns from the visitor's behavior and determines what should be presented. This adapting of a site to the customer's needs usually occurs via a small data file, called a **cookie**, that the program stores on the user's computer. Amazon.com is among the best known of the adaptive sites. It tries to discover what type of reading material and music the visitor likes so that it can recommend books, CDs, DVDs, and so on in a special Web page that appears the next time you visit the company's Web site. It uses a cookie to remember you when you return to the site. Web sites such as Amazon.com have proved very successful in terms of mass customization, because every visitor feels as if the site has been customized just for him or her through a personal welcome and a suggested list of books. This is true even though thousands of people visit the same site each day.

Web Strategies for FarEast Foods

FarEast Foods has created a Web site from which it can sell its products to consumers, but it now wants to increase the number of visitors to the Web site. For this reason, it is trying to link its catalog mailing list into an outgoing e-mail system that will alert prior customers who subscribe to the service to special deals that are available only on the Web site. The company is also working on using CRM to learn more about its customers and believes that this information can be used to customize the Web site to each visitor. Eventually, the plan calls for offering MyFareastfoods.com Web sites to customers on which only the types of foods typically purchased by a particular customer are shown and where customers can exchange recipes for these types of food products.

Quick Review

1. What are the two stages of an attractor Web site?
2. How do an adaptable Web site and an adaptive Web site differ?

CaseStudy

Using the Internet at Boeing

With the demand for jets slowing as airlines consolidate and the European conglomerate Airbus puts pressure on it, one might wonder why the management and investors of Boeing were so upbeat prior to the terrorist attacks in September 2001. Although revenue was down, earnings and operating margins were strong, thanks to the Internet. Even though Boeing cannot sell airliners over the Internet, the Internet can help Boeing handle a much higher-margin business—aircraft maintenance. Boeing obtains a 9 percent margin from building planes, but a 20 percent margin from servicing them, so the math supporting the company's changed emphasis is self-evident.

To compete in this market, Boeing has equipped its maintenance crews with wireless devices that enable them to connect to the company's databases directly from the tarmac of an airport, instead of having to drive back to a hanger to check the availability of a part. It has also created a Web site, myboeingfleet.com, that allows its customers to research design and maintenance issues on almost all of an airline's fleet of Boeing planes and order parts online. As a result, online sales of Boeing spare parts have increased 10 percent and the number of employees required to handle telephone calls and faxes from customers has declined from 60 to 12. Ultimately, the largest airplane manufacturer in the world hopes to implement an Internet strategy that may be so ambitious it surpasses even that of General Electric (see Chapter 1) or the automobile industry.

At Boeing, everything about the company—even deciding whether building airplanes is really what it should be doing—is being reconsidered, and the Internet is a part of every decision being made. One important effort seeks to establish a way to purchase airplane parts online from suppliers. To accomplish this task, Boeing had to standardize and reduce the number of different types of parts used by its 18 divisions and digitize many of the billion pages of paper it generates each year. In fact, streamlining this effort alone increased Boeing's margin on manufactured planes from 5 percent to 9 percent. In another effort to reduce costs, Boeing used an online reverse auction (in which the winning participant has the lowest bid rather than the highest bid) to purchase natural gas. The auction resulted in a 10 percent lower cost for gas, and the deal was completed in a few hours rather than several weeks.

Think About It

1. Using the concepts of creative destruction and extension from Chapter 1, discuss Boeing's use of the Internet to extend its business.
2. Discuss Boeing's use of the myboeingfleet.com Web site in terms of this chapter's discussion of Web sites.
3. Compare Boeing's approach to using the Internet with the approach GE is using (see Chapter 1).
4. What other ways might Boeing consider to further its use of the Internet?

Even though the events of September 11, 2001 have reduced Boeing's revenue from aircraft sales, it is using the Internet to increase revenue from services.

Source: Fred Vogelstein, "Flying on the Web in a turbulent economy." *Fortune*, April 30, 2001.

SUMMARY

To summarize this chapter, let's answer the questions posed at the beginning of the chapter.

What is electronic commerce, and how does it relate to the networked economy? Electronic commerce is the process of carrying out business transactions over computer networks. Increasing profitability, gaining market share, improving customer service, and delivering products faster are some of the organizational performance gains possible with electronic commerce. Its emergence will affect all organizations in the world will be affected in one way or another. Electronic commerce is limited to some extent by the fact that physical goods cannot be shipped over the Internet, thus requiring an efficient back-office operation. Electronic products can be shipped over the Internet and do not have this limitation.

What types of Web sites exist, and how do they relate to electronic commerce? Web sites can be classified in several ways, ranging from business versus personal Web sites, to the technology of the Web sites, to the types of applications for which they are used. Business-related Web sites can be categorized based on their Web topology: Internet, intranets, and extranets. The Internet has a global scope and is used for business-to-consumer electronic commerce. Intranets have an organizational scope and are used for business-to-employee purposes. Extranets have a business relationship scope and are used for business relationships. Business models provide another way to categorize Web sites: e-tailers/e-malls, portals, auctions, communities of interest (COI), informediaries, process/services improvement, value chain service providers, and value chain integration.

How is the nature of electronic commerce changing, and what are Internet effects? Originally experts thought that electronic commerce would change business operations by increasing profits and cutting costs while keeping prices up. Companies believed that they simply had to create a Web site and the customers would immediately gravitate toward it. They also believed that customers would not change. Many experts predicted that the so-called dot-com companies would drive the more established companies out of business. Today, bricks-and-mortar companies are making widespread use of electronic commerce to increase profits and market share. The Internet and its multimedia application, the World Wide Web, can have Internet effects on manufacturers and retailers through price transparency, communications flip-flop, and customer demand for perfect choice.

What electronic commerce strategies should be used to deal with price transparency and customer demand for perfect choice? Electronic commerce strategies should deal with price transparency and perfect choice through innovation because innovation is highly difficult to imitate. To counteract price transparency, a company needs an innovative pricing strategy that enables it to increase revenues without encountering customer resistance. This effort includes increasing the value of goods and services, using tiered pricing and revenue management, bundling goods and services, implementing customer relationship management (CRM) to create one-to-one marketing, and creating a Web-based experience for the customer that transcends price. To combat perfect choice, a company needs innovative strategies that provide customers with the choices they demand in the networked economy. These approaches include virtual showrooms, increased channel choices, wider component choice, and use of mobile technology.

How can electronic commerce strategies be used to take advantage of the communications flip-flop? Electronic commerce strategies to deal with the communications flip-flop rely on an understanding of the opportunities provided by Web advertising and the various types of Web sites that can be used. Once a company has selected the type of Web technology and type of site, it must find ways to attract customers to its Web site. This effort can involve the use of an influence filter, which brings in targeted stakeholders, and a target refractor, which determines the degree of customization. Three degrees of mass customization are possible: broad, specialized, and personalized. A broad customization attempts to communicate with several types of stakeholders or many of the people in one stakeholder category. In contrast, a specialized attractor appeals to a narrower audience. A personalized attractor is aimed at a single person. Two types of personalized attractors exist: adaptable and adaptive. The customer carries out the customization on an adaptable attractor, whereas an adaptive attractor customizes itself based on the customer's tastes and purchases.

REVIEW QUESTIONS

1. Why do you need to include all networks, rather than just the Internet, in the definition of electronic commerce?
2. What type of sales over the Internet is forecast to grow the most between now and 2006?
3. What are the three Web technologies? Which is used for business-to-business electronic commerce?
4. List the eight types of Web sites commonly used in electronic commerce. Provide examples of the first three types.
5. What are the limitations of electronic commerce?
6. Which Internet effect has changed the direction of advertising?
7. Which Internet effect has made price competition more difficult for companies? Why?
8. Which Internet effect exacerbates demand risk? Why?
9. Which Internet effect accentuates consumer power? Why?
10. In what ways have the Internet and Web made information about retail and manufacturer prices more transparent? In what ways does price transparency tend to affect prices?
11. What is mass customization? How can a Web site appeal to a mass audience and be customized at the same time?
12. What is disintermediation? How have companies reacted to it?
13. In what ways are consumers demanding greater choice?
14. List electronic commerce strategies that can be used to deal with price transparency.
15. What are some electronic commerce strategies that can give consumers greater choice?
16. What is a pop-up window, and what does it have to do with Web advertising?
17. What is mobile commerce?
18. What is U-commerce, and how does it relate to consumer choice?
19. What are attractors? Why are they important in choosing an electronic commerce strategy?
20. What is an influence filter? A target refractor? How do they relate to electronic commerce?

DISCUSSION QUESTIONS

1. Choose three businesses with which you are familiar and discuss the ways in which Internet effects influence them. How might these businesses use electronic commerce to deal with the Internet effects?
2. Use the Internet and Web to research costs for an automobile of your choice. Write a short (one-page) paper providing these results to a prospective buyer.
3. Assume you are an automobile manufacturer. Discuss ways in which you can deal with the fact that your customers know the cost of the automobile before visiting your dealers.
4. Discuss the concept of disintermediation as it applies to two businesses and professions other than those discussed in the text.
5. Select three electronic commerce Web sites and analyze them in terms of topology, type of site, and success as an attractor.

RESEARCH QUESTIONS

1. Use the Web to research the current level of online sales for Dell and other major computer manfacturers. Write a short report comparing the online sales for each of those manufacturers. As a part of this report, compare their Web sites for purchasing computers.
2. Use the Web to research the current state of digital mobile phones in the United States and the movement toward implementing the G3 technology. Write a two-page paper on your findings.
3. Research the current status of Tesco.com in the United Kingdom, and suggest ways in which its approach to online grocery deliveries could be applied in the United States. Create an electronic presentation, with at least 10 slides, on your findings.
4. Research how the Internet has changed Boeing beyond those effects discussed in the text. Specifically, find other Internet-based applications implemented by Boeing. Write a two-page paper on your findings.
5. The article used as the source for the Boeing case includes four suggestions of ways that organizations can use the Internet. Read this article in its entirety and apply those four suggestions to a business or organization of your choice. Create an electronic presentation, with at least 10 slides, on your findings.

WildOutfitters.com

Claire has just returned from leading Wild Outfitters' first "Corporate Bonding Adventure," an adventure weekend designed specifically for business teams. The three-day package is designed to improve team dynamics by providing a unique setting and challenging outdoor activities. So far, the orders for this new service have been encouraging. Alex and Claire began discussing the events of the weekend to assess the lessons learned, with the hopes of applying these lessons to future outings.

"I was pleasantly surprised," Claire said. "Them thar city folk larn't good in them woods," she added in her *Jethro* twang.

Smiling, Alex responded, "I'm glad that it went well. What do you suggest for the next time? Any changes?"

"I have a couple of pages of notes about that," she replied. "But first, they gave me a few ideas that we can use with the Web site."

The ideas for the Web site that Claire mentioned came from several encounters on her trip. First, she continuously sought feedback through observation and direct inquiries about how the trip was going and what the customers were feeling. This feedback enabled her to respond more effectively to their individual needs on the trip. In addition, she took some time on the last night around the campfire to ask them as a group what they thought about the trip and their suggestions for improvement. Claire also questioned what had attracted the campers to the trip in the first place and what would make them come back. These encounters gave her the content for several pages of notes specific to the "Corporate Bonding Adventure" trips.

On the way home from the trip, a light bulb went off in Claire's head. Why couldn't techniques similar to the ones she had used to seek feedback from the trip participants be used with Web customers? If the Campagnes could get good information about their online customers' preferences, then they could respond more effectively to each customer on an individual basis. They could also use this information to set up special services for attracting new customers and retaining existing customers.

Alex and Claire concluded that a sure way to attract customers was to offer special deals and price cuts on their products. It would be nice if they could somehow combine this approach with the collection of customer information. One solution that they decided to implement involved sending targeted e-mail about specials to selected customers. For example, if a customer had a special interest in water sports, the Campagnes would send the customer information about deals on water sport equipment or river trips. The e-mail would both remind the customer about the Wild Outfitters site, and serve as an attractor to the site to make a purchase. To sign up for the service, the customer would have to fill out an online survey about his or her outdoor hobbies and activities. The survey information would prove valuable to the Campagnes in several ways. First, it would provide them with the data needed to send e-mail to the appropriate targets. Second, the data could be used to gather aggregate statistics on the types of customers who are interested in and shop at their site. Finally, they could use the information for more individual customization of the site in the future.

"Wow! You really learned a lot on this trip," Alex exclaimed proudly. "Maybe we have this whole thing backward. Those city folk should have a trip to guide us around the big city."

Of course, this remark got Claire going again. "Reckon so, Pa," Claire cried, as she slapped her knee. "We kin do some corprit bondin' by the cement pond."

Think About It

1. In which category of companies affected by electronic commerce does Wild Outfitters fit best? Which business model are the Campagnes following?
2. What is your opinion of the Campagne's plans? What opportunities are available? Are there any potential problems they should work to avoid?
3. Which level of mass customization are the Campagnes focusing on in this case? Is this emphasis appropriate? Why or why not?

Hands On

4. Search the Web for other examples of commercial sites for the business model used by WildOutfitters.com. List and describe the features on these sites designed to be attractors for customers. Evaluate the items on your list and rate them according to those that would most likely attract you back to the site. Which would you suggest for WildOutfitters.com?

CHAPTER 8

ELECTRONIC COMMERCE TECHNOLOGIES

LEARNING OBJECTIVES

After reading this chapter, you will be able to answer the following questions:

- What are the layers in the electronic commerce infrastructure?
- How is a transaction carried out using the Web?
- What problems are associated with using the Web for e-commerce, and how are they handled?
- How are electronic commerce transactions protected from criminal activity?
- What electronic commerce payment methods are in use today, and what methods are predicted to emerge in the future?

Delivering Electronic Commerce with UPS

Although attracting customers to an electronic commerce Web site is key to the success of any organization's electronic commerce strategy, no matter how attractive the Web site, or how well it handles transactions, if the purchased item does not arrive at the customer's office or home within a few days, the customer will become unhappy. Today, the vast majority of packages shipped are delivered by United Parcel Service (UPS), the U.S. Postal Service, or Federal Express (FedEx). Of the three, UPS handles the largest volume of packages, 55 percent of the total number shipped, with the Postal Service second at 32 percent, and FedEx third at 10 percent. There is every reason to assume that e-commerce deliveries follow this pattern. In 2001, UPS delivered nearly 3.4 billion packages in 200 countries, earning $2.4 billion on revenues of $30.6 billion.

Although most people are aware of the big, brown trucks and brown-uniformed individuals who deliver packages to their doors, many do not recognize the information technology that makes all of this activity possible. A wide variety of network technology enables UPS to provide a wealth of services to its business partners and customers, including real-time package tracking, just-in-time shipment coordination, customs clearing, financial transactions, and Internet access. In addition, it has a telecommunications arm that connects more than 900,000 users in 100 countries, and its Web site (www.ups.com) averages 21 million hits per day with 2 million online requests as it tracks the movement of 13 million packages. Supporting this package delivery and tracking system requires a massive information technology infrastructure: 14 mainframe computers with 70 trillion bytes of secondary storage, 713 mid-size computers, and 245,000 PCs connected by 3500 LANs.

To improve its tracking system that feeds into this Web site, UPS recently moved to a new type of handheld tracking system that transmits package information in three-tenths of a second instead of the current 10 seconds. It has aggressively adopted mobile devices, with almost 6 million packets being sent daily over wireless networks. In addition, it has sought to address a key issue with electronic purchases—returns—by setting up a return service for online merchants. To further aid its customers, UPS has added logistics and consulting services that provide customers with suggestions on new ways to use supply-chain management and electronic commerce to increase profits. This effort provides further evidence that UPS believes that the information about a package is almost as important as the package itself.

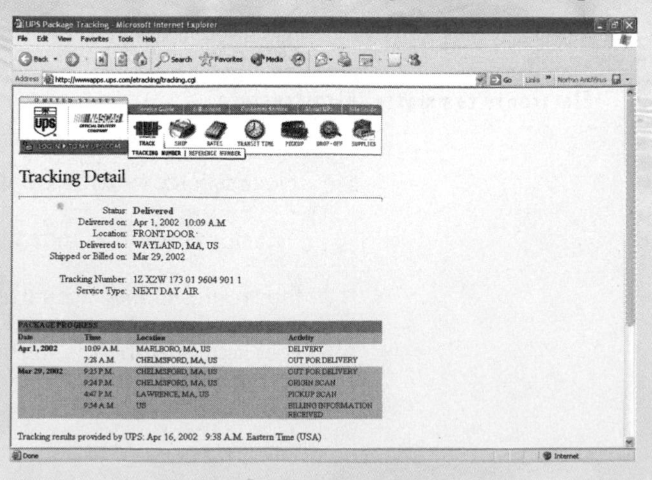

The UPS Web site allows customers to track shipments to their eventual destination.
Source: http://www.ups.com.

Electronic Commerce Technology

Chapter 7 defined electronic commerce and discussed its importance and limitations in the networked economy. It also discussed a number of electronic commerce strategies—the *what* of electronic commerce. This chapter considers the technologies needed to implement electronic commerce strategies—the *how* of electronic commerce. As discussed in the opening case on United Parcel Service, technology is an important part of any organization's overall strategy for success in the networked economy.

This chapter's discussion of electronic commerce technologies begins by looking at a layered model of electronic commerce technologies. Next, it examines the process of using the Web to carry out a transaction and the technologies for processing transactions on both the client (browser) and the Web server. The problems associated with using the Web for electronic commerce will then be discussed. Finally, methodologies for securing electronic commerce and paying for electronic commerce purchases will be examined.

Electronic Commerce Infrastructure

Electronic commerce relies on a number of different technologies. These technologies work together to create a layered, integrated infrastructure that permits the development and deployment of electronic commerce applications, as shown in Figure 8.1. Each layer is built on the layer below it and cannot function without it.[1]

Global Information Infrastructure Layer

global information infrastructure (GII) layer
The infrastructure layer composed of various national information infrastructures, in which some components may differ depending on the country.

national information infrastructure (NII)
Communication networks and protocols, including satellite and cable television networks, telephone networks, mobile communication systems, computer networks, EDI, and Internet protocols (TCP/IP).

The **global information infrastructure (GII) layer** is the bedrock of electronic commerce because all traffic must be transmitted by one or more communication networks and protocols. The GII consists of the **national information-infrastructure (NII)** for each country, which include satellite and cable television networks, telephone networks, mobile communication systems, computer networks, EDI, and Internet protocols (TCP/IP). Note that some components of a country's NII may differ from those of another country's NII. For example, radio and television standards and mobile telephone protocols in the United States tend to be different than those employed in much of the rest of the world. In contrast, the Internet protocol is the same around the world, and the trend is toward standardization of other infrastructure components. For example, the United States is moving toward adoption of the GSM mobile telephone protocol. Another trend in many countries involves increased competition among the various elements of the NII to increase its overall efficiency because an NII is critical to the creation of national wealth.

Figure 8.1 Electronic commerce infrastructure

Electronic commerce applications
Business services infrastructure
Electronic publishing infrastructure
Message distribution infrastructure
Global information infrastructure

1. This section is based on P. G. McKeown and R. T. Watson, *Metamorphosis: A Guide to the World Wide Web and Electronic Commerce*, V2.0 (New York: John Wiley, 1997), pp. 127–128.

Message Distribution Infrastructure Layer

message distribution infrastructure layer
The software layer of electronic commerce that sends and receives messages.

The **message distribution infrastructure layer** consists of protocols for sending and receiving messages. Its purpose is to deliver a message from a server to a client using the GII or NII. Electronic data interchange (EDI), e-mail (SMTP), File Transfer Protocol (FTP), and Hypertext Transfer Protocol (HTTP) are all examples of messaging protocols. The SMTP e-mail protocol is used to move e-mail messages over the Internet. Similarly, transfer of Web files from a server to a browser relies on the HTTP protocol. Messages can be unformatted (such as an e-mail over SMTP) or formatted (such as a purchase order over EDI).

Electronic Publishing Infrastructure Layer

electronic publishing infrastructure layer
The layer that permits organizations to publish a full range of text and multimedia over the message distribution infrastructure.

The **electronic publishing infrastructure layer** permits organizations to publish a full range of text and multimedia over the message distribution infrastructure. The Web offers a good example of the electronic publishing infrastructure layer. Recall that the Web has three key elements:

- The uniform resource locator (URL), which uniquely identifies any server
- A message composed in Hypertext Markup Language (HTML), which makes it possible to deliver plain text plus various formatting tags and hypertext links
- Associated multimedia files linked to the message

The electronic publishing layer is based on the GII layer because it requires the latter layer to deliver the message. The Web also requires TCP/IP for the Internet part of the GII layer. In addition, the Web requires a method of addressability (the URL) to find resources and use of a common language across the network (HTML). In some cases, such as the URL, these standards are built on the previous layer.

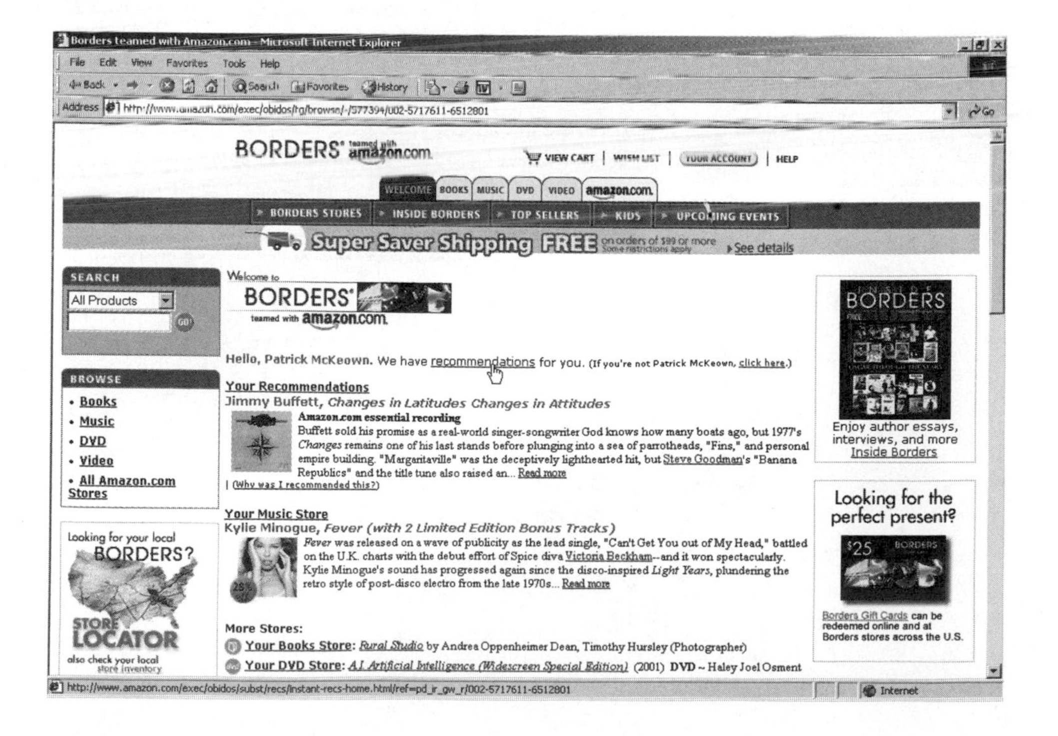

Borders combined with Amazon.com to provide their customers with a combination of bricks-and-mortar and Web site access.

Business Services Infrastructure Layer

business services infrastructure layer
The software layer of electronic commerce that handles the services required to support business transactions (for example, encryption).

The **business services infrastructure layer** comprises a number of common business processes that are used by most electronic commerce applications. Nearly every business, for example, seeks to collect payment for the goods and services it sells. Thus, the business services layer supports the secure transmission of credit card numbers between a customer and an online merchant. In general, this layer includes facilities for encryption and authentication (see the discussion of electronic commerce security later in this chapter).

Electronic Commerce Application Layer

Finally, on top of all the other layers sits an application layer with which customers interact. Transactions take place in this part of the electronic commerce infrastructure, and any Web site where you have purchased a product or compared products has an electronic commerce application. Chapter 4 discussed the electronic commerce application layer in detail, considering transaction processing from both a consumer and a business point of view. Well-known business applications on the Web include the extensive catalogs of books and CDs available at many Web sites, the listings of automobiles available at previously owned automobile sites such as AutoTrader.com, and the catalog of Asian foods on the FarEastFoods Web site. Each of these applications involves a great deal of programming in a variety of computer languages.

Although all of the other layers of this model reside on the server, the application layer represents a classic example of the four-tiered client/server model discussed in earlier chapters. That is, it entails a browser on the customer's computer communicating with a Web server, which in turn communicates with an application server and a database server. For very simple applications, the database, applications, and Web servers exist on one and the same computer; for high-volume or complex applications, the database and application servers reside on separate computers. In any case, the Web browser on the client computer interacts with the Web server software to complete the transaction, as shown in Figure 8.2.

Figure 8.2 Client/server model for electronic commerce

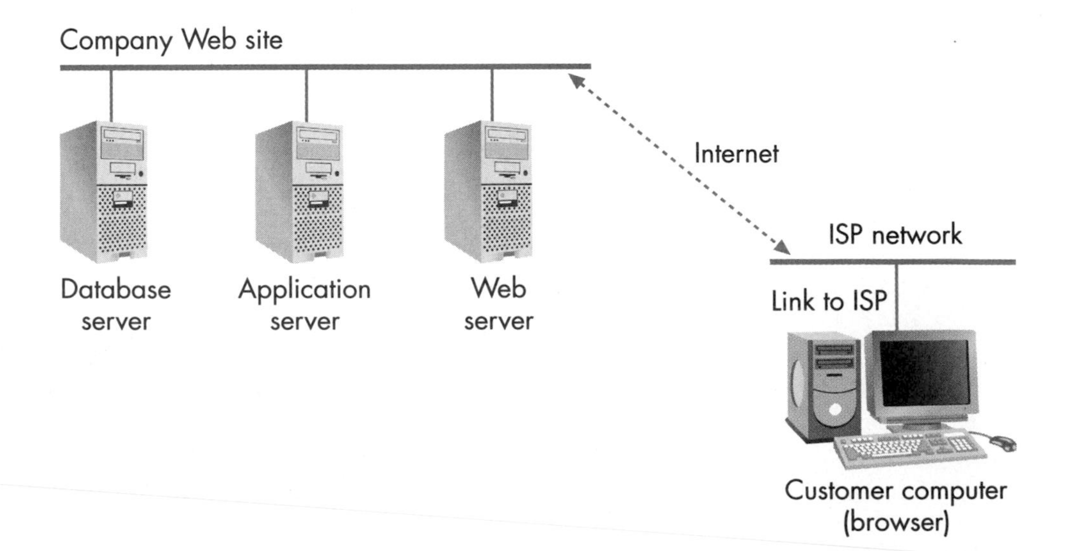

Application to FarEast Foods

As an example of the electronic commerce infrastructure, consider again FarEast Foods, with its online catalog of Asian food items. Figure 8.3 shows the electronic commerce infrastructure for this example. The application consists of FarEast Foods' Asian food item catalog and back-office functionality that receives orders from customers and requests that those orders be fulfilled. The business services infrastructure uses encryption to protect a customer's credit card number. Because this application involves a Web site, the electronic publishing infrastructure uses HTML, and the message distribution infrastructure uses HTTP to deliver the Web page to the customer and to return the completed order form to the company. Finally, the GII uses the Internet to send the messages between the customer and FarEast Foods.

Eric Allman

IT INNOVATORS

Although the World Wide Web has received the most attention as an electronic commerce technology, e-mail remains the killer application of the Internet and the one on which almost all electronic commerce sites depend. Whether it be for in-bound e-mail with orders, queries about prices, complaints, or requests to return an item, or out-bound e-mail with special offers or sales to online customers, e-mail is critical to the success of any electronic commerce strategy. Until the creation of the Web in the early 1990s, e-mail served as the driving force behind the growth and commercialization of the Internet, and its use continues to grow. By 2005, 35 billion e-mail messages are expected to be exchanged each day between 1 billion e-mail accounts. These numbers may actually underestimate the eventual use of e-mail, as organizations and companies look for ways to make wider use of e-mail in the wake of the distribution of anthrax via postal mail in the United States in late 2001.

The success of e-mail and the ultimate success of the Internet as a medium for communication can be traced to a single individual in 1982—Eric Allman, who was trying to make it possible for colleagues to send messages over the new Internet without having to use his computer. At that time, his computer was the only one connected to the Internet, so colleagues had to use it to send e-mail. To understand this issue, consider that when Ray Tominson invented e-mail in 1971, it was specialized for each different network. Mail on a LAN could not be sent over ARPANET (a predecessor to the Internet), and mail could not be sent between networks. Because Allman had a computer with a connection to ARPANET at the University of California-Berkley, his friends and colleagues often borrowed it to send mail to other ARPANET sites. To solve this problem, he wrote a program called Sendmail that transmitted mail between networks by resolving the differences among the network protocols. Allman essentially gave away his program and worked to help others use it. As a result, it became a huge success and is credited with the ultimate success of the Internet. Allman formed a volunteer group to support Sendmail, but the commercialization of the Internet in the mid-1990s so overwhelmed them that Eric co-founded a company, Sendmail, Inc., for this purpose. Today, Sendmail technology powers more than 60 percent of all Internet domains.

Eric Allman's Sendmail program enabled e-mail to become a widespread communication protocol.

Source: "Sendmail, Inc.," http://www.cwheroes.org.

Figure 8.3 Electronic commerce infrastructure for FarEast Foods

Electronic commerce applications	Food item catalog
Business services infrastructure	Encryption
Electronic publishing infrastructure	HTML
Message distribution infrastructure	HTTP
Global information infrastructure	Internet

1. List the five layers of the electronic commerce infrastructure.
2. What is the purpose of the message distribution infrastructure layer?

Web-Based Electronic Commerce

As noted throughout this book, the World Wide Web is the primary technology for electronic commerce. This section discusses the process of carrying out a transaction on the Web. This process begins with the customer finding a Web site by using a search aid or by responding to one of the attractor strategies discussed in Chapter 7. Next, the URL of an electronic commerce site is sent over the Web, the user receives one or more Web pages, and he or she makes a purchase from the Web site. The process ends with the customer receiving verification of his or her order.

A number of problems arise when using a protocol created for exchanging information—the Web—for carrying out transactions and counting visitors. These problems will be discussed in a subsequent section. Workarounds, in the form of Web-based shopping carts developed to solve these problems, will also be discussed. The FarEast Foods Web site will continue to serve as an example of Web- based transactions.

Finding Web Sites with Search Sites

As discussed in Chapter 7, for any electronic commerce transaction to occur on the Web, the merchant's Web site must attract visitors. One of the most important ways that prospective customers find a Web site is through a search site. In some ways, search sites serve as the eyes and ears of the Web because they help the customer find items to purchase. With the Web containing 7 *billion* pages at last count, without some type of search aids, you would never find the exciting new stuff in the 38 pages added to the Internet *every second*. The majority of Web users remain unhappy with the performance of the search sites they use; in one study, more than 70 percent of users expressed dissatisfaction with these sites. The problems are daunting. The Web is huge, containing a great deal of junk. People are not very adept at using search aids, and the software that runs them is not very good at determining what the user actually wants. Furthermore, some search sites have admitted biasing search results in favor of products for which they received payment. Table 8.1 shows the top five search sites in terms of number of unique visitors for July 2001.

Companies continue to improve the search sites, however, and their work is starting to pay off with better search results. To understand why, you need to understand how search sites work. In general, they take one of two approaches: computer or human.

Table 8.1 Most Popular Search Sites

Search Site	Number of Visitors
Yahoo!	64.3 million
Lycos	37.6 million
Excite	28.6 million
Google	15.1 million
Alta Vista	7.2 million

Computer-Driven Search Sites

search engines
Web sites that use technology to find as many Web sites as possible that match the user's request.

spiders
Computer programs that crawl the Web by following links from one page to the next, then send information back to a database of visited pages.

meta tag
An invisible HTML tag that describes the contents of the Web page.

Computer-driven search sites are commonly referred to as **search engines** because they employ technology to increase the breadth of the search. Computer programs called **spiders** crawl the Web by following links from one page to the next, and send information back to a database of visited pages that users then query. The largest search engines claim to have more than 1 billion pages in their databases.

The kind of information returned by the spider depends on the information that the search site is attempting to gather. Some spiders send back only the URL of the site, the title of the site, and maybe a little bit about the general subject. Others index some or all of the text contained on an individual site, the URLs to which the site links, multimedia elements, and site descriptions or subject information contained in hidden fields in the site's HTML coding. These hidden fields, called **meta tags**, do not appear on a Web page, but rather are placed in the coding of a document specifically for the purpose of increasing the chances that a search engine will locate the document. For example, to attract visitors to the FarEast Foods Web page, the company might use the following meta tag:

```
<meta name="keywords" content="food, Asian, Asian foods">
```

Most search engines, such as Lycos, Excite, Google, and Northern Light, take this approach.

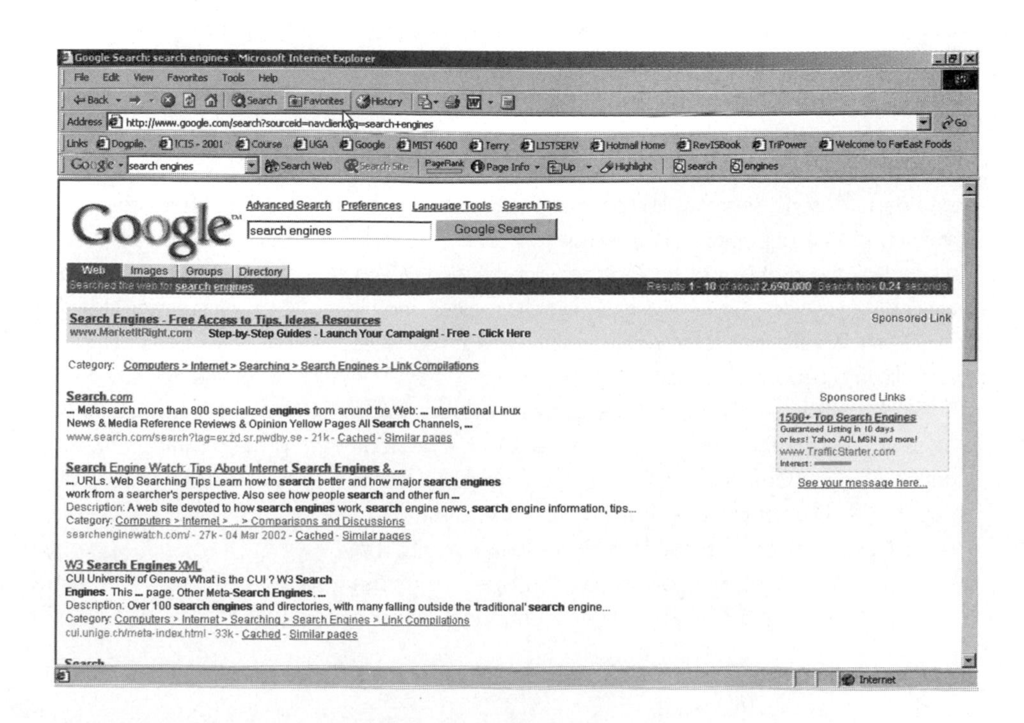

Google, although it looks simple, is one of the more popular search engines.

Figure 8.4 Use of search engine spider

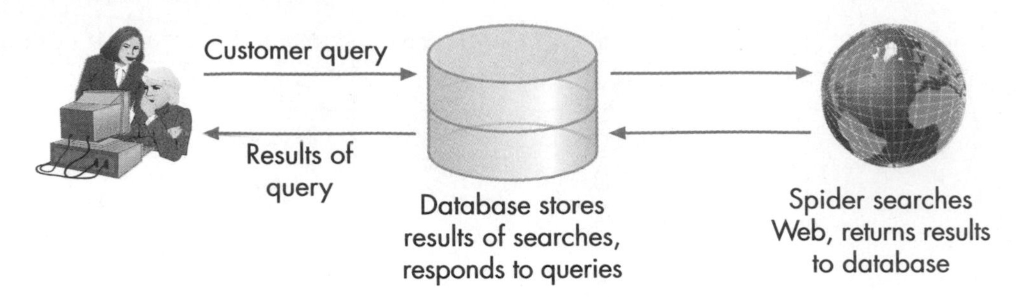
Customer query
Results of query
Database stores results of searches, responds to queries
Spider searches Web, returns results to database

Web page authors often add as many words as possible to the meta tag in hopes of increasing their Web page's visibility to search engines. Some even include meta tags that have nothing to do with the Web page, but will increase the number of hits by a spider. Figure 8.4 shows this methodology. Note that some search engines penalize sites that use irrelevant meta tags—a practice known as spamdexing—by dropping them lower in the results list than they might otherwise be.

Search engines use different measures to rank the pages they find. For example, DirectHit bases its measure of popularity on the amount of time that Web surfers spend at a page. In contrast, Google uses the number of pages with links to the page

Smarter Search Tools

INTERNET IN ACTION

Even though search engines and indexes are popular ways to find Web sites with desired information, products, and services, they remain a source of frustration for many users. Entering a word or phrase often returns a large number of matches unrelated to the object of the search, matching only one word in the search phrase, or matching the search phrase but in a foreign language. Given this frustrating performance, it is not surprising that new search engines, such as Wisenut (www.wisenut.com), or meta-search engines, such as Vivisimo (www.vivisimo.com), are being developed and released by frustrated searchers or companies with what they believe is a better idea. Wisenut searches like other engines, but adds the capability to look at pages without leaving Wisenut. This technique solves the problem of having to switch between Web pages to review a result. Vivisimo searches other search engines, shows the ranking in each engine, and categorizes the results in a separate window.

Smarter search tools also include software add-ons to the browser or operating system, such as the Google Toolbar, Atomica, and the Comet Smart Cursor. The Google Toolbar is an add-on to Internet Explorer that increases the efficiency of searches and that adds some features to found pages. For example, with the Google Toolbar, you can type in a phrase and search for it without leaving your current page. You can then search for the same term in matching pages, get information on the page, determine the validity of a match, have the matching term highlighted in a page, and so on. A neat feature is the capability to translate a foreign-language page into English.

Atomica and Smart Cursor do essentially the same thing: They enable you to position the cursor on a word and carry out a number of operations with that word, such as get a definition, get an encyclopedia description, or search for that word just as if you were working in a search engine. Atomica works with any type of document, but Smart Cursor works only on Web pages. You can download the Google Toolbar from google.toolbar.com, Atomica from www.atomica.com (only the personal version), and the Smart Cursor (along with a number of fun cursors) from www.cometsystems.com/sc.

Given the current instability of Web companies, any or all of the search engines or search tools mentioned here might not be available when you read this book. Given the growing importance of searching on the Web, however, others will undoubtedly spring up to replace them.

Sources: Stephen H. Wildstrom, "Smarter tools to scour a wider Web." *Business Week*, March 26, 2001, p. 26; and Thomas E. Weber, "New Web search tools offer useful shortcuts and some nice twists." *Wall Street Journal*, October 1, 2001, p. B1.

as its measure of popularity. Another search approach uses a meta-search engine, such as the aptly named Dogpile, which searches over a number of search sites and returns the top 10 results from each. This type of search approach allows you to look at the best hits on the widest variety of search engines.

For electronic commerce, one of the most useful search engines may be RealNames. The RealNames search engine allows users to enter a product name and find the related Web site, bypassing the need to enter a URL in the browser address box.

Human-Indexed Search Sites

Some Web search sites use the human brain to categorize Web pages into directories, which may provide better results than a general search conducted by a spider. These directories categorize sites into a hierarchy of topics, with subtopics under each main topic. Yahoo! is the best-known example of this approach, employing a large number of editors and Web surfers to do the categorizing. Yahoo! provides both new and seasoned Web users with a structured view of hundreds of thousands of Web sites and millions of Web pages. If a search argument doesn't lead to a topic page, it will still lead to results from the six or seven popular search engine sites to which Yahoo! links. Yahoo! began as the bookmark lists of two Stanford University graduate students, David Filo and Jerry Yang. After putting their combined bookmark lists, organized by categories, on a college site, the list began to grow into an Internet phenomenon. It became the first such directory with a large following. Other popular directories include AOL Search and MSN Search, and most portals now use the directory approach to organize information.

Getting a Web Site Indexed

How does a company get its Web site onto a search site? Although you could wait for a spider to stumble upon your Web site or a directory editor to index it, the best way is to submit the site yourself. For example, Yahoo! has a link that allows you to suggest a site for inclusion. You can submit your Web site's URL to most search sites, rather than just waiting for them to find you, or you can agree to pay a fee to have them include your site.

MSN uses humans and technology to create its directory.

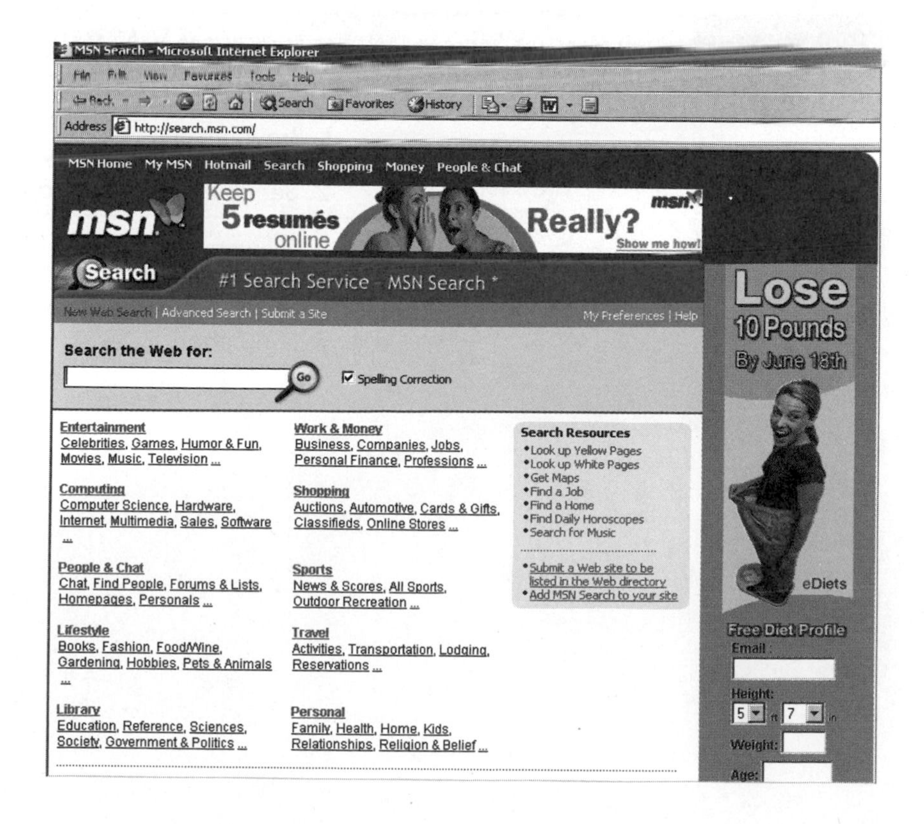

Transacting an Order over the Web

Once customers have found the Web site with which they wish to do business, the next step is to carry out the interactions between the customer's Web browser and the Web server of the merchant necessary to purchase the items. This interaction involves five steps, shown in Figure 8.5, some of which are repeated multiple times:

1. The customer sends a URL over the Internet to the Web server.
2. The Web server responds by sending a Web page to the customer.
3. The customer responds by sending data back to the Web server.
4. The server processes the data, often querying a database.
5. Based on the processing, the server sends a new Web page to the customer.

Note that during the five steps of this interaction, the customer's browser always sends data to the Web server, either in the form of a URL or as data regarding the item being purchased. In contrast, the Web server always sends Web pages back to the browser because browsers can only accept Web pages.

As customers seek additional information, they may repeat Steps 1 and 2 several times by clicking hyperlinks on the current page. Clicking a hyperlink sends another URL to the Web server, resulting in another page plus any associated multimedia files being sent back from the original Web server or other Web servers. For example, when you submit the URL for the FarEast Foods Web site (www.fareastfoods.com), you receive the home page for this Web site (number 1 in Figure 8.6). From the home page, you can select a type of Asian food—for example—Thai, by clicking the corresponding link (*Thailand*), and see a catalog page listing this type of food product (number 2 in Figure 8.6). Clicking the *Prawn Cracker* link on page 2 displays a page from which the customer can select the number of units of Prawn Cracker to order (number 3 in Figure 8.6). Finally, clicking the *Checkout* link displays page 4. The bidirectional arrows in Figure 8.6 indicate that the user can easily go back from any individual page to a previous page.

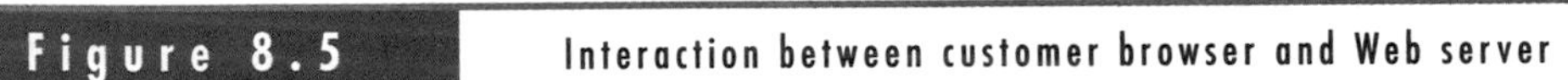

Figure 8.5 Interaction between customer browser and Web server

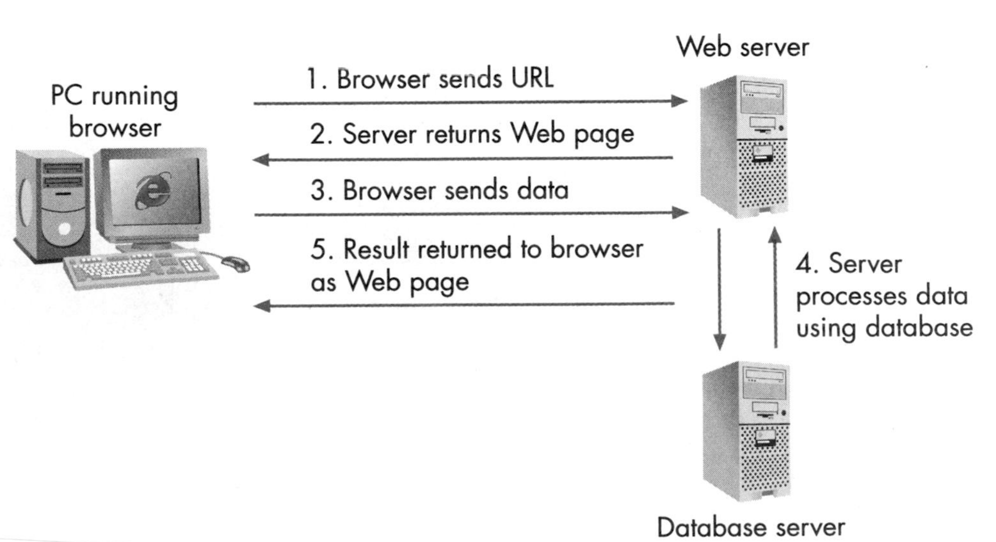

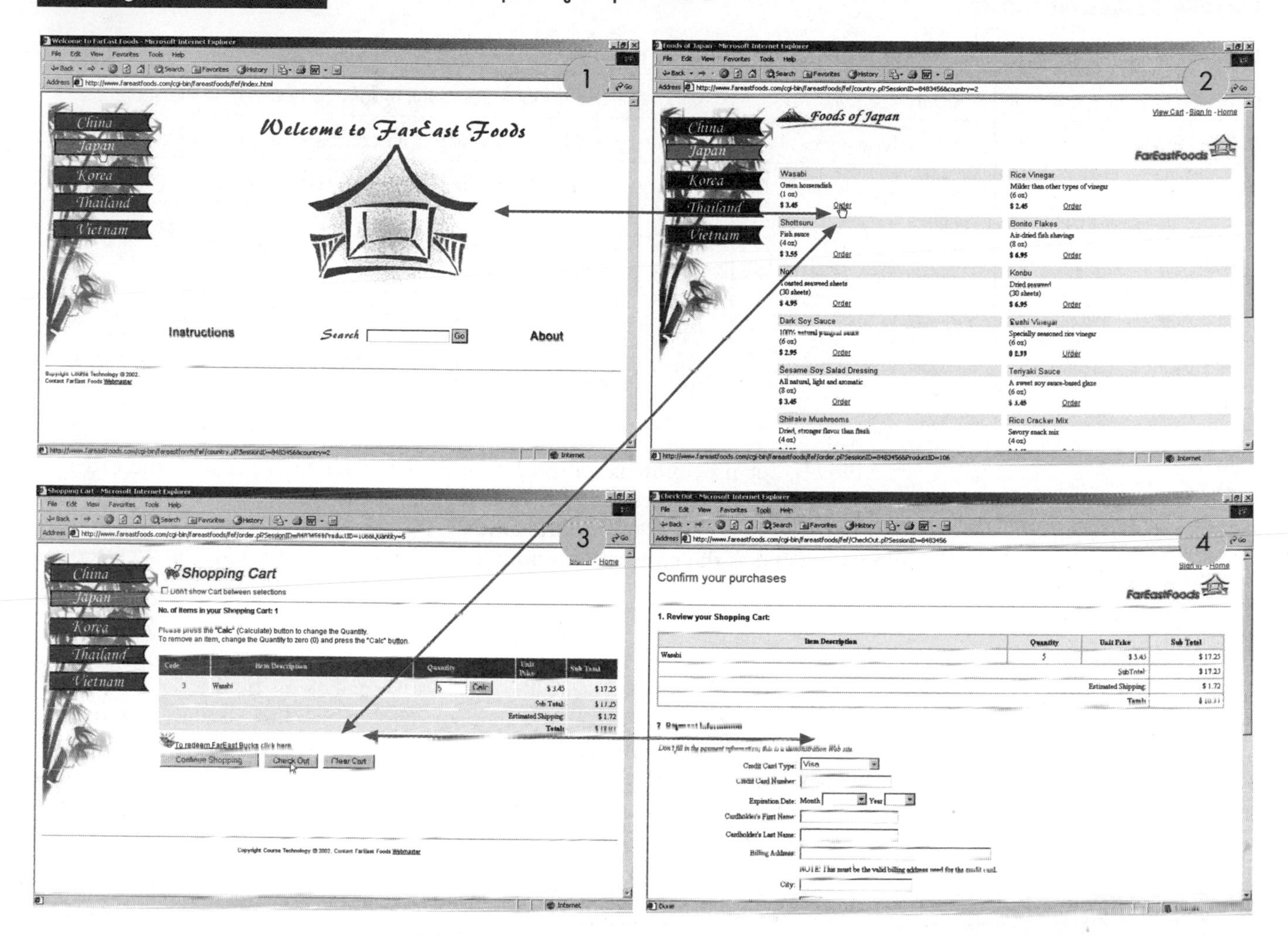

Figure 8.6 Customer repeating Steps 1 and 2

form page
A type of Web page that sends data to a Web server.

Once the customer reaches the checkout page page (number 4 in Figure 8.6), the third step of the process—sending data back to the server—will occur. In Step 3, the customer fills out a special type of Web page called a **form page**, which sends data to the server when the customer clicks the submit button. Developers create form pages with the HTML <form> </form> tag pair. They also create various input elements, such as text boxes, radio buttons, and list boxes, with HTML tags and place them inside the form tag pair. Users submit the data by clicking a submit button or clear the data by clicking a reset button. Table 8.2 shows some of the tags needed to create a form page, and Figure 8.7 shows the bottom portion of a completed FarEast Foods form page ready to be submitted. In Figure 8.7, a series of text boxes is combined with list boxes, a submit button, and a reset button. Clicking the submit button sends the data to the Web server.

In most cases, database access is an important part of the online purchase process. If the purchase involves an item possibly available in limited quantities, such as packages of wasabi at FarEast Foods, clothing at Lands' End, or seats on an airline, accessing a database can ensure that the item is actually available. After all, selling unavailable goods or services is not a good way to build customer happiness. Even when an item has unlimited availability, such as when a user downloads a copy of software or a copy of an audio file, the purchaser's name and pertinent information are usually stored on a database for warranty purposes, future upgrades, or marketing campaigns.

Table 8.2 HTML Form Tags

Object Created	Example	Purpose
Form	<form> ... </form>	Creates the form page from which data can be sent to the server
Text box	<Input Type=text>	Creates a text box into which data can be entered
List box	<Select> ... </Select>	Creates a list box from which an option can be selected
Radio button	<Input Type=radio>	Creates a radio or option button, only one of which can be selected
Check box	<Input Type=checkbox>	Creates a check box, several of which can be checked
Submit button	<Input Type=Submit>	Creates a button that, when clicked, submits data to server
Reset button	<Input Type=Reset>	Creates a button that, when clicked, clears all data from input elements

Finally, during confirmation of the order, if a problem arises with the order or other communication between the Web server and the customer becomes necessary, another Web page is sent to the customer's Web browser. Steps 3 through 5 are often repeated as the customer communicates back and forth with the Web server.

Client- and Server-Side Processing

client-side processing
Processing that takes place on the browser itself before data are actually sent to the server.

When a user submits a form to the server for processing, some processing may take place on the browser itself before the data are actually sent to the server. Termed **client-side processing**, this processing can involve checking for missing data or for making simple calculations prior to the data being submitted in Step 3 of the Web transaction. For the example shown in Figure 8.7, client-side processing could verify that all necessary input elements (the name, address, credit card

Figure 8.7 Form Web page with data entered

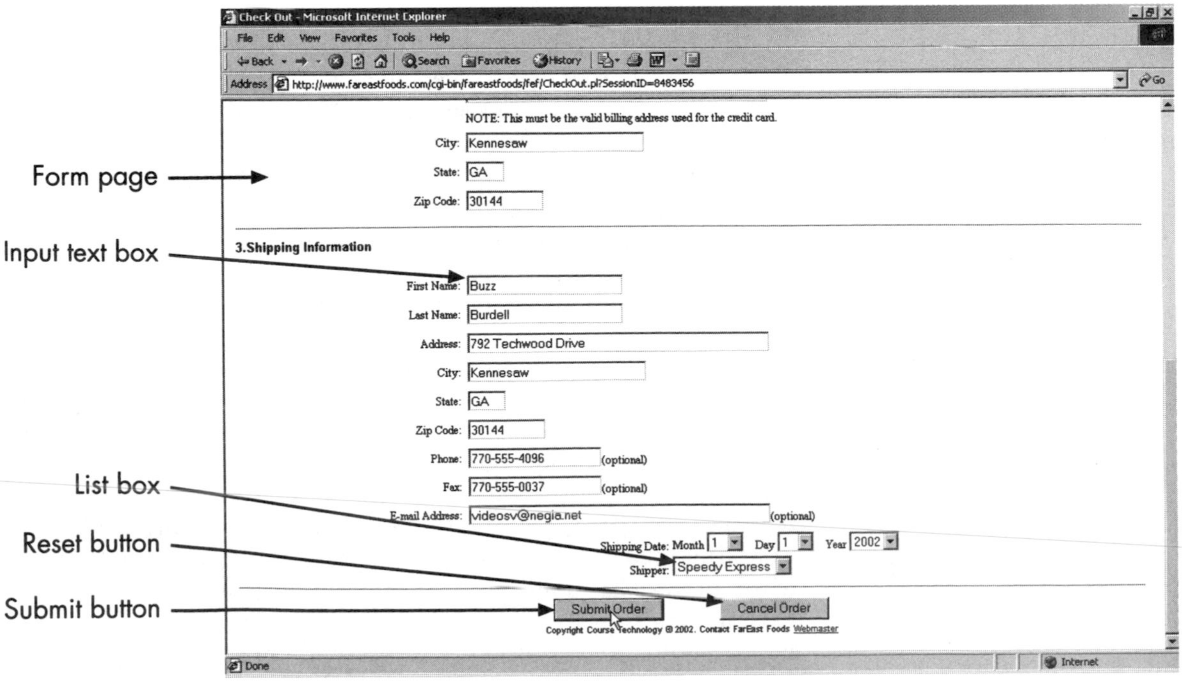

This Java applet can be used to determine mortgage payments.

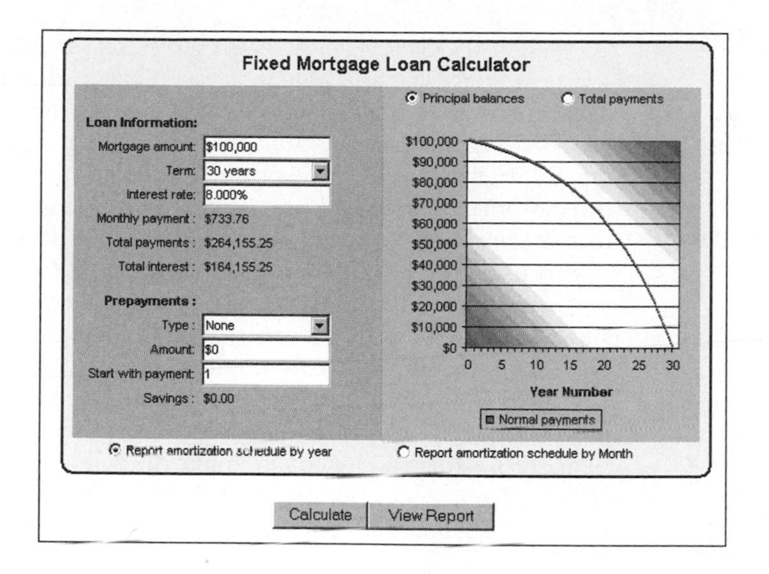

information, and so on) are completed, prior to sending the data to the server. Client-side processing is also used to make intermediate calculations on a Web page. For example, it might be used in the last of the four Web pages shown in Figure 8.6 to ensure that the user has filled out all required fields before the data are submitted to the Web server.

Client-side processing is useful because it can keep often over-burdened Web servers from trying to process data from incomplete form pages. It can also reduce the processing burden on Web servers by carrying out simple calculations directly on the Web page. Client-side processing is typically handled by one of three programming languages: JavaScript, Java, or ActiveX. Developers build programs written in JavaScript, called scripts, directly into a Web page, and the browser then executes these scripts. Java programs are self-contained computer programs on Web pages, known as **applets**, which are written by the Web developer and stored on the server. The server sends applets separately from the Web page; the applets are then referenced by the Web page through an <applet> HTML tag. If you can see programming statements such as those shown in Figure 8.8 when you select the View Page Source or View Source option on your browser, then the program uses JavaScript. If you see the <applet> tag in the source code, then it uses a Java applet. ActiveX controls are Microsoft's answer to Java applets and carry out the same functions on Web pages. You can recognize an ActiveX control by the <object> tag in the HTML code.

applets
Self-contained computer programs on Web pages written in Java.

Figure 8.8 Client-side processing with JavaScript

```
<HTML><HEAD><TITLE>Check Out</TITLE>
<SCRIPT language=JavaScript>
<!–
function cmdSubmitOrder_Click() {
        window.alert ("Purchase confirmed.\nThank you for
        shopping FarEast Foods.");
}
function cmdCancel_Click(strProductID) {
        strNewUrl = "order.pl?SessionID=8483456";
        document.location = strNewUrl;
}
//—->
</SCRIPT>
```

server-side processing
Accepting data from a user's browser, processing it, and generating and returning a Web page to the user.

When the data reach the Web server, server-side processing takes place on the server in Step 4 of the Web transaction. **Server-side processing** involves accepting data from a user's brower, processing it, and generating and returning a Web page to users. Taking the data sent from the form page and processing it often requires some type of database access to determine the price and availability of an item selected by the customer. Server-side processing typically results in the creation of a Web page that is sent back to the browser. For example, server-side processing created Web page 4 in Figure 8.6, based on user selections in the previous Web pages.

common gateway interface (CGI)
A method of communication between the Web server software and a computer program that processes data sent from the user.

Server-side processing can occur in two ways: the common gateway interface and server includes. The **common gateway interface (CGI)**, the older of the two methods, supports communication between the Web server software and a computer program that does the processing. The computer program that does the processing is usually stored in a folder named cgi-bin, and it is accessed through a URL or a <form> tag. For example, the following URL would access a program called CheckOut.pl, located in subfolders in the cgi-bin folder on the FarEast Foods Web server and execute it using data from the user:

http://www.fareastfoods.com/cgi-bin/fareastfoods/fef/CheckOut.pl

Developers can write CGI programs in a variety of computer languages, including Perl, C++, and Java. In the preceding example, the checkout program is written in Perl, as indicated by the .pl file extension. Figure 8.9 depicts the process of using CGI programming for server-side processing.

Figure 8.9 Server-side processing using CGI

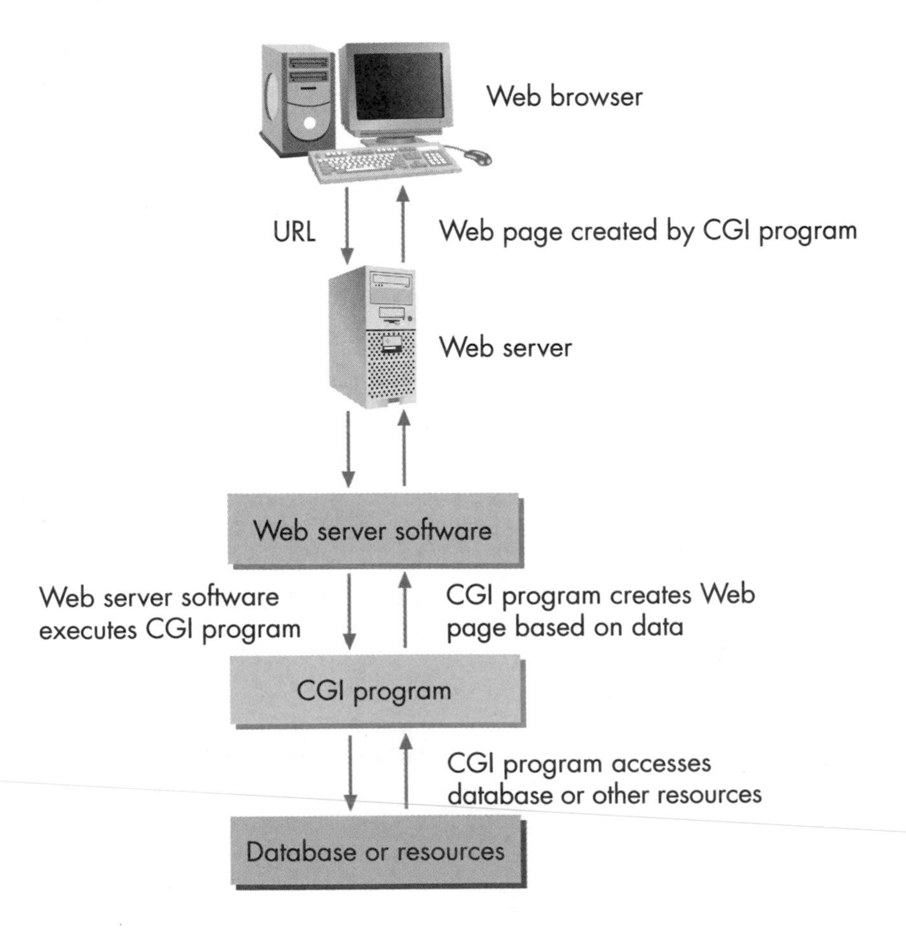

server includes
A way of processing data from the user that involves integrating the programming code into the Web page that is being sent to the server, but in such a way that users do not see the code.

Active Server Pages (ASP)
Web pages that include one or more embedded programs, which are processed on a Web server to generate and return a Web page to the user.

Server includes provides a way to handle server-side processing that avoids adding the CGI layer to the process. Instead, server includes work by integrating the programming code into the Web page that is being sent to the server, but in such a way that users do not see the code. The most popular approaches to server includes are Active Server Pages and Java Server Pages. **Active Server Pages (ASP)** are Web pages that include one or more embedded programs, which a Web server processes to generate a Web page that is sent back to the user. ASP was originally developed to work with Microsoft's Internet Information Server (IIS) Web server, but other companies have since modified their Web servers to support ASP as well. In Figure 8.9, a system using server includes would omit the CGI layer. Instead, the server includes programming statements would be integrated into the Web server software layer. Java Server Pages (JSP) use Java on the server rather than a form of Visual Basic *.NET used with ASP .NET.

The Jazz on Java

TECHNOLOGY ON THE EDGE

This book has often mentioned the Java computer language. Just what is Java? This object-oriented, cross-platform language is similar to C++. Java was developed by Sun Microsystems to resolve the technical morass of working with different operating systems and machines. Programs written in other computer languages for one operating system or for a specific type of computer are often difficult to move to a different operating system or computer. Because Java is designed to be compatible with any computer platform, it is portable. This portability makes it a good language to use to write Web applications, which need to run on many types of computers and browsers.

The basic element of Java Web applications is the applet, a small application that can be downloaded over the Internet when needed. Applications created in Java can be deployed without modification to any computing platform, thereby saving the costs associated with developing software for multiple platforms. Because Java applications reside on centralized servers, users do not need to insert disks and companies do not need to ship CDs to update software.

A key part of using Java is the Java Virtual Machine. The **Java Virtual Machine (JVM)** is a piece of software that is built into operating systems or can be downloaded and added to them; it interprets the code in the Java applet and executes it within the browser. As a consequence, a developer can create an applet in Java that will run on Netscape on a Windows or Linux-based computer as well as on Internet Explorer on a Macintosh, or vice versa. With Java and the JVM, you can do much more complex client-side processing than is possible with other approaches. For example, suppose you want to use a Java spreadsheet applet to analyze your investments. To do so, you would use your browser to download both the applet and financial data from your financial services company and then use the applet to run what-if analyses on the data. Microsoft's competitors to Java applets are ActiveX programs, which also run in browsers.

Not surprisingly, some confusion has arisen regarding the difference between Java and JavaScript. Although the names are similar, both share some similar language structures, and both run on browsers, the similarity ends there. Whereas Java is a true computer language that can run on its own or on browsers as applets, JavaScript is a scripting language that runs only on browsers and works only with HTML. In fact, the name *JavaScript* is a marketing ploy on the part of Netscape to piggy-back on the popularity of Java; its original name was NewScript.

Microsoft or Open-Source Web Server

Apache
An open-source Web server.

open-source software
Software that is created and supported by volunteers who make it freely available to users who can then add any features desired to it.

A major issue with electronic commerce technology is the type of Web server software used. As shown in Figure 8.9, the Web server software is a key element in server-side processing because it must accept URLs and data from browsers and return appropriate Web pages. Today, you have a number of choices for Web server software, but two packages control close to 90 percent of the market: Apache and Microsoft IIS. **Apache** Web server software, which runs on 60 percent of Web servers, is open-source software similar to Linux. **Open-source software** is created and supported by volunteers who make it freely available to users, who can then add any features desired to it. The Apache Web server software runs on almost all types of server operating systems, including Linux, UNIX, and Windows.

In contrast, Microsoft IIS was developed by Microsoft and is controlled by it. Running on the Windows server operating systems, IIS has close to 30 percent of the Web server market.

1. How do client-side processing and server-side processing differ?
2. How does the use of CGI differ from the use of server includes?

Problems with the HTTP Protocol

Although the World Wide Web has become the centerpiece of electronic commerce, problems with its underlying protocol, HTTP, require workarounds to make it work for electronic commerce. This section discusses the two biggest problems: the stateless nature of HTTP and the way it records information about visitors to Web sites.

stateless protocol
A client/server protocol in which the server has no memory of an interaction with the client other than logging some information about it.

HTTP is a **stateless protocol**. That is, when a Web server receives a URL, the server sends back the requested Web page and then forgets that this event ever occurred, other than logging the address from which the request originated. This stateless quality explains why trying to return to an earlier page during a Web transaction often does not work; the Web server has forgotten who you are between pages.

When you visit a Web site, HTTP does not count you or your machine, but rather counts your IP address. Recall that an IP address consists of four groups of numbers between 0 and 255 and that it uniquely identifies machines that are connected to the Internet. For example, 192.225.56.33 is the IP address for the Web server for FarEast Foods (www.fareastfoods.com). Because HTTP counts only IP addresses and many users share the IP addresses of their ISP, a Web server will miscount the number of visitors.

The Stateless Nature of HTTP

session
A client/server protocol in which a continuous sequence of transactions occurs between client and server.

Because HTTP is a stateless protocol, the browser does not carry on any sequence of communications with the Web server. Contrast this situation with how FTP works. With FTP, the client software engages in a **session** with an FTP server for a sequence of transactions. The stateless nature of HTTP is similar to calling someone on the telephone, but having the call disconnect immediately after the other person answers. To exchange information, you must redial and connect with the person again. When you reconnect the call, however, the person who answers does not remember the previous call.

You might ask why Tim Berners-Lee chose to use HTTP as the protocol behind the World Wide Web given its forgetful nature. To understand this decision, you need to remember that he was interested only in making information widely available—he was not thinking about the commercial implications of the Web. For sharing information, HTTP offers a number of important advantages:

- It requires only a modest amount of code or resources to operate
- Because the connection stays open for only one operation, it efficiently links from one object (Web page) to another and then to another
- It returns results from the server based on the URL, regardless of any previous operations performed by the client
- It allows for retrieval of an unrestricted set of formats, avoiding a need for standardization of formats
- It provides a degree of privacy, because, without a workaround, the Web server does not track user visits

HTTP allows the transfer of information with little programming or requirements for high-powered computers—key factors in its use in information sharing. In addition, HTTP enables users to jump to other Web pages or Internet resources quickly and easily because they need not disconnect from one page or resource before going to another. Also, the Web server can send the requested Web page or other Internet resource without worrying about previous interactions with the user's browser. Finally, the Web server can send both text and multimedia in whatever form is necessary because it does not have to worry about formats, as does EDI.

On the other hand, HTTP has some definite disadvantages for electronic commerce purposes. The biggest disadvantage is that Web sites cannot handle multiple purchases without some sort of modification. Because HTTP is stateless, it does not remember previous visits to the Web site so it does not have a record of earlier selections.

Counting Web Site Visitors

Virtually every electronic commerce Web site needs to know how many unique visitors it has. Knowing how many unique visitors have come to a Web site gives the individual, organization, or company supporting the Web site an idea of just how effective an attractor the site is. Also, the rate charged for third-party advertising on a site is often tied to the number of hits to that site. In addition, the number of unique visitors represents an important starting point for developing a complete picture of the audience served by the Web site.

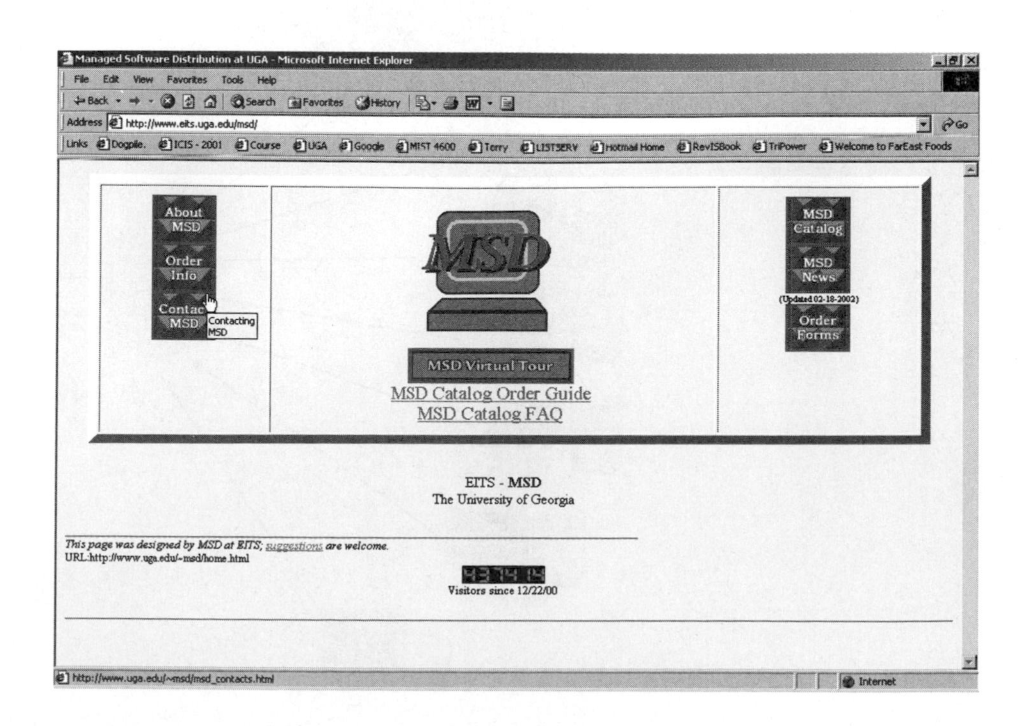

The counter on this Web page tracks the number of hits, but not necessarily the number of unique visitors.

HTTP does not handle counting unique visits to a Web site very well. The reason for its ineffectiveness in this area relates to the unknown number of visitors who come from behind an organizational firewall or from a dial-up or cable ISP. Web clients within an organization on a local network are often protected from outside attack by a computer known as a *firewall*, which resides between an organization's network and the Internet to control data access. It works by filtering every message going into the organization. The only IP address recorded by a Web site is the IP address of the firewall computer, no matter how many different people and machines exist behind the firewall. Similarly, because IP addresses are *dynamically* allocated to each dial-up or cable user as he or she accesses the ISP computer, a count of IP addresses coming to a Web server from an ISP can prove misleading. Figure 8.10 illustrates both of these problems.

Figure 8.10 Problems in counting visitors

a. Local network behind firewall

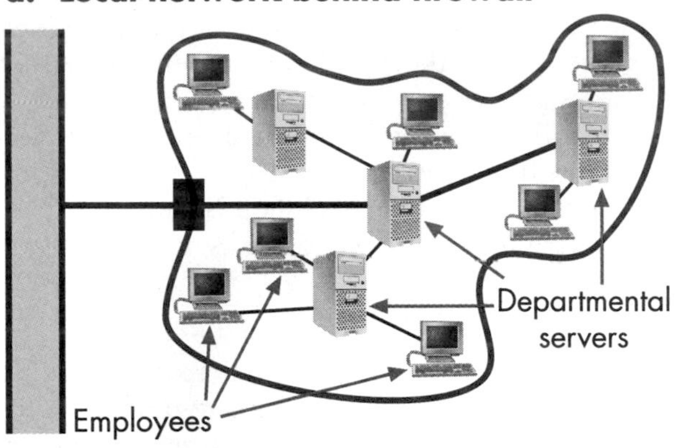

b. Dial-up or cable users

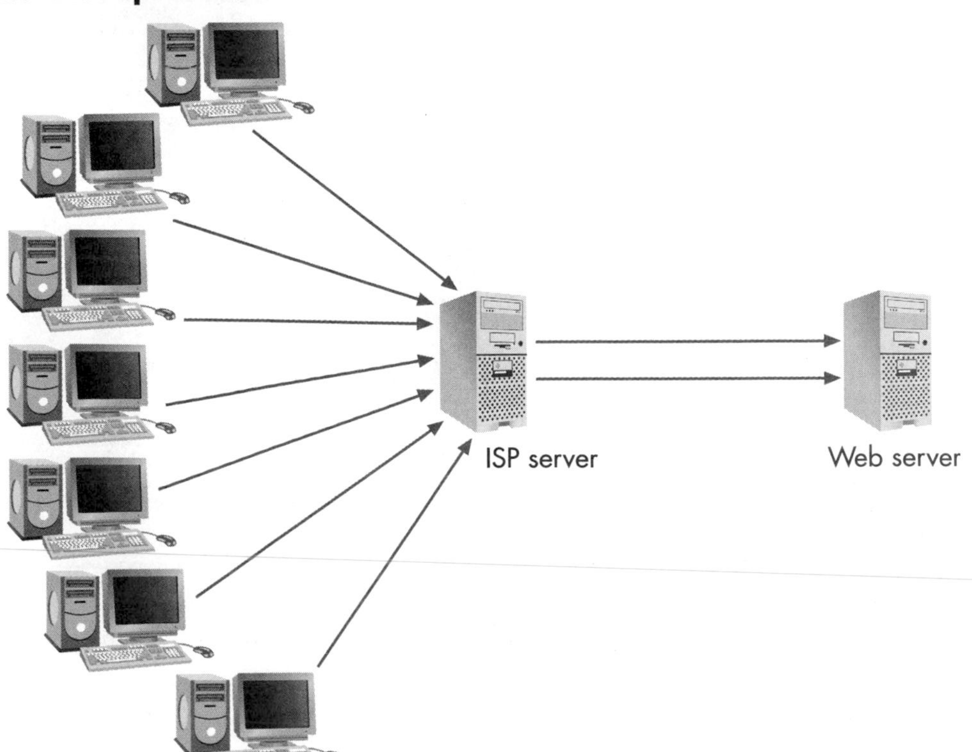

In Figure 8.10(a), numerous machines and users are located behind the firewall, but the Web server counts only a single IP address—that of the firewall computer. In the case of the ISP, even though there are multiple users, not all of them access the ISP at the same time, and the ISP server has far fewer IP addresses than customers. It dynamically allocates these IP address as customers dial in. For the example, in Figure 8.10(b), even though there are seven users, the Web server counts only two IP addresses.

Using Cookies for Shopping Carts and Counting Visitors

As it turns out, the problems of customers making multiple purchases from an electronic commerce site and counting visitors to a Web site can both be solved with a single approach—a cookie. A *cookie* is information that a Web site stores on a computer's hard drive to enable it to identify the computer at a later time. In the simplest form of a cookie, a unique identifier is created for each visitor the first time he or she visits a Web site and is stored in a database on the Web server. This identifier is sent back to the visitor's computer as a part of the URL of a Web page and stored on the user's hard drive. When the visitor returns to the Web site, the browser automatically sends the cookie information along with the URL, letting the Web server know the identity of the visitor. Figure 8.11 illustrates this process.

A typical cookie is stored on the visitor's hard drive as a line in a text file or as a separate text file, depending on the type of browser used. With Netscape, all cookies are stored in a file called cookies.txt in the Netscape folder. With Internet Explorer, each cookie is stored as a separate text file in the Cookies folder. This folder appears in different locations in different versions of Windows. (To look at your Internet Explorer cookies folder, use the Windows Start | Search or Find operation to search for *cookies*).

session cookie
A cookie that exists only during the current series of interactions between the browser and Web server.

persistent cookie
A file that exists indefinitely on the user's hard disk and that the browser uses to identify the user to the corresponding Web site.

Two types of cookies exist: session cookies and persistent cookies. A **session cookie** exists only during the current session of operations between the user and the Web server. It is not saved to the user's hard disk, but stays in memory during that time. In contrast, a **persistent cookie** lives on indefinitely as a file on the user's hard disk; the browser uses it to identify the user to the corresponding Web site. Figure 8.12 shows part of an Amazon.com cookie file for Internet Explorer. Note that Amazon.com uses the customer's ID as the session ID and that the time of the session is included as well. Because this cookie is stored on your hard disk, it is an example of a persistent cookie.

Figure 8.11 Use of a cookie

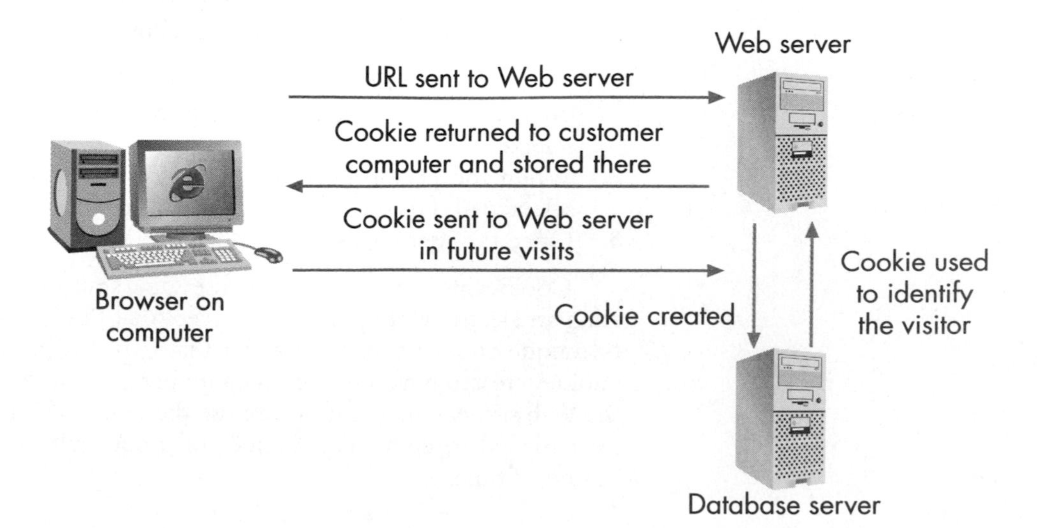

Figure 8.12 Cookie for Amazon.com

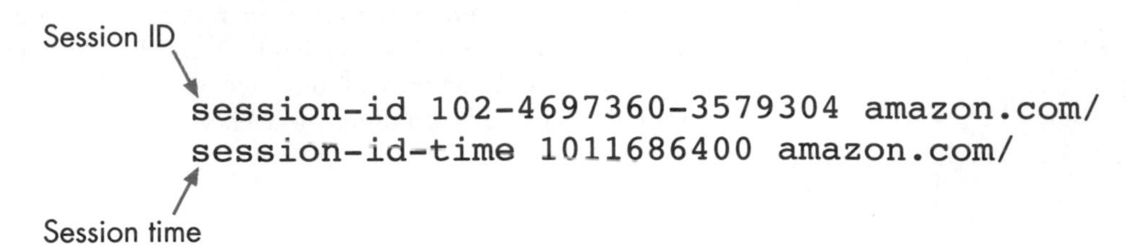

When a Web site is designed for making multiple purchases, developers use a cookie to create an electronic shopping cart. You may have noticed in Figure 8.6 and in actual online purchases references to a shopping cart. **Electronic shopping carts** on Web sites serve the same purpose as physical shopping carts in grocery and other stores—they enable you to keep shopping without having to check out after selecting each item. An electronic shopping cart is not really a physical shopping cart that contains purchases; rather, it offers a way to hold information about the customer and his or her purchases in the data that is passed back and forth between Web server and browser.

electronic shopping cart
Software on a Web server that enables the customer to keep shopping without having to check out after selecting each item.

To create an electronic shopping cart, you can use a session cookie that is temporarily stored in a a database, along with the inventory identifier for each item selected. The cookie is attached to the URLs of any pages sent back to the user as well as to any data returned from the user to the Web server. In essence, the cookie helps the Web server remember who the user is and what he or she has selected to purchase. For example, when you visit the FarEast Foods Web site and select a country (say, Thailand), the URL of the returned page might be changed from

http://www.fareastfoods.com/

to

http://www.fareastfoods.com/cgi-bin/fareastfoods/country.pl?SessionID=9111541&country=4

where the part after the question mark indicates that a cookie session ID (9111541, in this case) has been assigned to the session and that the country selected is Thailand. The session ID uniquely identifies the customer while he or she continues shopping. Because the FarEast Foods Web site is a demonstration site, it does not create a persistent cookie.

When the customer chooses to check out, the application uses the same session cookie ID to identify all of the items he or she has selected and added to the shopping cart. This process is summarized in the following steps and illustrated in Figure 8.13:

1. When an item is selected and added to the shopping cart, the cookie is saved on the company database along with the inventory identifier of the item selected.
2. When the person checks out, the cookie is used to identify all of the items selected and to show their prices on the checkout page.
3. Once the user checks out, the session ID cookie is retired.

Cookies are used for counting users in a similar way to how they are used for creating an electronic shopping cart, except that the Web site simply counts the number of unique cookies that it creates or that return to it on repeat visits. The number of cookies created measures the total number of unique visitors since the origination of the Web site. An organization can use the combination of cookies created and cookies returned to count the number of unique visitors to its Web site over a given period of time.

Figure 8.13 Use of cookie for electronic shopping cart

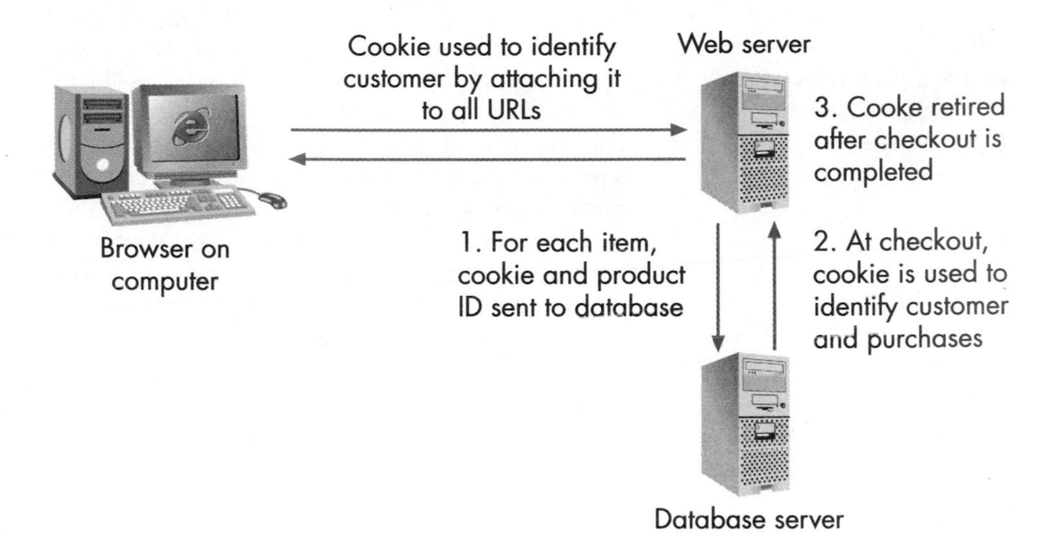
Cookie used to identify
customer by attaching it
to all URLs
Web server
3. Cooke retired
after checkout is
completed
Browser on
computer
1. For each item,
cookie and product
ID sent to database
2. At checkout,
cookie is used to
identify customer
and purchases
Database server

Because cookies are used to identify and track visitors to a Web site without their permission, quite a bit of controversy has arisen regarding privacy issues and the use of cookies. Chapter 12 will discuss these concerns.

Quick
Review

1. What are the disadvantages of HTTP for the Web?
2. What are the two types of cookies?

Securing Electronic Commerce Transactions

If any one thing has retarded the growth of electronic commerce, it is perceived security problems. Many would-be customers remain afraid to use electronic commerce because they worry about theft of their credit card numbers. People cite a variety of reasons for this concern. First, because the intent of the Internet is to give people remote access to information, it is inherently open. Traditional approaches to restricting access via physical barriers are less viable here, although organizations still need to restrict physical access to their servers. Second, the same technologies that form the basis of electronic commerce—computers and networks—can be used to attack security systems. People believe that hackers can easily use computers to intercept network traffic and scan it for confidential information, or find useful information on a server by breaching its security (for example, running repeated attacks, such as trying all words in the dictionary, to discover an account password). In fact, server security is generally not easy to breach, and most people stand a far greater chance of having their credit card information stolen at a local restaurant or other business. Nevertheless, the mere perception of a threat has caused many would-be buyers to avoid making purchases via the Web.

To help you understand the ways in which electronic commerce transactions are secured, a brief discussion of security follows; Chapter 11 provides a more-detailed discussion. This section focuses on only one issue: using encryption to protect the contents of an Internet message and to verify the sender of a message.

Security Issues

sniffer
A computer program on an intermediate Internet computer that will briefly intercept and read a message.

Internet messages can pass through many computers on their way from sender to receiver, and the danger always exists that a computer program called a **sniffer**, operating on an intermediate computer, will briefly intercept and read a message. For most e-mail messages, this possibility does not cause great concern, but what happens if your message contains your name, credit card number, and expiration date in an unprotected form? The sniffer program, looking for a typical credit card number format of four blocks of four digits (for example, 1234 5678 9012 3456), will copy your message before letting it continue on its way. The owner of the sniffer program can then use your credit card details to purchase products in your name and charge them to your account.

Without a secure means of transmitting payment information, customers and merchants will remain reluctant to place and receive orders, respectively. When the customer places an order, the Web browser should automatically secure the order prior to transmission. Providing this security is not the customer's task.

Credit card numbers are not the only sensitive information transmitted over the Internet. Because it is a general transport system for electronic information, the Internet can carry a wide range of confidential information (financial reports, sales figures, marketing strategies, technology reports, and so on). If senders and receivers cannot ensure that their communication is strictly private, they will not use the Internet. Secure transmission of information is necessary for electronic commerce to thrive.

Encryption

encryption
The conversion of readable text into characters that disguise the original meaning of the text.

decryption
The conversion of an encrypted, seemingly senseless character string into the original message.

private-key encryption
A form of encryption in which a single key is used to both encrypt and decrypt the message.

key
An algorithm used to encode and decode messages.

The most widely used method of protecting Internet messages from being read by a computer along their path between sender and receiver is encryption. **Encryption** transforms messages or data to protect their meaning. It scrambles a message so that it becomes meaningful only to the person knowing the method of encryption and holding the key to deciphering it. To everyone else, the message is gobbledygook. The reverse process, **decryption**, converts a seemingly senseless character string into the original message. Two primary forms of encryption systems exist: private-key and public-key encryption.

Private-key encryption uses the same private key to encrypt and decrypt a message. A **key** is an algorithm used to encode and decode messages. Although private-key encryption may sound like the simplest method, it has significant problems. For example, how do you securely distribute the key? You can't send the private key with the message, because if the message is intercepted, the key can be used to decipher it. You must find another secure medium for transmitting the key. Do you fax the key, or send it via telephone? Neither of these methods is completely secure, and each is time-consuming to use whenever the key changes. In addition, how do you know that the key's receiver will protect its secrecy? Another problem with private-key encryption is that you must create a separate private key for each person or organization with which you exchange encrypted messages.

Figure 8.14 Public-key encryption

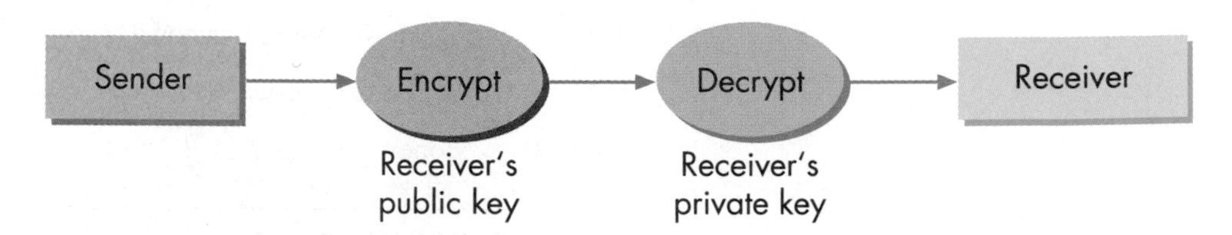

public-key encryption
An encryption system with two keys—one private and one public—where the public key is used to encrypt a message and the private key is used to decrypt it.

public key
In a public-key encryption system, a key that is freely distributed to encrypt messages.

private key
In a public-key encryption system, the only key that can decrypt the message.

In contrast, a **public-key encryption** system has two keys: one private and the other public. Your **public key** is freely distributed and used to encrypt messages coming to you. In contrast, your **private key** remains secret; you use it to decrypt the messages encrypted with your public key. For example, you would distribute your public key to anyone who might need to send encrypted messages to you. The sender would encrypt a message with your public key. Upon receiving the message, you would apply the private key, as shown in Figure 8.14. Your private key—the only key that can decrypt the message—must remain secret to permit secure message exchange.

The public-key system avoids the problem of secure transmission of keys. Public keys can be freely exchanged. Indeed, a public database might contain each person's or organization's public key. For instance, if you want to e-mail your credit card details to a catalog company, you could simply obtain the company's public key (probably from its Web site) and encrypt your entire message prior to transmission.

Of course, you may wish to transmit far more important data than your credit card number. Consider the message shown in Figure 8.15; the sender would hardly want this message to fall into the wrong hands. After encryption, the message is totally secure, as shown in Figure 8.16. Only the receiver, using his or her private key, can decode the message. A free form of public encryption is Pretty Good Privacy (PGP).

Figure 8.15 Message before encryption

To: Pat McKeown <pmckeown@dogs.uga.edu>

From: Rick Watson <rwatson@dogs.uga.edu>

Subject: Money

G'day Pat

I hope you are enjoying your stay in Switzerland. Could you do me a favor? I need $50,000 from my secret Swiss bank account. The name of the bank is Aussie-Suisse International in Geneva. The account code is 451-3329 and the password is 'meekatharra'. I'll see you (and the money) at the airport this Friday.

Cheers

Rick

Figure 8.16 Message after encryption

To: Pat McKeown <pmckeown@dogs.uga.edu>

From: Rick Watson <rwatson@dogs.uga.edu>

Subject: Money

-----BEGIN PGP MESSAGE-----

Version: 2.6.2

hEwDfOTG8eEvuiEBAf9rxBdHpgdq1g0galP7zm10cHvWHtx+9++ip27q6vl
tjYblUKDnGjV0sm2INWpcohrarI9S2xU6UsSPyFfumGs9pgAAAQ0euRGjzy
RgIPE5DUHG ultXYsnlq7zFHVevjO2dAEJ8oualX9YJD8kwp4T3suQnw7/d
1j4edl46qisrQHpRRwqHXons7w4k04x8tH4JGfWEXc5LB+hcOSysPHEir4EP
qDcEPlblM9bH6 w2ku2fUmdMaoptnVSinLMtzSqIKQIHMfaJ0HM9Df4kWh
ZbY0yFXxSuHKrgbaoDcu9wUze35dtwiCTdf1sf3ndQNaLOFiljh5pis+bUg
9rOZjxpEFbdGgYpcfBB4rvRNwOwizvSodxJ9H+VdtAL3DIsSJdNSAEuxjQ0
hvOSA8oCBDJfHSUFqX3ROtB3+yuT1vf/C8Vod4gW4tvqj8C1QNte+ehxg==

=fD44

-----END PGP MESSAGE-----

steganography
A form of encryption that hides messages within graphic or audio files.

Security experts think terrorist groups have used a graphical form of encryption known as **steganography** (Greek for hidden writing) to send messages to terrorist cells in Western countries. This technique, also known as digital watermarking, hides messages within graphic or audio files. More than 140 steganography tools are available over the Web, and experts suggest that the Al Qaeda terrorist group is using it. In fact, a man captured before he could blow up the U.S. Embassy in France had been instructed to use pictures posted on the Internet as a means of communication.

The photograph on the left shows a steganographic image based on a photograph of a flower. The photograph on the right, which shows a ship, is hidden within the photograph of the flower.

Figure 8.17 Creating a digital signature with the public-key system

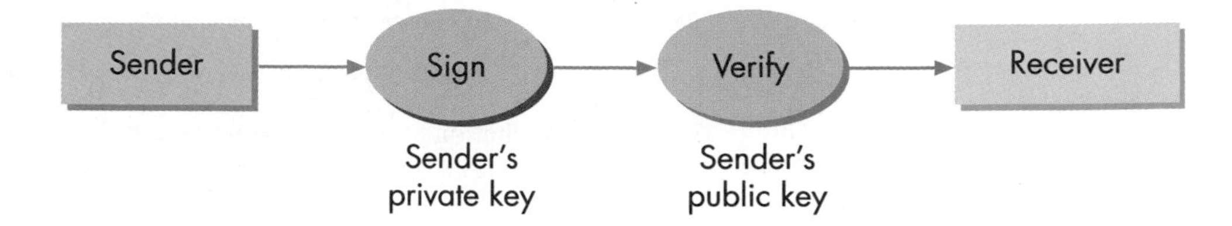

Digital Signatures

digital signature
A digital code that is attached to an electronically transmitted message and that uniquely identifies the sender.

With the growth of electronic commerce and increased interactions between people and organizations that have no physical contact has come the need to have some way of verifying identities. To solve this problem, many companies have come to rely on digital signatures. A **digital signature** is a digital code that is attached to an electronically transmitted message and that uniquely identifies the sender. For example, imagine that people pay $1000 per year for an investment information service. The provider might want to verify that any e-mail requests it receives are from subscribers by having them sign all electronic messages with a digital signature known to them and the company. As part of the subscription signup, subscribers have to supply their public key and, when using the service, sign all electronic messages with their private key. The provider is then assured that it is servicing only paying customers. Naturally, any messages between the service and the client should be encrypted to ensure that others do not gain from the information.

The most popular form of digital signature uses the same approach as public key encryption. With this system, to verify his or her identity, a sender uses his or her private key to create a digital signature for the message. The receiver then applies the sender's public key to verify the signature, as shown in Figure 8.17.

As an example of the use of digital signatures, assume that one of your friends attempts to have some fun at your expense by sending you the e-mail message shown in Figure 8.18. If the president were actually in the habit of communicating electronically, it is likely that he would electronically sign his messages so that the receiver could verify them as shown in Figure 8.19. When the purported sender's public key is applied to this message, the identity of the sender can be verified (it was not the president).

Figure 8.18 Message before signing

To: Pat McKeown <pmckeown@dogs.uga.edu>

From: President@whitehouse.gov

Subject: Invitation to visit the White House

Dear Dr. McKeown

It is my pleasure to invite you to a special meeting of Internet users at the White House on April 1 at 2 pm. Please call 212-555-7890 and ask for Mr. A. Phool for complete details of your visit.

The President

Figure 8.19 Digitally signed message

To: Pat McKeown <pmckeown@dogs.uga.edu>

From: President@whitehouse.gov

Subject: Invitation to visit the White House

Dear Dr. McKeown

It is my pleasure to invite you to a special meeting of Internet users at the White House on April 1 at 2 pm. Please call 212-555-7890 and ask for Mr. A. Phool for complete details of your visit.

The President

-----BEGIN PGP MESSAGE-----

Version: 2.6.2

iQCVAwUBMeRVVUblZxMqZR69AQFJNQQAwHMSrZHWyiGTieGukbhPGUNF3a
B +qm7E8g5ySsY6QqUcg2zwUr40w8Q0Lfcc4nmr0NUujiXkqzTNb+3RL41w5x
fTCfMp1Fi5Hawo829UQAlmN8L5hzl7XfeON5WxfYcxLGXZcbUWkGio6/d4r
9Ez6s79DDf9EuDIZ4qfQcy1iA==G6jB

-----END PGP SIGNATURE-----

Digital signatures were made legally binding in the United States in June 2000. Under the E-Sign law, no contract, signature, or record can be denied legal effect solely because it appears in electronic form. For example, legal experts and financial services companies hope that customers can open mutual fund accounts online using a digital signature, without the common paper requirements and delays. To protect people without computers, consumers must explicitly choose to have their records stored elecronically, ensuring that paper records could still be maintained.

1. What is a sniffer, and how does it affect electronic commerce?
2. What is *public* about public-key encryption?

Electronic Commerce Payment Systems

When commerce goes electronic, the means of paying for goods and services must also go electronic. Paper-based payment systems cannot support the speed, security, privacy, and internationalization necessary for electronic commerce. This section discusses three methods of electronic payment: credit cards, electronic funds transfer, and digital cash.

electronic funds transfer (EFT)
Any transfer of funds from one bank account to another without paper money changing hands; also, the transfer of payments between consumers or between organizations engaged in business-to-business electronic commerce and businesses.

digital cash
The storage of value in a digital format in one of two broad forms: card-based or computer-based.

Although you are already familiar with the use of credit cards, you may not be familiar with electronic funds transfer or digital cash. **Electronic funds transfer (EFT)**, in its broadest definition, refers to any transfer of funds from one bank account to another without any paper money changing hands. In this book, the term will also refer to the transfer of payments between consumers or between organizations engaged in business-to-business electronic commerce and businesses. **Digital cash** involves the storage of value in a digital format; it is an electronic parallel of notes and coins. Two forms of digital cash exist: card-based and computer-based.

card-based digital cash
The storage of value on a plastic card, such as a prepaid telephone card or a smart card, that can have value added to or removed from it.

computer-based digital cash
The storage of value on a computer, usually linked to the Internet, allowing for payment directly between the customer and merchant computers or for a transfer of funds between individuals.

Card-based digital cash is the storage of value on a plastic card, such as a prepaid telephone card or a smart card that can have value added or removed from it. **Computer-based digital cash** is the storage of value on a computer, usually linked to the Internet, allowing for payment directly between the customer and merchant computers or for a transfer of funds between individuals.

Concerns with Electronic Money

Four fundamental concerns are cited regarding electronic money: security, authentication, anonymity, and scale of purchase. Security of electronic money means that consumers and organizations know that their online orders remain protected from theft or manipulation and that large sums of money can be transferred safely. For any type of electronic money to be useful, it must be possible to authenticate it—that is, verify that it is real. Otherwise, firms and consumers will not have faith in electronic currency and will avoid using it. In addition, transactions using electronic money should retain anonymity; in other words, these transactions should remain invisible to persons who have no reason to see them.

The scale of purchase issue is a new issue that is very closely associated with the rise of electronic commerce. Traditionally, people have thought of making purchases no smaller than the smallest denomination of a national currency (for example, one cent in the United States). In contrast, electronic commerce allows people to make purchases using smaller denominations. Now a real need arises to make micro purchases (for example, a purchase for less than $1) or even nano purchases (less than one cent). These capabilities will permit high-volume, small-value Internet transactions, such as purchasing individual newspaper, magazine, or encyclopedia articles; renting software for an hour; or accessing a technical support area. To automate these types of transactions, it becomes necessary to adapt current payment systems. For example, overhead considerations might make a merchant reluctant to let you use your credit card to make a purchase of $0.01. On the other hand, some form of digital cash or e-cash system might work very well for small-value transactions.

Any money system, real or electronic, must have a reasonable level of security, a high level of authenticity, and a substantial degree of anonymity, or people will not use it. Although all electronic money systems are potentially divisible to any degree, some forms of electronic payment lend themselves better to small-value transactions than others. The various approaches to electronic money vary in their capability to solve these concerns, as shown in Table 8.3.

Although not all of the technical problems of electronic money have been solved, many people and companies are working on their own solutions because electronic money promises efficiencies that will reduce the costs of transactions between buyers and sellers. Given that counting, moving, storing, and safeguarding cash account for an estimated 4 percent of the value of all transactions, ample reasons exist to pursue such systems. In the next few years, electronic currency could potentially displace notes and coins for many transactions. Let's now consider the types of electronic money in more detail.

Table 8.3 Characteristics of Electronic Money

	Security	Authenticity	Anonymity	Scale of Purchase*
Credit card	High	High	Low	Small to medium
EFT	High	High	Low	Small to large
Card-based digital cash	Medium	High	High	Nano to medium
Computer-based digital cash	High	High	High	Nano to medium

*Large: > $10,000; medium: < $10,000; small: < $1,000; micro: < $1; nano: < $0.01.

Credit Cards

Credit cards are a safe, secure, and widely used remote payment system. Millions of people use them every day to order goods by telephone, by mail, and over the Internet. Furthermore, people think nothing of handing over a credit card to a restaurant server, who could easily find time to write down the card's details. In the case of fraud, banks already protect consumers to some degree, depending on the national jurisdiction. For example, in the United States, credit card holders are typically liable for only the first $50 of any purchases made with a stolen credit card.

Web-based credit card systems now universally include real-time authorization, and the use of secure servers and clients makes transmitting credit card data extremely safe. These systems include Netscape's Secure Sockets Layer (SSL) system and the Secure Electronic Transaction (SET) system supported by MasterCard, Visa, Microsoft, and Netscape. These systems enable online purchases using encrypted credit card numbers.

electronic wallets
A digital form of storage that enables the user to electronically store multiple credit cards in a combination of software and data.

Credit card systems have been expanded through the use of **electronic wallets**, which enable users to store multiple credit cards in an electronic form as a combination of software and data. When the customer purchases an item from a merchant that supports the electronic wallet system, an appropriate message is returned to the customer's computer, and the user can then select a credit card from the wallet.

Credit cards have some major shortcomings. For example, they do not support person-to-person transfers, and they do not have the privacy of cash. In addition, consumers cannot now use them for micro and nano purchases, although the threshold for the required amount of a purchase is declining.

Electronic Funds Transfer

Electronic wallets like this one enable users to select a credit card to use for an e-commerce purchase.

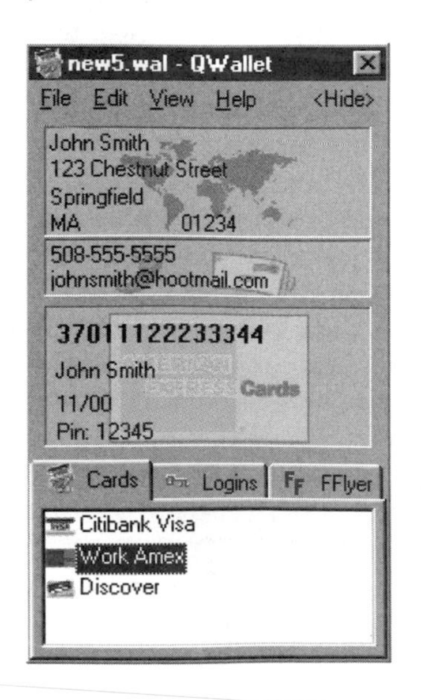

Electronic funds transfer, introduced in the late 1960s, uses the existing banking structure to support a wide variety of payments. For example, consumers can establish monthly checking account deductions for utility bills, and banks can transfer millions of dollars. EFT is essentially electronic checking. Instead of writing a check and mailing it, the buyer initiates an electronic checking transaction (for example, using a debit card at a point-of-sale terminal). The transaction is then electronically transmitted to an intermediary (usually the banking system), which transfers the funds from the buyer's account to the seller's account. A banking system has one or more common clearinghouses that facilitate the flow of funds between accounts in different banks. Online banking systems fall into this category, whether they send a paper check or create an EFT arrangement with the payee. Figure 8.20 shows this process for a consumer purchase using a debit card.

Electronic checking is fast; transactions occur virtually instantaneously. Paper-handling costs are substantially reduced, and bad checks cease to be a problem because the system verifies the buyer's account balance at the moment of the transaction. EFT is flexible; it can handle high volumes of small consumer payments and large commercial transactions, both locally and internationally. The international payment clearing system, consisting of more than 100 financial institutions, handles an average of $3.5 *trillion* per day.

Figure 8.20 Use of EFT for a debit card purchase

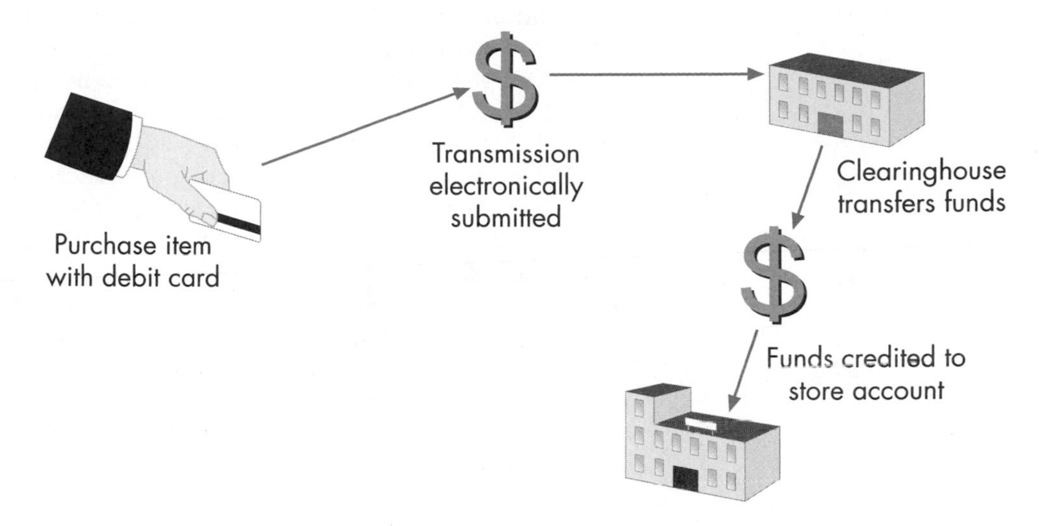

Internet bill payment/presentment (IBPP)
A system through which both the bill itself and the payment of the bill are presented in an electronic fashion over the Web.

A fairly new use of EFT that is especially appropriate for electronic commerce applications comprises Internet bill payment/presentment. With **Internet bill payment/presentment (IBPP)**, not only is the payment electronic via your bank or one of the bill payment services, but the bill itself is presented in an electronic fashion over the Web. The primary problem with IBPP is the lack of a single Web site at which you can receive and pay all of your bills. Often you must set up a separate account to receive and pay your bills at each company's Web site. However, companies such as CheckFree have begun to form alliances with large billers with an eye toward making it possible to pay all bills from one site.

The major shortcoming of EFT is that all transactions must pass through the banking system, which is legally required to record every transaction. This lack of privacy can have serious consequences for those interested in keeping their transactions private. In addition, EFT does not handle micro and nano transactions well. The scale factor keeps EFT from being used for very small consumer purchases.

Bill presentment and payment software is becoming a popular way to pay bills electronically.

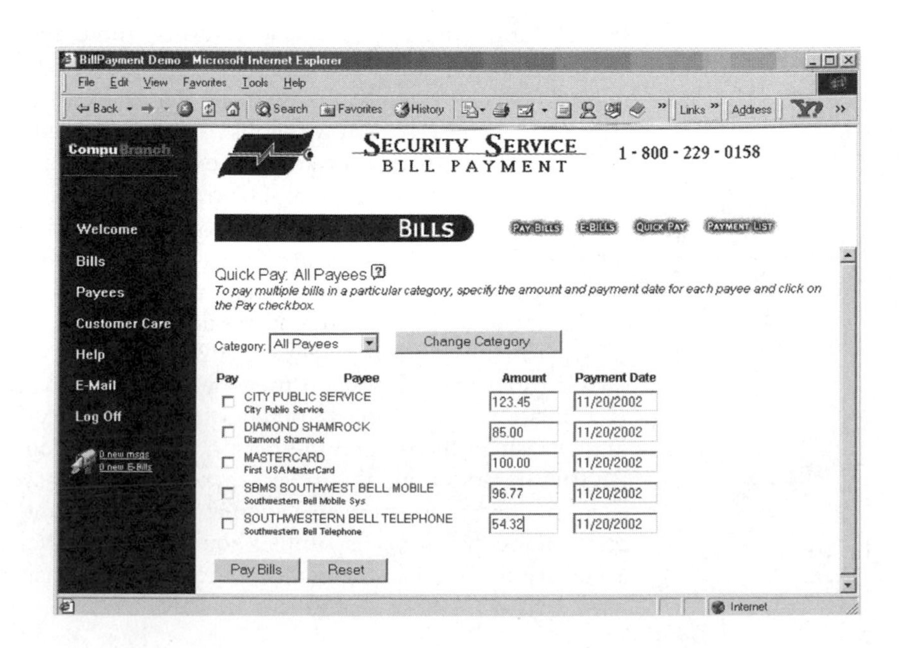

Card-Based Digital Cash

As mentioned earlier, digital cash is the electronic parallel of notes and coins. Two forms of card-based digital cash are presently available: prepaid cards and smart cards. The phone card, the most common form of prepaid card, was first issued in 1976 by the forerunner of Telecom Italia. In addition, you may be familiar with prepaid cards for making copies at the library or for other special purposes.

When people use prepaid special-purpose cards, such as phone and photocopy cards, they can end up with a purse or wallet full of cards. In contrast, a *smart card*, which consists of a computer chip with built-in memory and microprocessor, can simultaneously serve as a personal identification card, a credit card, an ATM card, or a telephone credit card. It might also hold critical medical information and serve as cash for micro to medium-sized transactions. Standard storage on a smart card is 16 KB, but higher capacities are becoming available up to 128 KB. In addition, the microprocessor can be programmed to carry out a variety of activities.

Consumers can use the stored-value card, the most common application of smart card technology, to purchase a wide variety of items (for example, fast food, parking, and public transportation tickets). Consumers buy cards of standard denominations (for example, $20, $50, or $100) from a card dispenser or bank. To use the card to pay for an item, the customer inserts it into a reader. The amount of the transaction is transferred to the reader, and the value of the card is reduced by the transaction amount.

The problem with card-based digital cash, as with physical money, is that you can lose it or it can be stolen. Although this form is not as secure as the other electronic money alternatives, most people are likely to carry only small amounts of digital cash on their cards, so security is less critical. Because smart cards often have unique serial numbers, consumers can limit their loss by reporting a stolen or misplaced smart card to invalidate its use. Adding a personal identification number (PIN) to a smart card can raise its security level.

France, where smart cards were introduced more than a decade ago, and the rest of Europe make heavy use of smart cards. In 2000, more than 1.6 billion cards were in use worldwide, but only 18 percent of them were being used in *all* of the Americas. The United States is the one area of the world where smart cards have not become widely popular, but banks and credit card companies continue to work to change this attitude.

Computer-Based Digital Cash

For making purchases over the Internet, most people currently use credit cards. However, these payment options do not work on auction sites such as eBay because the seller is usually an individual with no capability to process a credit card transaction. Payment by check or cash is a problem if the seller fails to send the item to the buyer. A variety of methods are now available to store value in the form of digital cash on either the user's computer or a server. The user can send this digital cash to pay for everyday Internet transactions, such as buying software, receiving money from parents, or paying for a pizza to be delivered without using a credit card. Digital cash can provide the privacy of cash because the payer can remain anonymous under some systems.

Currently, the hottest company in the digital cash field is PayPal. Born from the need to handle person-to-person (P2P) payments for auctions such as eBay, PayPal served more than 12 million users by early 2002, with more than 2 million business accounts, and was adding 18,000 new accounts every day. To use the PayPal version of digital cash, both you and the person or business you are paying need accounts with PayPal. In addtition, the payer's account must have a source of funds—a bank account, credit card, or even a money market account. Once the payer's account is set up, making payments is simple: You just enter the recipient's e-mail address and account number. The recipient will receive an e-mail message that says "You've Got

PayPal is the most popular application for transferring funds via e-mail.

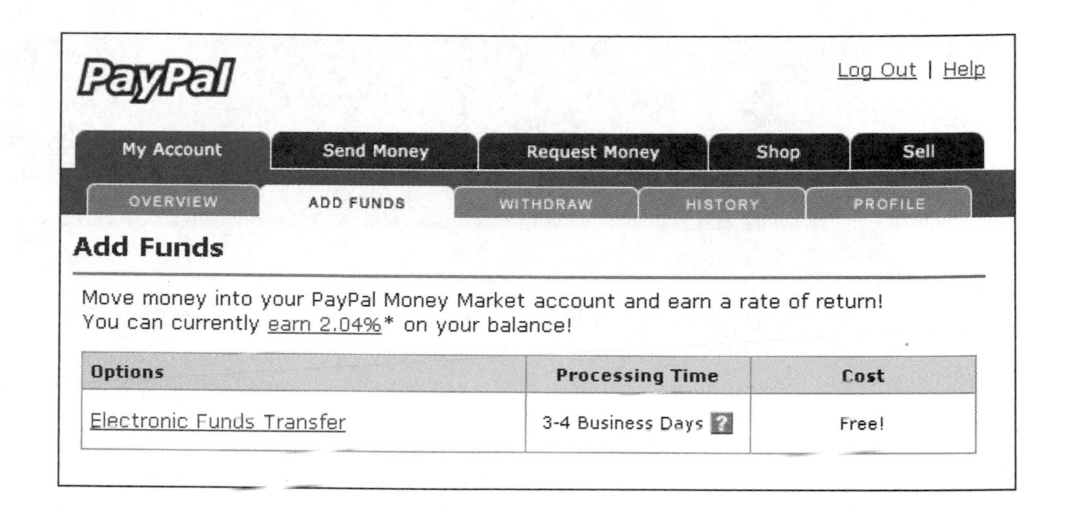

Cash!" The recipient can then go to www.paypal.com, log in into his or her account, and have a check sent out, send the funds to somebody else, or just leave the funds in the account for future use. If the recipient does not have an account, he or she must create one before the transaction can be completed; this setup can be done on the spot.

Transactions of virtually any size can be handled in this manner. Although PayPal does not charge a fee to send money, individuals and merchants wishing to use an electronic shopping cart or accept credit card payments must pay a modest fee. CheckFree and CitiBank are moving aggressively into this market as competition to PayPal. Figure 8.21 explains how a system such as PayPal works for electronic commerce transactions, using the purchase of a piece of antique clothing from an individual as an example.

1. What three electronic payment methods are currently in use?
2. Why is a credit card a safe way to purchase goods and services over the Web?

Figure 8.21 Use of digital cash for individual purchase

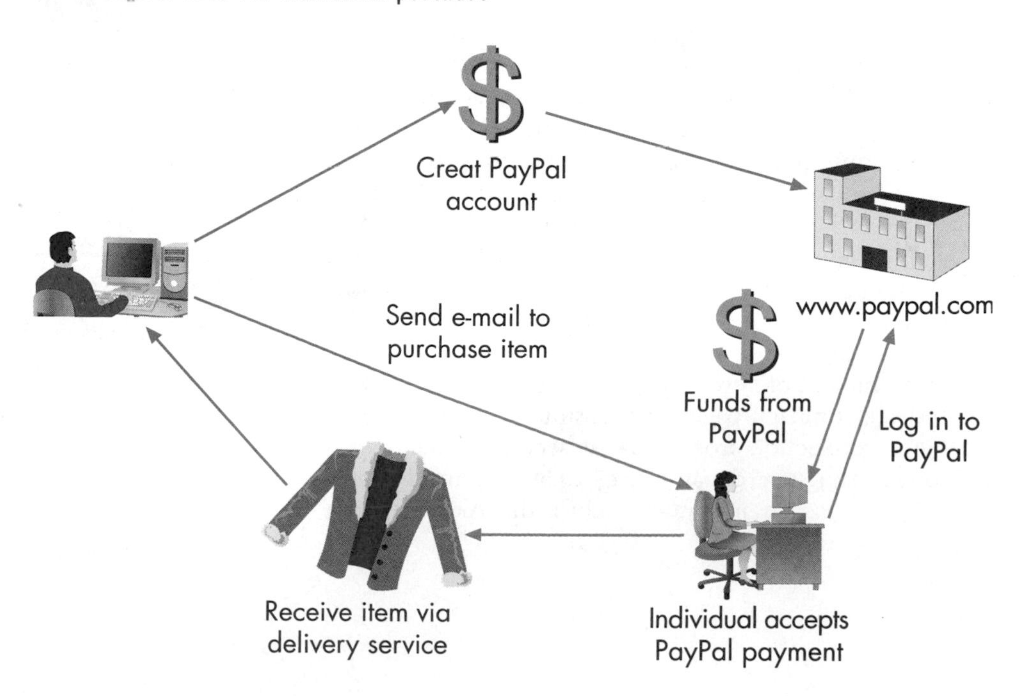

CASE STUDY

Case*Study*

Akamai Technologies

When loading Web pages with a high graphical content, you may have noticed that the URL displayed at the bottom of the screen includes a host name of "akamai.net." If you haven't seen this name yet, you may want to take a second to load the home page of one of Akamai's clients—they range from Apple to Yahoo! and include BET, Lands' End, and Monster.com—and note the URL in the bottom of the browser while graphics load. The akamai.net host name appears because the graphic images of the Web pages for these companies are stored on one of the more than 12,600 servers on more than 1000 networks in 60 countries that are a part of Akamai Technologies' globally distributed network. Located in Cambridge, Massachusetts, Akamai has more than 1000 customers who depend on it to serve up their Web content at a much faster rate than the customers' servers could because Akamai servers are closer to the end user in terms of download speed. Akamai provides this service through intelligent routing and mapping technology that determines in real time how to serve the end user via the optimal network path.

Akamai servers are also widely used for streaming video. Although live video represents an exciting part of the interactive Web, it can slow down traffic significantly. Webcasts can also become degraded due to Internet congestion, inconsistent servers, and interrupted streams. Akamai technology works to ensure that only the highest-quality streams reach the audience. For live broadcasts, it sends multiple streams that are recombined into their original quality format.

Despite a loss of customers following the failure of many dot-com companies and the tragic death of chief technology officer and co-founder, Daniel Lewin, in the terrorist attacks of September 11, 2001, Akamai is continuing to roll out new technology. One such technology is EdgeSuite, which lets companies handle visits from millions of new users without buying new servers. Instead, company pages are transmitted over the much faster connection from Akamai servers. When a customer requests a dynamic page that may involve new data such as an inventory record, the Akamai server checks whether it has a page with that information. If not, it goes back to the company server, downloads the small bit of new information needed, and then sends the updated page to the customer. The over 185 EdgeSuite clients are finding that they can toss out computers and reduce office space with the new system and, even after paying monthly EdgeSuite fees, save big bucks. For example, the Motley Fool financial news and information Web site reduced its Web farm from 25 to 5 servers for a significant cost savings.

The various Akamai servers, which provide graphics for many popular Web pages, are controlled from this control room.

Source: Daniel Lyons, "Living on the edge." *Forbes.com*, http://forbes.com/forbes/2001/0709/134.html.

Think About It

1. Why would a company choose to store its graphic images on Akamai servers rather than its own servers?
2. How does the existence of a company such as Akamai help customers?
3. For what other reason beyond that which you gave in Question 1 would companies choose to use Akamai?

SUMMARY

To summarize this chapter, let's answer the questions posed at the beginning of the chapter.

What are the layers in the electronic commerce infrastructure? Electronic commerce technologies are layered one upon the other. The lowest layer, the global information infrastructure (GII), includes television and radio broadcast networks, cable television networks, telephone networks, cellular communication systems, computer networks, and the Internet. The next layer up, the message distribution infrastructure, consists of software for sending and receiving messages between a server and a client. Electronic data interchange (EDI), e-mail (SMTP), and Hypertext Transfer Protocol (HTTP) are all examples of messaging protocols. Above the messaging layer is the electronic publishing infrastructure, of which the Web is a very good example. This layer addresses problems of having addressability (a URL) and having a common language across the network (HTML or XML). The next-to-the-top layer, the business services infrastructure, supports common business processes—for example, it supports secure transmission of credit card numbers by providing encryption and electronic funds transfer. The top infrastructure layer is an electronic commerce application where the transaction takes place.

How is a transaction carried out using the Web? The steps necessary to carry out an electronic commerce transaction using the Web begin with the customer finding a Web site by using a search site or bu responding to an attractor strategy. Search sites take one of two general approaches: computer and human. Computerized approaches use search engines that crawl the Web seeking out new pages, whereas human approaches create directories of pages in various categories. Computerized search engines often use meta HTML tags, an important way of identifying the subject of a Web page. The URL of an electronic commerce site is sent over the Web, and HTML pages are returned from which the user can make selections and fill in forms to make purchases. In addition, the customer receives verification that his or her order has been accepted. The data from an HTML form can be processed on the browser via client-side processing or at the Web site via server-side processing. Client-side processing may occur through JavaScript, Java applets, or ActiveX controls. In contrast, database access can be handled at the Web server only through server-side processing using approaches such as ASP, JSP, or CGI. Web server software can be either Microsoft or open source.

What are the problems associated with using the Web for e-commerce, and how do you handle them? Because the Web is based on HTTP, which is a stateless protocol, a Web server has no memory of a user's previous Web site visits. To facilitate eletronic commerce, customers need to purchase multiple items at a Web site but check out only once. In addition, a problem occurs with counting unique visitors because HTTP counts only individual computer IP addresses. Because knowing the number of unique visitors is important for assessing the success of a Web site and raising advertising revenue, you need some way to identify each visitor. Both problems are handled through cookies—text files that are written on the customer's hard disk and then used to identify the customer on return visits. For multiple purchases, cookies are used to create an electronic shopping cart. For counting visitors, they are used to count only unique visitors, regardless of the IP address associated with the visit.

How are electronic commerce transactions protected from criminal activity? Encryption is the most popular method for protecting Internet messages from being read by another computer along their path between sender and receiver. Encryption transforms a message so that it is meaningful only to the person knowing the method of encryption and having the key to decipher it. The reverse process, decryption, converts a seemingly senseless character string into the original message. A popular form of encryption is public-key encryption, which uses a pub lic key and a private key. With the growth of electronic commerce and increased interactions between people and organizations that have no physical contact has come the need to have some way of verifying identities. Digital signatures are used for this purpose, with the most common approach using public and private keys similar to those employed for public-key encryption.

What electronic commerce payment methods are in use today, and what methods are predicted to emerge in the future? Paper-based payment systems cannot support the speed, security, privacy, and internationalization necessary for electronic commerce. Four fundamental concerns arise regarding electronic money: security, authentication, anonymity, and scale of purchase. Security of electronic money means that consumers and organizations receive assurance that their online orders are protected from theft and that large sums of money can be transferred safely. For any type of electronic money to be useful, it must be possible to authenticate it—that is, verify that it is real. Transactions using electronic money should remain invisible to persons who have no reason to see them. Three commonly used methods of electronic payment include credit cards, electronic funds transfer (EFT), and digital cash. Credit cards are used for all types of transactions except those involving very large or very small amounts of money. EFT refers to the transfer of payments between consumers or between organizations engaged in business-to-business electronic commerce. Digital cash involves the storage of value in a digital format and comes in two broad forms: card-based and computer-based. Card-based digital cash is the storage of value on a plastic card, such as a prepaid telephone card or a smart card, which can have value added or removed from it. Computer-based digital cash is stored on a computer, usually linked to the Internet. It allows for payment directly between the customer and merchant computers or for transfers of funds between individuals. PayPal is a popular form of digital payment used for electronic commerce.

REVIEW QUESTIONS

1. List the five layers of the electronic commerce infrastructure model.
2. In which layer does encryption occur? In which layer would SMTP be found?
3. What is the first step in carrying out an electronic commerce transaction?
4. What are the two types of search sites? Which crawls around the Internet looking for new Web sites?
5. What is open-source software, and how does it relate to Web servers?
6. What is a meta HTML tag, and what is its purpose?
7. How do the data sent from a Web server to a browser differ from the data sent from a browser to a Web server?
8. What kind of HTML tag do you use to create a Web page from which data can be sent to the Web server?
9. What kind of HTML tag do you use to submit data? What kind of HTML tag do you use to clear the Web page of data?
10. Describe the steps in the interaction between a browser and Web server to carry out a electronic commerce transaction.
11. What are the two methods of processing data sent from a Web page? Which type occurs on the browser?
12. Which method of processing Web page data do you use to query a database?
13. What is the difference between using CGI and server includes?
14. What is meant by the stateless nature of HTTP?
15. Why is the stateless nature of HTTP considered a problem for electronic commerce?
16. What is the most popular solution to the problems associated with the stateless nature of HTTP?
17. How does an electronic shopping cart work?
18. What is the difference between a private key and a public key in encryption?
19. How are digital signatures different from encryption?
20. What are the available payment systems for electronic commerce?

DISCUSSION QUESTIONS

1. Explain why the term *global information infrastructure* is more appropriate than *national information infrastructure* when discussing the electronic commerce infrastructure.
2. What is your favorite search site? Explain why you like it and what problems you have experienced with it.
3. Discuss why search sites are an important attractor strategy for electronic commerce.
4. How do you feel about having a cookie placed on your hard disk by a Web site? Discuss the pros and cons of this solution to the problems associated with the stateless nature of HTTP.
5. Discuss how the E-Sign law combined with other events during 2001 might change the handling of legal affairs.

RESEARCH QUESTIONS

1. Research how the Internet and Web have changed the way UPS conducts its business. Create an electronic presentation on your findings with at least 10 slides.
2. Go to the Web site for Eric Allen's company, Sendmail, Inc., and research how it hopes to make money on what had previously been a volunteer project. Write a two-page paper on your findings.
3. Download the three smart search tools described in the "Internet in Action" feature and try out each one. Discuss the one you like best and provide reasons for your choice. Write a two-page paper on your findings.
4. Use the Web to compare Java applets with ActiveX controls. List the pluses and minuses of each approach. Create an electronic presentation on your findings with at least 10 slides.
5. Compare the approach taken by PayPal to P2P transactions to that taken by CheckFree and CitiBank. Write a two-page paper on your findings.

"Can I get extra whipped cream on that?" asked Alex, indicating the Caramel Mocha Latte that he had just ordered.

When her husband had finished, Claire ordered. "I'll have a Tall Coffee of the Day. Black."

After having their Frequent Drink card punched and retrieving their order, they settled into a comfortable booth in the corner. With their growing business becoming increasingly hectic, the Campagnes occasionally found it necessary and enjoyable to retire to a local coffee shop/bookstore to discuss their plans for Wild Outfitters. As usual, they commenced with a few preliminary questions about each other's day as Etta Jones played on the café stereo. Eventually, they turned the conversation to business.

Claire started a discussion of their plans for the WildOutfitters.com site. "Let's see," she began. "I made a 'To-Do' list. Let me read it to you." Her list was something like the following:

- Provide a registration option with a survey of customer preferences
- Track customer preferences when visiting the site
- Send targeted e-mails about deals and promotions based on customer preferences
- Increase the site's transaction processing capability
- Provide more customer payment options (credit cards)
- Enhance catalog searching and browsing features
- Develop customized views of the site based on customer preferences

Alex stared at the list. Confidently, he said, "That seems rather ambitious. But I've learned a lot about maintaining our Web site, and I think that with a little work I can figure out how to do these, too."

"Hold your horses, hotshot!" Claire replied. "I did some browsing in the computer book section before you got here. The number of technologies and computer languages that are available today is overwhelming."

Thoughtfully, Alex nodded. "You're probably right," he admitted. "There is a lot to learn." He paused a moment, then added, "I do think that we know enough to develop a plan. I can do some browsing through the books and on the Web and at least determine some alternatives for getting this list completed."

"Good idea," Claire agreed. "We can also begin prioritizing the list. Some of these capabilities would be nice to have yesterday, but we can wait on some others for a little while."

"With level-headed plans like these, I think that we're getting to be quite the computer-age professionals," Alex said and smiled as he took a sip of his latte.

Claire grinned. "You're cute when you're being professional," she replied. Then she leaned over and wiped the whipped cream mustache off of her husband's upper lip.

Think About It

1. Which technical components of electronic commerce would Wild Outfitters use at each level of the electronic commerce infrastructure, and how?

2. Would you recommend that the Campagnes develop the components on Claire's list or hire someone else to develop the components for them? Why or why not? Which components, if any, should they control on their in-house servers?

Hands On

3. As described in the case, the Campagnes want to send targeted e-mail to alert their good customers about special deals and tours. Sketch out a new form on the WildOutfitters.com Web site that would allow customers to sign up for this service. Include boxes on the form for soliciting the customer's contact information as well as questions about his or her product and travel interests. Add a new table to your Wild Outfitters database to store the information from your form.

Hands On

4. If you have access to a Web editing package, try to create the form that you designed in Question 3. If you do not have a Web editing package available or lack the knowledge to use it, a form page is available for downloading from the textbook Web site at www.course.com. Look at your page in a browser and then view the source code. Identify the HTML tags that were discussed in this chapter. Are there other tags that you have not seen before? Where could you go to learn about these tags?

DESIGN AND DEVELOPMENT OF INFORMATION SYSTEMS

For information systems to benefit organizations, they must first be developed. Organizations can develop systems in a number of ways, including developing them internally, having someone else develop them specifically for the organization's needs, or acquiring them from a commercial developer. If an organization decides to develop an information system internally, it can use a number of approaches, including ad hoc programming, structured systems development, end-user development, and rapid application development (RAD). Structured systems development, also known as systems analysis and design, is a popular approach that has been in use for nearly 30 years. RAD, a newer approach, attempts to speed up the often lengthy structured approach. When internal development is not feasible, the project can be outsourced to an outside developer or acquired.

Part IV analyzes approaches to development of information systems. It provides detailed coverage of the popular structured systems development approach, including the planning, analysis, design, and implementation stages. It also examines the use of RAD as a way to speed up the process, as the networked economy often requires that systems be up and running in record time. In addition, Part IV discusses outsourcing and acquisition as alternatives to accelerate the development process.

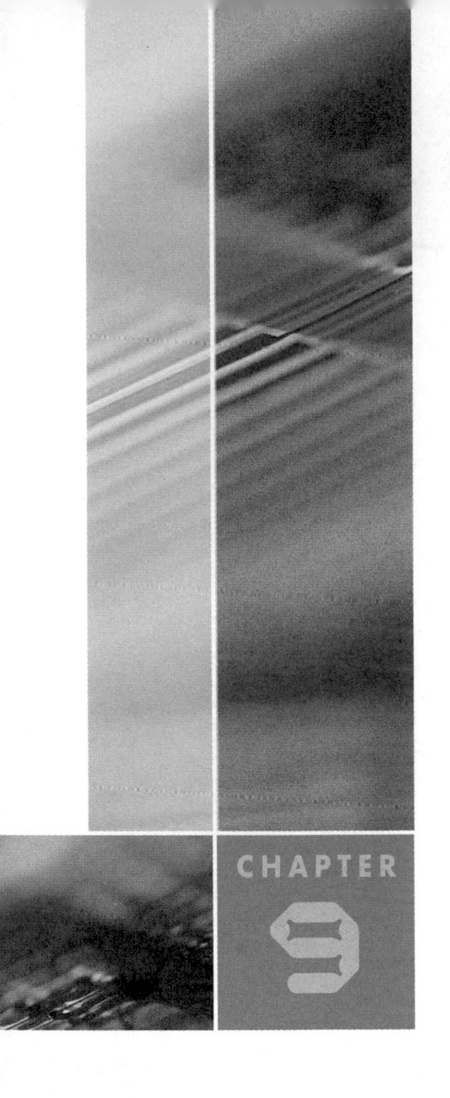

CHAPTER 9

DEVELOPING INFORMATION SYSTEMS 1

LEARNING OBJECTIVES

After reading this chapter, you will be able to answer the following questions:

- Why is information systems development or acquisition important to all types of organizations?
- What are the most commonly used approaches to systems development?
- What are the stages in the structured approach to development?
- What are the steps in the planning stage of development?
- What are the steps in the analysis stage of development?
- How do data modeling and process modeling fit into the analysis of a replacement or new system?

Developing the Manheim Online Web Site

The end-of-chapter case in Chapter 4 discussed Manheim Online and its Web site, where the company sells used cars directly to automobile dealers and to end buyers. Manheim began developing this Web site in early 1996 because of a perceived threat from AUCNET, a company that had successfully auctioned cars in Japan using a television network. Wishing to blunt the threat from AUCNET, Manheim experimented with its own television-based system, but decided to pursue an Internet-based system instead. To test the idea, it approached Intellimedia Commerce, an Atlanta-based Internet development company, about creating such a car auction system. Based on preliminary discussions, Intellimedia Commerce quickly developed a prototype Web site that demonstrated how dealers would be able to purchase automobiles directly from Manheim Auctions.

This initial prototype proved so successful that Manheim asked Intellimedia Commerce to create a working version of the sales system. Early on, Manheim decided the purchase system should use a set wholesale price for selling program cars (off-lease or company executive cars), rather than an online bidding system; the goal was to offer a different approach from that provided by the existing physical auctions. Manheim also felt a set wholesale price would give purchasers access to a larger inventory pool and provide sellers with potentially a higher return on their vehicles.

Interestingly, Manheim's information technology (IT) department had only minimal involvement in developing the new sales system. Instead of having the IT department develop the new system, the vice president in charge of the venture chose to use an outside developer to create both the demonstration prototype and the final working version of the system. Although the developer had to interact with the IT people at Manheim to create a system that would access the data on Manheim's AS400 midrange computer, the IT department had little to do with the development of the final system.

Shortly after the 1998 interview with Ralph Liniado, vice president of development, that served as the basis for this discussion, Manheim Auctions brought the operation and further development of Manheim Online inside the organization. In addition, it added AutoTrader.com, a retail used-car system, to its online system. This operation gave the company access to both the retail and the wholesale markets.

Source: Author's interview with Ralph Liniado, vice president of development of Manheim Auctions, Atlanta, January 15, 1998; updated February 28, 2002, in an interview with Ben Dyer, CEO of Intellimedia Commerce, Inc.

Systems Development

As discussed throughout this textbook, information systems that are used to handle the present, remember the past, and prepare for the future are an integral part of all organizations. They process transactions quickly; store data, information, and knowledge, and help managers and employees make decisions. Without information systems, most modern organizations would cease to operate in just a few days, if not hours. All organizations face a problem, however, in deciding how to obtain the information systems they need. Whereas some systems can be acquired, others must be designed and built to meet the specific needs of the organization. Properly building or acquiring information systems is critical to the long-term well-being of an organization. James Martin, one of the best-known authors in the field of information technology, has said: "The building of systems for unique [competitive] capability is often the single most important activity for an IT organization."[1] Given the importance of acquiring or building information systems, this book will devote the next two chapters to this topic.

When building or acquiring an information system, IT professionals must carefully analyze the organization's requirements so that the resulting information system accomplishes the desired purposes. The process of an organization building an information system internally, contracting out the development, or acquiring it is called **systems development**. This broad term encompasses the processes of acquiring an information system and contracting out the development because, even when a system is acquired or contracted out, staff members must still do a great deal of the problem determination and analysis work inside the organization. Because it involves analyzing a problem and designing a solution for it, systems development is also often referred to as **systems analysis and design (SAD)**.

systems development
The process of developing a system design to meet a new need or to solve a problem in an existing system.

systems analysis and design (SAD)
Another name for the systems development process.

When Is Systems Development Needed?

The process of systems development can take place for a variety of reasons, including to capitalize on a new opportunity to use information technology, to satisfy a request for an enhancement to an existing system, or to solve a problem with an existing system not meeting organizational needs. For example, consider the Manheim Auctions case, in which potential competition and new technology (the Internet) prompted the development of a new information system to carry out an existing operation—selling program cars to dealers—in a different way.

In addition, existing information systems are frequently modified through enhancements that enable them to carry out additional operations. Individuals or departments often request such enhancements to help them do their jobs better.

To understand the last situation, in which the existing system fails to meet the organization's needs, you need to understand that all systems go through an **information system (IS) life cycle**, which has three phases:

information system (IS) life cycle
The various phases in the life of an information system—systems development, operational use, and decline in usefulness.

Phase 1: Systems development
Phase 2: Effective operational use
Phase 3: Decline in usefulness

Phase 1 of the IS life cycle includes all of the elements of systems development that are carried out to create a new information system, including recognizing the need for one, developing the new system, and implementing it in the organization. This phase of the IS life cycle will be the subject of much of this and the next chapter. In Phase 2, the information system is (ideally) effectively carrying out the purpose for which it was designed. However, virtually from the end of the systems development phase until the new system is implemented, the information system must be modified to meet changing business conditions and technology. Not surprisingly, after several

1. James Martin, *Cybercorp: The New Business Revolution* (New York: AMACOM Books), p. 104.

system degradation
The point in the IS life cycle in which the performance of the system drops off markedly and the quality of information provided by the system suffers.

modifications, a system may begin to resemble a patchwork of subsystems that do not work well together, resulting in **system degradations**. In Phase 3, system degradations cause the performance of the system to drop off markedly, and the quality of information provided by the system suffers. At this point, the old system must to be discarded and a new one either acquired or built.

Figure 9.1 illustrates the three phases of the IS life cycle. In the figure, required resources appear on the vertical axis, and time is graphed on the horizontal axis. Because the axes are not to scale, you should not try to interpret the graph as showing actual resource use or time spent in each phase.

Note that Figure 9.1 shows two versions of the IS life cycle: a theoretical version and an actual version. In theory, except for an occasional blip, the use of resources should decline as the information system grows older until its phase-out and the installation of a new system. In actual practice, although the need for resources decreases during the later stages of systems development, continued changes required by any information system tend to require addition of more resources at various points in operational use. An information system will definitely require additional resources during the period while its usefulness is declining.

The rate at which this process of development, use, and decline progresses depends on the nature of the business environment, changing business conditions, and technological advances. Rapid changes like those produced by the networked economy may dramatically shorten the length of the typical information system life cycle.

In general, however, the length of use for information systems ranges from 5 to 15 years or longer. For example, at the turn of the twenty-first century, organizations continued to use information systems that had been developed as long ago as the 1960s and 1970s. At the time of the systems' creation, no one thought that they would still be in use 30 years later. As you may be aware, because many of these old systems saved dates in a two-digit format with the first two digits assumed to be 19 (originally done to save disk storage space), they would not work in the new century. A great deal of work was done leading up to January 1, 2000, to ensure that these systems could correctly handle years beginning with 20.

Figure 9.1 Information systems life cycle

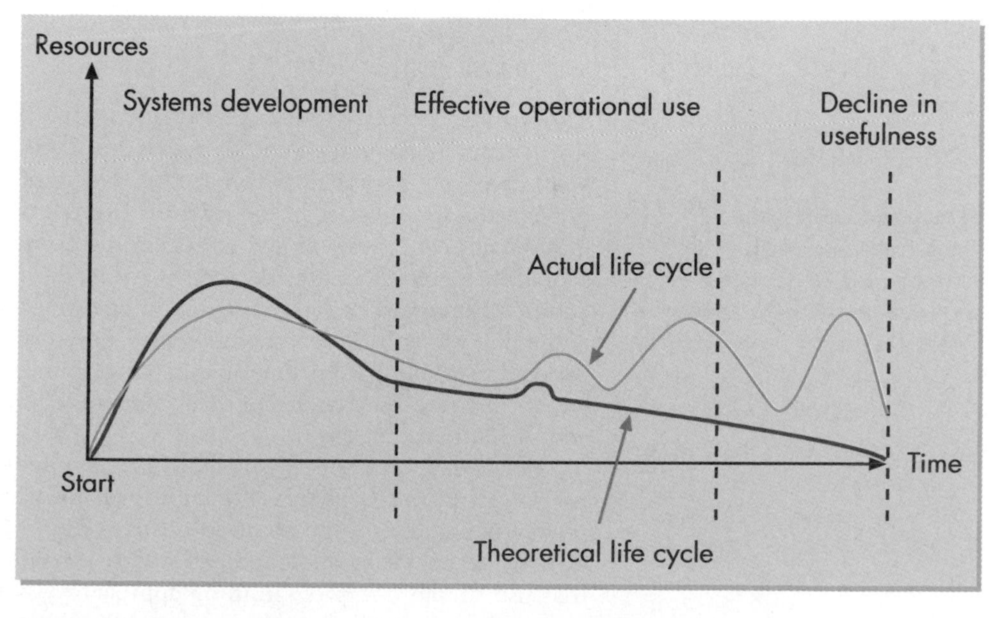

Source: Guy Fitzgerald, "Addressing Information Systems Flexibility: Theory and Practice," Working paper, Brunel University, UK, 2001.

In any case, all information systems must continually be monitored to ensure that they satisfy current business needs and to determine when they must be upgraded, maintained, or scrapped in favor of new systems. In this chapter and the next, you will look at various approaches to the first phase of the IS life cycle—systems development.

Systems Development at FarEast Foods

After scanning the environment and noting the actions taken by other Internet e-tailers, the management at FarEast Foods is considering enhancing its ordering system to give customers rewards based on their level of spending with the company—something similar to a frequent purchaser plan. Management hopes that this enhancement will give the company a competitive advantage in the sale and delivery of Asian foods using the Internet. Thus the FarEast Foods information system will be modified so as to enhance the existing system needs. This example will be used throughout this chapter and the next.

1. Why is information systems development so important to organizations?
2. What are the four stages of the information system life cycle?

Approaches to Systems Development

All approaches to systems development proceed through four stages: planning, analysis, design, and implementation. The planning stage determines the problem to be solved. The analysis stage determines what must be done to solve the problem. The design stage determines how the problem will be solved. The implementation stage actually solves the problem. These four stages of systems development all fit into the development phase of the information systems life cycle depicted in Figure 9.1.

Individuals and organizations use a variety of approaches to systems development, including ad hoc programming, a structured approach, rapid application development, end-user development, outsourcing, and acquisition. Although these six approaches share some similarities, they also have many important differences. For example, some approaches emphasize one or two stages of systems development over the others. Many systems development projects use a combination of approaches. Table 9.1 profiles the six approaches to systems development. Each will be described in more detail in the following sections.

Ad Hoc Programming

ad hoc programming process
A process in which an individual or a group that needs a new or revised system meets with a programmer, and together they decide what should be done.

Although the benefits of using a systematic approach to systems development have been apparent for many years, some organizations still do not formally go through all four phases typically associated with systems development—planning, analysis, design, and implementation. Instead, in the **ad hoc programming process**, an individual or a group that needs a new or revised system meets with a programmer and decides what should be done. Once this decision is made, the programmer works on the various computer programs that make up the information system, meeting with the person(s) requesting the work when questions arise. In essence, the analysis and design stages are handled in an informal manner by the programmer consulting with the user group to ensure that he or she is solving their problem. This approach emphasizes the planning and implementation stages.

Ad hoc programming can result in a usable information system for the initial user or group, but it often yields an information system that is poorly documented, poorly planned, not integrated into the overall organization information plan, and difficult to maintain and modify. Because these systems just evolve, organizations that use the ad hoc programming approach typically wind up with an uncoordinated, hodgepodge information system that cannot deliver the information needed for the organization to operate in today's competitive, networked economy.

Table 9.1 Approaches to Systems

Approach to Systems Development	Description	Location of Development	Comments
Ad hoc programming	The user and programmer go directly from the planning stage to the implementation stage	Internal by user and programmer	Fast, but very risky; the system may not be usable when the programmer leaves
Structured systems development	The development team goes through all four stages systematically	Internal by formal team	Slow, but accounts for all details
Rapid application development	The development team abbreviates the first three stages to speed up the development process	Internal by formal team	Faster than the structured approach, but also more risky
End-user development	The user develops the system, often skipping the first three stages	Internal by individual	Can be much like ad hoc programming, but is done to the user's own requirements
Outsourcing	The organization engages an outside vendor to develop the system	External	Does not require internal expertise, but control of system can be lost
Acquisition	The organization purchases a system from an outside vendor	External	Fastest approach of all, but may not provide a competitive advantage

Structured Systems Development

structured systems development
An approach to systems development in which each stage must be completed in a specific order after certain objectives are achieved in the previous stage, including a specified set of deliverables and management approval.

To deal with the problems with ad hoc programming, structured systems development techniques emerged in the 1970s and 1980s. This approach is called **structured systems development** because completion of each stage occurs in a specific order only after certain objectives have been achieved in the previous stage. The development team must prepare a specified set of deliverables at the end of each stage, and team members must obtain management approval before they initiate the next stage. The structured approach to systems development is particularly useful when the organization needs large transaction processing systems (TPS) and data-based decision support systems and when users can specify information requirements and the necessary data for these requirements. It is also appropriate when the organization must integrate the major components of the system into a comprehensive system.

Because it succeeds so well ensuring that nothing is left out of the system, over the last 30 years the structured approach has become the standard for information systems development. Major disadvantages of the structured approach include the time required to develop a new system, the need to identify all information requirements at the beginning of the project, and the fact that users are not always sure what they are getting. The slowness of the structured approach has become more of a problem as events in the networked economy are occurring at a faster pace than earlier eras. Even though the speed of development is slow, structured systems development remains the standard with which anyone studying information technology should be familiar. For this reason, this chapter and the next will discuss structured systems development in detail.

Rapid Application Development

rapid application development (RAD)
Methods and tools that allow for faster development of application software.

prototype
A version of the system that contains the bare essentials and that can be used on a trial basis.

Often an organizaion needs a new system quickly to meet a competitive threat. This speed becomes especially essential in the networked economy, where business conditions and technology are evolving rapidly. When an organization needs a new system as soon as possible, it can expedite the development process through **rapid application development (RAD).** A variety of approaches to RAD attempt to shorten the planning, analysis, and design stages. Many rely on prototyping, the approach described in the opening case. With this technique, the organization develops a **prototype** system that contains the bare essentials and uses it on a trial basis. After some experience with the prototype, users further refine their requirements or try a completely different approach to solving the problem. The organization then designs a revised or new prototype for use.

Prototyping is an iterative process. An information system is briefly analyzed, designed, used, reanalyzed, redesigned, used again, reanalyzed, and so on, until management has a system that meets its needs. Usually, the system becomes more comprehensive with each iteration. When the users are satisfied, the organization implements the prototype.

Prototyping also proves useful when needs change rapidly, users find it difficult to articulate requirements without seeing the system first, the risk of developing the wrong system is high, or the organization must review several alternative systems prior to accepting any particular project design. Because developers can easily create prototypes for Web applications, prototyping has been widely used for Web development. For example, Manheim's management did not have a clear idea of what it wanted when it began developing Manheim Online. The Web development firm created an initial prototype to show the possibilities. It then used this prototype as a starting point to create the final product. Chapter 10 will discuss RAD in more detail.

End-User Development

end users
Non-IT professionals who use computers to solve problems associated with their jobs.

end-user development
Development of an information system by an end user.

With computing power migrating more toward users, organizations are increasingly using a type of home-grown development referred to as end-user development. **End users** are non-IT professionals who use computers to solve problems associated with their jobs, and **end-user development** entails the development of an information system by the eventual user. End users are usually interested only in doing their own jobs better, rather than in creating application software for other users. When you complete computer projects as a part of this course or other courses, you are acting as an end-user developer. The end-user developer short-circuits the often-lengthy structured development process by going directly from the planning stage to the implementation stage. Thus, the end-user development approach has many of the same advantages and disadvantages of the ad hoc programming approach to development, with the difference focusing on who does the work and which tools are used in development.

graphical applications development environment
A software package that allows end users to create menus, boxes, and so on, and then to write just the instructions needed for a specific menu or box.

Typical end-user development tools include spreadsheets and database management systems. With spreadsheets, end users can use macro languages to create very sophisticated applications. For example, an instructor could create a spreadsheet macro that allows projects to be graded electronically. Similarly, end users can work with database software to write applications by using either a programming language built into the software or database tools to create screens. The newest form of end-user tool is the **graphical applications development environment**, such as the Microsoft Visual Studio .Net suite of programming languages, which enables end users to create menus, boxes, and so on, and then write just the programming statements needed for the specific menu or box.

Outsourcing

outsourcing
A process that involves turning over some or all of the responsibility for the development or maintenance of an information system to an outside group.

A popular alternative to internal development of an information system is **outsourcing** the development process to an outside organization. Outsourcing can range from having an outside company, such as IBM, handle the entire information technology operation of an organization to hiring contractors and temporary office workers on an individual basis, and every combination in between these two extremes. A company may choose to outsource a systems development project because its internal information technology group does not have the skills needed to create the new system or because management does not want to distract the group from its ongoing projects. Manheim outsourced its new project for several of these reasons. Its internal IT group had no experience with Web development, and members were involved in the mission-critical work of processing data on existing auction sales. In outsourcing, the actual systems development process followed usually involves either the structured approach or some form of RAD.

application service provider
An IT company that runs one or more application servers, which a customer can access to process data, to query data warehouses, or for a host of other purposes.

In another form of outsourcing, organizations contract with companies known as application service providers. An **application service provider (ASP)** runs one or more application servers, which a client can access to process data, to query data warehouses, or to pursue a host of other purposes. ASPs are a rapidly growing solution to the problem of high setup and maintenance costs for specific applications. Chapter 10 discusses outsourcing, including the use of ASPs, in detail.

Acquisition

Enterprise resource planning (ERP)
A multi-module application software that helps an organization manage the important parts of its business, including managing the supply chain, maintaining inventories, providing customer service, and tracking orders.

Acquisition involves purchasing an information system from an outside vendor, rather than developing it internally. In addition to personal productivity packages, such as word processors, spreadsheets, presentation software, and so on, organizations use a variety of other packaged software. Typical packages include accounting, payroll, order entry, and Internet shopping cart software. Companies such as SAP, PeopleSoft, and J. D. Edwards offer an all-in-one type of software called enterprise resource planning systems. **Enterprise resource planning (ERP)** refers to multi-module application software that helps an organization manage the important parts of its business, including managing the supply chain, maintaining inventories, providing customer service, and tracking orders. Like RAD and outsourcing, acquisition will be discussed in Chapter 10.

Choosing Among Internal Development, Outsourcing, or Acquisition

Determining which approach to use—internal development, outsourcing, or acquisition—is an important decision. Each approach offers both advantages and disadvantages. For example, internal development often provides a competitive advantage for the organization, but it can cost more, take longer, and require more skills on the part of the internal information technology staff than either outsourcing or acquisition. Outsourcing represents a good option when the internal IT professionals lack the skills needed to build the desired systems. Many products, such as word processing or spreadsheet applications, for which standard products exist, are almost always acquired. Other cases are less clear-cut, and an organization must carry out an analysis of the options. Chapter 10 describes the decision of whether to develop internally, outsource, or acquire an information system.

A Look Ahead

The remainder of this chapter focuses on the first two stages of the structured systems development process to create customized information systems, using the FarEast Foods customer reward project as an example. Chapter 10 discusses the last two stages of the structured systems development process, outsourcing, acquisition, and the use of RAD.

1. List the approaches taken to information systems development.
2. In which type of development are the analysis and design stages usually skipped?

Structured Systems Development

Recall that structured systems development arose as a way to solve the problems associated with the ad hoc programming approach, including poor planning, lack of integration into the overall organization information plan, and difficulty in maintaining and modifying the system. These problems occurred because the programmer

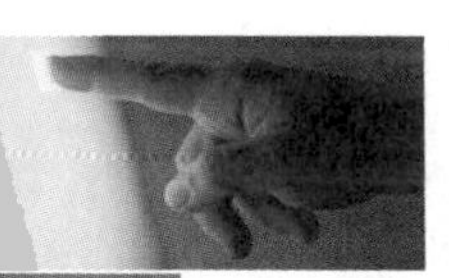

IT INNOVATORS

Bill Gates

Very few companies develop more software systems than Microsoft. Products from Microsoft include its Windows operating system, which runs on more than 85 percent of all personal computers and a wide range of servers; Micrsoft Office Suite, which dominates the personal productivity market; popular programming languages included in the Visual Studio .Net suite; Internet Explorer and other Internet tools; and a variety of games and other products. In 2001, revenue from these products amounted to $25.3 billion, and the company had the second highest total corporate value in the world. As you probably know, the chairman of Microsoft is Bill Gates. Although he no longer serves as the company's CEO, very little happens at Microsoft without Gates's knowledge.

Gates got his start with computers as a high school student in Seattle. After a year at Harvard, he joined with Paul Allen in 1975 to form Microsoft to develop software using the BASIC computer language. Gates and Allen were successful in this market, but their real break came in 1979, when they became aware that IBM was working on a personal computer and needed an operating system. They knew that another Seattle company had developed an operating system called DOS (Disk Operating System) that would work with the chip used in the IBM PC, so Microsoft purchased it. In turn, Microsoft licensed DOS to IBM for its new PC. By licensing rather than selling DOS to IBM, Microsoft retained control of the software and could sell other versions to IBM's competitors. DOS emerged as the primary operating system for PCs, providing much needed revenue for Microsoft until it released the first of the Windows operating systems (Windows 3.1) in the early 1990s. Windows 3.1 was followed by several versions of Windows, culminating with the latest edition, Windows XP, which was released in October 2001. All through Microsoft's more than 25 years of growth, Gates has remained very involved in the development of new software, meeting with programmers to test and critique their work on new products, as well as providing new ideas for the company to pursue.

Although computers and software remain at the center of his interests, Gates is also a father of two, philanthropist, and writer. The Bill & Melinda Gates Foundation tops the world's philanthropic organizations, focusing on education and world health. Gates has also written a best-selling book, *Business@the Speed of Thought*, and a number of essays, which have appeared in the popular press.

Bill Gates, although no longer CEO of Microsoft, is still very involved in developing a vision for its future.

often moved directly from a fairly simple planning stage, consisting of a conversation with the group initiating the new system, to the implementation stage, in which the computer program was written. Little analysis or design activities occurred in between these stages.

The ad hoc programming approach to developing information systems is similar to a builder talking with you about your visions for a house and then starting the building process the next day. Although this approach to building would work reasonably well for a dog house, a residence built this way most likely would *not* fit your requirements. On the other hand, structured systems development involves using a carefully thought-out process of planning, analyzing, designing, and implementing. It goes from an idea to a set of preliminary sketches to preliminary plans to blueprints to the actual construction process. In both the construction of a building and the structured approach to systems development, a step-by-step transition from a logical concept to a physical system occurs.

systems analyst
A person who carries out the systems analysis and design process.

project manager
A person who ensures that the project is completed on time and within budget and who takes responsibility for bringing other people onto the team.

This systems analysis and design process has spawned an entirely new occupation whose practitioners are known as systems analysts. **Systems analysts** work as part of a development team carrying out a problem-solving process to determine the cause of a current system's problem, suggest solutions to this problem, and then work with the programmer and end users to implement one of these solutions. Every team includes a **project manager** who ensures that the project is completed on time and within budget and who takes responsibility for bringing other people onto the team. Project managers often have extensive experience working as systems analysts and are the primary point of contact for outside people. Depending on the size of the project, in addition to the project manager and one or more systems analysts, the team may include other specialists, including programmers, database designers, and technical writers.

The Systems Development Life Cycle

As mentioned earlier, structured systems development involves four development stages, with each stage depending on the successful completion of the previous stages:

1. Planning
2. Analysis
3. Design
4. Implementation

systems development life cycle (SDLC)
Another name for the structured approach to systems development.

These four stages are commonly referred to as the **systems development life cycle (SDLC)** because they describe the conception, birth, and growth of the system. Figure 9.2 depicts the four SDLC stages. Note that the figure shows the process as a closed loop because systems that degrade over time present problems that must be resolved by repeated applications of the SDLC approach. The SLDC refers only to the first phase of the IS life cycle and should not be confused with it, even though both refer to life cyles.

As you move from the planning stage through the analysis stage to the final implementation stage of the SDLC, you move from a broad, logical understanding of the problem to a detailed solution to the problem. Each stage of the process consists of steps to generate deliverables in the form of written documentation or computer files, which provide information about what has been accomplished and what is planned for the next stage. In each stage, IT professionals and other members of the development team use different techniques to generate the deliverables.

Figure 9.2 Stages of the systems development life cycle

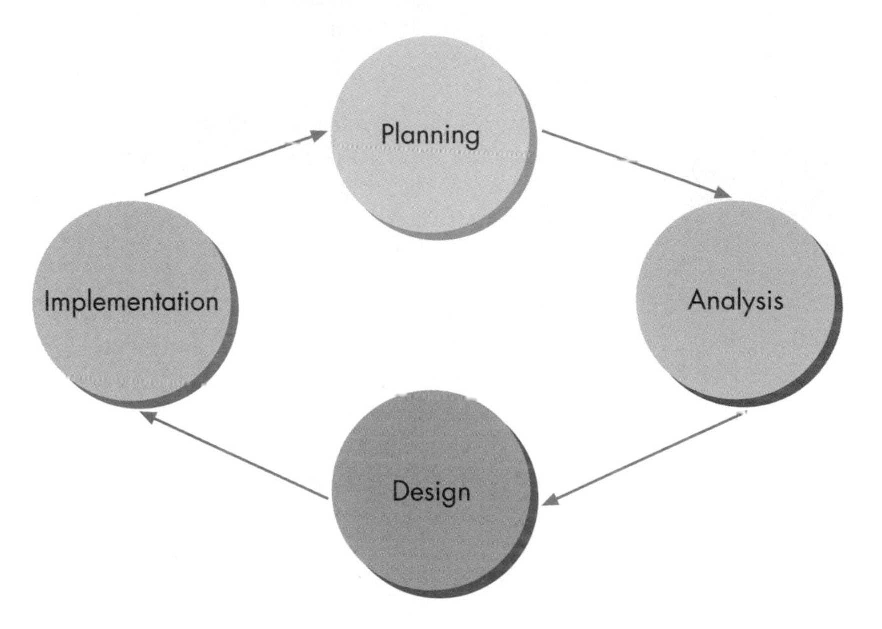

The following discussions will use the FarEast Foods Web-based customer ordering system as an example. You might ask why an organization would use the structured approach for a Web-based project such as the www.fareastfoods.com system. The structured approach is slow and methodical, and Web development usually needs to be done quickly using prototyping, as Manheim did. Although prototyping is the preferred way to build a Web site, it may not result in a system that can adequately handle numerous or complex transactions. Often, a development effort must use elements of the structured approach to ensure that the Web site can correctly handle all transactions. The consequences of an incomplete or too-rapid development process can be severe. For example, USAA, a large insurance company, found it necessary to halt access to its Web-based bill-payment service for four days in February 2002 to fix problems. Because the modification to the FarEast Foods Web site involves complex transactions that affect the entire customer ordering system, it provides a suitable example of the structured approach.

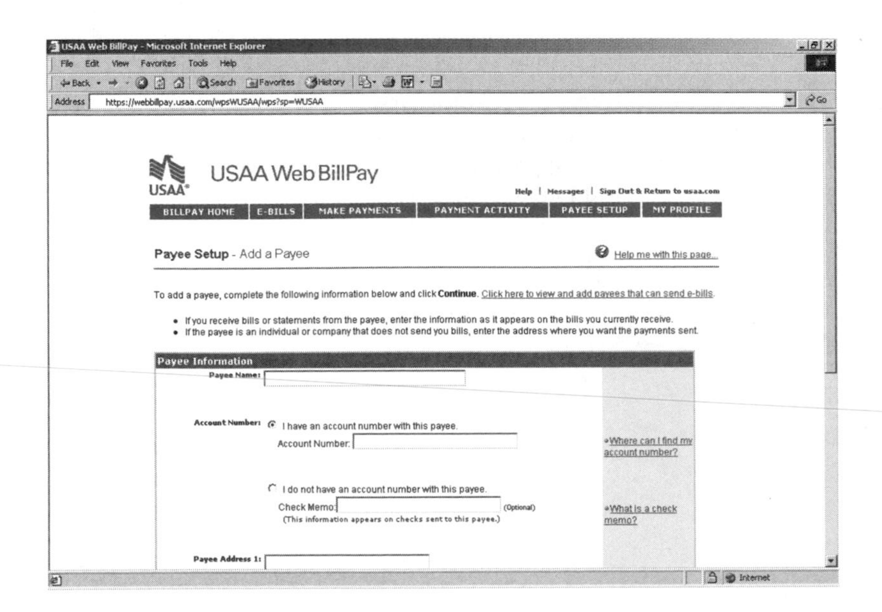

USAA took this Web site offline for four days to make changes to it.

Stages of the Structured Approach

The structured approach begins with the planning stage, in which someone in the organization becomes aware that either an existing system is not working correctly or a new system is needed to meet an opportunity. The planning stage encompasses a number of steps, including project initiation, feasibility analysis, and so on. For now, assume that the organization has already carried out the project initiation step, identifying a problem or need and initiating a project to solve it. The next step is to determine the feasibility of solving the problem or creating the new system. If the organization judges the project to be infeasible, perhaps because of technical, economical, or organizational reasons, then it can terminate the project without going any further. The key deliverable of the planning stage is a document called the **project plan**, which describes the desired information system. The project plan includes a clear statement of the scope of the desired project.

project plan
A document that describes the desired information system, including a clear statement of the scope of the desired project.

In the analysis stage, the goal is to gather data that will answer the question of *what* the system will do. Only when this question has been answered does the project proceed to the next stage—design. The key deliverable from the analysis stage, the **system proposal**, describes what the new system should look like.

system proposal
A document that describes what the new system should look like.

In the process of answering the question of what the system will do in the analysis stage, it may become clear that this system should not be developed or acquired, either because the questions cannot be clearly answered or because the answers cause management to rethink the need for a modified or new system. As with the previous stage, stopping here will save a great deal of money and time compared to continuing with the development of a wrong or unnecessary system.

In the next stage—design—the *how* questions are answered. The first *how* question is: *How* will the system be developed—internally, through outsourcing, or by acquiring it? The analysis stage should have provided enough information to answer this question. If the organization decides to pursue outsourcing, then it must select a vendor, and turn the project over to the vendor. If the organization decides on acquisition, then it must develop systems requirements to use in selecting a system to be acquired. On the other hand, if the organiztion elects to develop the system internally, then the design stage must answer another question: *How* will the system operate?

In the design stage, the organization creates the physical design to specify the details of the system. This **system specification** is a document that is used in the implementation stage to program the new system, outsource it, or acquire it, depending on the decision made at the beginning of the design stage.

system specification
A document that is used in the implementation stage to develop the new system internally, outsource it, or acquire it, depending on the decision made at the beginning of the design stage.

In the implementation stage, if the system is being developed internally or outsourced, programmers turn the system specification into working programs. If the new system is being acquired, the organization uses the system specifications to identify a system for purchase. Whether developed internally, outsourced, or acquired, the new system is installed and tested extensively at this stage, and the users are trained. The deliverable from this stage consists of an installed working system. Note that, even when the planning, analysis, and design stages are accomplished on schedule, the development process can become dramatically delayed or even aborted during the implementation stage. A number of factors can lead to the delay or termination of the project, including poorly completed early stages or programming problems. All too often, however, the problems relate to people in the organization subverting the implementation process when they feel threatened by the changes associated with the new information system. These people may include employees who fear that they will be forced to learn a new system or those who think the new system will put them out of a job.

Table 9.2

Systems Development Deliverables

Stage	Deliverable	Percentage of Time
Planning	Project plan	18
Analysis	System proposal	24
Design	System specification	9
Implementation	Working system	49

After installation of the project, it must be maintained by solving any day-to-day problems. During the implementation stage, you can discover that so much maintenance work is necessary to keep the system running that it becomes obvious that the system is not working as required. This problem can occur because of changing information requirements, dynamic business conditions, or technological innovations. If the system fails to work as required, you move back to the planning stage and start the systems development life cycle over again.

Table 9.2 shows the four stages of structured systems development and the key deliverable for each stage. It also shows the results of a survey of software developers regarding how much time they spend on each activity.[2] Note that the last stage takes almost 50 percent of the development time. This fact is not surprising given that the system must be programmed, tested, and installed during this stage. Often, developers do not set aside sufficient time to carry out the implementation stage, which results in new information systems being delivered late.

waterfall approach
Another name for the structured approach to systems development.

The structured approach to systems development is often referred to as the **waterfall approach** to development because the process resembles a series of waterfalls, with the deliverable from each stage *falling down* to the next stage, as shown in Figure 9.3. And, as with a waterfall, it is often difficult (but not altogether impossible) to go backward in the structured approach. Ideally, each step of the structured approach should be completed before going on to the next stage. In practice, situations often arise in which it becomes necessary to backtrack to a previous stage to add a missing element or to fix a problem. Although returning to a previous stage (shown as a dashed backward arrow in Figure 9.3) should be avoided, it is a reality in systems development.

Figure 9.3

Waterfall development

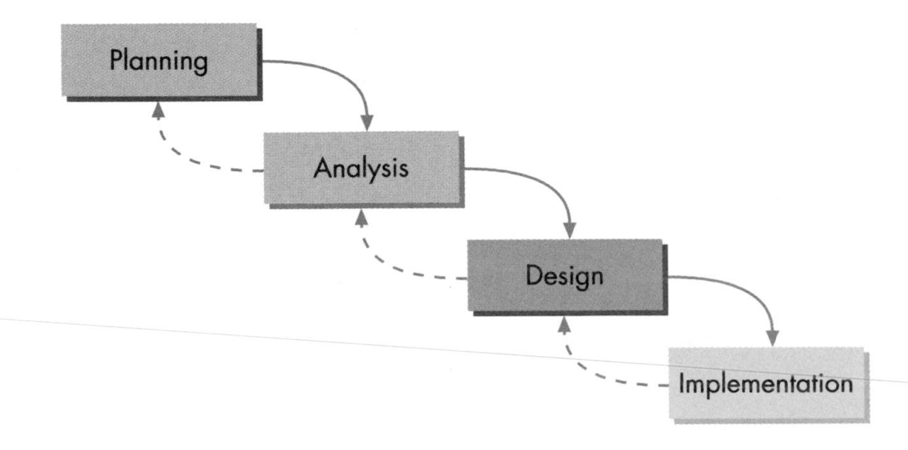

2. Guy Fitzgerald, "Addressing Information Systems Flexibility: Theory and Practice," Working paper, Brunel University, UK, 2001.

Systems Development in the Networked Economy

Like every other facet of global business, structured systems development has been changed by the networked economy. No longer can the process be slow and tedious—it must move fast to keep up with the speed of the networked economy. As a consequence of this demand for speed, systems development professionals have combined the best features of structured systems development with those of RAD prototyping in the analysis and design stages, and adopted computer languages that lend themselves to faster development. In addition, systems development professionals are emphasizing thinking about the implementation stage at the beginning of the process, rather than waiting until the end. This effort often involves including implementation problems as a part of the planning process.

The next sections will consider the planning and analysis stages in more detail, using the customer rewards enhancement to the FarEast Foods customer ordering system as an example. Chapter 10 will cover the design and implementation stages.

Quick Review

1. What are the four stages of structured systems development?
2. Why is the structured approach referred to as the *waterfall approach* to systems development?

Going from Windows to Mac Through Kiev

TECHNOLOGY ON THE EDGE

One problem facing Apple Computer in its competition with Windows-based machines is providing enough software to its users. One way to develop Apple software is to write original software specifically for the Apple Macintosh. Another option is to convert Windows-based software so that it will run on the Macintosh. For example, The WorldBook Encyclopedia on CD ROM was developed first for the Windows platform and then converted into a Mac version. Similarly, the popular SPSS statistics package had to be converted from Windows before Macintosh users could benefit from it. As with all such conversions, these translations from Windows to Macintosh required a great deal of work on the part of a large number of computer programmers. The surprising part of the conversion of these two software packages and more than 200 other such packages is that it did not take place in Silicon Valley. Rather, it was accomplished in Kiev, Ukraine, by a company called Software MacKiev.

Kiev has become a hotbed of Apple programming for a number of reasons. An estimated 15 percent of the city's population has a computer science degree due the concentration of computing power there during the Soviet era. Combine this fact with the creation of a center, in cooperation with Apple's development group, for the retraining of nuclear scientists and engineers for peaceful purposes, and you have a definite interest in Macintosh programming. To take advantage of this supply of Macintosh programmers, Software MacKiev was founded to convert Windows software programs to run on Macintosh computers. With a staff of more than 100 programmers, it is the largest development company in the world devoted exclusively to converting Windows titles to the Macintosh. Whenever possible, the firm tries to add features to the Mac titles not found in the Windows version. Jack Minsky, founder of Software MacKiev, which also has offices in Cupertino, California, happily reports that former Soviet nuclear scientists are now writing *Sesame Street* programs and other peaceful projects.

Programmers at MacKiev specialize in converting Windows software to run on Macintosh computers.

Source: David Graham, "The Ukrainian connection." *Hot News*, http://www.apple.com/hotnews/features/mackiev/ as updated by the author.

Planning Stage

The planning stage of the systems development process involves five steps:

1. Identifying the project
2. Initiating the project
3. Performing a feasibility study
4. Creating a workplan
5. Staffing the project

Through these five steps, the development team creates a project plan containing the best estimate of a project's scope, benefits, risks, and resource requirements. This project plan will be reviewed by the steering committee, which will decide whether to go forward with it.

Identifying the Project

The project identification step begins when someone in the organization—a manager, salesperson, or customer service representative, for example—recognizes that an existing information system has a problem or that a new one could enhance the organization's ability to achieve its goals. Regulatory or tax law changes can also trigger a need for a modification to an existing system or the acquistion of an entirely new one. Problems with an existing system become apparent through complaints from users, an audit (which may show the current system to be too expensive even though it is working), or output from the system (which may not match organizational needs). Management may recognize that the organization needs a new system as the result of strategic plans, from a drive to meet competitive demands, from an employee's idea for a new way to carry out an existing process, or from an idea to use information technology to move into a completely new area of business. Many of the dot-com companies mentioned in Chapter 7 were inspired by ideas that led to a new type of information system. In many cases, a company creatively destroyed itself and then rebuilt its operations around the new information system.

Like the idea for Manheim Online discussed in the opening case, ideas for new systems often come from individuals outside the IT area. People in marketing, sales, production, or almost anywhere in the company can see a need for a new system that will help the organization compete more effectively. Although some ideas turn out to be impossible to implement, the feasible ones can bring huge benefits to the company. The best case occurs when a person outside IT works with the IT group to develop an idea into a workable project.

project sponsor
A person who has an interest in seeing the system succeed and who will provide necessary business expertise to the project.

system request
A document that lists the business need for a systems development project, its expected functionality, and the benefits that would likely result from its completion.

Regardless of where the recognition of a problem or the idea for a new system originates, the new system must have a **project sponsor** (also known as a *champion)* who has an interest in seeing the system succeed and who will bring business expertise to the project (or appoint someone to provide the needed expertise). If a project does not have a sponsor sufficiently high up in the organization to push it, no matter how good the idea is, the project will probably never reach completion. At Manheim Auctions, Ralph Linado, vice president of development, was the sponsor who pushed the project forward.

Initiating the Project

IS steering committee
An internal group composed of management, users, and developers that reviews proposals for information system development.

Once a sponsor has identified a project, the project initiation step begins. In project initiation, the project sponsor usually creates a **system request** that lists the business need for the project, its expected functionality, and the benefits that would likely result from its implementation. The system request goes to an **IS steering committee**, composed of management, users, and developers, for review. If the steering committee approves the project for further study, the IT department assigns a systems analyst to investigate the project and work with the sponsor on the next step of the process, the feasibility study.

Performing a Feasibility Study

The feasibility study that is carried out by the project sponsor and the systems analyst aims to answer the following questions:

- Is the problem worth solving?
- Is a solution to the problem possible?

If the answer to either question is *no*, then the structured approach is terminated.

In a feasibility study, the analyst does not attempt to find a solution to the problem or opportunity. Instead, the objective is to reach an initial understanding of the problem or opportunity and to decide whether it is feasible to proceed with a full-scale study of it. The feasibility study looks at technical, economic, and organizational aspects of the problem. **Technical feasibility** means that technology exists to solve the problem or to meet the need. **Economic feasibility** means that solving the problem will offer financial benefits to the organization. **Organizational feasibility** means that the problem can be solved or the need can be met within the limits of the current organization.

technical feasibility
An indication that the technology exists to solve the problem or develop the system.

economic feasibility
An indication that solving the problem or developing a system will offer financial benefits to the organization

organizational feasibility
An indication that the problem can be solved or a system developed within the limits of the current organization.

A project that is judged to be technically infeasible cannot be developed with current technology, or the development group does not have the necessary skills to develop it. To make this judgment, the systems analyst must have knowledge of the available technology and the skill set of the development group. Developers should avoid using experimental technology because it can result in working on the *bleeding edge* of technology, in which case a technology might not work as advertised. This type of failure can prove catastrophic when the application is a part of the central mission of the organization. For example, in the late 1980s, the state of California tried to convert an antiquated information system for checking outstanding loans on property to one that used the latest PC-based technology. Unfortunately, the technology had not been thoroughly tested and failed to live up to expectations. The fiasco resulted in the loss of millions of dollars—not just to the state, but also to almost 500 banks that depended on the system for information.

Determining economic feasibility involves running a cost/benefit analysis of the project. This analysis should take into account both monetary and nonmonetary costs and benefits of solving the problem or taking advantage of the opportunity by creating a new system. At this stage in development, costs and benefits are merely estimates because the actual solution is not yet known. If this cost/benefit analysis shows that the costs will likely outweigh the potential benefits, then the systems analyst will recommend scrapping the project. On the other hand, if the cost/benefit analysis yields positive results, then the organization may further consider the project. Economic feasibility did not represent an issue in the Manheim Online case, because the company was responding to a competitive threat to its dominant position in the market. However, if the costs of setting up and running a Web site to sell cars to dealers turned out to exceed any benefit that might result, then the project would be considered economically infeasible.

Determining organizational feasibility involves assessing how well the organization will accept the new system. No matter how good the system is, if the members of the organization are unwilling to accept it, it will *not* work! As mentioned earlier, if not addressed at this stage, organizational feasibility can become a problem at the implementation stage, after the organization has already spent a large amount of money on the project. To address organizational feasibility, the firm can undertake a **stakeholder analysis** to determine how the new or revised system might affect any particular person or group. Typical stakeholders include the project sponsor, the IS development group, users, employees in other departments, customers, and suppliers. Failure to analyze the potential reaction of a stakeholder to the new system can result in severe problems upon implemetation of the system. For example, at Manheim, stakeholders included the vice president (the sponsor); the IS group, which was left out of the development process; the dealers who would use it, and the people who worked at the traditional auction sites, who may have been threatened by the new system.

stakeholder analysis
An analysis to determine the effect of a new or revised system on a particular person or group.

If the new system appears to be feasible on technical, economic, and organizational grounds, then the analyst will recommend that the project continue.

Creating a Workplan

workplan
A document that lists the tasks that must be accomplished to complete a project, along with information about each task, the number of persons required, and an estimated time to complete each task.

If the project is approved for continuation, management typically assigns a project manager and formally launches the project. The project manager's first responsibility is to create a document called a **workplan**, which lists the tasks that must be accomplished to complete the project, information about each task, the number of people required to complete each task, and an estimated time to finish each task. Estimating the time to complete each task requires a great deal of experience on the part of the project manager, and even then the actual time required to complete the project could potentially differ dramatically from the estimates in the workplan. Late projects are so common in systems development that they are almost an accepted part of the process. This fact explains why redoing systems to handle the Year 2000 problem was so different from the typical project: The new or revised systems could not be late or else the results could have been catastrophic.

In creating a workplan, the project manager must consider the three competing demands of every information systems development project: time, functionality, and cost. It is not possible to reduce the time to complete the project, increase functionality, and reduce cost at the same time. In fact, improving on any one of these elements will usually affect one or both of the others negatively. For example, adding more functionality to a project in the form of more features will almost always lengthen the project's development time and increase costs. Figure 9.4 depicts this relationship.

Staffing the Project

Once the project manager has created the workplan and management has approved it, the next step is to bring the team members on board. Depending on the size and complexity of the project, the team will include a variety of skilled professionals. The degree of experience will also vary as employees who are new to the systems development field work alongside veterans to provide the new people with much-needed experience. The project manager takes responsibilty for staffing the project and assigning duties to each person or subteam based on the tasks listed in the workplan.

Note that a structured systems development project is an organizational project. To develop the best system possible, users from all functional areas must take part in the development effort. Failure to do so could cause serious morale problems as well as design and implementation problems later in the process.

Figure 9.4 Competing IS project demands

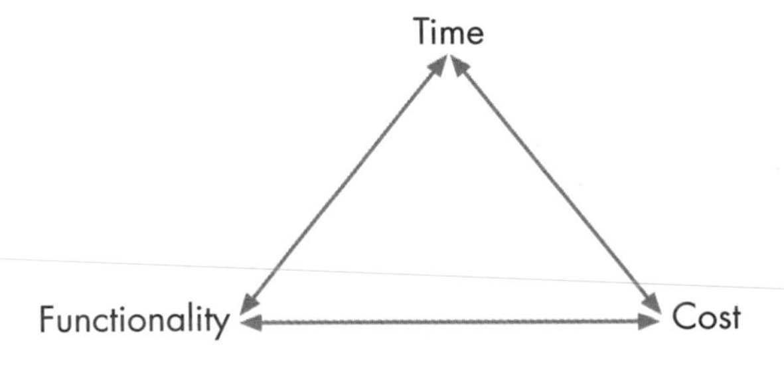

Cost/benefit analysis for FarEast Bucks customer reward program

	A	B	C	D	E	F	G
1	Item	Current Values		Projected Values			Comment
2	Average Price	$7.50		$7.13			Assume 5 percent discount on all sales
3							
4	Average Cost	$4.50		$4.50			Average cost does not change
5							
6	Average Unit Profit	$3.00		$2.63			Unit profit = price - cost
7							
8	Monthly Unit Sales	70,000		87,500			Sales increase by 25 percent each month
9							
10	Monthly Profit	$210,000.00		$230,125.00			Monthly profit = unit profit x unit sales
11							
12	Monthly Increase/Decrease>>			$ 20,125.00			Monthly increase = current profits - project profits
13							
14				Development Cost	$100,000		Assumed development cost of $100,000
15							
16				Months to Pay Off	4.97		Months to pay off = development cost / monthly profit increase
17							

Table 9.3 uses a worst-case scenario for the customer discount. *All sales* in this analysis are subject to a 5 percent discount when, in actuality, customers will receive the discount only after spending at least $200 to obtain 10 FarEast Bucks. Under this more rigorous assumption, if the project pays for itself in a few months, it will do even better under actual conditions. The marketing manager has researched the effect of similar customer discounts in the past and has found that monthly unit sales typically increase by *at least* 25 percent with this kind of plan. Finally, the systems analyst has estimated that the customer rewards system will cost *no more* than $100,000 to develop and can be completed in 13 weeks. The questions now become: Will the increased unit sales combined with decreased unit profits result in increased monthly profit? If so, how long will it take to recoup the cost of development? Because FarEast Foods has a policy that a new project must pay for itself in six months to win approval, this question is important.

Figure 9.7 shows these data and assumptions entered into a spreadsheet. You can see that the combination of 5 percent smaller unit profits with 25 percent higher unit sales results in an increased monthly profit of over $20,000. Dividing the $100,000 in development costs by the increased profit results in a payoff period of a little less than five months—good enough to continue the development process.

Next, the analyst considered organizational feasibility. This new system does not appear to affect any stakeholders other than customers, and it affects them in a positive way. Thus it would appear to be organizationally feasible.

Developing a Workplan and Staffing the Customer Rewards Project

Based on the feasibility study, the steering committee at FarEast Foods gave its go-ahead to the project and management assigned a project manager. This project manager will immediately begin to develop a workplan for the project and to pull together a team to build the new system. Figure 9.8 shows some of the tasks to be performed, along with the project manager's best estimate of the time to complete each task. Note in Figure 9.8 that two tasks can be performed in parallel, but the last two tasks must be performed in sequence. If the design task runs late, it will therefore push back the start of programming, testing, and installation, potentially resulting in a late project, especially if the implementation also runs late, (a very

The final deliverable of the planning stage is the project plan, which is a compilation of all of the work that has gone into this stage. The project plan includes the systems request, the feasibility studies with the cost/benefit analyses, the workplan, and the staffing assignments. It represents the team's best estimate of the project's scope, benefits, costs, time requirements, and expected results. The project plan serves as the "highway map" for the project, and it will be revised as the team learns more about the project during the analysis stage.

Figure 9.5 shows the five steps of the planning stage. In the figure, the rectangles indicate actions and the triangles indicate decisions that the steering committee makes. Deliverarables or outcomes are shown on the arrows.

Figure 9.5 Steps in the planning stage

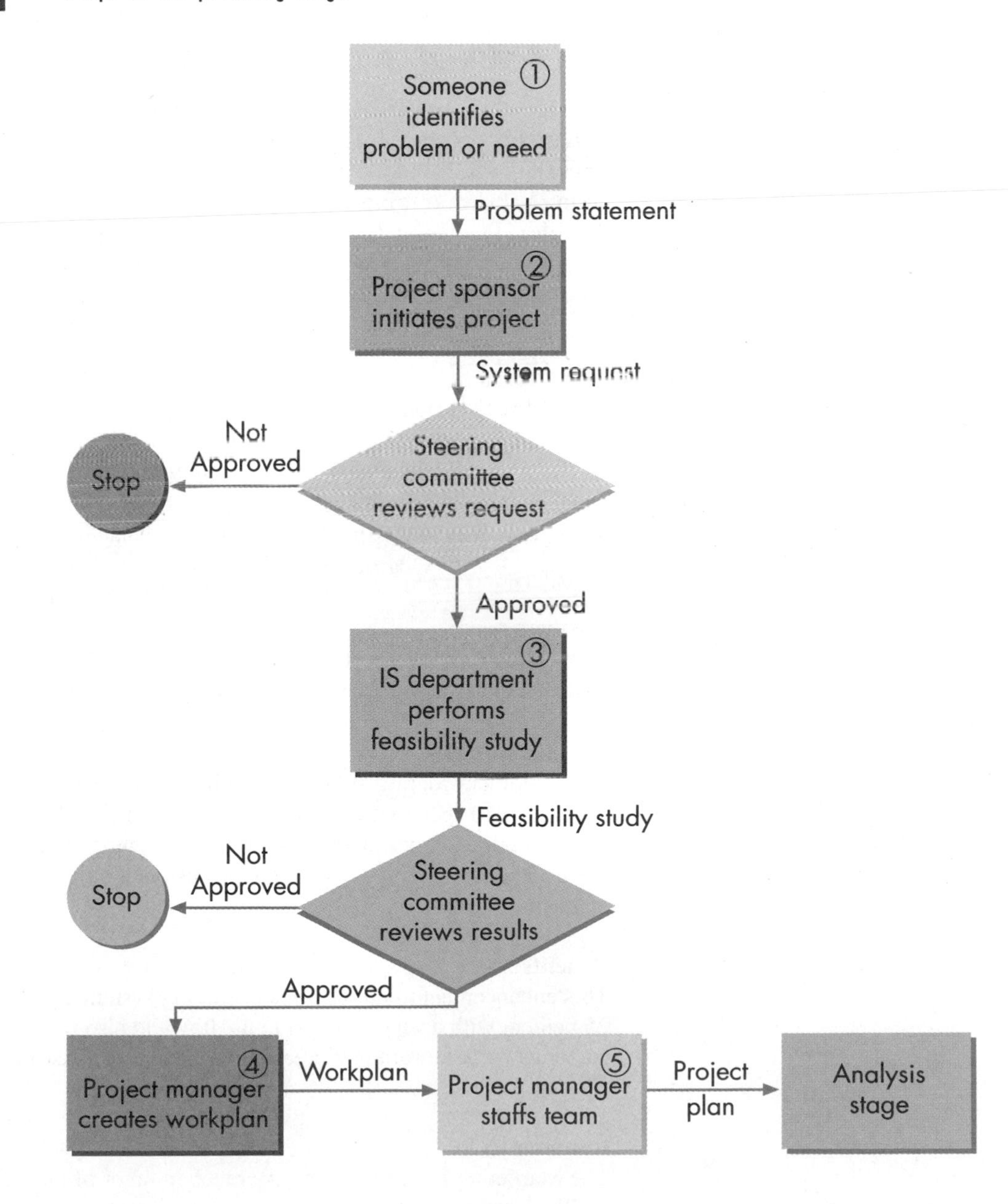

Magnitude of Effort

Although the FarEast Foods project considered in this chapter is fairly small, involving only a few people, much larger projects are common. Projects such as creating Windows XP involve hundreds, if not thousands, of people working for several years to complete the many elements of the new system. Rewriting code in thousands of information systems to fix the Year 2000 problem required a great deal of systems development effort. Experts estimate that a typical large company spent 400 programmer-years to fix this problem (a programmer-year is equivalent to one programmer working full-time for one year).

1. What are the five steps in the planning stage of systems development?
2. What types of feasibility must be considered in the planning stage?

Planning Stage for FarEast Foods Project

Management at FarEast Foods has proposed that the existing customer ordering system be enhanced to provide incentives to customers to spend more at the www.fareastfoods.com Web site. A marketing employee came up with the idea after taking a vacation using frequent flier miles, and then talked with her manager about the idea. Realizing the potential of this suggestion, the marketing manager put together a system request for the project. Figure 9.6 shows a shortened version of the system request.

Figure 9.6 FarEast Foods customer rewards system request

System Request

Project name: FarEast Bucks
System Sponsor: Chris Patrick, Marketing Manager
CPatrick@fareastfoods.com

Business need:
To increase sales and customer loyalty by providing frequent customers with FarEast Bucks, which they can use to purchase items from our catalog.

Functionality:
The customer ordering system would be enhanced to give customers one FarEast Buck for every $20 spent with us. When they have at least $10 in FarEast Bucks, customers could use them to purchase items from our catalog. The FarEast Bucks would be stored electronically on our database, and the customer's total would appear each time he or she visits our Web site.

Benefits:
This enhancement to our customer ordering system could increase sales by as much as 25 percent with a very low cost to us. It would also increase customer loyalty, because customers would want to increase their FarEast Bucks and use them to purchase items from us.

Comments:
The Marketing Department is very much in favor of this project and views it as a way to build a loyal customer base.

Customer Rewards Feasibility S

The IS approval committee
program, and the IS depar
feasibility study. In stud
technically feasible to
each transaction, tra
to use the bucks to ma
tomer rewards program w
well as adding functionality t
reduce the amount due from the

Because the customer reward pr
system, FarEast Foods is adding new fun
information systems today, sections of pro
can be added to enhance the existing systems.
existing system by passing messages and data, bu
task for which they are designed. The development
existing system proceeds much the same way as that
Although the customer rewards program will actually requ
the existing customer ordering system, the current discussion
development of the new modules.

module
A separate program that performs a specific task and shares data with other modules to lead to an integrated system.

To check economic feasibility, the analyst and sponsor did a cost/
Organizations can perform such analyses using a variety of methods,
present value (NPV), return on investment (ROI), and break-even anal
With the speed at which information technology is changing in the
economy, one way to carry out a break-even analysis is to look at how *long* it
for a project to pay for itself. If a project won't pay for itself in 6 to 12 mo
smaller company such as FarEast Foods might not want to invest in it because
low return on investment. Consider the values shown in Table 9.3 to see how t
type of analysis would work for the customer rewards system at FarEast Foods.

Table 9.3 Data and Assumptions for Customer Rewards Project

Item	Value	Assumption
Current average price	$7.50	Historical data
Current average cost	$4.50	Historical data
New average price	$7.13	All sales discounted 5 percent
Old monthly unit sales	70,000	Historical data
New monthly unit sales	87,500	Unit sales will increase 25 percent based on marketing manager's research
Development cost	$100,000	Analyst's estimate

Figure 9.7

Figure 9.8 Portion of workplan for customer rewards program

Workplan for customer rewards project

Task	Time Estimate
Analyze system and determine requirements	2 weeks
Create data flow diagram	1 week*
Build data model	1 week*
Design system	4 weeks
Program, test, and install system	7 weeks

*Can be done together.

common occurrence). Note also that, because two tasks can occur in parallel, an estimated 14 weeks will be required to complete the project—only one week longer than the time estimated by the systems analyst in the feasibility study.

Although this list of tasks and expected times is adequate for this project, a number of project management tools are available for developing workplans and schedules for more complex projects. For example, the **Gantt chart** lists project tasks along the vertical axis and time along the horizontal axis. It uses shaded boxes to indicate completed tasks and unshaded boxes to indicate uncompleted tasks. When a task runs late, tasks that depend on it automatically move back, vividly displaying the effect of the delay on the completion of the entire project. The Gantt chart for the customer rewards project in Figure 9.9 shows that the first task has been completed on time.

Gantt chart
A graphical project management tool used for developing workplans and schedules.

Figure 9.9 Gantt chart

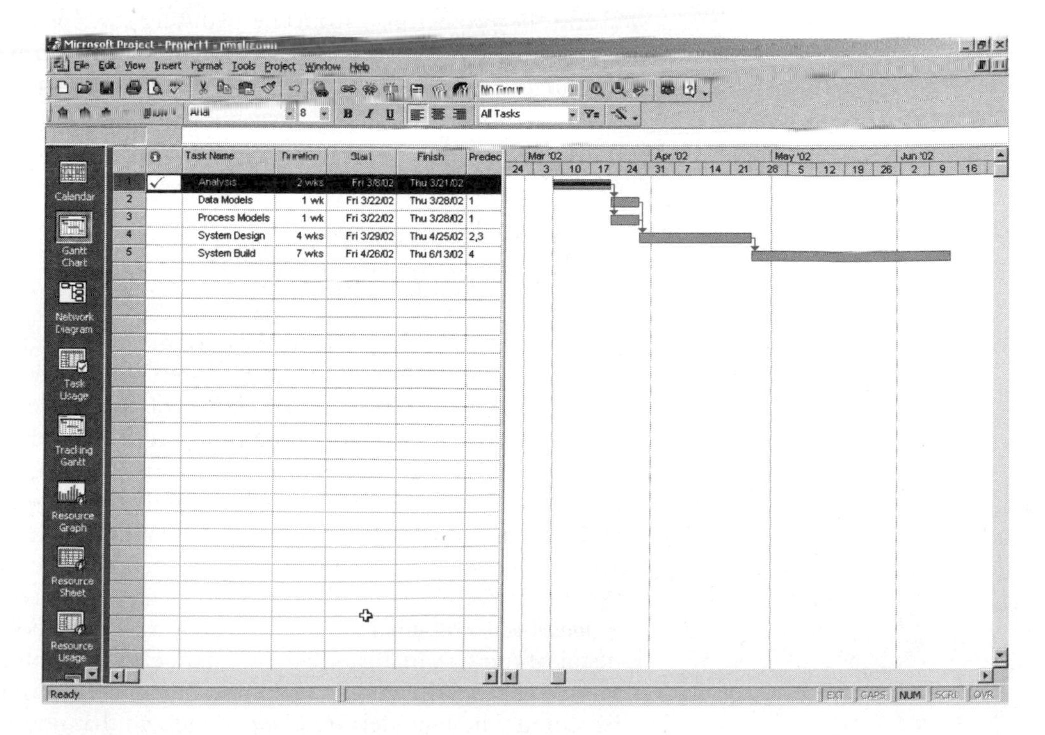

Table 9.4 Staffing for Customer Rewards Project

Name	Title	Responsibility	Time Allocation (%)
Heather	Project manager	Oversees the entire project	100
Rick	Systems analyst	Works with Heather to analyze, design, and implement the project	75
Eleanor	Lead programmer	Takes a leadership role in implementing the design	50
David*	Analyst/programmer	Supports the work throughout the project	100

*New hire with very little experience.

With the workplan and staffing completed, the project manager can combine all of the documentation into the project plan. This project plan will be the basis for the analysis stage of the development process.

Table 9.4 shows the staff members, along with the percentage of their time they will likely allocate to the project. The staffing plan, in addition to the project manager and systems analyst already on the project, includes a lead programmer and programmer/analyst. The programmer/analyst is an inexperienced new hire who is expected to work on all stages of the project as a part of his training experience.

1. Create the spreadsheet shown in Figure 9.7, and change the increase in sales from 25 percent to 20 percent. Will the plan still meet the company's requirement of a six-month payoff?
2. Why is Figure 9.7 considered a *worst-case analysis?*

Analysis

Using the project plan from the planning stage as a guide, the analysis stage of the process begins. In this stage, the development team works closely with the sponsor to understand the existing system (if one exists) and the new system needed. Once team members understand what the existing system does and what the new system must do, they can develop a design for the new system. Included in the analysis is determining the *who, where,* and *when* of the system:

> *Who* will use the system?
> *Where* will the system be used?
> *When* will it be used?

The analysis stage includes two main steps: determining requirements and building logical models. In the first step, the development team members learn as much as possible about the problem or opportunity at hand. In particular, they learn about the *requirements* for the system they are building or acquiring. In the process of carrying out the requirements analysis, the development team usually identifies both good and bad features of the existing system. The team members can carry the good features forward into the new system, with a reduction in the design effort, even as they avoid repeating the bad features in the new system.

In the second step, team members create models of both the old and new systems. A *model* is a simplified version of reality; as such, it does not attempt to capture every detail of the system. Instead, the model helps conceptualize the way in which the system works and determine what must be done to solve the problem or what data must be stored. These models are *logical* models, in that nothing physical takes place.

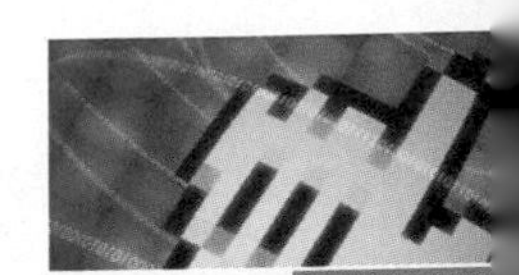

Developing Web Services

With the continuing dramatic growth of the Internet and Web has come a transition to a new type of application known as Web services. **Web services** are distributed computer applications that can be easily located, accessed, and used over the Internet. For example, the eMRWeb computerized patient records application discussed in Chapter 5 is a Web service. Users can access Web services through a peer-to-peer arrangement rather than by going to a central server. In addition, Web services can communicate with each other.[3] An October 2001 survey of more than 800 software developers revealed that more than 90 percent spent some of their time building applications for the Internet, up from 80 percent six months earlier. Of those who developed for the Web, 37 percent built Web services for their company. This number is expected to nearly double by 2002.

As an example of this trend toward Web services, consider Trans World Entertainment Corporation (www.twec.com). The company wanted to establish a single, solid brand and a consistent experience for customers visiting its Web site or its 700 bricks-and-mortar stores, which include the Strawberries, Coconuts, and Record Town chains. Trans World's newest chain, FYE—an acronym for "For Your Entertainment"—is using Web services to launch a smart application that will allow customers to catalog their music collections, share listings with friends, create wish lists, listen to music samples, and even view movie trailers. Customers can access the application from a variety of locations, including home-based PCs, more than 2500 listening and viewing stations, or 1000 kiosks in the stores. The application is available on CD-ROM or can be downloaded from the Web site. The new system uses Microsoft's .Net services, Windows XP operating system, and media technologies.

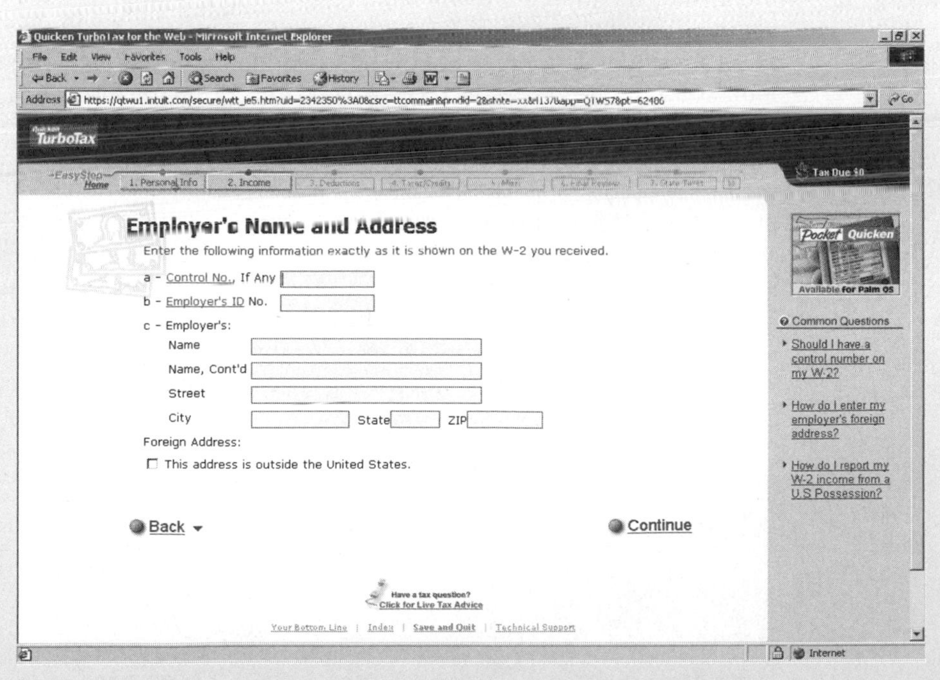

TurboTax is a Web service that many people use to economically prepare their federal and state income taxes.

Sources: Matt Berger, "Study: Web development reigns." *ComputerWorld,* http://www.computerworld.com/itresources/rcstory/0,4167,KEY11_STO64992,00.html; and Carol Silva, "Media retailer turns to Web services, Windows XP. *ComputerWorld,* http://www.computerworld.com/ storyba/0,4125,NAV47_STO65115,00.html.

---

3. John Hagel III and John S. Brown, "Your next IT strategy." *Harvard Business Review,* October 2001, pp. 105-113.

At this stage, the team attempts to understand what the current system does and what the replacement or new system must do—not *how* it will be done. Understanding how the new system will work must wait until the design stage. Note that this stage is not merely analysis—rather, it is analysis of the existing system or need *plus* logical design of the new system.

Requirements Determination

To determine the requirements of the system under construction, the team members must gather facts and information to answer a variety of questions, including the following:

- How does the current system function, or what need will the new system meet?
- What data are needed for the revised or new system?
- What reports does the current system generate (if a current system exists)?
- How should the replacement or new system operate?
- What reports or results should the replacement or new system generate?
- How would the new system affect employees' jobs?

Answering these questions requires that the team use a number of tools and techniques, including interviews, surveys, observations, joint application development meetings, and reviews of output generated by the existing system. Interviews with the project sponsor and other stakeholders closely associated with the project provide firsthand ideas about approaching the problem or opportunity. Surveys of staff can provide less-detailed information from a wider audience regarding problems with an existing system. In **joint application development (JAD)** meetings, the development team meets with the project sponsor and the users to discuss the project at all stages of development. At the analysis stage, such meetings provide useful information on what is required to fix an existing system or develop a new system. Reviews of output from an existing system can reveal why the system fails to satisfy the organization's needs. In addition, the development team may want to look at information systems used by competitors, if available.

joint application development (JAD)
A process in which the development team meets with the project sponsor and the users to discuss the project at all stages of development.

For a new system, another useful requirements analysis tool is a throw-away prototype of the system. A **throw-away prototype** helps in developing the system requirements but is *not* meant to be used as an actual system. The Manheim Online project used this tool very successfully.

throw-away prototype
A prototype that is not meant to be kept but rather used for exploratory work on critical factors in the system. It is used just for interface designing or for demonstrating an idea to a client.

Requirements analysis is a critical step in the process of systems development because it serves as the source of all information that the development team uses. The development team should not rush the completion of this step. A hurried approach to requirements analysis can cause a team to miss an important piece of information, which might lead it to create the wrong system in later stages of the process.

JAD meetings bring clients, developers, and management together to discuss a systems development project.

Figure 9.10 DFD symbols

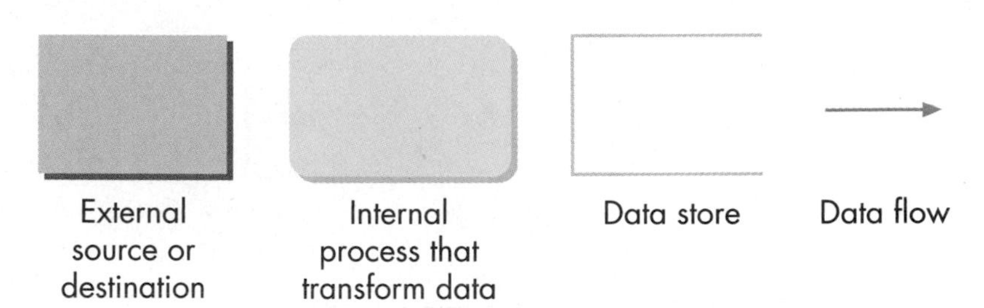

Model Building

Once the development team members believe that they have a clear understanding of the problems with an existing system or the new system that is needed to capitalize on an opportunity, they can move ahead to the model building step. Two types of models are used in this step: data models and process models. Recall from Chapter 5 that a data model is a graphical description of the columns, tables, and identifiers in a database (usually relational). Because data and databases serve as the basis for all processing in any information system, it is absolutely necessary that the team construct a correct data model before attempting to go forward.

data flow diagram (DFD)
A pictorial representation of the flow of data into and out of the system.

In contrast, process modeling involves describing the flow of data within the system. It uses a **data flow diagram (DFD)**, which is a graphical representation of the flow of data into and through the system and information out of the system. Only the four DFD symbols shown in Figure 9.10 are used in process modeling.

In a data flow diagram, the rectangular external source or destination symbol represents parties external to the system who interact with it, such as customers, vendors, and other systems. The oval internal process symbol represents processing that takes place, such as calculating how many FarEast Bucks should be added to a customer's account based on the amount of the sale. The data store open rectangle represents the storage of data, and the data flow arrow represents the flow of data into and within the system and information out of the system. An example of a data flow diagram will be shown later in the discussion of the analysis stage for the enhanced FarEast Foods customer ordering system.

As an example of using a data flow diagram to model a process, consider a personnel placement firm that matches job candidates with companies seeking qualified employees. Job candidates have submitted their résumés, which are stored in a résumé file, and companies have submitted their skill requirements, which are stored in a job file. The interview matching system, which we wish to model, aims to match résumés and skill requirements so as to set up interviews that may lead to job offers. When a match occurs, both the job candidate and the company are notified and an interview time is set up. Figure 9.11 shows this process.

One question always arises at this stage: Which system is being modeled—the existing system or the proposed system? The answer is both. If the organization plans to replace an existing system, then it must understand what this system is doing before it can build a replacement system that accomplishes the same (and, usually, additional) purposes. Without understanding the existing system, there is no way to build a replacement. The replacement system should also be modeled to ensure that it accomplishes the desired purposes. Conversely, if the organization plans to create a completely new system, then the new system must be modeled to understand what it must do.

Figure 9.11 DFD For interview matching system

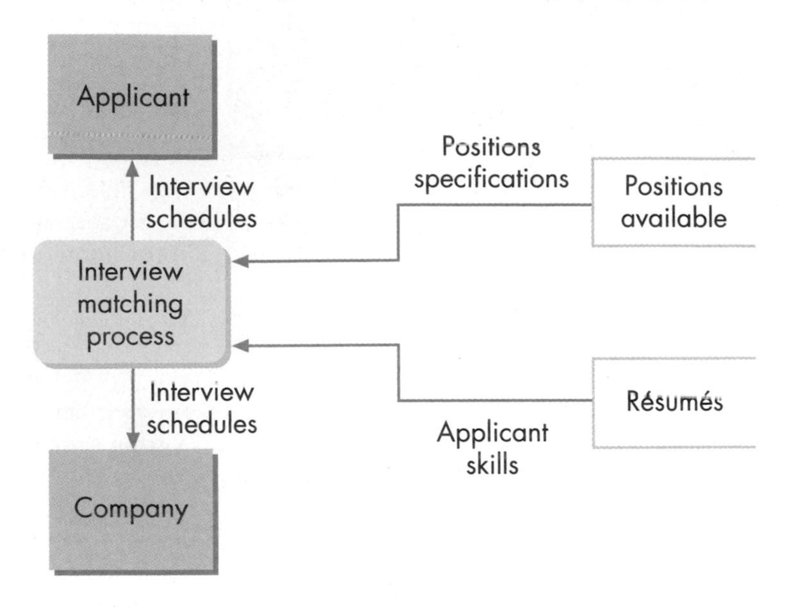

The System Proposal

After modeling the information system, the development team writes a system proposal that goes back to the steering committee for another review. This proposal describes what the team has learned about the problem or need and indicates the proposed solution. Such solutions may include acquiring a system, internally developing one, or outsourcing the development. After reviewing the system proposal, the steering committee decides whether to accept the team's proposal, to instruct them to consider yet more alternatives, or to terminate the project. Figure 9.12 shows the steps in the analysis stage of the development process using the same symbols as were used earlier in Figure 9.5.

1. What are the steps in the analysis stage?
2. What is the purpose of a process model?

Figure 9.12 Analysis stage of systems development

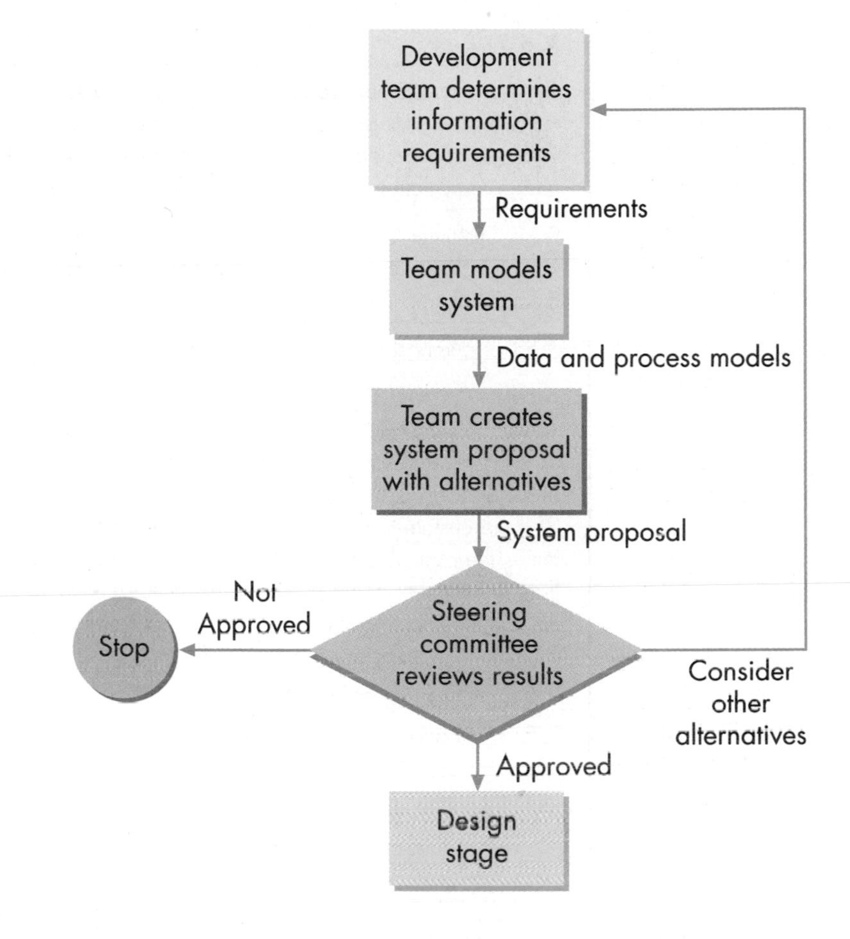

Analysis of FarEast Foods Customer Rewards Project

To carry out the analysis of the customer ordering system enhancement, you need to begin with the system requirements step. Recall that this step involves using a variety of tools and techniques, including interviews, surveys, and analyses of competitors, to answer a series of questions. Table 9.5 poses those questions again and presents the answers that the project development team found.

Note that the answer to the question of how the new or enhanced system should operate is a list of steps that the system must follow. These steps represent the *logic* of the system, and the development team must carefully consider them to ensure that the logic matches the system proposed by the sponsor. This logic will become even more important in the design and implementation stages because it stipulates the manner in which the system will operate.

Table 9.5 Questions and Answers from Analysis

Question	Answer
How does the current system function? What need will the new system meet?	The current customer ordering system does not provide for any type of customer rewards. To make customer rewards possible, the system requires new modules that award FarEast Bucks to customers, who then spend them on food items. The new modules will communicate with the existing order entry system to update the number of FarEast Bucks as customers spend them or make new purchases.
What data are needed for the revised or new system?	From Customer Order system: Customer ID, Sales Amount, whether FarEast Bucks are being used on a purchase, or whether FarEast Bucks are being created by a credit card purchase. From database: FarEast Bucks available for customer to spend.
What reports does the current system generate (if a current system exists)?	A list of purchases by item, but no lists by customer.
How should the replacement or new system operate?	1. If the customer is making a purchase, the amount of the purchase is converted to FarEast Bucks and added to the existing number for that customer. 2. If the customer wants to use FarEast Bucks to reduce a purchase amount, a new module determines the available FarEast Bucks. If the customer has at least 10, the number used is subtracted from the amount of the purchase (up to the amount of the purchase). The number of FarEast Bucks used is subtracted from the existing number of FarEast Bucks. 3. If the number of existing FarEast Bucks is less than 10, a message is sent to the order system that they cannot be used for a purchase.
What reports or results should the replacement or new system generate?	Total sales and FarEast Bucks spent by a customer.
How would the new system affect employees' jobs?	No effects that can be ascertained.

Model Building: The FarEast Bucks Data Model

Once the development team has answered the questions listed in Table 9.5, the next step is to model the system. Recall that this step involves creating a data model and a process model for the system. Figure 9.13 shows the data model for the system. Note that it closely resembles the data model for the special request system discussed in Chapter 5: It includes the same Customer table along with Products and Purchase tables. Recall that the primary key for the Customer table was the customer ID. A new column for the number of FarEast Bucks in the customer's account has been added to the Customer table. The Purchase table contains all of the information on each transaction—a number that uniquely identifies the transaction (the primary key, denoted in Figure 9.13 by an asterisk), the customer ID, the product ID, the price paid for the item, the number of FarEast Bucks used, the quantity purchased, and the date of the purchase. The table includes columns for price paid and number of FarEast Bucks used in case a refund becomes necessary. The Price Paid column contains the original purchase price, which may differ from the current product price. Similarly, the column for number of FarEast Bucks used is necessary to ensure that purchasers are refunded FarEast Bucks (and not a credit to their charge card) if FarEast Bucks were used in the purchase. Finally, the Products table contains the product ID as a primary key, the product name, the

Figure 9.13 Data model for FarEast Foods customer rewards system

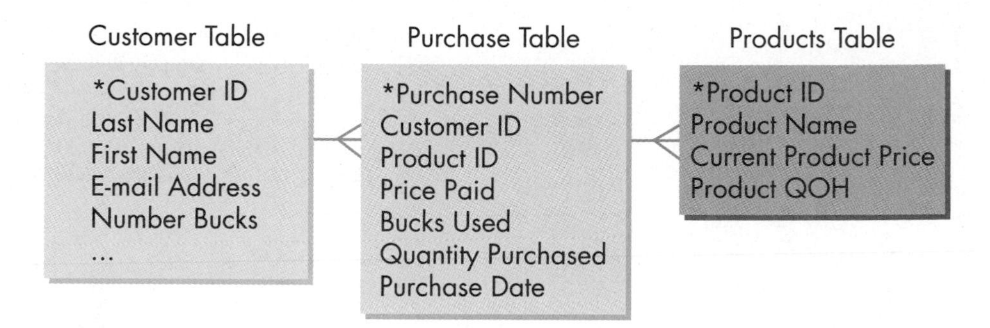

product price, and the quantity on hand (QOH). Although this table is not needed to work with FarEast Bucks, it is a key part of the existing customer order system and is shown for completeness.

Note that the data model includes the data required to answer the question about the data required by a new system (FarEast Bucks available for customers to spend). If this were not true, the team would need to revise the data model to match the data requirements. Note also the thought that goes into the data model to ensure that it can handle the various operations that might occur after the purchase.

Model Building: The Process Model

As mentioned earlier, the process model for any system is a data flow diagram that shows the flow of data into and through the system and information out of the system using the symbols shown in Figure 9.10. Because this textbook does not specialize in systems development, this very important technique will be covered only briefly here. Numerous articles on process modeling[4] and textbooks on systems development[5] cover this topic in detail. Instead, Figure 9.14 shows the appropriate data flow diagram for a purchase involving FarEast Bucks. Process modeling will be discussed using this diagram. To keep this example simple, it omits the case of a refund of a purchase using FarEast Bucks—something that the development team would need to consider for a complete system.

Figure 9.14 includes only one external object (the customer ordering system), two processes, and one data store for customer data. The existing customer ordering system is treated as an external entity because it already exists and operates separately from the two new processes. Process 1 occurs when customers want to use their FarEast Bucks to reduce the cost of a purchase, and Process 2 occurs when they make a purchase and create new FarEast Bucks, which are added to their account. For both processes, the input consists of the customer ID and the sales amount. Also, the customer data store is queried and updated as a part of this process. Note that the figure omits the detail of the internal logic of each process, as that is handled in the design stage. The purpose here is simply to create a logical model of the overall process.

---

4. A good reference is J. Satzinger, "Essential systems analysis: Its use and implications," *Proceedings of the Seventh Annual Conference of the International Academy for Information Management*, Dallas, 1992, pp. 287-301.
5. See John W. Satzinger, Robert Jackson, and Stephen D. Burd, *Systems Analysis and Design in a Changing World* (Boston: Course Technology, 2000).

Figure 9.14 Data flow diagram for FarEast Foods customer rewards system

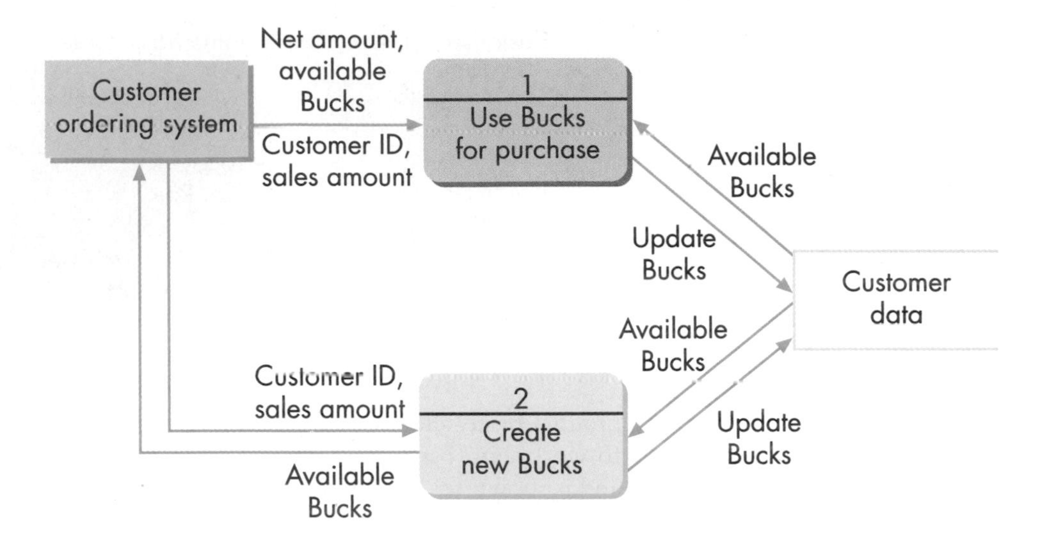

The models being built in this case focus on the *new* system, because there is no existing system to replace. These models will be crucial in helping the team design the new system in the next stage.

Customer Rewards System Proposal

After modeling the customer discount system to meet the systems requirements gathered in the analysis stage, the development team will write its system proposal recommending that the organization develop the customer rewards systems internally rather than trying to acquire one or outsource the development process to another company. The team will report this opinion to the steering committee, which will decide whether to accept the proposal, terminate the project, or instruct the team to look for other ways of meeting the requirements. In this case, the steering committee decided that the system proposed by the development team was worth continuing, and it directed the development team to begin the design process immediately. Chapter 10 dicusses this effort.

1. Why is it necessary to have both data models and process models?
2. Explain the logic associated with Process 2 in Figure 9.10.

Case *Study*

Building Systems at Corrugated Supplies

When Dave Pung was hired at the cardboard manufacturing company Corrugated Supplies (see the opening case in Chapter 3 for more information on the company) as the director of information systems, he found that all information systems were outsourced; the company had no information technology staff other than himself. Pung hired an IT staff of five people and immediately began working to combine three different computer systems, which handed off data to each other as jobs moved from one part of the manufacturing process to another. At first, Pung tried using off-the-shelf software. Because Corrugated Supplies creates cardboard sheets to meet customer orders and an almost infinite variety of sizes and styles of sheets are possible, however, he found that no off-the-shelf software fully met the company's unique needs. He also thought about outsourcing the development, but decided that outsourcing would not fit the company's needs, either.

Pung's first tasks were to build a centralized database from which a variety of applications could pull needed data and to build the applications themselves. He settled on Microsoft SQL Server and Visual Basic for much of the development process. Over the next three years, his staff built new systems for such important operations as order entry, scheduling, manufacturing, shipping, and inventory. In building these systems, Pung first tried to simply replace existing functionality. Later, his team members took a hard look at the underlying business processes and attempted to build better versions of the systems. Pung's extensive knowledge of the industry was a key element in the successful analysis and design of the new systems for Corrugated Supplies.

While building the replacement systems, Pung knew that he eventually wanted to move operations involving customer orders onto the Web to make it easier for customers to make purchases. Using Cold Fusion, he developed Web applications and made them available to customers. The 25 percent of customers who had been using a DOS-based dial-up system for 12 years were the first group to move to the Web, because the new system replaced the older and more-difficult-to-use system. Eventually more than 80 percent of Corrugated's customers moved to the Web-based order systems built by the IT staff.

Dave Pung oversaw the development of an entirely new information system at Corrugated Supplies.

Source: Interview with Dave Pung, Director of Information Systems at Corrugated Supplies, November 12, 2001.

Think About It

1. Discuss the options facing Dave Pung when he decided to install a new information system at Corrugated Supplies, and suggest reasons why he decided to go with internal development.
2. Examine the Corrugated Supplies Web site at www.csclive.com/, and discuss its purpose in terms of the company's overall goals.

SUMMARY

To summarize this chapter, let's answer the questions posed at the beginning of the chapter.

Why is information systems development or acquisition important to all types of organizations? Information systems used to handle the present, remember the past, and prepare for the future are an integral part of all organizations. They help process transactions quickly; store data, information, and knowledge; and help managers and employees to make decisions. A problem facing all organizations is deciding how to obtain the information systems they need. Whereas some systems can be acquired, others must be designed and built to meet the specific needs of the organization. Properly building information systems is critical to the long-term well-being of an organization. In designing and building or acquiring an information system, the organization must carefully analyze its requirements so that the resulting information system accomplishes the desired purposes. Systems development entails the process of analyzing information requirements, designing systems to meet these requirements, and building these systems.

What are the most commonly used approaches to systems development? Individuals and organizations can take any of several approaches to systems development: ad hoc programming, the structured approach, rapid application development, end-user programming, outsourcing, and acquisition. All of these approaches have four basic stages: planning, analysis, design, and implementation. The planning stage determines the problem to be solved, the analysis stage determines what must be done to solve the problem, the design stage determines how the problem will be solved, and the implementation stage actually solves the problem. Different approaches emphasize one or two of these stages over the others, and many development projects use a combination of these approaches.

What are the stages in the structured approach to development? The four stages of the structured approach to development (also known as the systems development life cycle) are planning, analysis, design, and implementation. The planning stage identifies a problem or an opportunity and determines whether it is possible to solve the problem or capitalize on the opportunity. The key deliverable of the planning stage is the project plan. In the analysis stage, the organization gathers data that will answer the *what* question of the system: *What* will the system do? The key deliverable is the system proposal. In the design stage, it answers the *how* questions: *How* will the system be developed, and *how* will it operate? In the design stage, the organization creates a physical design to specify the details of the system in a systems specification. In the implementation stage, the new system, whether developed or acquired, is installed and tested extensively; users also receive training at this stage.

What are the steps in the planning stage of development? The planning stage of development involves five steps:

1. Identifying the project
2. Initiating the project
3. Performing a feasibility study
4. Creating a workplan
5. Staffing the project

The project identification step begins with the organization determining that a problem exists in the current system or that a new system could capitalize on an opportunity. In the project initiation step, the project's sponsor creates a system request that lists the business need for the project, its expected functionality, and the benefits that would likely result from its completion. A feasibility study aims to answer the following questions: Is the problem worth solving, and is a solution to the problem possible? Three types of feasibility must be considered: technical, economic, and organizational. If the new system appears to be feasible, the IS department names a project manager who begins by creating a workplan that lists information on the tasks necessary to complete the project. The project manager also staffs the development team and develops a project plan that is submitted to the steering committee for review.

What are the steps in the analysis stage of development? This stage involves two steps: determining requirements and building logical models. In the first step, the development team members learn about the requirements for the system under construction by answering a series of questions about it. They use a variety of tools, including interviews and surveys, in this effort. This step is very important because all successive work will depend on its results. In the second step, team members create models for conceptualizing the way the system works and for determining what must be done to solve the problem. Data models and process models are commonly used in this phase. The models can either re-create the existing system that is being replaced or represent a new system.

How do data modeling and process modeling fit into the analysis of a replacement or new system? Data models help to clarify the relationships between the columns and tables in a relational database. Because data serve as the basis for all processing in any information system, it is absolutely necessary that the development team construct a correct data model before attempting to go forward. Process modeling describes the data flows within the system in the form of a data flow diagram, which graphically represents the flow of data into and through the system and information out of the system.

REVIEW QUESTIONS

1. What conditions make information systems development necessary?
2. What are the four phases of the IS life cycle?
3. At which phase of the IS life cycle does systems development occur?
4. Why do resource requirements increase in the systems life cycle after the development phase?
5. How does the theoretical IS life cycle differ from the actual life cycle?
6. What approaches to systems development are in use today? Which approach to systems development was used in the Manheim Online case?
7. What problems can occur with the ad hoc programming approach to systems development? Can you suggest a situation where it would be appropriate?
8. What are the stages of the structured approach?
9. What are the steps in the planning stage?

10. What does a systems analyst do? A project manager?
11. What role does the steering committee play in the systems development process?
12. Why is a team needed to develop an information system? Why might a new employee be included as part of the team?
13. What deliverables are expected from each stage of the structured approach?
14. Why is a feasibility study necessary? What types of feasibility should the organization consider?
15. Why would stakeholder analysis be used in the systems development process?
16. What is the purpose of the analysis stage? What questions must be answered?
17. What tools are used during the analysis stage?
18. What system is being analyzed?
19. What is a model? How does a data model differ from a process model?
20. What symbols are used to create a process model?

DISCUSSION QUESTIONS

1. Discuss the need for a sponsor for a project.
2. Discuss situations in which the structured approach might be appropriate for Web development.
3. Which approach to systems development might best fit the networked economy? Why?
4. Discuss the application of the structured approach to an information systems development situation with which you are familiar.
5. Discuss the rationale given in this chapter for omitting the detail for each of the processes in a data flow diagram.

RESEARCH QUESTIONS

1. Go to www.manheim.com and look at the sample interaction available there. How does it compare with that available on autotrader.com? In a two-page report, describe the integration of the two systems.
2. Use the Web to research the approach that Microsoft uses for its systems development. Write a two-page report about your findings.
3. Use the Web to research companies, other than Software MacKiev, that convert software from one platform or operating system to another. Compare one of these companies to Software MacKiev in a short report.
4. Look on the Web for examples of Web services, and write a short report on one of them.

WildOutfitters.com

Alex awoke from his nap when he heard the keys in the door, and began to rub his eyes.

"Wake up sleepy head," said Claire. "There are groceries to be brought in from the car."

Even though he was still woozy, Alex noticed the words CLARK'S GROCERY written in bold letters on the bags in Claire's hands.

"Why did you go there to shop?" he asked. "The Foodbasket and Qubert's are closer."

"Well, I've always liked the selection," she responded. "Besides, when they run my check cashing card through, I can collect points."

"Points? What can you use them for? Are you in a shopping league or something?"

"No, silly! You get points for every dollar that you spend and then when you get a certain amount you can get store coupons."

"Oh, it's kind of like that credit card we have that gives a little cash back every time we use it. They have really made loyal customers out of us...." With that, Alex stopped, and the couple looked at each other and smiled.

The Campagnes suddenly realized that incentive programs like those used at the grocery store and with the credit card are designed to bring customers back for repeat business. With this realization came the insight that a similar program might work for them on the Wild Outfitters Web site. The Campagnes have been struggling with the question of how to attract more customers and encourage repeat business from their current customers. Over the next week, they begin to plan a frequent-buyer program for their site. Their strategy will be to give away special items when buyers make six purchases worth $100 or more. Before spending too much time on the project, they must assess the project's feasibility, however.

Initially, the Campagnes feel that they will need to add functionality to both the Web page and their database. Additions to the Web page will, at the least, include links to incentive program information such as rules and prizes, the ability for a customer to register for the program, and possibly secure access to the customer's current total. They will also need to adjust the structure of their database to include the amount of sales for each customer.

Because it is already set up with a focus on customer service, Wild Outfitters should be able to administer the program with its current staff. The major hurdle is whether the company has the technical ability to implement and maintain the incentive program. The Campagnes will need to decide whether they will build the system in-house or if they will need to turn to outsourcing.

After several conversations with an IS consultant and a little research on their own, Alex and Claire conservatively estimated a few of the benefits and costs that would result from the project. They have estimated that an incentive program will increase their sales by 5 percent. With their current monthly sales of about $76,000, this boost would mean an extra $3800 monthly. A rough estimate of the costs includes an initial cost of $8500. The Campagnes would have to bear the extra costs of the prizes and the operational expenses of the program. Because they want the prizes to be meaningful, they estimate that together these costs would amount to about $1650 per month. A banker friend suggested that it would be reasonable to use a 6 percent annual discount rate. A discount rate is a value used to analyze an investment, and it reflects the time value of money in the analysis. In many ways, it resembles a savings account interest rate. With a savings account, the interest rate reflects the amount of money that the initial saved amount will earn from compound interest. In a cash-flow analysis of a project, the discount rate reflects the amount of money that the initial investment will earn over the project's lifetime.

With this information in mind, Alex and Claire have begun the initial stages of developing the project. After completing their initial analysis, they will be able to decide whether to continue.

Think About It

1. In which of the four stages of the SDLC is the customer incentive project at WildOutfitters.com? What deliverables should the Campagnes develop?
2. Does this project sound feasible for WildOutfitters.com? Describe the criteria that you would use to make this determination in terms of technical, economic, and organizational feasibility. What information do you need to make this determination?

Hands On

3. Using the information in the case, along with what you have learned so far about electronic commerce, write a system request for this project.

Hands On

4. A spreadsheet file containing a cost/benefit analysis of the project is available from the Course Technology Web site. Use the file to perform a what-if analysis. If the Campagnes are willing to take on the project only if it has a payout period of less than one year, what should they decide based on your analysis?

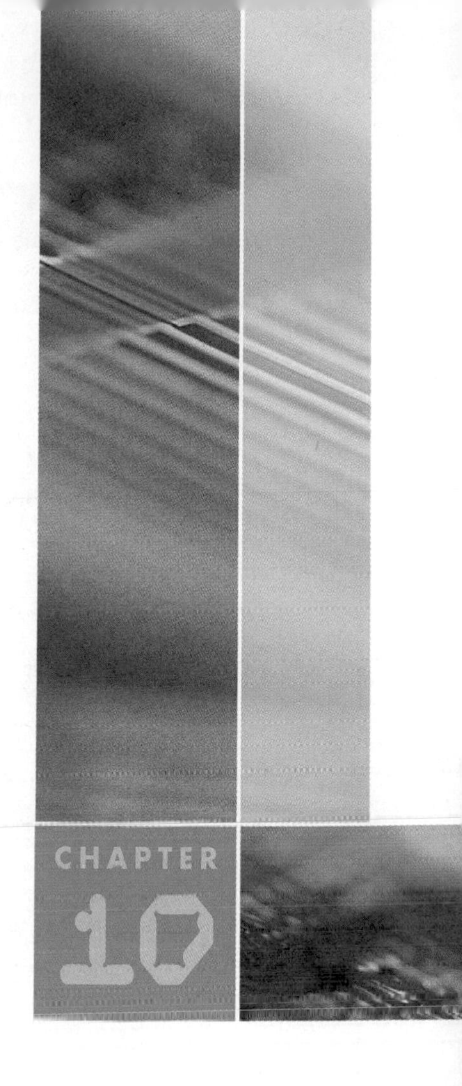

DEVELOPING INFORMATION SYSTEMS 2

CHAPTER 10

LEARNING OBJECTIVES

After reading this chapter, you will be able to answer the following questions:

- What is the purpose of the design stage of the structured approach to the systems development process?
- What operations must occur during the implementation stage, and how is computer programming involved?
- What is outsourcing, and what issues should be considered when deciding whether to outsource a project?
- What key steps are involved when an organization chooses to acquire an information system?
- What is RAD, and how does it differ from the structured approach to systems development?

TGA at Merrill Lynch

Merrill Lynch & Company has long been a leader in the financial services field, with almost $1.5 trillion in assets in the third quarter of 2001—more than the GDP of many developed countries. In 1997 Merrill Lynch realized that the growth of low-cost trading over the Internet represented a threat to its business. To combat this perceived threat, the management at Merrill Lynch decided to compete on the Internet by concentrating on improving its customer service. Specifically, the company wanted to improve information access for its financial consultants who spent their time helping clients accrue wealth. To do so, Merrill Lynch needed a completely new information system, one that centered on providing financial consultants with the analytic tools and information they needed to develop, implement, and monitor financial plans for clients.

The resulting Trusted Global Advisor (TGA) system was completed by October 1998 at a total cost of $850 million. To implement this system quickly, Merrill Lynch set a goal of upgrading 10 offices per week. Two weeks before the conversion from the existing system to TGA at each office, all employees in the office underwent mandatory training on the new system. This training focused on how to use the system's basic functionality as well as how to use the extensive online help system consisting of online cue cards and multimedia demonstrations. The Friday before the conversion, an installation team arrived at the office to remove the old technology and install the new version. On the Sunday before the new system went live, the trainers provided a three-hour review session. On Monday morning, the new system went live. The trainers stayed for a week after the conversion to provide solutions to any problems that arose. Merrill Lynch found the problems with the conversion to TGA to be fewer than expected, primarily due to the intuitive quality of the new system and the thoroughness of the training.

The Merrill Lynch Trusted Global Advisor software helps its employees provide trading advice on the New York Stock Exchange to its customers.

Continued

Immediately upon completion of the proprietary version of TGA, Merrill Lynch began a two-year project to leverage TGA into other systems. The result of this effort was its online investing system called ML Direct. Released in December 1999, ML Direct was named by *Barron's* as one of the three best such sites on the Web, and it was named the second best such system by *Money* magazine. So much of this system is based on TGA that it could be called TGA-online "lite." Going further, Merrill Lynch has used its experiences with TGA to create MLOL, an investor's research system. All of these applications are linked back to the TGA database, a data warehouse that is linked to Merrill's deep legacy systems, which consist of the mainframes that connect to the various stock exchanges.

Sources: Bill Gates, *Business @ the Speed of Thought.* New York: Warner Books, 1999, pp. 80-86; and Andrew Rafalaf, "Can Merrill overcome its legacy?" *Wall Street & Technology,* August 8, 2000, http://www.wallstreetandtech.com/story/WST20000808S0011.

Continuing the Development Process

Chapter 9 discussed the importance of the information systems development process to an organization and the most commonly used types of systems development, including ad hoc programming, structured systems development, rapid application development (RAD), acquisition, and outsourcing. The chapter then went into some detail regarding the use of structured systems development because it is so widely used. Also known as the waterfall approach to systems development, it incorporates four stages: planning, analysis, design, and implementation (see Figure 10.1).

You learned about the first two stages of this approach—the planning and analysis stages—and applied them to the development of an enhancement to the customer ordering system for the FarEast Foods Web site. Recall that the planning stage involves determining that a need or an opportunity exists. The planning stage results in the development of a project plan containing the best estimate of a project's scope, benefits, risks, and resource requirements. The second stage in the structured approach, analysis, concentrates on determining *what* must be done to solve the problem or capitalize on the opportunity determined in the planning stage. This effort involves the development team learning what the existing system does

Figure 10.1 Four stages of structured systems development

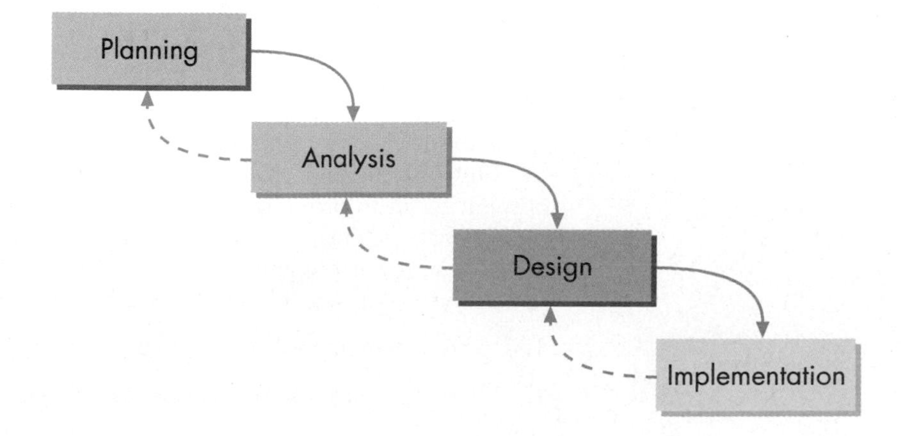

and what the new system must do. The analysis stage includes two steps: requirements determination and building logical models.

When the planning and analysis stages are complete, the development team knows the problem to be solved or opportunity to be met and knows what it must do to address the problem or realize the opportunity. The team members are ready to begin the design and implementation stages of systems development. As with the planning and analysis stages, the customer rewards system for www.fareastfoods.com will serve as the example throughout this discussion.

Design Stage

The design stage focuses on the *how* questions: How will the new system be developed, and *how* will the new system work? The development team must answer the question of how the system will be developed first. If the team decides to develop a custom system, it can do the work internally or outsource the job to another firm. When a company outsources a project, an entity outside of the organization develops all or part of the project. If the team decides to acquire the information system, it must find and purchase an appropriate system. If the decision is to outsource or acquire, then the team skips the remainder of the design stage and begins the implementation stage.

If the team decides to develop the system internally, the design stage must answer the question of how the new system will work. It does so by moving from the logical models and system proposal document of the analysis stage to a physical design that specifies *all* of the details of the system. For example, in the analysis stage, the team used process modeling to determine what the system should do, but made no effort to specify how the actual processes would work. In the design stage, the team must specify the manner in which all of the processes in the data flow diagrams will work. The design stage results in a system specification that is used in the implementation stage to program the new system. The **system specification** comprises a complete and detailed group of deliverables, including the physical data model, physical models of each process in the process model created in the analysis stage, and the interface screens.

system specification
A complete and detailed group of deliverables, including the physical data model, physical models of each process in the process model created in the analysis stage, and interface screens.

Implementation Stage

Once the team has completed the design stage, it must implement the design. This process involves either building and testing the system or acquiring it, installing the new system, converting from the existing system (if one exists) to the new system, training the employees who will use the system, and supporting and maintaining the system. Failure to appropriately carry out the installation step can doom even the best system to failure. The success of the new TGA system at Merrill Lynch described in the opening case was ensured by a well-thought-out installation process.

Because programming is essential to all information systems development, this chapter will discuss computer languages and programming briefly. Although this discussion is not meant to teach you to program, it should help you understand more about this very important process.

Outsourcing and acquisition represent important alternatives to internal development of an information system; these two topics will be covered in more detail in this chapter. Finally, because the networked economy often requires high-speed development, various approaches to RAD, which results in faster development than structured systems development, will be discussed.

1. What is the purpose of the design stage of the structured approach?
2. When the implementation stage is completed, what should be the result?

Design Stage

After the analysis stage of the structured development process, the development team knows *what* must be done, but has not yet worked out *how* it will be done. The design stage requires that the development team investigate the three development alternatives—develop internally, outsource, or acquire and present a recommendation to the steering committee.

Develop, Outsource, or Acquire?

How does an organization decide whether to develop an information system internally, outsource it to an outside developer, or acquire a system? Making this decision is a function of five variables: cost, customization, time, competitive advantage, and in-house skills. The organization must consider the trade-offs for each variable. Developing a system internally or having one developed by an outsourcer leads to a high degree of customization and provides a competitive advantage because the organization can control exactly which features go into the system, how it looks and feels, and so on. Internal development or outsourcing comes at a price, however, because it takes much longer and costs more to create a new system than it does to acquire one.

Conversely, acquired systems may not fit the organization's needs exactly and probably will not provide a competitive advantage because other companies can readily acquire the same system. Acquiring a product with the hopes of modifying it to fit organizational needs is another approach, but can lead to many problems when the customization fails to achieve its goals. Note that the maker of the product has no obligation to solve the problems caused by customization changes.

In terms of competitive advantage, internal development is typically superior to outsourcing or acquisition because the resulting system will be unique to the organization. Of course, it also requires that the organization have employees with the skills necessary to carry out this process. If the firm lacks those skills, then it can use outsourcing to bring in a company that has exactly those needed abilities. Note that companies often resort to outsourcing to develop an information system, but then use their own people to actually run and maintain it. The reverse can also occur. In addition, a compromise between internal development and outsourcing is possible in which an organization complements the existing staff with consultants to develop the system and uses the internal staff to maintain the system. Table 10.1 compares the advantages and disadvantages of the three approaches to systems development.

Sometimes an organization can decide whether to develop internally, outsource, or acquire immediately after the planning stage because it realizes that it can easily solve the problem with acquired software or that it does not have the skills in-house to develop the system. In other cases, the organization makes this decision after the analysis stage, when it knows what must be done. For example, it appears that the FarEast Foods' project to create a system that will reward customers based on their purchases on www.fareastfoods.com will require that a system be developed either internally or through outsourcing. The degree of customization required far outweighs the time it may take to develop the system. In addition, this system is meant to give FarEast Foods a competitive advantage over other companies that provide the same product or service. Only custom development can provide that type of competitive advantage.

Now consider a different development project that FarEast Foods might undertake. Assume that the company has discovered that its use of employees from a temporary firm is not working out very well. The temporary employees combine too many orders incorrectly, and FarEast Foods believes that permanent employees who can earn promotions and other benefits will have more pride in their work, resulting in fewer mistakes. To handle this new group of employees, the company

Table 10.1 Comparison of Systems Development Methods

Development Method	Advantages	Disadvantages
Internal development	› Provides a competitive advantage for the organization › Provides the organization with complete control over the final system › Builds the technical skills and functional knowledge of the developers	› Requires the dedicated effort of an in-house staff › Can lead to slower development › May have higher costs than other approaches › Can result in a system that may not work when completed or that does not provide the desired functionalitiy
Outsourcing	› Exploits the outsourcer's more skilled and experienced programmers › Does not divert internal staff from their current work	› May lead to loss of control over the project › May prevent internal developers from learning the skills necessary to maintain the system › May lead to higher costs than with acquisition › Creates problems if the outsourcer does not deliver on its claims or the final system does not provide the desired functionality
Acquisition	› Becomes available sooner and has a high probability of working › Leads to lower costs because development is spread out over many users	› Offers little competitive advantage › Must accept funtionality of purchased system › May not integrate well with existing systems › May require modification to meet needs

needs a payroll system that can deal with these hourly employees as well as the existing salaried employees. In contrast to the customer rewards project, this project appears to require little customization. In addition, it does not provide any competitive advantage, and FarEast Foods needs the new system quickly to make the move to permanent employees as quickly as possible. For these reasons, FarEast Foods' best bet is probably to acquire an appropriate payroll system.

The next section will delve into the internal development process. Outsourcing and acquisition will be discussed in detail later in the chapter.

Internal Development

If the development team recommends to develop the system internally, then the team members must answer the second *how* question in the design stage: *How* will the system operate? To do so, they must develop plans for a new system and present those plans to potential users for comments. At this stage, close contact between the team and users is extremely important to ensure that the replacement system will solve the problems in the existing system without introducing any new ones or to prevent the new system from failing to capitalize on the opportunity identified in the planning stage.

Although the design stage has many aspects, the three most important ones are converting the logical database model into a physical database specification, converting the logical process model into physical forms that programmers can use to write the necessary computer programs, and developing the interface screens with which users will work.

The deliverables from the design stage include a complete and detailed specification of the physical data model, physical models of each process in the process model created in the analysis stage, and the interface screens. Let's discuss each of these design elements in turn and apply them to the FarEast Foods example in the next section.

Physical Database Specification

Once the development team decides to go with internal development, the next step in the design process involves creating a physical database specification. Design of the physical database specification begins with the data model that the development team created during the analysis stage (see Chapter 9). Recall that this data model includes the tables, the columns in each table, the data type for each column (for example, currency), and the relationships between the tables. Although this information is essential for understanding the system, you need to go further to design the new system by providing data about the data in the database, which are called **metadata**. Metadata include the type of database being used (Oracle, Access, and so on), the names of the tables and the fields in the tables, the primary key for each table, and the foreign keys in each table. Although the table and field names may match those shown in the data model, you may have to change them to meet the database system's naming requirements or the organization's standard naming conventions.

metadata
Data about the data in a database, including the type of database being used, the names of the tables and the fields in the tables, the primary key for each table, and the foreign keys in each table.

As a part of specifying the database, you also need to ensure that the database tables provide the data specified by the data stores in the data flow diagrams (DFDs). Although the tables in a database often have a one-to-one correspondence with the data stores, this is not always the case. Sometimes, you may need multiple tables to provide for the data in one data store.

Converting Process Models to Physical Forms

To design the processes that make up the information system, you must convert the logical process models to a form that programmers can use in developing the software portion of the system. That is, you must convert all of the processing ovals in the DFDs into two physical forms: input/processing/output tables and pseudocode procedures. These two physical design elements are critical to creating the computer programs that implement the information system.

As their name implies, **input/processing/output (IPO) tables** show the inputs to a process, the required outputs for that process, and the logic that is necessary to convert the inputs into the desired outputs. In many cases, you know the inputs and outputs from the process models, so the important work involves determining the processing needed to convert inputs to outputs. For example, assume that you have a simple Web order form in which the name, purchase price, and number of an item being purchased serve as inputs; required outputs include the name, price, number, subtotal, tax, and total cost. You compute the subtotal by multiplying the purchase price times the number purchased, and you compute the total cost by adding 7 percent sales tax to the subtotal. Figure 10.2 shows the resulting IPO table.

input/processing/output (IPO) table
A table showing the inputs to a process, the required outputs for that process, and the logic needed to convert the inputs into the desired outputs.

Figure 10.2 IPO table for computing total price

Input	Processing	Output
•Item name •Purchase price •Number purchased	1. Subtotal = price × number 2. Tax = subtotal × 0.07 3. Total = subtotal + tax	•Item name •Purchase price •Number purchased •Subtotal •Tax •Total

Figure 10.3 Pseudocode for computing total price

```
Begin procedure
 Input item name and purchase price
 Input number purchased
 Subtotal = price X number
 Tax = subtotal X 0.07
 Total = subtotal + tax
 Output item name, purchase price, number purchased, subtotal, tax, and total
End procedure
```

pseudocode
A way of expressing the logic of processing in structured English rather than in a computer language.

Once you have created IPO tables for all of the processes in the DFD, you should create corresponding pseudocode procedures. **Pseudocode** provides a way to express the logic of the processing in *structured English* rather than in a computer language. It enables you to describe the procedure for the DFD process very clearly in English without worrying about the special syntax and grammar of a computer language. The pseudocode should consist of a set of clearly defined steps that enables a reader to see the next step to be taken under any possible circumstances. Also, the language and syntax should be consistent so that the programmer can understand the designer's pseudocode. A programmer can then easily implement the DFD process as a computer program. Figure 10.3 shows the pseudocode corresponding to the IPO table from Figure 10.2.

Once you have developed an IPO table and the pseudocode for each process, it is often straightforward for programmers to write a computer program in the next stage that will carry out the necessary processing.

Creating Interface Screens

interface
The design of the screen that users will see when they access an information system.

Although converting logical models to physical models is crucial for communicating the design of the information system to the programmers who will actually build it, these physical models are often not a very good way of communicating with the project sponsor or potential users. Instead, the design stage usually must incorporate examples of the **interface**, the design of the screens that users will see when they access the information system. The project sponsor and potential users are usually better able to understand the design by viewing and working with the interface than with the various physical models. The interface screens can range in complexity from a screen with little or no interactivity to one that allows the user to input values and click buttons to see results. Figure 10.4 shows an interface screen for the example Web application described in the IPO and pseudocode discussions. Notice that the screen is a skeleton and doesn't have the graphic elements and visual appeal that a customer screen ready for presentation on the Web would have.

Using CASE

computer-aided software engineering (CASE)
Software used to help in all phases of the systems development process so as to improve the productivity of systems development.

CASE repository
A database of metadata about the project that is used to automate much of the paper flow typically associated with structured development.

A growing trend in developing large software systems is the use of computer-aided software engineering. With **computer-aided software engineering (CASE)**, developers use software in all phases of the systems development process to improve their productivity. Development teams can use a **CASE repository**, or a database of metadata about the project, to automate much of the paper flow typically associated with structured development. Team members can store information about the project in the CASE repository on a server accessible to all team members. In addition, they can use CASE-based diagramming software to reduce the effort involved in creating process and data models and applications, or they can use code generators to automate the actual writing of the programs. Similarly, CASE can automate the documentation

Figure 10.4 Interface screen for computing total price

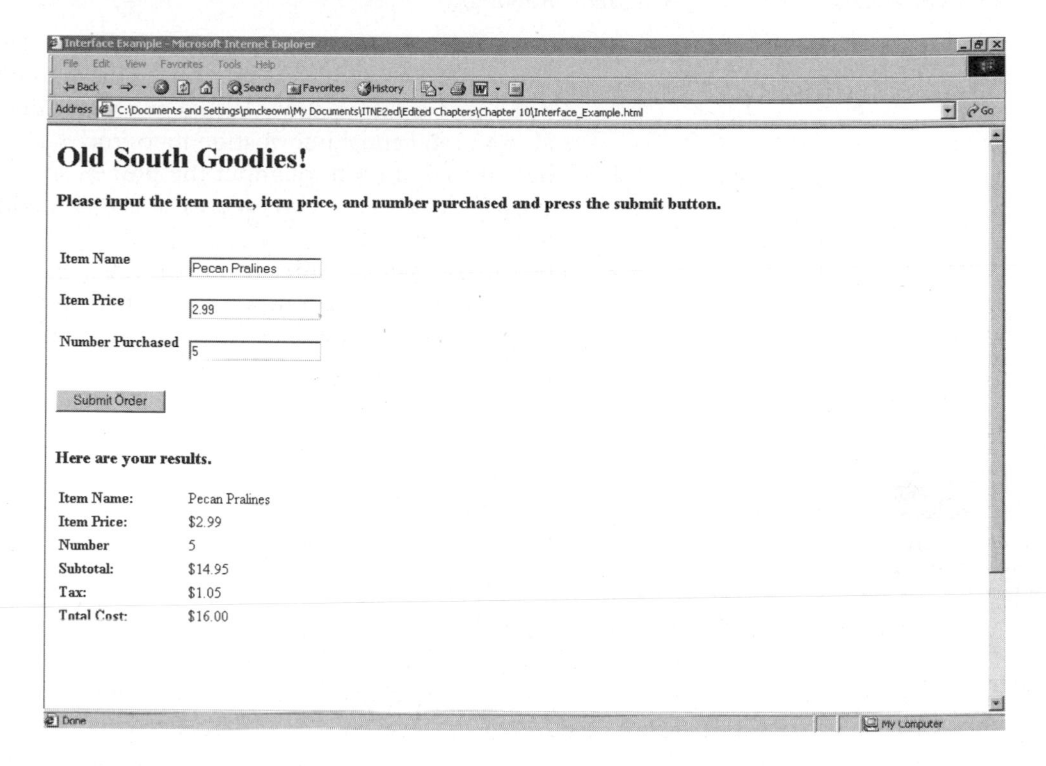

process so that as programmers develop new versions of the program, they simultaneously create new documentation. The objective of the CASE process is to have rapid production of program instructions that can be reused with other applications.

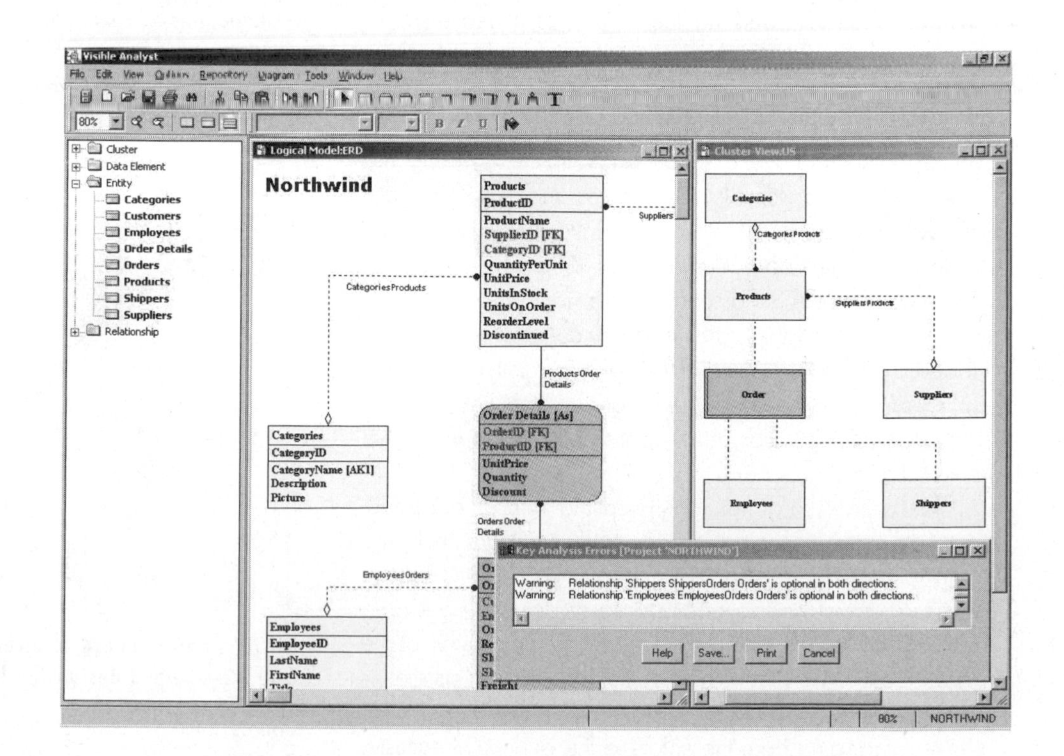

Developers can use CASE software to create process models such as this one for a complex catalog ordering system.

CASE tools
CASE software packages.

The benefits of using CASE software packages—called **CASE tools**—include the following:

- They reduce the length of the overall systems development life cycle by compressing the various stages and causing them to overlap.
- They improve information sharing, as paperbound methods become automated and stored in central information repositories, and reduce duplicated effort.
- They involve users throughout the process so that more effort is expended in the early design stages, when critical errors might arise that could prove difficult to correct later.
- They make systems easier to modify by storing system parameters and functions in a central location. When one part of a system changes, all elements of the system affected by the change automatically access the new value.

INTERNET IN ACTION

Kids Designing Kids' Web Sites

When it comes to creating electronic commerce Web sites, failure to develop the best design can result in a marketing failure, regardless of how well the other three development stages are carried out. A poor or difficult-to-navigate design will turn users off to the Web site, causing them to leave quickly and possibly never return. This loss of viewership can occur faster for Web sites aimed at children than for Web sites aimed at any other age group. When you consider that almost 18 million children and teenagers go online each month, and that they spend or influence spending of more than $500 billion every year, making a Web site attractive to this group is obviously critical to the success of companies that hope to market to it. Research shows that online marketers have 8 seconds to capture a young visitor's attention—or lose it. Companies such as Nintendo, Disney, or Crayola do not want to repeat mistakes like those made by *Sports Illustrated for Kids*. The director of new media for the SIKids.com Web site thought kids would be attracted to trading cards that they could download and added a link to the Web site to make this activity possible. Kids visiting the site completely ignored the link; when it comes to trading cards, they want to be able to hold them in their hands.

To avoid such gaffes, an increasing number of companies that want to market to kids and teens are including them in their design process. For example, Mattel has an online board of directors composed of 26 girls aged 6 to 12 who have a significant influence on the design of the barbie.com Web site. When Mattel redesigned this site, its online board of directors helped pick key elements of the new site, ranging from the colors to the pulse of the music. Surprisingly, the first thing to go was the traditional Barbie pink; it was replaced by purple and lime green. The group also told Mattel to avoid including a chat room because they had been warned away from such forums by their parents. For another design issue, the group of girls selected an icon of a Nintendo-like hand control to represent a new game on the site, rejecting the suggestion of a board game icon—many of the networked economy girls had no idea what a board game was.

Other companies are also listening to their younger customers. For example, Crayola has revised its site to make it easier to navigate. Disney has redesigned its site to add more pictures of the well-known Disney characters with fewer words.

Mattel's online board of directors—a group of girls aged 6 to 12—helped design the barbie.com Web site.

Source: Bruce Horowitz, "Marketers call on kids to help design Web sites." *USA Today*, June 5, 2001, pp. B1-B2.

A side effect of using CASE technology is that it encourages developers to concentrate on the front end of the software design procedure: the analysis and design of new applications. The applications produced in this manner require less debugging or redesigning and go to the customer sooner.

1. What are the deliverables from the design stage?
2. What tools are used to convert the logical process models into physical models?

Designing the FarEast Foods Customer Rewards System

After researching the various alternatives for the new rewards systems for the FarEast Foods Web site, the development team has recommended to the steering committee that the company use internal development. The team members could not find an off-the-shelf system that met their particular needs and worried that outsourcing might result in a loss of competitive advantage. After considering the development team's recommendation, the steering committee decided to proceed with internal development.

Upon receiving this approval, the development team began the design process, creating a complete and detailed specification of the physical data model, developing physical models of each process in the process model created in the analysis stage, and generating interface screens.

Physical Data Model for the FarEast Foods System

To create the specification of the physical data model, you need to start with the logical data model developed in the analysis stage (see Chapter 9) and then convert it to a physical data model for specific database management software. Figure 10.5 shows the data model for the proposed www.fareastfoods.com rewards system, and Figure 10.6 shows the corresponding Access design screen for the Purchase table with key metadata labels highlighted.

In the www.fareastfoods.com rewards system, the Customer data store matches the Customer database table, so no change is needed in the database structure.

Converting FarEast Foods Process Models

Once you have created the physical data model in database management software, the next step is to convert the processes in the data flow diagram into IPO tables and pseudocode. To demonstrate this step for the www.fareastfoods.com rewards system, consider again the process model created for it in Chapter 9. Figure 10.7 shows the data flow diagram. Recall that Process 1 corrresponds to customers using

Figure 10.5 Data model for rewards system

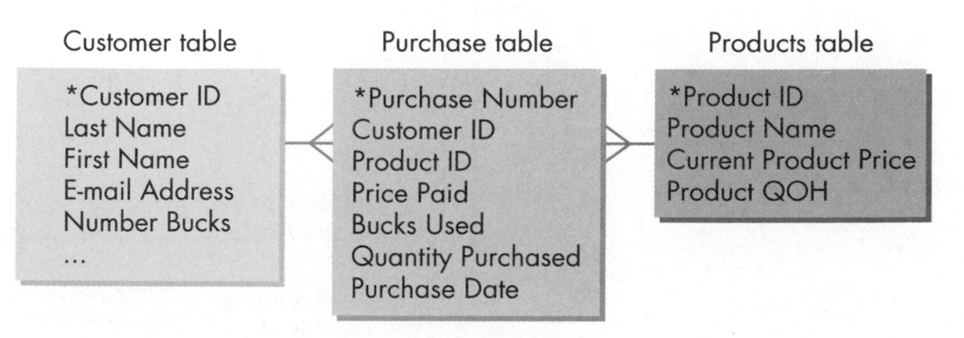

Figure 10.6

Access design screen

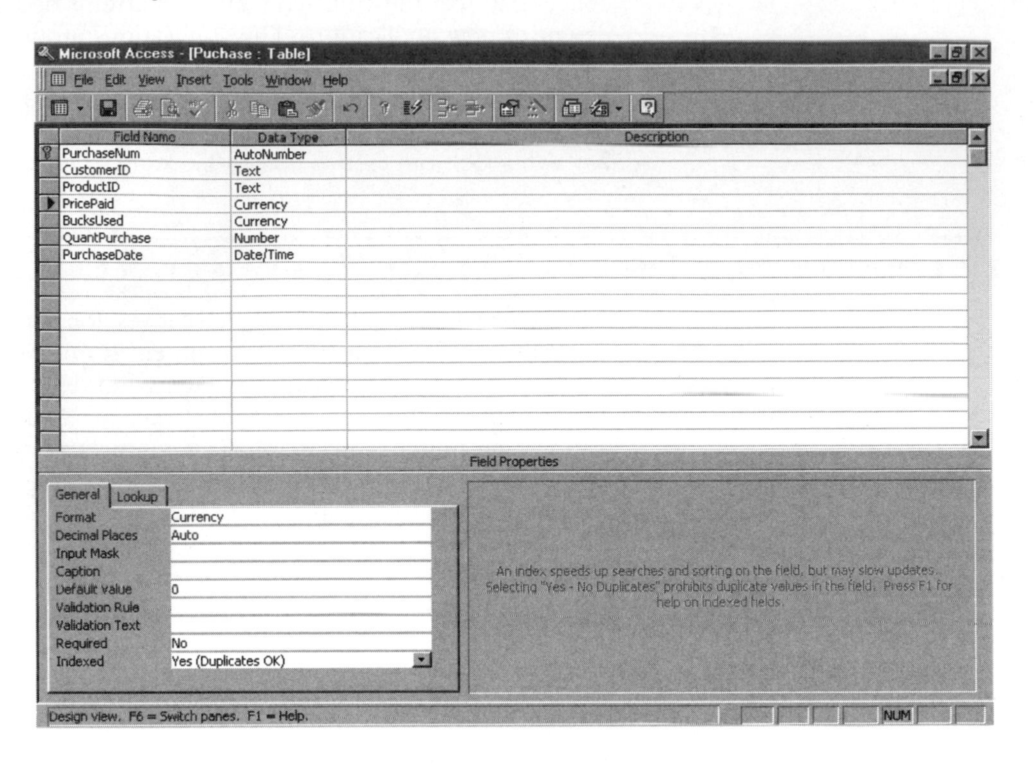

FarEast Bucks to help purchase an item, and Process 2 describes how they create new FarEast Bucks by purchasing an item with their credit card.

The logic for Process 2 is very simple—compute the number of new FarEast Bucks by dividing the purchase amount by 20 and adding them to available FarEast Bucks. In contrast, the logic for Process 1 involves checking the number of available FarEast Bucks to determine whether the customer has at least 10 and then computing the net purchase price after subtracting the FarEast Bucks. The number of FarEast Bucks is then updated.

Figure 10.7

Data flow diagram for rewards system

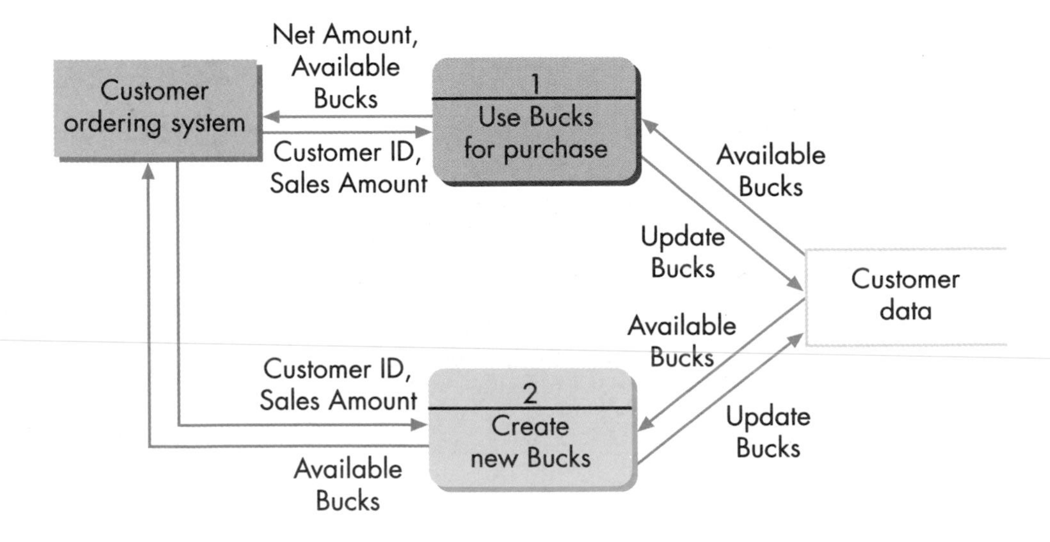

Figure 10.8 IPO table for Process 1 of rewards system

Input	Processing	Output
•Customer ID •Sales Amount •Available Bucks •Bucks Used	1. Use Customer ID to query database for Available Bucks 2. If Available Bucks > 10 then Net Amount = Sale Amount – Bucks Used; Set New Available Bucks = Old Available Bucks – Bucks Used 3. If Available Bucks <= 10 Then Net Amount = Sale Amount; New Available Bucks = Old Available Bucks + Sale Amount / 20 4. Update Available Bucks	•Net Sale Amount •New Available Bucks

The IPO table for Process 1 should include the inputs to this process, the logic of the process, and the outputs from the process. The Process 1 inputs include the customer ID, the sale amount, the number of FarEast Bucks the customer wants to use for the purchase from the customer order system, and the number of available FarEast Bucks from the Customer data store. Process 1 outputs consist of the net sale amount, which goes to the customer order system, and an updated number of FarEast Bucks, which goes to the Customer data store. The logic involves three steps:

Step 1: Use Customer ID to find Available Bucks.
Step 2: If Available Bucks is greater than 10, calculate Net Sale Amount by sub tracting number of Bucks Used from Sale Amount; set Available Bucks equal to old number of Available Bucks minus the number of Bucks Used.
Step 3: If Available Bucks is less than 10, then Net Sale Amount is equal to the Sale Amount, and Available Bucks is increased by the Sale Amount divided by 20.

Figure 10.8 shows the resulting IPO table. Note that if the number of available FarEast Bucks is less than 10, the logic matches that for Process 2; that is, increase the number of Available Bucks by the Sale Amount divided by 20.

The next element of the design process is developing the pseudocode for Process 1, corresponding to the the IPO table shown in Figure 10.8. Figure 10.9 shows this pseudocode. Note that you can easily follow the logic in this pseudocode procedure.

Creating the Interface for the FarEast Foods Rewards System

The interface screen for the FarEast Food enhanced customer ordering system might appear as shown in Figure 10.10. The designers have added some functionality to the screen. The sponsor or users can enter a specific customer ID and sale amount and see the result of using the system. Note that the interface does not actually query the database, but rather uses values that are *hard-coded* into the program and that work only for the specified ID values.

Figure 10.9 Pseudocode for Process 1 of rewards system

```
Begin Procedure
   Input Customer ID, Sale Amount, Bucks Used
   Query database for Available Bucks for this Customer ID
   If Available Bucks > 10 then
        Net Amount = Sale Amount - Bucks Used
        Available Bucks = Available Bucks - Bucks Used
   Otherwise
        Net Amount = Sale Amount
        Available Bucks = Available Bucks + Net Amount / 20
    End decision
   Output Net Amount and update Available Bucks in database
End procedure
```

1. If you have Access database software at your disposal, create the structure for the Products table of the FarEast Foods data model.
2. Create an IPO table and the corresponding pseudocode for Process 2 of the www.fareastfoods.com rewards system.

Figure 10.10 Interface screen for rewards system

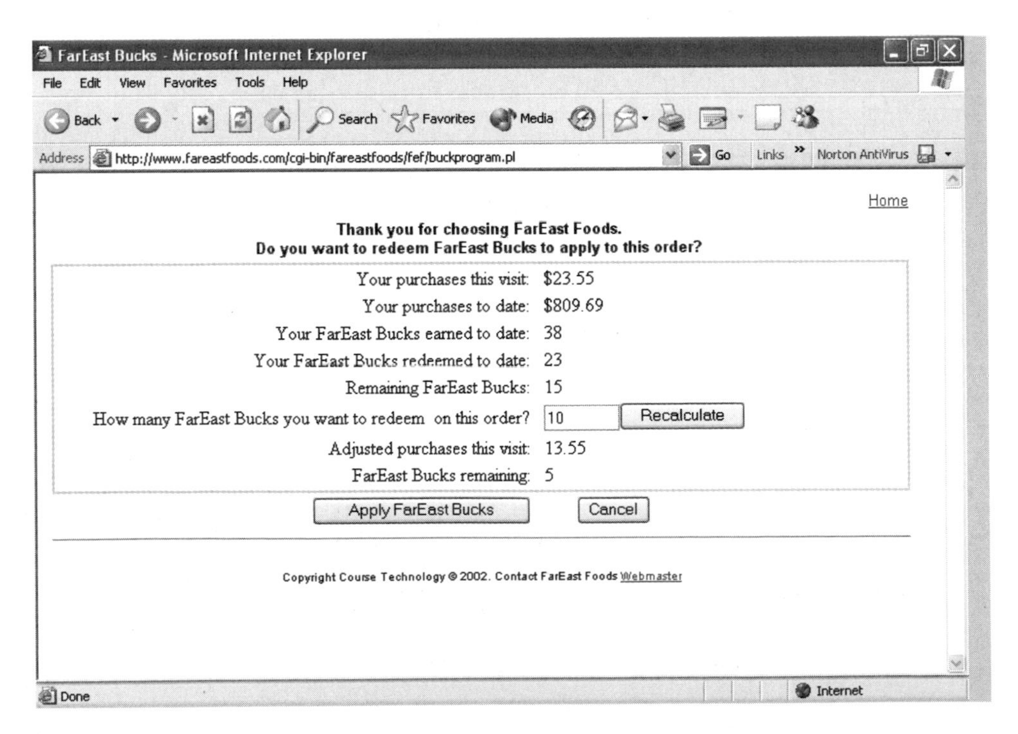

Implementation

implementation stage
The stage of the structured systems development process in which the information system design is built, tested, installed, and maintained. Training also takes place during this stage.

If the steering committee decides to implement the design recommended by the development team, then the final stage of the development process—the implementation stage—can begin. The **implementation stage** involves turning the information system design into a reality. This section assumes that the design team has recommended internal development. Later sections will consider outsourcing and acquisition.

The implementation stage for internal development of an information system involves four steps:

Step 1: Build and test the system.
Step 2: Install the system.
Step 3: Train users on the system.
Step 4: Maintain the system.

Ed Yourdon

IT INNOVATORS

Along with the increasing use of computers to handle large projects in the late 1960s and early 1970s came the need for a more organized way to handle the development of such projects. One of the leaders in developing the structured approach for large information systems was Edward Yourdon, creater of the popular "Yourdon method" of structured systems development. Educated at MIT, Yourdon began his career with Digital Equipment Corporation (DEC, which is now a part of Compaq Computers) in 1964. After working for DEC and General Electric, where he became involved in the development of several pioneering computer technologies, such as time-sharing operating systems and virtual memory systems, Yourdon started his own consulting firm, Yourdon, Inc., in 1974. Over the next 12 years, his company grew to a staff of more than 150 people who trained some 250,000 people around the world in the pioneering concepts developed by Yourdon. In 1986, Yourdon sold his consulting firm (it is now a part of IBM). The publishing division, which has published more than 150 titles on computer science topics, eventually became part of Barnes and Noble.

Yourdon followed up his work on structured systems development methods with pioneering work on the use of object-oriented methods for systems development in the late 1980s and 1990s. His efforts included the development of the Yourdon/Whitehead method of object-oriented analysis/design and the Coad/Yourdon object-oriented methodology. In 1997, Yourdon was inducted into the Computer Hall of Fame. In 1999, he was named as one of the 10 most influential men and women in the software field.

As the author of 26 computer books and more than 500 technical articles, Yourdon's most recent books include *Managing High Intensity Internet Projects* and *Prepping*, a discussion of what we have learned from the Year 2000 problem and the September 11, 2001, attacks on the World Trade Center and the Pentagon. *Prepping* lays out ways of adjusting to a new world in which significant, "unthinkable" risks are likely to be a *permanent* part of our lives.

Ed Yourdon has pioneered many of the approaches to systems development in use today.

Source: http://www.yourdon.com/bio.html.

In the first step, the development team builds and tests the system by writing appropriate computer programs or assembling programming objects that implement the design created in the design stage. Next, the programs are thoroughly tested—not only by the programmers, but also by other members of the development team and potential users. Only after a complete testing and correction process can installation of the new information system on the organization's computers take place. The installation process also involves testing it in that location and converting from the old system to the new. The third step is to train the staff who will use the new system. Training not only prepares employees to use the new system, but can also help avoid "people problems" that arise whenever any kind of change occurs in an organization. Finally, once the system is installed and running in a production mode, the organization must support and maintain it. All too often, organizations give this last step insufficient weight, even though the cost of maintaining an information system often greatly exceeds its development cost.

Building the System

Building an information system involves writing the computer programs that provide the functionality specified in the first three stages of the systems development process. In writing computer programs, the programmer gives the computer a sequence of instructions in a computer language, such as COBOL, FORTRAN, C, C++, Java, or Visual Basic .Net. If an organization is building a Web application, then programmers typically use programming languages and approaches to development such as such as Java, JavaScript, PHP,[1] ASP.Net using Visual Basic .Net, or CGI using Perl. Regardless of the high-level language in which the program is written, the programmer must initially translate the logic developed in the design stage into a working program, first by learning the vocabulary and syntax (grammar) of the language, and then by studying how to handle various logical situations in that particular language.

Although programmers can use all of the languages listed to write computer programs, COBOL and FORTAN are quite different from some of the others. Languages such as C++, Java, Visual Basic .Net, and JavaScript are **object-oriented languages** to a greater or lesser degree, whereas COBOL and FORTRAN are not. In programming, an **object** is a self-contained module that combines data and instructions and that cooperates with other objects in a program by passing strictly defined messages to the other objects. Programmers find it easier to work with an objected-oriented methodology than with other programming techniques because the former is more intuitive than traditional programming methods, which divide programs into hierarchies and separate data from programming code. Programmers can combine objects with relative ease to create new systems and extend existing ones. Around us is a world made of objects, so the use of objects to create information systems provides a natural approach to programming.

object-oriented languages
Computer languages that use objects to carry out the required logic of a program.

object
A self-contained programming module that combines data and instructions and that cooperates with other objects in a program by passing strictly defined messages to the other objects.

The Programming Process

The most important concept to learn about programming is that it is a form of *problem solving* in which programmers convert the design created in the design stage into a computer program that implements the design. If the development team has done the work in the design stage correctly, the programmer does not need to know very much about the original problem: He or she need simply follow the design. The data model, IPO tables, pseudocode, and interface screens, if correctly done, should suffice to guide the programming process.

1. PHP stands for a very early version of the language called Personal Home Page Tools. Today, only the letters are used.

Java is a very powerful object-oriented language that developers are using in many Web projects.

Q2.java - Notepad

File Edit Format Help

```
//Uses command button to increment counter in text box
import java.awt.*;
import java.awt.event.*;
import javax.swing.*;
public class Q2 extends JFrame implements ActionListener {
   private JTextField top;
   private JButton bottom;
   private Container container;
   public Q2() {
          super("Enjoying Quiz 2");
          g1 = new GridLayout(2,1);
          container = getContentPane();
          container.setLayout(g1);
          top = new JTextField("");
          container.add(top);
          bottom = new JButton("Increment");
          container.add(bottom);
          bottom.addActionListener(this);
          setSize(150,150);
          setVisible(true);
   }
   public void actionPerformed( ActionEvent event) {
               String temp;
               int it;
                        temp = top.getText();
                        if (temp.length() == 0)
                               top.setText("0");
                        else {
                               it = Integer.parseInt(temp);
                               top.setText(Integer.toString(++it));
                   }
   }
   public static void main(String args[]) {
          Q2 application = new Q2();
          application.setDefaultCloseOperation(JFrame.EXIT_ON_CLOSE);
   }
```

Assuming the development team has worked through the design stage correctly, ideally programmers and the development team must carry out six programming steps to build the system:

Step 1: Programmers incorporate the step-by-step logic from the design stage in a computer program.
Step 2: Programmers test the program extensively and correct any errors.
Step 3: Development team members test the program, and programmers correct any new errors.
Step 4: Development team members write documentation for the program.
Step 5: Development team members release the program to users for additional testing in a work environment.
Step 6: Users report errors, which programmers fix, leading to a revised release of the program.

In Step 1, the programmer creates or builds the program. in Steps 2 and 3, the programmers and the development team test the program and correct any errors that they find. Testing is an extremely important aspect of the programming process that must not be taken for granted. Once the programmers and development team believe the program is correct, they write documentation for it and then release it to the designated users for further testing in the work environment. When (not if) users find errors, they report them to the programming team; team members then correct the errors and send out a new version of the program. This process can be ongoing, with versions numbered 1.0, 1.1, and so on, until an entirely new version of the software results.

bugs
Errors in a computer program.

beta testing
Engaging the services of users to test pre-release versions of software.

Users frequently find errors or **bugs** in popular commercial software programs even after the development team has done extensive testing. Commercial software developers, such as Microsoft, Adobe, and Apple, try to avoid releasing software that contains errors or bugs by engaging the services of users to test pre-release versions of software in a process called **beta testing**. The beta testers report any bugs or errors in exchange for receiving free copies of the software. The commercial software developers also release revised versions of software as they correct errors.

Writing documentation that explains to users and other programmers how a program works is an essential part of the development process. Unfortunately, development teams often ignore this step in their rush to get new systems up and running. However, without proper documentation, both users and support personnel

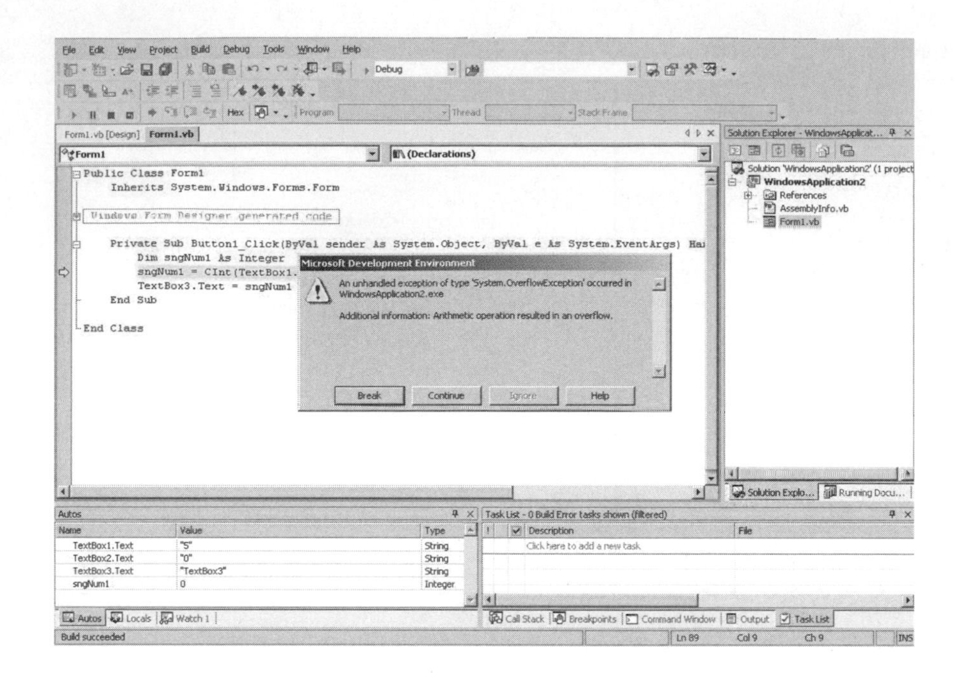

VB .NET, the newest version of Microsoft's extremely popular Visual Basic programming language, comes with powerful debugging tools.

can quickly become frustrated when they cannot get the system to work as planned or solve problems with the system. In addition, documentation is essential to the maintenance step, which keeps the system running after the development team has moved on to other projects.

A key concept in programming is that program development steps take place only *after* a great deal of planning, analysis, and design work has been done. It is important because no matter how well the program is written, the organiztion will not achieve its objective if the program is written to solve the *wrong* problem.

Installation

Regardless of whether an organization uses commercial software or specially developed software, once the new information system is acquired or developed, the development team must install it in the organization. Installation includes adding any necessary hardware, loading the software on new or existing hardware, testing the system in place, converting data files to suit the new system, and switching operations over to the new system. In the opening case, Merrill Lynch dedicated an entire weekend at each office to removing old hardware and wiring, installing new computers and Internet connections, and installing the TGA system on the new hardware.

In system testing, the team extensively checks the new system's processing capabilities for throughput, turnaround time, and access time. The developers must also test the system to determine its capacity to handle both a normal volume of transactions and abnormally high volumes. In addition, they must determine the system's capacity to restart and recover after abnormal system termination events (sometimes called *crashes*). The development team must analyze the results of system testing to determine the magnitude of any problems and to figure out how to fix them.

Once the development team members have completely tested the system, they may have to convert database files to the new system. File conversion is typically necessary if the previous system utilized several types of files or proprietary software. The development team must give careful attention to the integrity of new and old files; otherwise, an organization can make poor decisions because of bad or missing data.

After testing and data conversion are complete, the final installation steps involve system conversion. **System conversion** entails the process of changing over from the old system to the new one. Four approaches to system conversion

system conversion
The process of changing over from an old system to a new one.

Table 10.2 Comparison of Approaches to System Conversion

Conversion Aproach	Characteristics	Advantages	Disadvantages	Risk
Direct	Simultaneously shut down old system and start up new system	> Conversion is fast > Costs are lower	New system may not work	High
Parallel	Run old and new system at same time	Developers can fix problems with the new system while the old system is still in operation	> Conversion is slower than with the direct approach > It is expensive to run both versions > Performance problems may crop up from running two versions	Low
Pilot	Install and test system in one part of organization before installing it everywhere	Developers can find and fix problems without affecting the entire organization	Problems with high volume of transations may not be found	Moderate
Phased	System is installed sequentially in different locations	> Technique requires fewer number of installation staff > Problems at one location can be fixed before installing the system elsewhere	Different locations do not use the same version of the system	Moderate

are commonly used: direct, parallel, pilot, and phased. Table 10.2 summarizes these approaches and gives the advantages, disadvantages, and risk level of each.

Table 10.2 shows that the direct method is faster, but much more risky than the more conservative parallel approach, in which both systems run side-by-side for a period of time. More than one IS management job has been lost when managers tried to speed up the conversion process by going with a direct conversion, only to have to revert back to the old system when the new one failed to work as promised. The phased and pilot approaches represent compromises between these two extremes. Notice that Merrill Lynch used a phased installation of its new TGA system because the same system was being installed at hundreds of offices around the world.

Training

Merrill Lynch gave a great deal of emphasis to training the potential users of its new TGA system both by giving mandatory training sessions two weeks before it installed the new system and then by providing a review session the day before the system went live. This example indicates the importance that organizations should place on training their employees on how new systems will help them perform their job responsibilities. Without appropriate training, even the best system can quickly frustrate users and eventually cause them to reject it.

What training should organizations provide to users? The answer may surprise you. The training should focus not only on how to use the system, but also on how the system will help users do their jobs better. This type of training involves helping users understand how the new system fits into the organization's overall mission. If users do not understand the business aspects of the new system, they may fail to see why learning to use it is important.

Training should highlight the user's actions rather than the features of the new system. For example, most users of word processing or spreadsheet packages use only a small fraction of their capabilities; the same is true of many new information systems. Unfortunately, development teams usually become so excited about their new "baby" that they want to show off all its capabilities, overwhelming the users with features during training. Training should concentrate on the 20 percent of the system's features that 80 percent of the users will use, thereby enabling the vast majority of users to become confident in their ability to work with the new system. Users can learn about more advanced features when they need them by reading the documentation or by talking with members of the development team or the technical support staff.

Types of training include classroom training, one-on-one training, and computer-based training (CBT). Classroom training is a popular choice when a large number of users must be trained simultaneously, but often proves less effective than one-on-one training or CBT. However, one-on-one training and CBT are more expensive than classroom training *unless* the same training will be repeated frequently. When training will be used in many situations and with many users, the organization can spread the start-up costs of creating a CBT program over many trainees. Note that the movement toward using the standard browser interface for software applications has reduced the amount of user training required because most people are already quite comfortable with this interface.

Training is critical not only to provide users with the knowledge necessary to use the new system but also to avoid problems related to resistance to change. A well-known aspect of organizational behavior is that any change will bring about resistance. Unless the change process is well planned and includes adequate training, employees who are being asked to change the way they do their jobs can resist to the point of ensuring that a new system fails. To minimize or avoid resistance, organizations should start the change process much earlier than the training stage and involve potential users in the earlier planning, analysis, and design stages of the systems development process. If users become involved early on, they often feel that they had a big part in designing the new system and can be enthusiastic about the change rather than resist it.

Maintenance

Any system, no matter how well designed, must be continually modified to handle changes in input, output, or logic requirements. These alterations occur through **maintenance**, the ongoing process of keeping a system up-to-date. Although it might seem like a minor part of the analysis, design, and implementation process, maintenance is a critical aspect of keeping information systems running as designed. In fact, a recent study shows that programmers spend more than 45 percent of their time maintaining information systems.[2] Although very high, this estimate is actually significantly lower than the traditional estimate of spending *70 percent* of programmer time on maintenance.

maintenance
The ongoing process of keeping a system up-to-date by making necessary changes.

Maintenance consists of two important steps: determining what changes need to be made and then making the changes. The change identification step closely resembles the analysis stage of the systems development process. That is, a systems analyst studies a situation and then pinpoints the problem or responds to a request from a user for an upgrade. Once the analyst has identified the problem or change, changing the existing system occurs via a small-scale version of the design and implementation stages.

The ease with which an organization can maintain an existing system depends a great deal on the system documentation and the quality of the computer code. Documentation consists of the descriptions and instructions provided with the

2. Guy Fitzgerald, "Addressing information systems flexibility: theory and practice." Working paper, Brunel University, UK, 2001.

Tracking Bugs in Real Time

Imagine that you are the CEO of a small company that depends on software from a software company to run your business. Unfortunately, the software is not working correctly, and no one seems to know what the problem is. If you are lucky, the software company may be willing to send some of its staff to try to figure out what is going wrong, but they may need to take the software out of production to track down the problem. For your company, this effort means losing production time trying to find the bug using various debugging tools, even if the company's programmers have written code into the software that will help them search for the problem. Although all modern computer languages include some type of debugging tools, many of them are meant only for working with the code before it is compiled into 0s and 1s and installed on customers' machines.

Software from a small software company located in Cambridge, Massachusetts, called InCert, provides one solution to this problem. InCert's system can generate snapshots of diagnostic information on command or when certain events are likely to occur while the software is running—or, in computer lingo, in *real time.* As one customer said about looking for errors in a software program, "Before, it was basically shooting in the dark and, hopefully, if you shot enough times, you hit something." Another customer said that his software company can write 50 percent less code with InCert's products because they don't have to include code to trace the execution of the software when errors occur. According to the same person, InCert's products can reduce the time to fix problems encountered by end users by as much as 80 percent. When downtime means that your company is out of business until its software is fixed, you can only hope that InCert's products are built into your software.

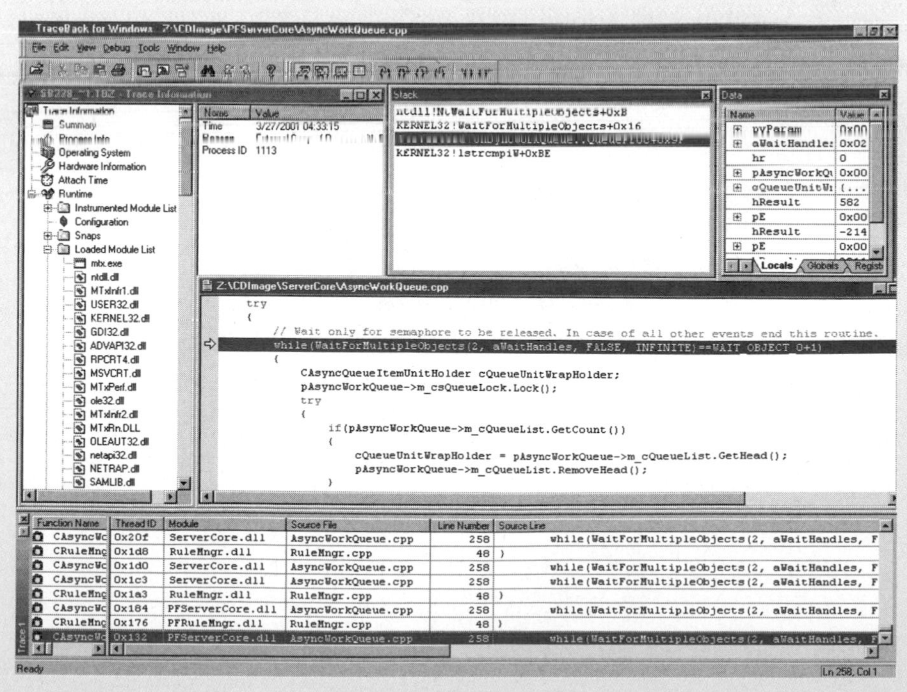

Programmers can use software from InCert to detect problems in programs after they have been installed and are in use at a customer's office or factory.

Source: Gary H. Anthes, "InCert locates bugs in real time." *Computerworld*, August 27, 2001, http://www.computerworld.com/storyba/0,4125,NAV47_STO63277,00.html.

hardware and commercial software or the documentation for internally written software. Without this documentation, changing the system may be virtually impossible and the existing system may have to be junked in favor of an entirely new system—a potentially expensive process.

1. What are the four steps in the implementation stage?
2. Why can the quality of training on a new information system make or break the new system?

Implementation at FarEast Foods

Let's take a look at the implementation process to enhance the www.fareastfoods.com purchase system to award one FarEast Buck to customers for every $20 they spend with the company. Recall that when customers have more than 10 FarEast Bucks, they can use them to purchase items from the company. With the design stage already completed, the next step is to implement the design by building the system, installing it, training the staff who will use it, and maintaining it.

Programmers will build the Web-based rewards system in the Perl language, which is commonly used to carry out processing on a Web site using CGI. Other languages that they might have chosen for this purpose are Java, C++, ASP.Net, or PHP. Figure 10.11 shows a portion of the Perl code used to implement the design corresponding to the IPO table and pseudocode for Process 1 shown earlier in Figures 10.8 and 10.9.

Once the FarEast Foods programming staff builds and debugs the system, the next step is to test it extensively before having it go live. The initial testing of this system involved having FarEast Foods employees simulate the Web-based ordering process, receive FarEast Bucks, and then use them to simulate purchasing goods. After extensive testing by employees, the company undertook beta testing of the system with a small group of customers who were offered discounts for using it. When the beta testing was completed, and the programmers corrected all errors, the system was ready to be installed on the application server at FarEast Foods

Figure 10.11 Portion of Perl code for FarEast Foods rewards system

```
sub IsSubmitOK
{
        if ($cart->isEmpty($db))
        {
                return 1;
        }
        my $shopperid       = ($us_lngShopperID eq "")?"null":$us_lngShopperID;

        my $lngRedeemNum    = $session->getValue($db, "usRedeemNum");
        if ($lngRedeemNum eq "")
        {
                $lngRedeemNum = 0;
        }
        if ($lngRedeemNum == 0) {
                $bucksearned = $dblSubTotal / 20;
        } else {
                $bucksearned = 0;
        }
        my $strUpdShopper  = "";
        if ($us_lngShopperID ne "")
        {
                $strUpdShopper = qq{Update Shoppers SET
                      PurchaseAmountUTD = (PurchaseAmountUTD \+ $dblSubTotal)
                      , BucksRedeemedUTD = (BucksRedeemedUTD \+ $lngRedeemNum)
                      , BucksEarnedUTD = (BucksEarnedUTD \+ $bucksearned)
                      ,LastOrderID = LAST_INSERT_ID()
                      Where ShopperID = $us_lngShopperID};
        }
```

Figure 10.12 FarEast Bucks being redeemed

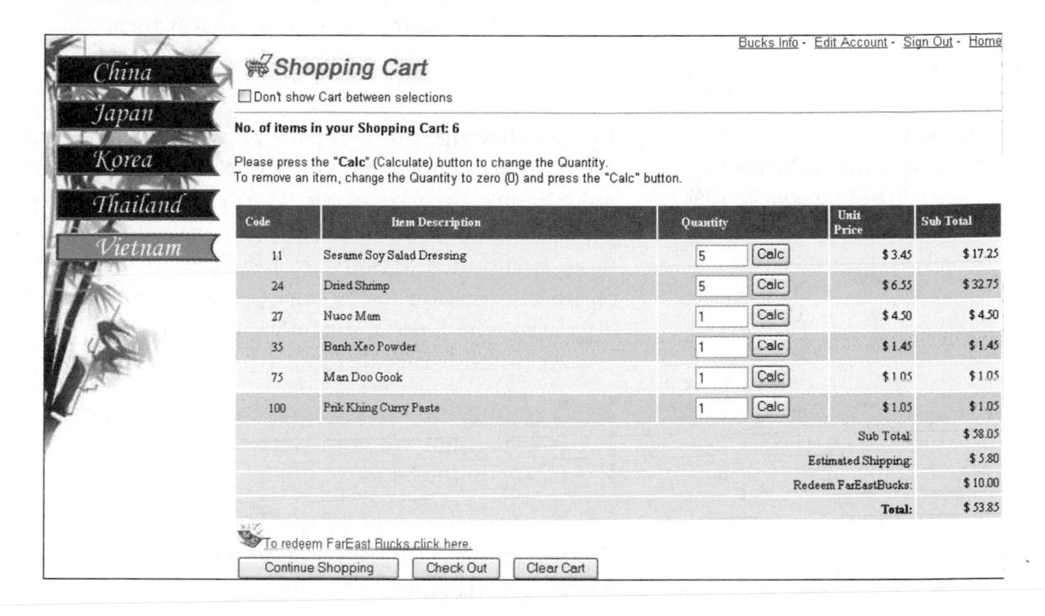

Code	Item Description	Quantity	Unit Price	Sub Total
11	Sesame Soy Salad Dressing	5 Calc	$ 3.45	$ 17.25
24	Dried Shrimp	5 Calc	$ 6.55	$ 32.75
27	Nuoc Mam	1 Calc	$ 4.50	$ 4.50
35	Banh Xeo Powder	1 Calc	$ 1.45	$ 1.45
75	Man Doo Gook	1 Calc	$ 1.05	$ 1.05
100	Prik Khing Curry Paste	1 Calc	$ 1.05	$ 1.05
			Sub Total:	$ 58.05
			Estimated Shipping:	$ 5.80
			Redeem FarEastBucks:	$ 10.00
			Total:	$ 53.85

along with the rest of the customer purchase system, which resides on a number of physical computers in a server farm. It then became available to the general customer population after completion of the training step.

At FarEast Foods, a customer service department answers questions from customers, handles returns, and, in general, tries to help customers. Because this group would handle questions about the new FarEast Bucks program, they needed to be trained in its use. This training involved classroom and one-on-one training on the system and problems that might arise.

The company must constantly monitor the new FarEast Bucks system for problems and, when found, immediately remedy them. One problem that FarEast Foods must watch for is response to a large demand. Although development team has tested the new system with employees and some customers, they have not tested the system among the entire population of potential customers. The possibility exists that the hardware and software could be overwhelmed by too many customers wishing to use their FarEast Bucks. You can see and use the final customer rewards enhancement to the FarEast Foods Web site at www.fareastfoods.com by clicking the About link and then the link to the enhanced Web site. You will be able to register to earn and use FarEast Bucks in a simulation of purchasing Asian food items. Figure 10.12 shows the checkout Web page, where a customer redeems 10 FarEast Bucks to reduce the amount of the purchase.

1. Some experts say that it is very difficult to test a Web site to determine how well it will handle high levels of demand. Why might this difficulty arise?
2. Why did FarEast Foods use employees to test its system first?

Outsourcing and Acquisition

As discussed earlier, an important outcome of the design stage is a decision on how the organization will develop the information system: by an internal development team, by an outside development team (outsourcing), or by a commercial software company (acquisition). The preceding section discussed internal development, and this section will cover outsourcing and acquisition. Although each technique offers advantages relative to internal development, each also has definite problems. Both the outsourcing and acquisition approaches require that an organization give special attention to selecting a vendor. In addition, the organization must carefully consider the type of outsourcing contract that it uses.

Outsourcing

outsourcing
An arrangement in which one company provides services for another company that could otherwise have been provided in-house.

In **outsourcing**, one company provides services for another company or organization, which chooses not to provide those services internally. Outsourcing is a growing trend, representing a market expected to exceed $177 billion by the year 2004. It can range from the use of contractors, temporary employees, or off-shore programmers, such as those mentioned at Software MacKiev in Chapter 9, to outsourcing of a company's entire information technology operation. The extent of outsourcing depends on the degree to which the company believes IT contributes to its competitive advantage.

As noted earlier in Table 10.1, outsourcing offers the following advanatages:

- The outsourcer has more skilled and experienced programmers.
- Internal staff are not diverted from their current work.
- Costs might be lower.

The development of the Manheim Online used car sales system, discussed at the beginning of Chapter 9, provides a good example of the use of outsourcing. Manheim management realized that their internal development staff had little or no experience or skills in Web development. They also did not want to divert the attention of the internal IT staff from their primary purpose of processing auction sales data into usable information. At the same time, management wanted to add value to the company. For these reasons, Manheim outsourced the development of the online sales system until it became a part of the company's competitive advantage. When it became clear that Manheim Online represented an important part of the company's future, Manheim hired IT people knowledgeable in Web development and brought the system inside the company. The reverse situation can also occur: An organization can develop a system internally and then outsource its maintenance. The choice all depends on the way the company views its use of information technology.

Also as listed in Table 10.1, outsourcing has a number of disadvantages:

- The company may lose control of the project.
- Internal developers may not learn the skills necessary to maintain the system.
- The outsourcer may not deliver on its claims.

Of particular interest in systems development is the third point: The outsourcer may not deliver on its claims. Although late software projects are common regardless of where the software is developed, this problem becomes more critical when the project is handled outside the organization, because management may not be able to visit with the development team to ask questions, view outputs, and judge the true state of project completion.

Once an organization has decided to outsource a project, it must select a vendor and negotiate a contract. Important considerations in vendor selection include the outsourcer's record on meeting its commitments with similar projects, current and prior customers' satisfaction levels, and the technical competence of the outsourcer's employees as evidenced by prior work.

Organizations typically use three types of contracts when engaging an outsourcer: time and materials, fixed price, and value added. Table 10.3 describes these contracts, their pricing plans, and their advantages and disadvantages.

Table 10.3 Types of Outsourcing Contracts

Contract Type	Pricing	Advantages	Disadvantages
Time and materials	Payment based on time spent on the project and any materials involved	Good when the time required for the project is difficult to estimate	Can result in a large bill
Fixed price	Fixed price for development of the system	The company knows the costs prior to commencement of the work	Can result in the outsourcer cutting corners to keep costs down
Value added	Payment based on benefits resulting from the new system	Enables the outsourcer to share in the risks and benefits of the project	The company has to share the benefits with the outsourcer

Application Service Providers

application service providers (ASP) Companies that offer individuals or enterprises access over the Internet to applications and related services that would otherwise have to be located in their own personal or enterprise computers.

Recall that over the last few years, a new type of outsourcing firm has become increasingly popular. Called **application service providers (ASP)**, these companies provide computing applications, data storage, reporting tools, and technical support over the Web. Many of these services would otherwise have to be located on personal computers or enterprise servers. The ASP market is expected to grow from $770 million in 2000 to $15 billion by 2005.

You can classify ASPs into four categories: enterprise ASPs, general business ASPs, specialist ASPs, and vertical ASPs. Enterprise ASPs offer applications that require customization, such as data warehousing, customer relationship management, and enterprise resource planning for multiple industries. In contrast, general business ASPs are relatively simple applications that require little customization to provide users with access to standard business tools (such as word processing and spreadsheet applications) over a network. Specialist ASPs focus on a single area of business. For example, the eMRWeb system for hospitals and clinics discussed in Chapter 5 is a specialist ASP; it captures patient records, stores them on their servers, and makes them available to health care workers through the Web. Finally, vertical ASPs provide packaged or specialized applications for a specific industry, which an organization can modify slightly to run on most standard business tools. The automotive, computer manufacturing, health care, or insurance industries, for example, might have need for a vertical ASP with services tailored to that particular industry. For example, an automotive ASP might provide specific design services for an automobile manufacturer.

The key feature that separates ASPs from other service providers is the use of the Internet to make applications available to customers. This approach provides clients with affordable access to technology with improved cost and performance over installed systems, thereby allowing them to focus on their core business competency. The primary problem with using an ASP relates to the danger that it will go out of business, leaving the client company with no applications for carrying out its key operations. One estimate by the Gartner Group found that on an annual basis, 60 percent of ASPs will drop out of the ASP business, go out of business altogether, or be acquired.

Acquisition

Whereas projects requiring a high degree of customization or competitive advantage often rely on outsourcing and internal development, *acquiring commercial off-the-shelf (COTS) software* can be very useful when speed or cost is the primary consideration. For example, it would probably be better to acquire a new payroll system for FarEast Foods, because a payroll system rarely requires customization and does not generally add to a company's competitive advantage.

You have probably noticed that the discussion of systems development has not mentioned hardware; this omission reflects the fact that organizations almost always acquire hardware instead of developing it. The various hardware companies, such as IBM, Compaq, and Dell, now specialize in creating hardware systems to meet almost any company's needs. However, when an organization acquires hardware and software together, the acquisition needs to follow a particular sequence, as shown in Figure 10.13. Note that the software is specified and selected *before* the hardware is specified and selected. The acquisition process must follow this general order; otherwise, an organization may acquire hardware that won't run with its new software.

The acquisition process starts with the design stage, in which the development team develops specifications for the software; that is, the development team decides

Figure 10.13 Hardware and software acquistion process

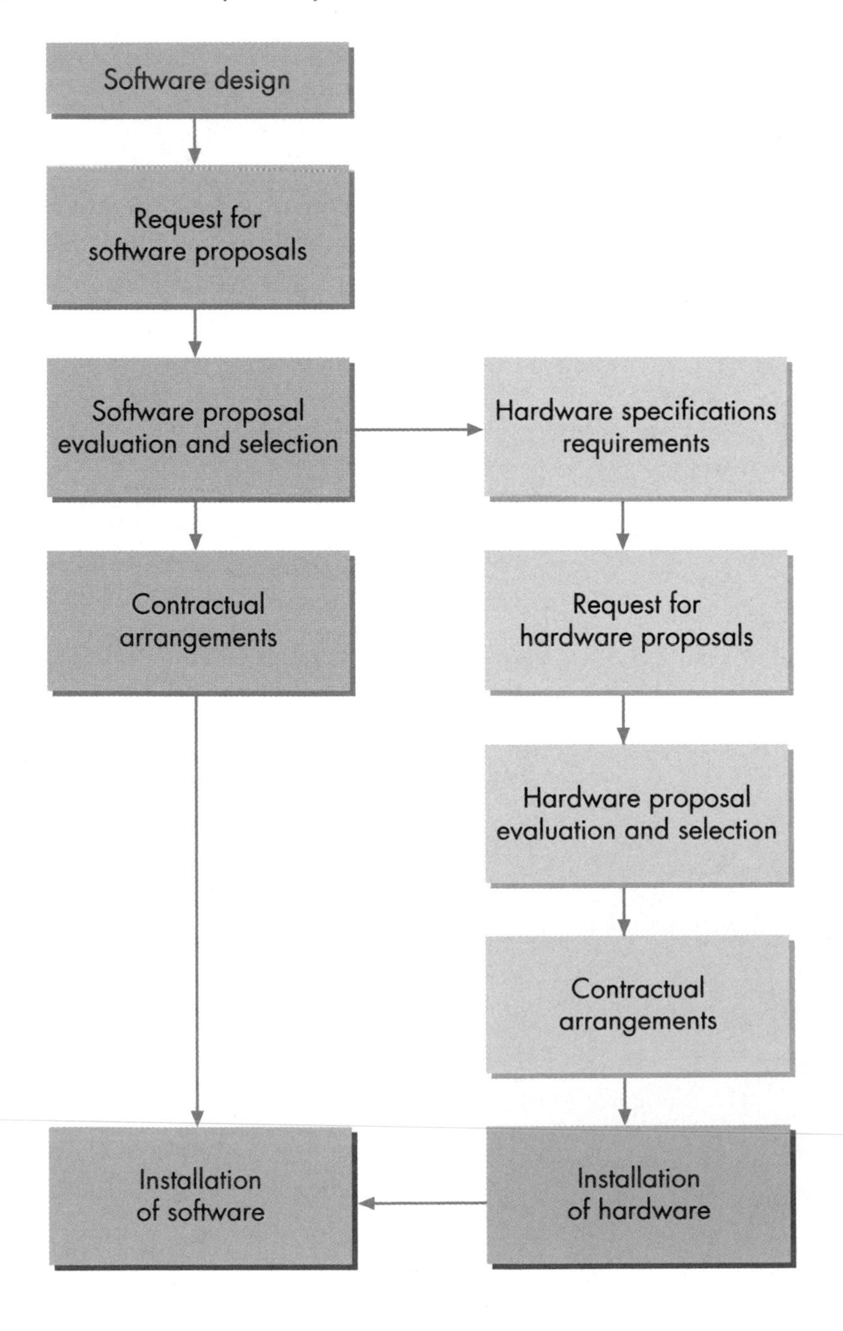

request for proposals (RFP)
A complete list of specifications used by vendor or contractors to prepare a bid on a project.

benchmark test
A test in which the development team compares competing products using programs and data typical of the actual conditions under which the proposed system will operate.

what the software must accomplish. Once the team members have completed a software specification, they send out a request for proposals to vendors that are known to offer software of the type under consideration. A **request for proposals (RFP)** should list the specifications for the software as completely as is needed for a vendor to prepare a bid.

When the development team receives software proposals, it must evaluate the proposals against the software design. To do so, the staff can measure the proposals against a weighted set of criteria, giving more weight to criteria that have greater importance. Another way to compare competing software packages is to benchmark them. In a **benchmark test**, the development team compares competing products using programs and data typical of the actual conditions under which the proposed system will operate. The development team can combine these two methods by using the weighted comparison to reduce the list of proposals and then conducting benchmark tests on the best two or three alternatives.

Once the team selects the best software option, the next step in the acquisition of software is to work out the contractual arrangements with the vendor. This contract must include financing of the cost of the software, arrangements for technical support, and conditions for acquiring updates to the software.

Only after the team has evaluated software proposals and selected the best one should it give any thought to selecting and acquiring hardware that matches the software. Hardware selection follows the same process as acquiring software, except the contractual process demands additional consideration. Whereas software is typically purchased or leased, hardware can be rented, leased, or purchased, so the organization must consider these options in the hardware acquisition process. Once the team members have installed and tested the hardware, they can then install the software and test it on the new hardware.

Enterprise Resource Planning Systems

enterprise resource planning (ERP) system
A large, integrated system that handles business processes and data storage for a significant collection of business units and business functions.

A special case of the acquisition process involves a type of software known as enterprise resource planning systems. An **enterprise resource planning (ERP) system** is a large, integrated system that handles business processes and data storage for a significant collection of business units and business functions.[3] ERP systems comprise packaged software designed with a specific type and size of organization in mind. Some target human resources or manufacturing, whereas others are more general in nature. These systems permit some tailoring of the business processes to be used in any given organization, but only within fixed bounds, because all the business processes are designed to work together using a single database. SAP, the developer of the popular R/3 ERP system, is a leader in this field. Other ERP companies include PeopleSoft, Oracle, and J. D. Edwards.

Business activities typically included in an ERP system are product planning, purchasing, inventory management, supply chain management, customer service management, and order tracking. In addition, ERP can include modules for managing the finance and human resources sides of a business. Figure 10.14 shows a screen from a popular ERP package.

To understand the use of ERP, you must recognize its three key concepts: standardization, restrictions, and integration. ERP requires standardization of data and processes, which means that an organization must use a single database with common units of data throughout its departments, plants, and business units before implementing an ERP system. If two plants use different databases, they cannot be part of the same ERP system. Process standardization implies that all units within

3. Thomas Gattiker and Dale Goodhue. "Understanding the plant level costs and benefits of ERP systems: Will the ugly duckling always turn into a swan?", *Proceedings of the 33rd Hawaii International Conference on System Sciences*, 2000.

Figure 10.14 Screen shot from ERP Package

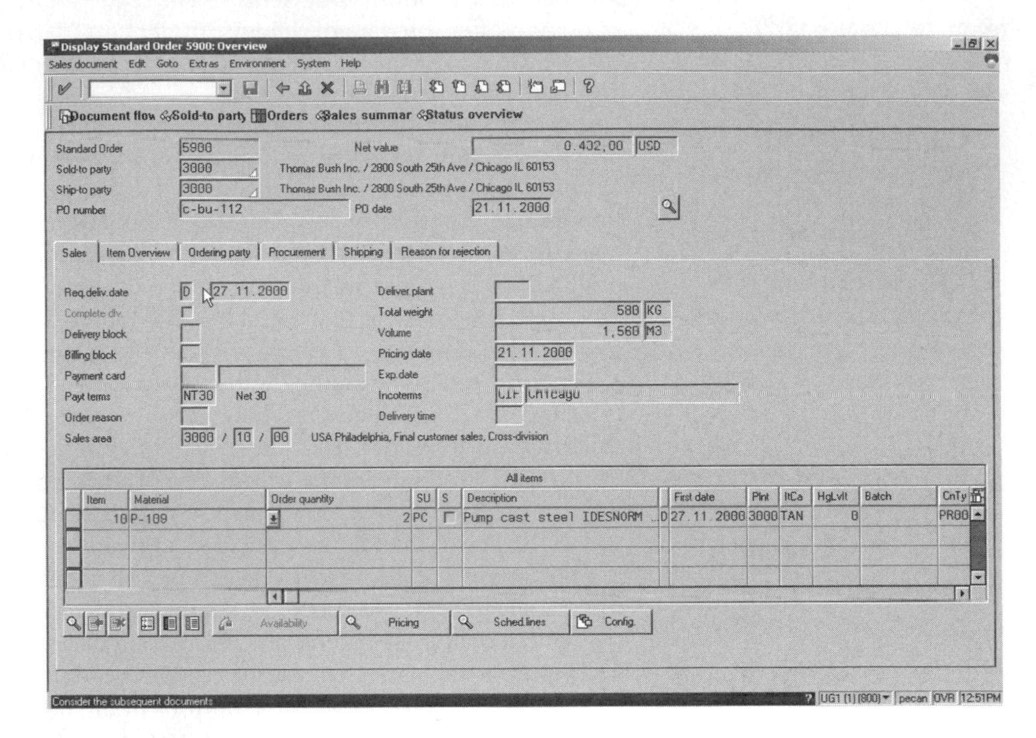

an organization using a single ERP system must carry out a process in the same way. Thus one unit could not use a system of payment for raw materials that differs from that used in other units. As you might imagine, in many cases data standardization forces process standardization on the organization.

ERP also restricts the type of business processes that an organization can use to a set of options that are built into the software. These options rely on a set of best practices—a set of techniques and methodologies that have proven to provide desired results—as selected by the software developer. This restriction feature means that many companies must revise the way they do business to be able to adopt an ERP system.

Finally, integration is a key feature of ERP in that it links many (if not all) of a company's locations and functions so they have access to all relevant information. As you see, using ERP is not just the installation of a new software product. In many cases it leads to a whole new way of thinking about the organization and its business processes.

ERP can benefit a firm by improving communication and coordination and increase organizational efficiency by forcing the firm away from inefficient processes for which no real rationale existed. The enforced standardization can also have the side effect of improving IT maintenance and the deployment of new IT systems. Before ERP, if a CEO wanted the big picture on the firm, he or she would have to obtain information from each department head and then integrate it. With ERP, the integration is automatic, allowing the CEO to spend less time determining what is going on and more time improving the firm's operations.

This enhancement of a firm's operations does not happen without a great deal of work. The work begins with the decision to adopt an ERP system and selection of the correct package. Numerous stories describe companies that decided to abort the ERP installation process after installing one module or none of the modules. This situation often occurs because the company did not completely understand its own business processes and recognize how ERP would work with them. Selecting and installing an ERP package must begin with a thorough study of the firm's business processes.

ERP systems are not always meant to be simply installed in the same way as a word processing package. In many cases, they are modified to fit the needs of the purchasing company. For that reason, they can be envisioned as a hybrid between COTS software and software that is designed and built specifically for a firm's needs.

Problems with ERP include costs, functionality, and implementation. ERP is expensive in terms of both time and cost. An ERP system from SAP can cost $4 million and take three years to install. In terms of functionality, a particular ERP system may not support a particular business process desired by the company. Finally, implementation problems can be associated with installing the ERP system and with changing business procedures. Installation of an ERP system often requires the use of consultants who specialize in the type of system under consideration. In the worst case, if the consultants do not transfer their knowledge to the in-house staff, they may never leave. In addition, a firm must change its business procedures to use ERP. As a consequence, employees must learn both the new business process and the new ERP system at the same time.

Although ERP is a type of software package, choosing to use this approach to business integration is much more involved than simply purchasing software. It requires a total dedication of the business to the practices built into the ERP system—something that all too many businesses fail to recognize until they become deeply involved in installing a system they don't understand. To make installation and use easier, some ERP companies, such as Oracle and SAP, are looking at using the ASP approach discussed earlier.

1. How do outsourcing and acquisition differ? How are they similar?
2. Why is ERP an important type of packaged software?

Rapid Application Development

Although the structured approach to systems development has been the most widely used methodology over the last 25 years, the speed at which things change in the networked economy and the need for Web sites to be built in a hurry has forced developers to look for faster ways of creating information systems. Even though the CASE methodology discussed earlier has speeded up the structured approach by computerizing much of the paper flow associated with the waterfall method, the process remains too slow for many applications, particularly those related to electronic commerce. By the time a firm completes the planning, analysis, design, and implementation stages for an electronic commerce project, the project may no longer be appropriate or the company may have gone out of business while awaiting the new system.

For example, the rewards system for the www.fareastfoods.com Web site is planned to take almost six months using the structured approach. Of course, by the time it is completed, there is no telling what kinds of changes will have occurred in the market or if another company will have beaten FarEast Foods to market with the idea. As one developer aptly put it, "We can't wait nine months to have a baby anymore; we have to do it in two. And it has to come out perfect and it has to run when it's two weeks old."[4] The ideas expressed in this quote are far more true today than when it was said *more than 13 years ago*, well before the development of the Web and the resulting commercialization of the Internet. The growth of the Web has dramatically increased the need for faster development.

To respond to the demand for faster development in the networked economy, a number of approaches have emerged. Two especially popular approaches to rapid application development are outsourcing and acquisition. Outsourcing can be

4. Quoted in Robert L. Schier, "Taking the quick path to systems design," *PC Week*, June 19, 1989, p. 65.

quicker than internal development because the company doing the work will have a more skilled development team that is already trained in Web development techniques. Acquisition can offer the fastest approach to systems development if the company is willing to accept the loss of flexibility and competitive advantage associated with using an off-the-shelf software package. However, if a company wants fast development combined with customization and competitive advantage, it may want to take a hard look at a type of RAD known as prototyping.

prototyping
A software development process in which the development team creates a quick-and-dirty version of the final product, often using special languages or software tools.

In **prototyping**, the development team creates a quick-and-dirty version of the final product, often using special languages or software tools. Prototyping can help the developer to short-circuit the often lengthy structured process by replacing some of the planning, analysis, and design steps with prototypes of the final system. Developers can create prototypes using HTML editors such as Microsoft FrontPage or Adobe PageMill. Although prototypes are not meant to provide the industrial-strength programming necessary to handle the higher volume of transactions in the final application, they do provide important user feedback about the system. Developers commonly use prototypes for creating Web applications to determine customer reactions. This approach was taken by the various companies marketing to children as described in the Internet in Action feature earlier in the chapter.

Types of Prototypes

Four basic types of prototypes exist: user-interface, demonstration, throw-away, and evolutionary.

user-interface prototype
A prototype that demonstrates an example interface for the information system.

In a **user-interface prototype**, the developer creates an example interface. An example interface, much like a test drive, enables prospective users to gain first-hand experience with the new systems interface. Adoption of the standard browser interface in Web-based electronic commerce systems, such as that being used at FarEast Foods, has greatly facilitated the creation of interface screens. Users readily understand how to use a browser, so the designers need to make sure only that their design fits within this standardized system.

demonstration prototype
A prototype that is used instead of a written proposal or a slide presentation to demonstrate what the final system will accomplish.

Demonstration prototyping represents a variation of user-interface prototyping. A company uses a **demonstration prototype** when it bids on an outsourcing or systems development project. Instead of creating a written proposal or a PowerPoint slide presentation, the company creates a user-interface prototype that demonstrates what the final system will accomplish. Intellimedia Commerce, Inc., used this approach to win the contract to create the Manheim Online system discussed in Chapters 4 and 9.

throw-away prototype
A prototype that developers use to carry out exploratory work on critical factors in the system. It is discarded after being developed.

The **throw-away prototype** tends to be more complex than a user-interface or demonstration prototype because developers use it to do exploratory work on critical factors in the system rather than simply for showing the interface design or demonstrating an idea to a client. This type of prototype is thrown away because it is not intended to be used in the final product, but rather to help determine information requirements.

Because these three types of prototypes will not be used as the final application, developers often create them using a special-purpose computer language that supports interface creation, but is inappropriate for creating a final working project. Occasionally, developers create user-interface prototypes in a language such as HTML, which they would also use for the final application. This tactic permits a developer to evolve a user-interface prototype into the actual application.

evolutionary prototype
A prototype that evolves into a working application.

The **evolutionary prototype** is meant to eventually become a working application. This prototype evolves over time as users have an opportunity to work with it and provide feedback to the developers. Evolutionary prototyping will be discussed in more detail in the next section.

Table 10.4 describes all four types of prototypes, along with their primary use and the development tools that help create them.

Table 10.4 Types of Prototypes

Type of Prototype	Primary Purpose	Development Tools
User-interface prototype	Used in the design stage of the development process to explore interfaces of final applications	Special-purpose computer language or final computer language Browser and HTML
Demonstration prototype	Used to show a potential client how the application will look and act	Special-purpose computer language
Throw-away prototype	Used in the analysis stage to determine specifications; also used to explore factors critical to the system's success	Special purpose computer language that results in much faster development than the final computer language
Evolutionary prototype	Development of actual system in an iterative fashion that can be modified to reflect users' feedback	Any of a number of computer languages

Source: Much of this table is based on Steve McConnell, *Rapid Development*, Redmond, WA: Microsoft Press, 1996.

Evolutionary Prototyping

The evolutionary prototype is worthy of discussion in more detail than the other types of prototypes because developers use it to build working information systems. Evolutionary prototyping has some key features that distinguish it from the structured approach to systems development:

- Development is an iterative process with short intervals between system versions and rapid feedback from users; the emphasis is on *speed*.
- Users are closely involved in the development process.
- The initial prototype has a low cost that does not require justification.
- With the iterative nature of prototyping and the feedback it provides, users develop information requirements while seeing what a new system can do.

Note that evolutionary prototyping emphasizes four things—speed, low cost, iterative development, and development of information requirements. Although speed and low cost are easy to understand, the other two features require more discussion.

iterative development
A development process in which the user tries out the latest version of the information system and provides feedback to the development team.

In **iterative development**, the user tries out the latest version of the information system and provides feedback to the development team. Based on this feedback, the team makes changes in the system and gets additional feedback from the user on the revised version. To keep this process moving as quickly as possible, the user must be willing to try out the system and provide rapid feedback. Compare this process to the waterfall method of development, in which backing up and changing the deliverable from a previous step can prove difficult. With the qualities of speed and low cost, and the ability to quickly respond to user feedback, it is easy to see why evolutionary prototyping has emerged as an important part of the Web development process.

In contrast to the waterfall approach, evolutionary prototyping does not assume that users can specify their information requirements up front. Instead, because they don't always know what they want until they have used a system for some time, users develop their information requirements as they work with the system and include them as a part of their feedback. This last feature makes evolutionary prototyping extremely useful in situations where the sponsor or users do not know precisely what they want, but just that they need *something* to remain competitive in the networked economy. Evolutionary prototyping is appropriate for creating many types of electronic commerce Web sites.

If you compare these features of evolutionary prototyping to structured systems development, you can readily identify several differences. First, whereas user involvement is mandatory in evolutionary prototyping, the user may not participate at all in the structured approach, other than to provide information in the planning and analysis stages and to sign off on the results of the various stages. Second, in evolutionary prototyping, users need not clearly define their information needs at the beginning of the process. By comparison, in the structured approach, users must be able to clearly define these needs at the beginning of the process. Finally, evolutionary prototyping emphasizes speed and low cost, whereas the structured approach focuses on completeness and ensuring that developers have considered every detail. Because evolutionary prototyping does not take a detailed approach, it may not be appropriate for creating complicated transaction processing systems. When creating complicated transaction processing systems, the lack of a planning, analysis, and design stage can result in developers missing or leaving out important elements of the system.

Figure 10.15 shows a diagram of the steps in the evolutionary prototype process. The step number appears in the upper-right corner of each box in the diagram. Note that when developers complete the prototype, an organization can actually use it as a production information system or as a design model for the implementation stage of the structured approach to development. When an organization implements the evolutionary prototype as a fully functioning new system, as

Figure 10.15 Steps in the evolutionary prototyping process

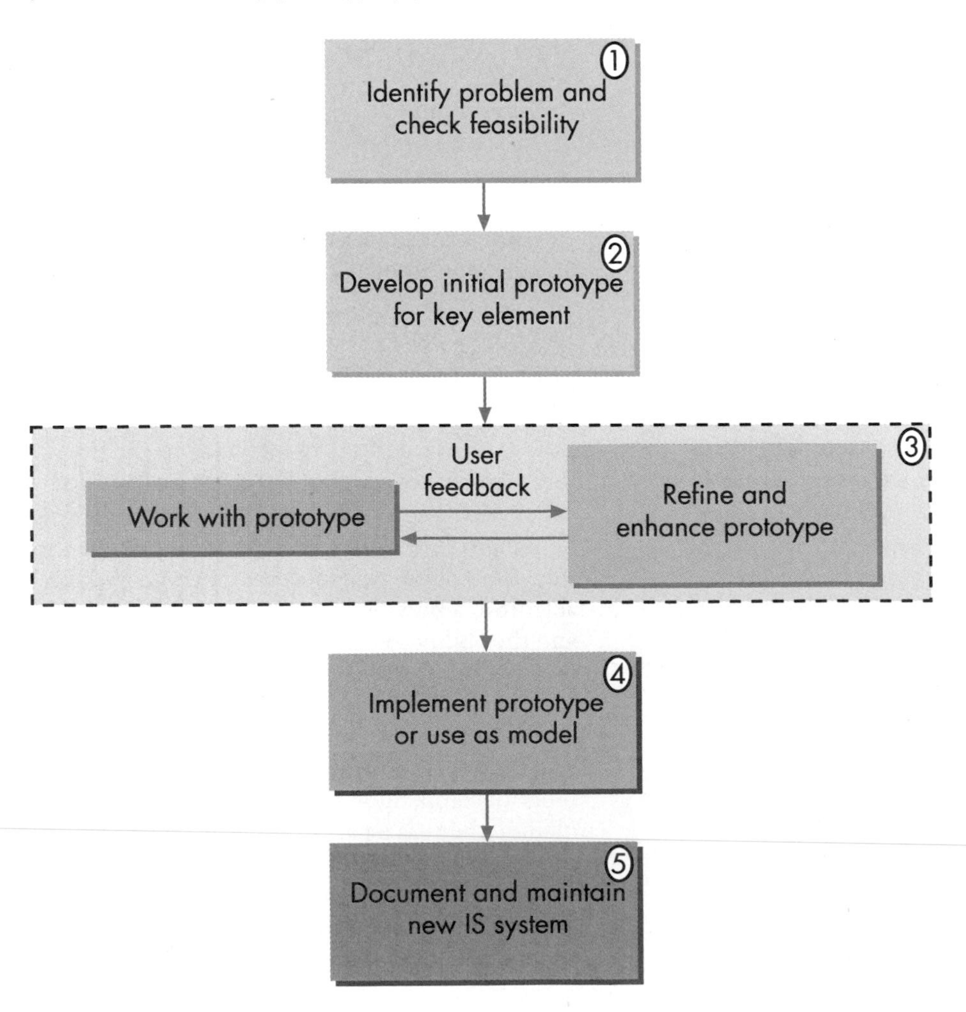

Table 10.5 Advantages and Disadvantages of Evolutionary Prototyping

Advantages	Disadvantages
> Users understand and react to prototypes far better than paper specifications. > It is usually quicker to build a prototype than create paper specifications. > Reality testing is introduced into the project at an early stage. > This approach can help avoid building systems with inadequate or wrong features. > It encourages creative input from users. > It enables errors and weaknesses to be caught before expensive design and programming are done.	> A quick, rough design may replace a well-thought-out design. > This approach may encourage users to continually change their minds about requirements, resulting in feature creep. > Users' expectations may be too high based on early prototypes. > Users may not want to go from the prototype to the production system; they may want to keep the prototype. > Users may not understand why the final cost for the full system is so high. > Users may not work hard enough to identify flaws in the prototype.

with any information system, it must maintain the evolutionary prototype by updating its features and correcting problems.

As with any approach to systems development, using evolutionary prototyping has both advantages and disadvantages, as noted in Table 10.5. In general, the advantages relate to the increased speed, lower cost of development, and greater user involvement. The disadvantages of evolutionary prototyping derive from the lack of clear planning, analysis, and design stages. They also result from users not understanding the purpose of evolutionary prototyping or being unwilling to do their work in testing the prototype and providing feedback to the development team. Another disadvantage is that users may want to add more features to the system as they use it, a process called **feature creep**. Both users and developers need to guard against feature creep.

feature creep
A process in which users continually want to add more features to the system as they use it.

1. What are the four types of prototypes commonly used in systems development?
2. List three ways that evolutionary prototyping differs from the structured approach to systems development.

Using ASP

Although taking a risk on new technology can sometimes initially cause more problems than it solves, at other times it can bring unexpected benefits. For example, San Francisco-based Putnam Lovell Securities, Inc., moved to an ASP to consolidate its customer information from nine separate databases. Now, Putnam Lovell's investment bankers go into a client meeting armed with a comprehensive view of the customer's situation. Prior to the installation of the ASP system, the different databases made it nearly impossible for one person to see the latest information in every system. An unexpected benefit of the ASP system was the $800,000 saved in paper costs in its first year of use. This saving came from the ability to rely on a single data source to contact customers electronically rather than having to send physical mail on a regular basis.

Canada's top computer-aided design integrator, CAD Resource Centre, located in Scarborough,

Ontario, determined that it needed to add an extra feature to its ASP-based online sales force automation software to generate proposals. The company had been using the ASP system for 18 months, but decided to add its own proposal system so it could create sales proposals within the online CRM and sales force automation systems. The resulting combination saved salespeople time in the process of creating a proposal and reduced errors because the data were take directly from the ASP software.

Because ASPs can have a low cost per user (as low as $50 per user per month), business units of some large companies can bring critical information technology resources into their organizations without first getting the approval of IT managers. For example, Thomas Cook Currency Services, Inc., in Toronto, Ontario, needed to automate its sales process, but did not have the resources to develop and support a major application. It turned to an ASP, skipping any review by IT management. Only when Thomas Cook planned a global installation of a similar application did the Currency division came clean about its use of ASP. The decentralized nature of Thomas Cook allowed the Currency group to keep its ASP at a much lower cost than the application planned by the corporate IT group.

Think About It

1. Compare the three uses of ASP discussed here in terms of the benefits to the companies. What are the similarities and differences?
2. From a corporate planning point of view, is the use of ASP by individual units always a good idea? Why or why not?
3. In each case, why would a Web-based ASP application be better or worse than a proprietary application?

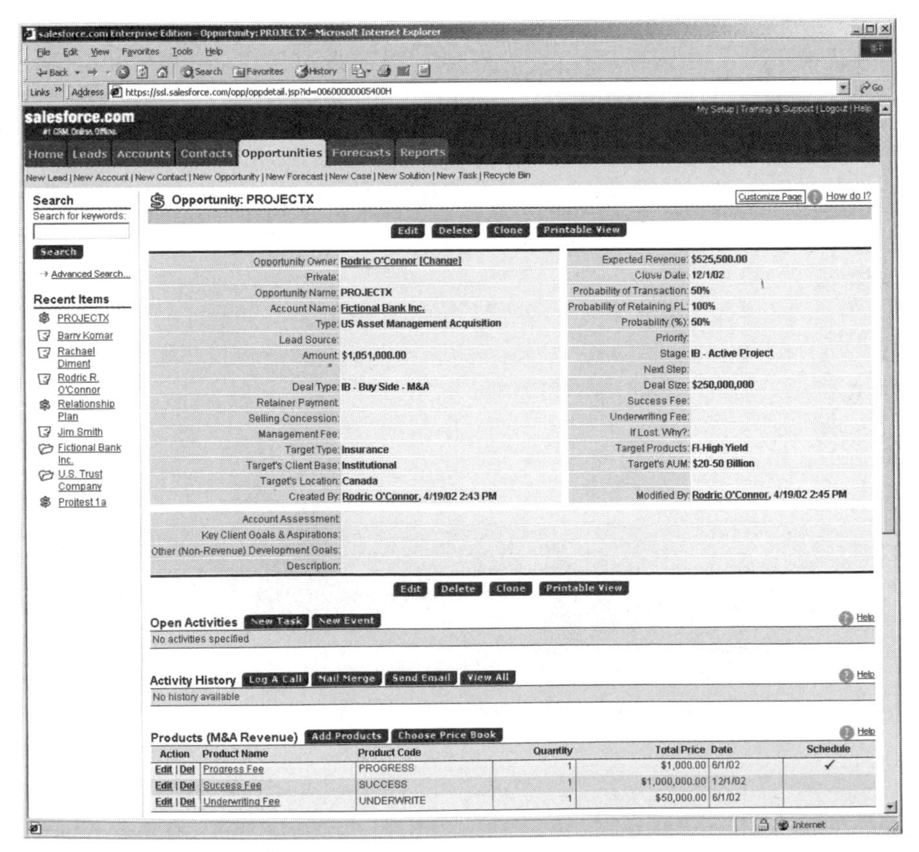

Putnam-Lovell has found that outsourcing to an ASP is a very effective way to manage its investment banking opportunities, allowing it to better meet its customers' needs.

Source: Mark Hall, "Unexpected benefits." *Computerworld*, August 20, 2001, http://www.computerworld.com/itresources/rcstory/0,4167,STO63113_KEY423,00.html.

SUMMARY

To summarize this chapter, let's answer the questions posed at the beginning of the chapter.

What is the purpose of the design stage of the structured systems development process? The design stage is dedicated to the *how* questions: *How* will the new system be developed, and *how* will the new system work? A firm must decide how it will develop the system first. If a firm chooses to develop a custom system, it can carry out development internally or outsource development to another firm. When a firm outsources a project, it uses an entity outside the organization to develop all or part of the project. If a firm decides to acquire an information system the system, it must find and purchase an appropriate system. If it decides to outsource or acquire the system, then it skips the remainder of the design stage and begins the implementation stage. If it decides to develop internally, the design stage must answer the question of how the system will operate. A development team makes this determination by moving from the logical models of the analysis stage to a physical design that specifies *all* of the details of the system.

The development team must convert the logical data models into physical database specifications and the process models into physical models using tools such as IPO tables and pseudocode. The development team also uses interface screens to specify the look and feel of the screens with which users will work.

What operations must occur during the implementation stage, and how is computer programming involved? The implementation stage involves four steps: building the system, installing the system, training users, and maintaining the system. In the first step, the development team builds the system by writing appropriate computer programs that implement the design created during the design stage. The next step is to install the system on the organization's computers, test it in that location, and convert from the old system to the new one. The third step involves training the staff who will use the new system. Training not only prepares employees to use the new system, but can also help avoid "people problems" that occur whenever an organization makes any kind of change. Finally, once the development team has installed the system and it is running in a production mode, the organization must support and maintain it.

Computer programming lies at the heart of the process of building the new system. Developers can use a variety of computer languages for this process, with object-oriented languages becoming more popular. Regardless of the language, developers must pay a great deal of attention to testing the programs to find errors, or bugs. Programming depends heavily on the design work that goes before it.

What is outsourcing, and what issues should be considered when deciding whether to outsource a project? Outsourcing is an arrangement in which one company provides services for another company that could also be, or usually have been, provided in-house. Advantages of outsourcing include access to more skilled and experienced programmers, avoidance of diverting internal staff from current work, and possibly lower costs. Disadvantages of outsourcing include losing control of the project, failure of internal developers to learn skills necessary to maintain the system, and failure of the outsourcer to deliver on its claims. An important aspect of outsourcing is the need to select the best vendor and write the best type of contract. Important vendor selection considerations include the outsourcer's record on meeting its commitments with similar projects, the satisfaction level of current and prior customers, and the technical competence of its employees as evidenced by prior work. Firms commmonly use three types of contracts when engaging a outsourcer: time and materials, fixed price, and value added. A relatively new type of outsourcing involves the use of an application service provider (ASP) to provide applications over the Internet from an application server. ASPs make it easier and cheaper to distribute applications across an entire organization.

What key steps are involved when an organization chooses to acquire an information system? During the acquisition process, it is important to always specify, select, and acquire the software *before* acquiring the hardware that matches the software; otherwise, the software may not run on the hardware. The acquisition process starts with the design stage, in which the development team creates the specifications for the software. Once the software specifications are complete, the company sends out a request for proposals (RFP) to vendors that lists the specifications for the software. The organization must evaluate the resulting proposals against the software design using a weighted set of criteria, benchmarking, or a combination of the two. Only after the organization has selected the software can it acquire the hardware via a similar process.

An important type of packaged software in use today is enterprise resource planning (ERP) software. An ERP system is a large, integrated system for handling business processes and data storage for a significant collection of business units and business functions. ERP systems permit some tailoring of the business processes to be used in any given organization, but only within fixed bounds, because all business processes are designed to work together using a single database.

What is RAD, and how does it differ from the waterfall approach to systems development? RAD is an acronym for rapid application development, and it incorporates a variety of approaches that organizations use to speed up the traditional approaches to systems development. Both outsourcing and acquisition are RAD approaches. For internal development, the most popular approach to RAD is prototyping, which involves creating a quick-and-dirty version of the final product, using either advanced languages or software tools. Prototyping can help the developer short-circuit the often lengthy structured process by replacing some of the planning, analysis, and design steps with prototypes of the final system. Types of prototypes include user-interface, demonstration, throw-away, and evolutionary. Development teams often use evolutionary prototyping to replace or supplement the waterfall method. It is a lower-cost, faster, iterative approach that does not mandate complete knowledge of information requirements; it also relies on heavy user participation in the process. Dangers of using evolutionary prototyping include the possibility of a quick, rough, design replacing a well-thought-out design, users never converging on a final set of requirements for the prototype, users having overly high expectations, and users not providing adequate feedback to the development team.

REVIEW QUESTIONS

1. What is the purpose of the design stage?
2. What three alternatives must be considered in the design stage?
3. What are the three key aspects of the internal development process in the design stage?
4. How are process models converted to a physical form?
5. What is another name for pseudocode? Why is it so named?
6. Why must you specify metadata for a data model?
7. What is CASE, and how do development teams use it in the development process?
8. Why are user interface screens a necessary part of the design stage?
9. What are the four steps in the implementation stage?
10. What are the steps in the programming process?
11. Why is the creation of documentation a key part of the programming development process?
12. During which steps of the implementation process does testing occur? Why does it take place at more than one step?
13. List the four types of conversion that development teams commonly use in installing a new information system. Which involve the highest and lowest risks?
14. What are the steps in the maintenance process? What factors affect the ease of maintenance?
15. When would outsourcing be a better approach to system development than internal development? When would outsourcing be a less desirable choice?
16. Why is ASP becoming a popular way to distribute applications across an organization?
17. Why should an organization select and acquire software *before* acquiring hardware?
18. Why is ERP becoming an important type of packaged software?
19. List the four types of prototypes used in RAD. Which is also used in the design stage?
20. What is feature creep, and how does it affect program development?

DISCUSSION QUESTIONS

1. Discuss the difference between the analysis and design stages.
2. Assume that your company needs software to start a business on the Web, and you have found several companies that offer turnkey shopping cart software for this purpose. Under what circumstances would you choose to purchase this software as opposed to developing it in-house?
3. Assume in Question 2 that your company decided to develop the shopping cart software in-house. Discuss the conditions under which your company would employ evolutionary prototyping rather than the waterfall approach to development.
4. Assume that a data warehouse would be useful for your business. Discuss the conditions under which you would turn to an ASP to provide this service. What type of ASP would you use?

5. Discuss how the development of the customer rewards system for FarEast Foods would have been different had the company used RAD with prototyping rather than the structured systems development approach.

RESEARCH QUESTIONS

1. Take the tour offered at the ML Direct Web site (www.mldirect.com). Based on the tour, write a description of the services that Merrill Lynch offers through this Web site.
2. Visit Ed Yourdon's Web site at www.yourdon.com. Write a short review of the first chapter of his new book, *Prepping*, which you can download from the site.
3. Visit two kids-oriented Web sites and look for features that would make these sites either attractive or unattractive to children. Discuss your findings in a short paper.
4. Visit the Web site for InCert, and research the capabilities of its Halo product. Use PowerPoint to present your findings.
5. Go to the FarEast Foods Web site at www.fareastfoods.com, click the About link, and click the link to the enhanced FarEast Foods site. Register with the Web site and then purchase a number of items worth at least $200 (you won't be asked to pay for them). Next, close and restart your browser, and then go to the enhanced site and sign in to redeem FarEast Bucks. You should now have at least ten FarEast Bucks to use in reducing the cost of future purchases. Try redeeming some FarEast Bucks, and then write a two-page paper comparing the final product you have just used to the system proposal (the deliverable of the analysis stage) and the system specification (the deliverable for the design stage).

Alex glided to a stop just short of the spot where Claire had settled for a rest on a felled tree. The two were breaking trail with their cross-country skis in the wooded hills behind the shop.

"Hey, slowpoke. What took you so long?" The chilly air imparted a visible quality to Claire's greeting.

Alex took off one of his gloves and grabbed the thermos that she offered. After a couple of deep breaths and a swig of the warm liquid, he replied, "Too many nights in front of the computer, I'm afraid. I'm a little out of shape and my herringbone technique is rusty. I nearly lost it on that hill."

"Speaking of computers," Claire said, "since you weren't around to talk to, I've been giving some thought to the customer incentive project for the Web site." Based on their initial feasibility studies and cost/benefit analysis, the Campagnes have decided to go forward with adding a customer incentive program to their Web site.

As he settled down beside her, Alex said, "You know what they say about all work and no play, don't you? So what are your thoughts?" The Campagnes' lives had become very busy with the success of their business, and they took the time to discuss important issues like this one whenever they could find it.

First, Claire explained the objectives of the project. The primary goal is to improve their market share and customer loyalty by providing an incentive system. The incentive system will keep track of the dollars that customers spend on the Web site and will then offer "freebies" to customers whose total purchases exceed a specified amount. The new system should be an enhancement to, and a fully functional part of, the current Web site. Finally, they want to have the new system up and running in three months from the current date.

Whether they develop the system themselves or decide to outsource it, the project team will need to fulfill several requirements. The team will need to plan the Web pages for content, functionality, and compatibility. This plan should be accompanied by a written specification. The team will be responsible for the look and feel of the pages as well as the client-side and server-side functionality. The new pages will be hosted on the current facility, necessitating an analysis of the capacity of the current facility. If changes to the current hosting facility become necessary, the team must provide a solution for them.

The Campagnes envision that customers who wish to take part in the incentive program will first register with the site. Upon registration, the system will make entries into the database designed to keep up with the customer's spending. When the customer's spending level exceeds the target level, he or she will be notified via e-mail and will be allowed to choose from the current free items available. The registration process should capture as much demographic data as possible. The team's proposal should focus on how this information will be captured and the specific modifications to the database required. Claire would also like for the team to explore the possibility of using a cookie system. A cookie could be used in a couple of ways. First, it could help registered users with automatic logon to the site and tracking of their spending levels. Second, it could allow WildOutfitters to analyze where site visitors go and how they interact with the site. The team's proposal should explain the solution for cookies and the features offered by the system.

The team will need to develop a detailed project report that describes the hosting requirements, maintenance, training issues, and all technical considerations. The report should include specifications for all resources needed for the system, including software, hardware, and personnel. Software specifications should discuss the programming languages used as well as any third-party software that will be used. Hardware specifications should describe the additional hardware needed, if any, to incorporate the new system into the Web site. Personnel specifications should define the number of people needed to develop and maintain the system and the appropriate skills that these people should possess.

With these thoughts in mind, the Campagnes now need to incorporate them into a formal document describing the project's mission. Only then will they decide how to complete the project.

"Well, enough mixing business with pleasure," Claire said vigorously. "Race you back to the house." With that she arose, and with a quick kick started glided down to the base of the next hill.

Alex groaned and forced himself up over his skis. He was thinking about the four miles, mostly uphill, back to the shop as he began to follow his wife.

Think About It

1. In which of the four stages of the SDLC is the customer incentive project at WildOutfitters.com currently? What deliverables should be developed?
2. Would outsourcing be a good choice for the development of the customer incentive system? Explain. What advantages and disadvantages will the Campagnes experience if they decide to develop the project through outsourcing?

Hands On

3. Assume that the Campagnes decide to pursue outsourcing for the project. Use word processing software to prepare a request for proposals (RFP). Recall that the RFP should list the specifications for the system as completely as possible. Feel free to add your own ideas to supplement the information provided in the case. A template with instructions for the RFP may be downloaded from the textbook Web site at www.course.com.

Hands On

4. Take the role of a company that provides outsourcing services. Prepare a presentation using a variety of software tools (word processor, spreadsheet, presentation software, and so on) that your company may use to convince the Campagnes to hire you for the job. Be sure to address the major points provided in your RFP.

ISSUES IN THE NETWORKED ECONOMY

The growth of the networked economy means that information technology is a part of everyone's life, in one way or another. Crime, security, privacy, ethics, economics, health, and international issues are just as much a concern in the networked economy as they were in the industrial economy, if not more so. Criminals are finding new and different ways to use information technology to carry out their illegal activities. For this reason, security has become an even more critical concern for organizations and individuals who buy or sell goods and services over the Internet or who use information technology to carry out some element of their daily life. In our online society, privacy has quickly emerged as one of the hottest issues for many of us. Similarly, because of our greater dependency on information technology, ethical issues related to its use have drawn our attention. Because of our widespread use of information technology, you should understand and be able to deal with economic, health, and international issues associated with IT.

Part 5 covers crime and security, privacy and ethics, and societal issues related to the networked economy.

CHAPTER 11

CRIME AND SECURITY IN THE NETWORKED ECONOMY

LEARNING OBJECTIVES

After reading this chapter, you will be able to answer the following questions:

- How has the face of crime changed in the networked economy?
- What types of crime are having the greatest effects on individuals and organizations in the networked economy?
- What kinds of attacks are being made on information technology and users?
- How can organizations protect themselves from crime in the networked economy?
- How can an organization protect its enterprise network from intrusions over the Internet?
- What nontechnical issues should you consider in dealing with IT security?

Combatting Fraud at PayPal

In the short time since PayPal was started in 1998, it has become the leading electronic payment network for online auction web sites, including eBay. But problems have come with its success, with the biggest issue being fraudulent use of credit cards. One scam was fairly simple: Criminals stole credit cards and, instead of using them to purchase and sell expensive items such as computers, sent themselves or others cash payments charged to the card. When the credit card company disallowed these charges, PayPal (and other similar electronic payment sites) were left holding the bag because the money had already been collected by the thieves. More than 1 percent of PayPal's credit card charges were being rejected (as compared to 0.07 percent of all transactions), which caused severe problems for the firm. Although some payment sites such as PayMe or PayPlace were acquired or went out of business altogether, because of this theft, PayPal decided to fight the fraud.

To fight credit card fraud, PayPal created a program named Igor (after a Russian mobster who had given it problems in 2000). Igor sifts through the PayPal database looking for patterns of activity that indicate fraudulent transactions. For example, payments that are consistently close to the maximum payment amount or payments to an account where the ZIP code does not match the customer's area code are flagged for investigation by PayPal's anti-fraud team. In one case, Igor flagged a number of $350 payments. The team tracked these payments to a person selling but not delivering 3,000 Sony PlayStation 2s in the middle of the 2000 Christmas season when these items were in high demand. PayPal put a hold on the account and turned the case over to local and federal authorities. In another case, PayPal's testimony and computer logs were instrumental in convicting at least one Russian criminal who had defrauded the company out of $100,000 using stolen credit cards.

PayPal CEO Peter Theil and board member Elon Musk display credit cards, which are often used to transfer funds via PayPal, but which also cause fraud problems for the company.

Source: Brad Stone, "Busting the Web bandits." *Newsweek*, July 16, 2001, p. 55.

The Changing Face of IT Crime

In most cases, information technology is used for completely legal purposes. Unfortunately, as illustrated in the opening case, individuals and organizations sometimes find ways to use information technology illegally for their own profit. You can expect the volume of crime associated with information technology to grow in tandem with the Internet and electronic commerce. The movement into the networked economy means that criminals will adapt their illegal tactics to this new environment. For example, instead of using the telephone to peddle worthless stocks, real estate, or cancer remedies, they may use the Web to do so. The fraud discussed in the opening case offers another example: Instead of using stolen credit cards to buy and sell physical goods, criminals used IT to steal money directly.

In this chapter, you will learn about crime and security as they relate to information technology (IT). An **IT crime** is an illegal act that requires the use of information technology, and **IT security** comprises the methods used to protect the hardware, software, data, and users from both natural and criminal damage. The need for security has assumed a much higher degree of immediacy since the terrorist attacks of September 11, 2001. In addition to producing tremendous loss of life, these events destroyed a massive amount of data and equipment, highlighting the need for IT security in addition to physical security.

IT crime
An illegal act that requires the use of information technology.

IT security
The methods used to protect hardware, software, data, and users from both natural and criminal damage.

The true extent of IT crime in the United States remains unknown because many companies hesitate to prosecute the people involved for fear that the general public will lose confidence in the companies. Companies also worry that discussing details of IT crime in open court may encourage others to try to replicate the crimes. Although we may not know the exact cost of IT crime, it undoubtedly involves a large and growing number of companies. For example, a joint 2001 Computer Security Institute/FBI survey revealed that 85 percent of respondents said they had suffered a computer security breach in the last 12 months. The same survey found that organizations that had experienced computer fraud reported losses averaging more than $2 million.[1] Compare this amount to the average burglary, which nets $3,500, and the average armed robbery, which claims approximately $8000. With such great potential losses, IT crime has become a critical consideration for organizations in the networked economy.

To understand the changing nature of crime, consider Table 11.1, which compares the industrial economy, represented by the year 1952, with the networked economy, represented by the year 2002. This comparison focuses on four aspects of crime: location, monetary impact, format, and risk. Location refers to whether the crime occurs locally or at some remote distance. Monetary impact refers to the amount of money (low or high) involved in the crime. Format refers to the manner (physical or electronic) in which the crime is carried out. Risk refers to the risk (high or low) facing the criminal.

Table 11.1 Changing Nature of Crime

Aspect of Crime	Industrial Economy (1952)	Networked Economy (2002)
Location	Local	Remote
Monetary impact	Low	High
Format	Physical	Electronic
Risk	High	Low

1. http://www.csi.com.

In looking at Table 11.1, you can see that because crime in the 1950s was usually face-to-face, it was local with low monetary impact—it was physically difficult to steal large amounts of money. It was also physical, and the risk to the criminal was high. In contrast, computer crime in the 2000s is remote with a potentially high impact in terms of amount of money stolen. The format is electronic, posing a low risk to the criminal. This comparison vividly shows why IT crime in the networked economy is a growth industry.

IT Criminals

Kevin Mitnick, one of the most widely publicized hackers, was sentenced to a 5-year prison term for his computer crimes and is on parole until 2003.

hackers
Individuals who seek unauthorized and illegal access to computers and computer networks for a variety of reasons, including "for the fun of it."

cyberterrorists
Organized groups of individuals who attempt to damage the IT infrastructure of a country or culture.

IT criminals may be classified in four major categories: employees, outside parties, members of organized crime, and cyberterrorists. These categories are shown in Table 11.2, along with their objectives.

Although crimes committed by outside parties often receive the most publicity, most surveys show that employees and ex-employees form the largest group of computer criminals, usually because they have the easiest access to a company's computers. The criminal may be a disgruntled or laid-off employee wishing to get revenge against the company either by attacking its information technology or by using that system to steal from the company. At other times, the employee finds a way to perform an illegal act with the organization's computers or other IT infrastructure, such as downloading child pornography or sending hate-related e-mail.

People outside the organization have used the Internet and other networks to break into computers for various reasons. The best known are those individuals who seek unauthorized access for a variety of reasons, including "for the fun of it." These individuals are often referred to as **hackers** because they hack away at code trying to gain illegal entry to computers. Numerous movies have glamorized this group, but breaking into somebody else's computer constitutes a criminal act regardless of how cool it seems to be. In addition, the process of hacking into a computer always has the potential to damage programs and files on the hacked computer.

Organized crime has discovered that IT can prove extremely useful in furthering its objectives. For example, Asian gangs in California have become heavily involved in counterfeiting Microsoft products—not just the floppy disks or CD-ROMs, but also the boxes, warranty cards, end-user licenses, and even the hologram that Microsoft uses to authenticate its products. Police raids on these operations have netted as much as $100 million of counterfeit computer software.

Cyberterrorists are organized groups of individuals who attempt to damage the IT infrastructure of a country or culture. This group has received a great deal of attention since the World Trade Center and Pentagon terrorist attacks. Because cyberterrorists are usually political in nature, Chapter 13 will discuss them in more detail.

Table 11.2 Types of IT criminals

IT Criminal Type	Objective
Employees	Steal money from the company or harm it in some way
Outside parties	Steal from the company or damage its infrastructure in some way
Organized crime	Use the company's information technology for monetary gain
Cyberterrorist	Damage the IT infrastructure of a country or culture

Although it may look real, this photograph shows an example of a counterfeit software certificate of authenticity.

IT Crime and FarEast Foods

Small companies like FarEast Foods need to be most concerned with the threat of IT crime from outside people and from their employees or ex-employees. First, FarEast Foods needs to work to keep intruders out of its network. Second, it must closely check the background of any employee who has access to the company servers and other IT equipment.

1. Why does the true extent of crime remain unknown? How does IT crime today differ from crime in the 1950s?
2. What are the four types of computer criminals?

Types of IT Crime

An IT crime can be committed in a variety of ways. In general, however, it involves theft, fraud, copyright infringement, or attacks on information technology or users' computers. Theft includes stealing computers, peripheral devices, company data, or private corporate information either directly or through networks. In fraud cases, individuals use information technology to carry out fraudulent activities involving credit cards like those discussed in the opening case. Other fraudulent activities include identity theft, investor fraud, travel fraud, Web auction fraud, and pyramid schemes. Copyright infringement, involving the illegal reproduction of software, music, or videos, represents a major problem for companies that develop and sell these products. Finally, attacks on information technology or users' computers are accomplished by the use of damaging software.

All of these types of IT crime involve computer networks and the Internet—not surprising in the networked economy. Table 11.3 summarizes the four types of computer crime and the purpose of each; each type is detailed in the sections that follow.

Table 11.3 Types of Computer Crime

Type of IT Crime	Purpose
Theft	Steal hardware, data, or information from individuals or organizations either directly or through networks
Fraud	Use computers and the Internet to steal money or other valuable items by deceiving victims
Copyright infringement	Use software, music, or trademarks, which in many cases are obtained over the Internet, illegally
Attacks	Damage hardware, data, or information

Theft of Hardware, Data, or Information

When most computers were mainframes, the theft of hardware constituted a relatively small problem involving a few terminals or other peripherals, because a high level of security surrounded a mainframe and the computer center. However, the trend toward almost universal use of PCs as network clients has been accompanied by the widespread theft of hardware items, including entire PC systems, laptops, and printers, or computer elements such as keyboards and monitors. According to a survey published by Safeware Insurance, Inc., almost 591,000 laptops were stolen in 2001, accounting for 96 percent of all computer hardware losses.[2] The FBI has estimated that 57 percent of all computer crime is associated with the theft of laptops. In 2001, the FBI itself admitted to Congress that the agency had lost 180 laptops, with at least one containing classified data. As a result of the dramatic increase in computer theft, especially of laptops, the National Computer Registry has been established to help recover stolen equipment.

Although it remains expensive to replace stolen hardware, the real cost of hardware theft comes from the loss of the software, data, and information stored on the computers. For example, Ontrack Data International has found that most companies value 100 MB of data at more than $1 million.[3] In many cases of laptop theft, the equipment is stolen with the intent of acquiring competitive information stored on the hard disk or to gain access to corporate servers and intranets through the passwords and remote access software stored on the computer.

Criminals also can steal data and information over a network. They can make their computers appear to be trusted computers on the Internet or other network, thereby tricking organizational computers into sending data and information to them. Another approach sends viruses to networked computers. The viruses create a secret entrance into the computer, called a **trapdoor**, through which criminals can access the computer or network even after the original virus has been removed.

trapdoor
A secret entrance into a computer through which criminals can access the computer or network.

Fraudulent Use of IT

Using IT and the Internet to deceive people is one of the fastest-growing types of IT crimes. The Internet Fraud Watch (run by the National Consumers League) reported that the average amount lost to Internet fraud rose by more than 50 percent from 2000 to 2001. Figure 11.1 shows the average size of monetary loss reported to the Internet Fraud Watch from 1998 to 2001.

2. http://www.safeware.com/losscharts.htm.
3. http://www.ontrack.com/.

Figure 11.1 Average Internet fraud loss

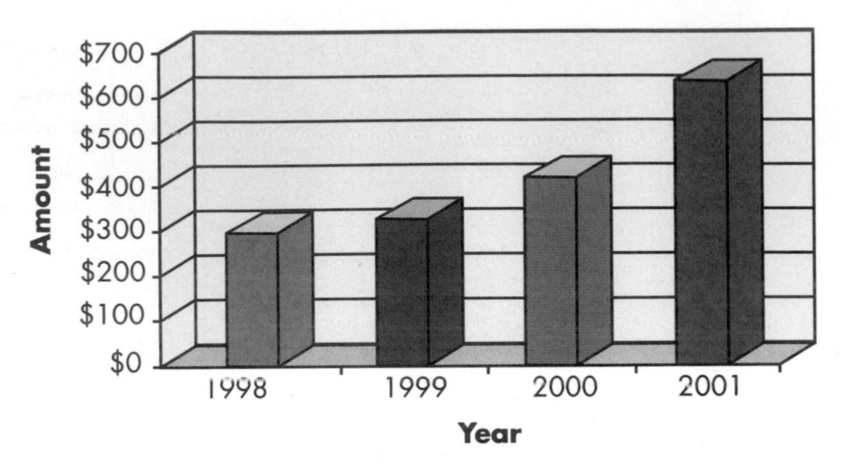

One reason for this increase is the low expense of running a fraud over the Internet compared to other fraudulent methods; because the Internet provides access to such a vast audience, criminals have to engage only a small fraction of that audience in their scheme to make the fraud profitable. The problem has become so widespread that these frauds have their own name: **dot cons**.

dot con
A fraud carried out over the Internet.

Many frauds run over the Internet constitute old frauds recast in a new format. Examples include Web sites that lure individuals into sending money for services that are normally free, for medical cures that don't work, or for nonexisting work-at-home opportunities. Other familiar frauds in a new guise are multilevel marketing (pyramid) schemes and the so-called Nigerian money offer scam. In the latter fraud, the potential victim receives an e-mail from Nigeria informing him or her that the sender needs to temporarily deposit several million dollars in a U.S. bank account. If the receiver will send his or her bank account number, the target is promised a sizable portion of this money. Of course, sending the bank account number merely allows the criminal to clean out the account. This last type of fraud is rising in number of hits simply because the criminals can reach a larger audience with e-mail than with postal mail.[4] The problems of all types of IT fraud are compounded by people who believe everything they read on the Internet, even when it is put there by people they do not know.

Other frauds use the Internet to scam users in ways that were difficult to implement before Internet use became so popular. These techniques include identity theft, travel fraud, telephone fraud, investor fraud, credit card fraud, illegal IDs, and online auction fraud. IT frauds can be categorized into those that primarily affect consumers and those that primarily affect companies. For example, travel fraud primarily affects consumers, whereas credit card fraud and illegal IDs have the greatest effect on companies. Some types of IT fraud, such as investor and online auction fraud, can affect both consumers and companies. Table 11.4 lists the various types of IT fraud, the users primarily affected by these activities, and a brief comment on how each type of fraud works. The types of fraud that target businesses are discussed in more detail here. Chapter 12 discusses identity theft in more detail, because it involves a loss of privacy as well as a potential loss of money.

4. Interview with Susan Grant, Internet Fraud Watch Director, December 11, 2001.

Table 11.4 Types of Internet Fraud

Type of IT Fraud	Primary Group Affected	Fraudulent Activities
Credit card fraud	Business	Web sites skim credit card numbers and use them to charge purchases or cash advances that are disallowed by the credit card company
Investor fraud	Consumer and business	Criminals use chat rooms to pump up the price of an almost worthless stock, which they then dump at a high price
Illegal IDs	Business	Underage individuals use IT to create IDs, which they use to purchase tobacco and alcohol
Online auction fraud	Consumer and business	The seller does not provide the goods after receiving payment or the buyer does not make payment after receiving the goods
Identity theft	Consumers	Information about an individual is stolen and used to set up a new identity for the criminal or other person
Travel fraud	Consumers	Web sites promise great ticket prices or trips and then do not provide them to the buyer
Telephone fraud	Consumers	Users are lured into unknowingly downloading software that calls an international number, resulting in huge telephone charges

Credit Card Fraud

The use of credit cards in electronic commerce has brought problems for both consumers and merchants. For example, consumers might be asked to give a credit card number as proof of age to visit adult-oriented Web sites, but the promoters behind the Web site could use those numbers to run up large charges on their cards. Fortunately, cardholders can dispute these kinds of false charges, and U.S. law limits a cardholder's liability to $50 if a card is misused.

For merchants accepting credit card payments over the Internet, the damages can be much greater. Because Internet credit card transactions lack a signature or magnetic stripe on the back of the card, merchants must agree to pay the full cost plus any penalty fees for sales involving invalid credit cards. Thus criminals with bogus or stolen cards can purchase items over the Internet, and the merchant is liable when the credit card company disallows the charges. Although the typical rate for credit card fraud is 0.7 percent of all transactions, new Internet merchants may experience a much higher fraud rate, quickly putting them out of business. International transactions are an even more attractive target for criminals. For example, Web transactions account for only 2 percent of Visa's international business, but almost 50 percent of disputed charges and fraud in that market. A Gartner study revealed that fraudulent credit card purchases cost Internet merchants more than $700 million in 2001—19 times higher than the amount at traditional stores. In one case, Flooz.com, a seller of online currency used as electronic gift certificates, filed for bankruptcy after it was hit by $300,000 in credit card fraud.[5]

A high percentage of Internet credit card fraud originates from free e-mail accounts, such as those offered by Yahoo! and Microsoft, which makes it very difficult to track down the person making the fraudulent purchase. Criminals use credit card generation programs available on the Web or set up Web sites that mimic those of legitimate merchants but with unbelievably low prices offered to customers whopay by credit card. These sites do not actually have any merchandise to sell; they

5. Amy Winn, "Cyber fraud rampart." *Atlanta Journal-Constitution*, March 5, 2002, p. D2.

Because the free Hotmail e-mail system requires no identification, criminals and others intent upon mischief can use it to carry out their objectives.

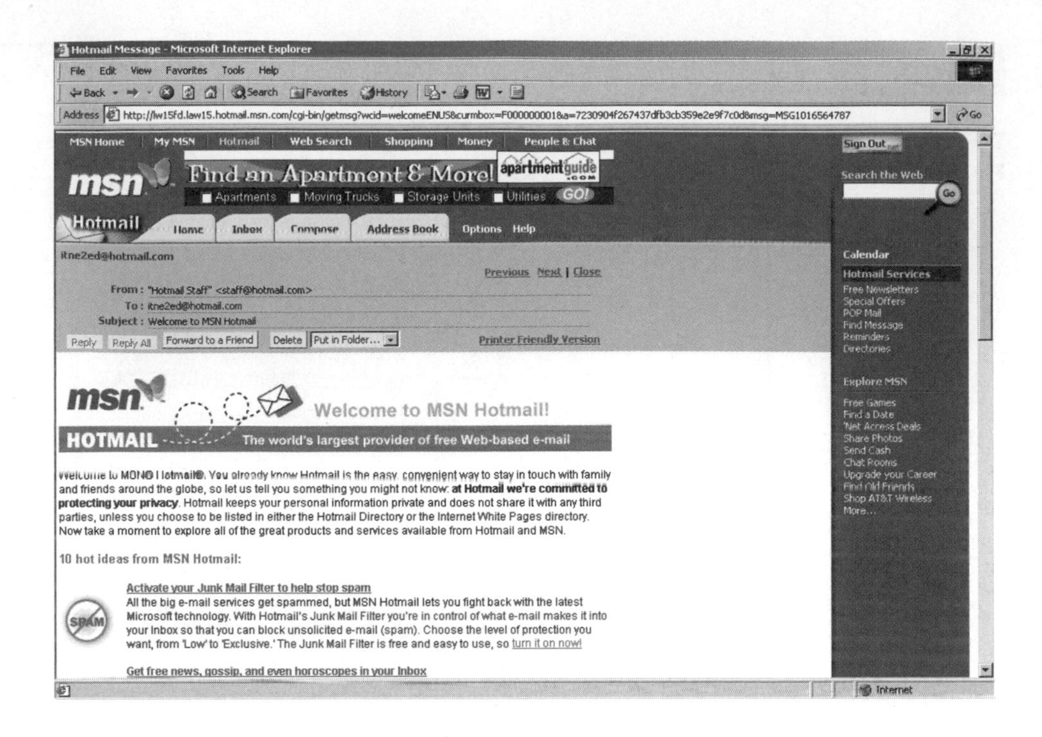

simply take orders that include valid credit card numbers. The thieves then use this information to purchase real merchandise or sell the credit card numbers on the Web to others with criminal intent.

secure electronic transaction (SET) A protocol that provides a way for buyers to transfer credit card information to the credit card issuer over the Internet without the seller seeing the credit card information.

A system called secure electronic transaction has been created to combat this type of credit card fraud. **Secure electronic transaction (SET)** provides a way for buyers to transfer credit card information to the credit card issuer without the seller seeing the credit card information, by requiring merchants to encrypt all credit card information. SET also defines minimum network and security requirements for companies that want to use the system. SET has been endorsed by Visa, MasterCard, American Express, and Japan's JCB Credit Card Corp. Figure 11.2 shows the process of using SET.

Visa has gone one step further by using smart cards along with a password to protect online businesses from Internet fraud. With this system, consumers obtain a smart Visa card and password from their credit card issuer and a smart card reader from a local consumer electronics store. Customers then use their password with the smart card to make a secure purchase. Figure 11.3 depicts this process.

Investor Fraud

The widespread use of the Internet for stock market trading has been accompanied by a dramatic growth in stock-related fraud that affects both consumers and businesses.

Figure 11.2 Use of SET to stop credit card fraud

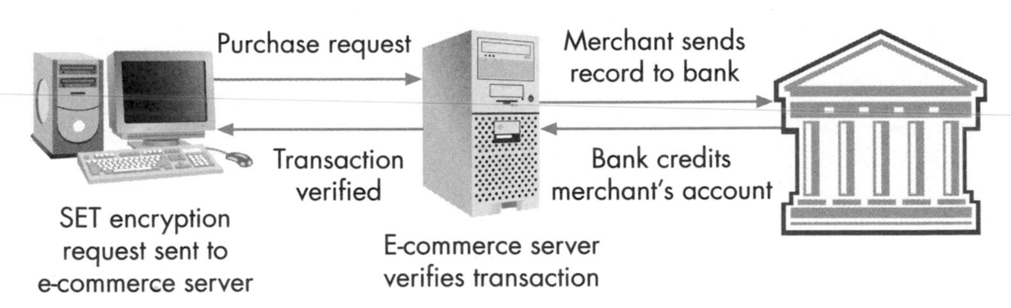

Figure 11.3 Use of smart cards to protect against Internet fraud

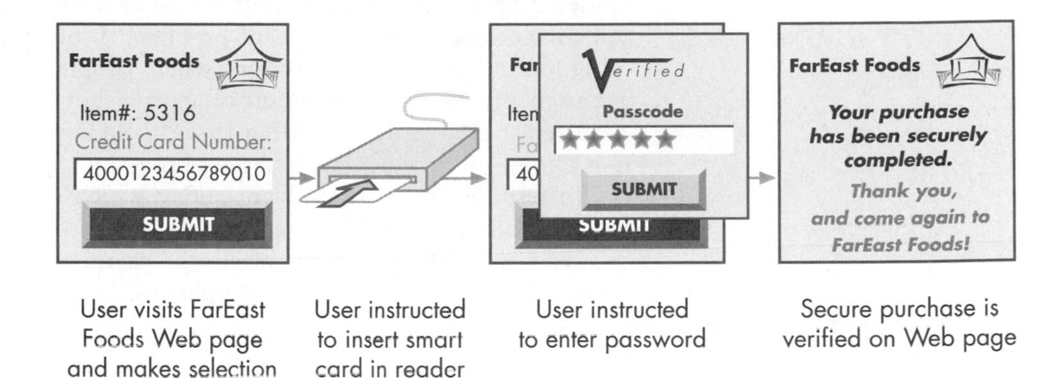

Investor fraud includes e-mail campaigns that improperly tout stocks, false information spread in chat rooms and newsgroups, and forgeries of legitimate brokerage Web sites. In the first two cases, promoters send a large number of junk e-mail messages to target groups or use chat rooms to sell stock in investment opportunities that do not exist or to pump up the value of a real stock so the promoters can then sell their shares at a profit. The latter approach is commonly known as "pump and dump." One of the worst cases of "pump and dump" involved a 15-year-old boy who made at least $285,000 using this approach.[6]

Sophisticated promoters can also create Web sites that appear to represent legitimate investment companies. They accept money for stock purchases but never buy the stock for investors. By the time investors figure out that they do not own any stock shares, the promoters have closed the Web site and opened another. This technique highlights an important point: Promoters can close down a Web site and move on much faster than they could in pre-Internet days. Before the advent of the Internet, they needed at least six months to restart their operation; today, it takes only a few days to be up and running again.

Illegal IDs

The development of scanners and high-quality color printers has made it possible to create illegal identification (ID). With the use of a template that is available from a number of Web sites, a scan of a photo, and a color printer, underage users can create copies of a real driver's license that makes the users appear to be older than they really are, complete with a hologram that will fool all but the best eye. Some Web sites even do all of the work for those with more money than time and expertise. Individuals then use these illegal IDs to purchase tobacco products (for those under age 18) or alcohol products (for those under age 21). This crime presents a problem for businesses, because they may lose their business or alcohol license if they serve underage consumers.

Driver's license age verification machines, such as the one shown here, enable merchants to check the validity of an ID.

Some states are creating databases against which driver's licenses can be verified, but this effort is far from universal and very few establishments that would need to check validity of the driver's licenses have the necessary reader to check such a database.

Auction Site Fraud

Since its launch in 1995, the eBay auction (and others like it) has drawn individuals and businesses to its Web site to buy and sell items. Unlike at a physical auction, the buyer and seller are not actually present at the auction, nor is the buyer's money or the item itself. This situation can quickly lead to problems when buyers fail to send the funds

6. Kevin Peraino, "A shark in kid's clothes." *Newsweek,* October 2, 2000, p. 50.

for a purchase or when sellers fail to send an item that has been purchased. In fact, this type of fraud causes by far the largest number of reports of Internet fraud to the National Consumers League, with 41 percent of all individuals using online auctions experiencing some type of problem. Another problem faced by eBay and other online auction sites occurs when people engage in the auction process although they have no intention of actually buying the items. This phenomenon occurred at eBay in 1999, when a 13-year-old boy bid more than $900,000 on various items—money he did not have to spend![7]

In response to these problems, eBay, which originally had a hands-off policy toward the bidding and payment process, has instituted new policies to discourage this type of fraud. In addition, electronic payment services like PayPal have sprung up to solve many of the payment problems.

7. "Online auctions: Fun $ profits, *Intrepid Solutions,* http://www.intrepid.net/newsletter/june1999/html/auctions.html.

Drive-by Hacking

TECHNOLOGY ON THE EDGE

When people think about hacking computer systems, they typically envision individuals trying to break into computer systems through the Internet. However, the growing popularity of wireless networks for organizations and homes has opened up an entirely new avenue for would-be hackers—breaking into wireless networks. With 6.2 million wireless systems sold in 2001, use of these networks is growing rapidly. They enable users to move about a building or campus without losing connection to their local area network. In homes, wireless networks permit users to connect computers without stringing wires through walls. Unfortunately, because the most popular protocol for such networks, WiFi 802.11b, uses the same frequencies as some cordless telephones, and because far too few wireless users install even standard security features, hackers can readily pick up wireless network transmissions with a laptop, an antenna, and a little computer coding.

To demonstrate how easily a person can hack into wireless networks, a computer scientist tapped into 1000 corporate networks in early 2001, including those of companies like Lucent and Cisco, which manufacture wireless networking technology. To do so, he simply set up his operation in the company's parking lot. Although he did nothing once he entered the corporate networks, the hacker could have done plenty, including obtaining bank account numbers and reading private e-mail. In another situation, a computer scientist reported driving the length of Market Street in San Francisco and using "drive-by hacking" to jump from one company's wireless network to another.

Even the Wired Equivalent Privacy (WEP) security standard for WiFi cannot stop hackers. Some of the same researchers who reported tapping into wireless networks were able to break this security code. The Wireless Ethernet Compatibility Alliance said the biggest problem is that users don't install security software correctly, but the group also admitted that problems exist and that it is working to develop a new security standard.

With just a laptop and an antenna, hackers can drive by businesses using wireless networks and capture data and information being sent over them.

Source: Bob Keefe, "Wireless boom in tech helps hackers, too." *Atlanta Journal-Constitution*, July 15, 2001, pp. F1-F4.

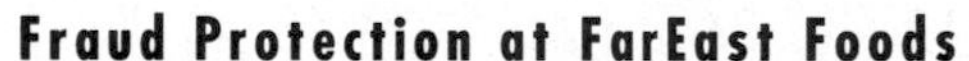

Fraud Protection at FarEast Foods

Because it handles a high volume of orders over the Internet that are paid for by credit card, a relatively small company like FarEast Foods could easily be put out of business if its credit card fraud rate exceeded 1 percent. To avoid this problem, the company has subscribed to the SET protocol to protect itself and its customers. It has also configured the Web site to accept passwords for those customers with smart cards. In addition, FarEast Foods does not allow its employees to save passwords on their laptops, which would allow entry into the corporate network if the laptops were stolen.

Quick Review

1. What motives drive the theft of laptop computers, besides their hardware value?
2. Why does credit card fraud affect merchants more than consumers?

Copyright Infringement

intellectual property
The ownership of any creation of the mind, including inventions, literary and artistic works, and symbols, names, images, and designs used in commerce.

Copyright infringement (using copyrighted material without paying the copyright holders a fee or having their permission to use the material) is a crime. Illegally copying or sharing software, commonly called software piracy, has become a very big business that costs companies such as Microsoft a great deal of money each year. Illegally downloading music and video files from the Internet also represents a copyright infringement. The problem of copyright infringement has spawned an entirely new branch of law, called intellectual property rights, to deal with it and related issues. **Intellectual property** refers to any creation of the mind, including inventions, literary and artistic works, and symbols, names, images, and designs used in commerce.[8]

Software Piracy

Before the introduction of PCs in the late 1970s, when mainframes and minicomputers served as the only sources of computing power, software piracy did not pose a significant problem because users could use only whatever software was available at their computer center or that they wrote themselves. In this environment, the need or the facility to copy commercial software from the computer seldom arose. Once PCs came into use, with software being distributed on floppy disks, some users began illegally copying or sharing software, rather than buying it.

Over time, software piracy has turned into a huge problem for software developers. A study by the Business Software Alliance (BSA) and the Software and Information Industry Association (SIIA), two software trade associations, estimated the worldwide cost of software piracy in 2000 at approximately $12 *billion*. The same study found that, worldwide, almost two out of every five new software packages installed today are illegal copies. In the United States alone, some 24 percent of all new software installations involve pirated software, resulting in an estimated 107,000 lost jobs, $5.3 billion in lost wages, and $1.8 billion in lost tax revenue in 1999.[9]

To counter this problem, the software industry has acted strongly against the worst domestic offenders and against retailers around the world. The industry has filed lawsuits or sent cease-and-desist letters to many offenders and distributed a software package that companies can use to audit their own computers (GASP, which can be downloaded from the BSA Web site at www.bsa.org/uk/freetools). Nevertheless, some countries are still referred to as "one-disk countries" because of the widespread software piracy practiced there. For example, the BSA-SIIA study found that more than nine out of ten business software applications are pirated in Vietnam, China, Indonesia, and Russia and other former Soviet countries.

8. World Intellectual Property Association, http://www.wipo.org/about-ip/en/.
9. http://www.bsa.org.

Software Piracy and the Law

site license
A license to use software that covers all of the computers in an organization.

To understand software copyright infringement, you must realize that when you purchase a commercial software package or a computer with software already installed, you do not become the owner of the copyright for that software. Instead, you simply buy the right to use that software under restrictions imposed by the software publisher or owner. Typically, you have the right to use it on one computer and to make one backup copy. Even when a single disk is used to install software on multiple machines, the organization must have a **site license** that covers all of the computers in it. Unfortunately, many people and some companies violate these restrictions by installing the software on multiple machines or making copies to give away or sell to others. Microsoft addressed the problem of a single copy of Windows being used on multiple computers in the design of Windows XP. Windows XP must be authenticated before each installation is complete, and it can be authenticated for licensed installations only. The default encompasses one license for one computer.

Although most software companies have not yet taken such a hard-line stand on "one license for one computer," the nature of the license remains the same. Thus, when a user installs a single-license software package on multiple computers, the user breaks a federal law. The possible penalties include payment of damages to the copyright holder—in this case, the software developer—as well as criminal prosecution. In cases of willfully copying and reselling of software, prison terms can be imposed. Enforcement of copyright law is usually reserved for the most blatant offenders because of the high cost of prosecution.

Software developers put a great deal of money and effort into creating software and expect a fair return on their investment. Every time someone gives away a copy of the program or uses it on multiple computers, the developer suffers a loss. If this happened enough times, companies would have no incentive to develop software.

In 1997, legislators passed a law targeting software and other types of piracy. The No Electronic Theft (NET) Act makes distributing illegal copies of online copyrighted material a federal crime if the value of the material is $2500 or more. The Digital Millennium Copyright Act of 1998 amended existing copyright laws to protect Internet service providers (ISPs) from liability relating to material posted on their servers by subscribers.

Music and Video Piracy

In recent years, copyright infringement for music and videos over the Internet has emerged as a hot topic. In 1997, the popular MP3 Rio player appeared on the market; this device allows users to download compressed music files in the MP3 format from the Internet and then play them. Software players such as Real Player and Windows Media Player also became available that allowed users to download and play MP3 files on their computers. The MP3 format (MP3 stands for Motion Picture Entertainment Group, or MPEG, version 3) enables music files that would ordinarily take as much as 50 MB of storage space to be reduced to 4 or 5 MB. This compression, which results in a very slight loss in sound quality, makes it feasible to download music files over the Internet.

Although individuals can legally create their own files and keep them for themselves, or download songs or videos with the permission of the copyright holder, the NET Act of 1997 makes it illegal to download or trade copyrighted songs without the permission of the publisher. When the trading of MP3 music files through the Napster server reached epidemic proportions, several industry groups successfully moved to shut it down as a free service (other services have since emerged to fill the gap left by changes in Napster). The same law may soon affect the trading of compressed versions of DVD visual files. It has already been ruled illegal to publish the code necessary to convert DVD files into compressed digital files. Content developers, such as Disney, are pushing for legislation to require new hardware to block illegal downloads of music and video.

1. Under what conditions can you legally make a copy of a software program?
2. What is a *one-disk country*?

Attacks on Information Technology

Although terrorists or other criminals can physically attack the devices that support information technology, attacks on information technology are much more likely to come through the Internet. In these cases, the targets typically consist of IT software or data or a server. Software and data attacks involve the introduction of destructive or malicious software into a computer to destroy software and data. In server attacks, the hacker takes control of a server or sends so many requests or messages to the servers that they collapse under the load. The cost of such attacks is huge and growing, with an attack by a single variety of destructive software possibly running into the billions of dollars.

Viruses, Worms, and DoS Attacks

virus
Malicious or destructive software that damages resources on a target computer.

worm
Malicious or destructive software that uses up resources on a target computer.

The most common types of destructive software are viruses and worms. Although formal differences exist between viruses and worms, it is becoming more difficult to distinguish between the two, and they have similar effects. **Viruses** and **worms** are malicious or destructive software programs that damage or use up resources on a target computer. They often come concealed within or masquerade as legitimate software (a software Trojan horse, as it were), sometimes spreading themselves to other computers via e-mail or through network connections. Hundreds of computer viruses and worms are active at any one time, and the number of infected computers remains unknown. The 2001 Computer Security Institute/FBI study found that 64 percent of the corporations surveyed had servers down for more than one hour due to viruses or worms, and 40 percent reported data loss due to these types of destructive software. Table 11.5 describes five of the most malicious viruses or worms that occurred between May 2000 and November 2001. By the time you read this chapter, you will undoubtedly have heard of others.

Table 11.5 Examples of Viruses and Worms

Virus/Worm	Month Discovered	Features
I Love You	May 2000	Delivered in an e-mail with "I Love You" in the subject line. Opening the attachment starts its execution. Sends itself to everyone in the host's address book. Writes over music and picture files. Created by a Filipino student.
SirCam	July 2001	Delivered in an e-mail with a subject based on a document in a previously infected machine. Requests advice from the recipient in an attachment that looks like a document or other file. Sends itself to everyone in the host's address book and hides in the Recycle Bin so it is not found by all antivirus software.
Code Red	August 2001	Attacks a flaw in Internet Information System (IIS) Web server software. Defaces Web pages and degrades system performance, and can cause overload on other servers.
Nimda	October 2001	Spreads via e-mail, Web sites, or shared hard disks on networks. Hits computers running Windows operating systems and repropagates periodically, reinfecting machines. Slowed the Internet down as it affected Web servers.
Goner	December 2001	Delivered in an e-mail that invites the recipient to look at a screen saver attachment. Attacks antivirus software on the host computer as well as firewalls on servers.

Figure 11.4 Distributing malicious software through e-mail

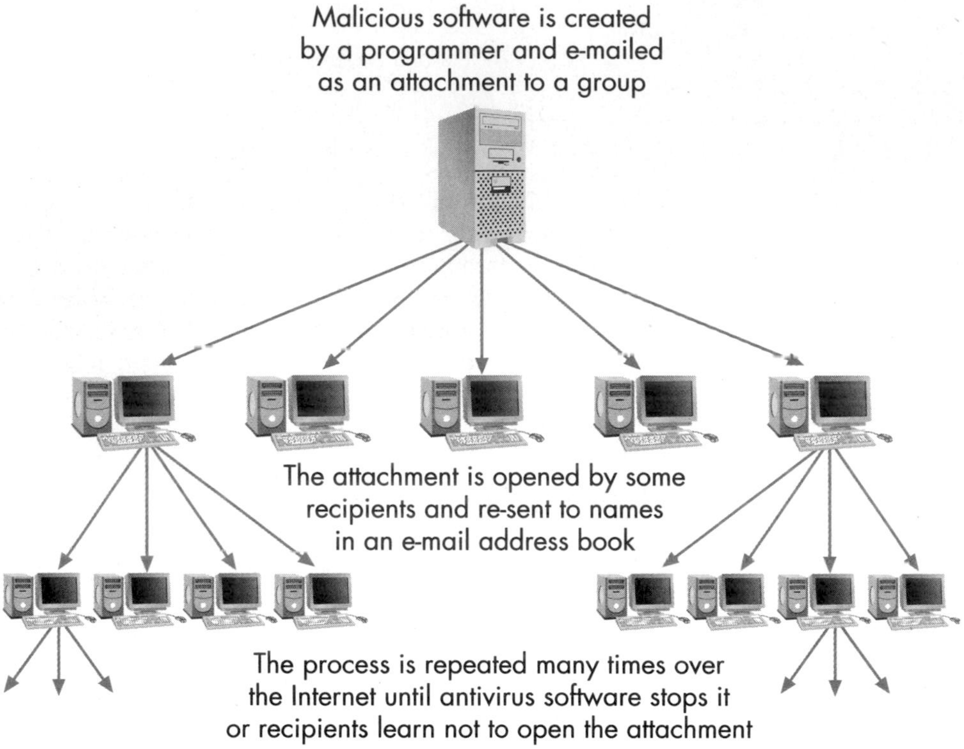

Malicious software is created by a programmer and e-mailed as an attachment to a group

The attachment is opened by some recipients and re-sent to names in an e-mail address book

The process is repeated many times over the Internet until antivirus software stops it or recipients learn not to open the attachment

Hackers often use Microsoft Outlook and Outlook Express, two popular e-mail software packages, to stage a malicious software attack by taking advantage of them as a platform to send destructive software to other users. Figure 11.4 shows the process used to create and distribute malicious software.

To combat viruses and worms, users should avoid opening any e-mail attachment that comes from an unknown source—and even those from known sources if you do not expect the attachment. The latter warning reflects the fact that these programs can make e-mail look like it is coming from a friend when it is not. You should also have up-to-date antivirus software running at all times that will check every file that is introduced into your computer and set off an alarm or automatically clean any infected files. Figure 11.5 shows an example of an e-mail containing the SirCam virus that was sent to the author. All SirCam e-mails had the same, strangely worded message and an attachment, which, if executed, released the software into your machine.

Figure 11.5 SirCam virus e-mail message

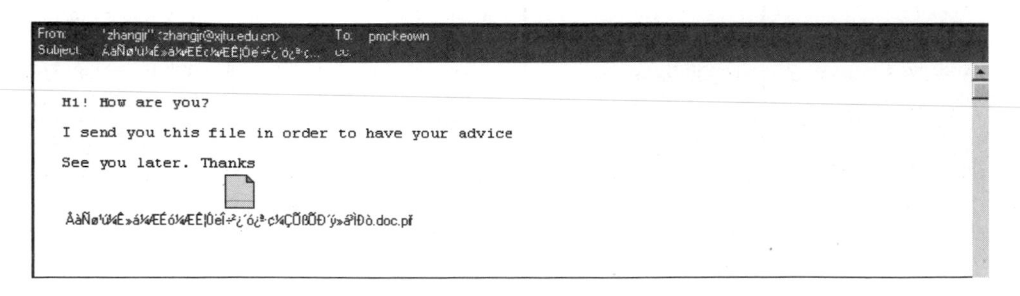
To: pmckeown
Hi! How are you?
I send you this file in order to have your advice
See you later. Thanks

denial of service (DoS) attack
A destructive use of software in which a Web site is bombarded with thousands of requests for Web pages, rendering the Web server unusable.

In addition to malicious software, other types of software attacks on IT are possible. One of the most destructive in terms of downtime for a Web site is a denial of service attack. In a **denial of service (DoS) attack**, a Web site is bombarded with thousands of requests for Web pages, putting the server out of action for several hours up to a few days depending on the overload. A DoS attack begins when a hacker installs software via the Internet on a large number of computers without the knowledge of their owners. Then, on a signal from the hacker, all of these computers send requests for information to a Web server at the same time. In almost all cases, the computers co-opted into sending the request are connected to the Internet via a LAN, cable, or DSL connection and have no protection from outside forces. Trapdoors left by viruses and worms also provide a way to co-opt a computer for use in a DoS attack. Hackers use software to search the Internet for these "open" computers and then install the DoS software on them.

Figure 11.6 shows that a DoS attack begins with the hacker locating unprotected Web servers in businesses or organizations. In step 2, the hacker uploads a "slave" program into the network servers. This program does nothing until the hacker gives it instructions. In step 3, the hacker sends a signal to the servers on which the slave program is installed. In turn, the slave programs direct those servers to query computers on the network as to whether they are online and working. In step 4, the network computers reply to this query, but the reply is redirected toward the target of the attack, say, the fareastfoods.com Web server. In step 5, all of the bogus replies arrive at the target computer at approximately the same time, making it impossible for legitimate messages to get through to the Web server and possibly causing it to crash. This process may be repeated any number of times until server administrators discover the problem and find ways to fix it.

Figure 11.6 A DoS attack

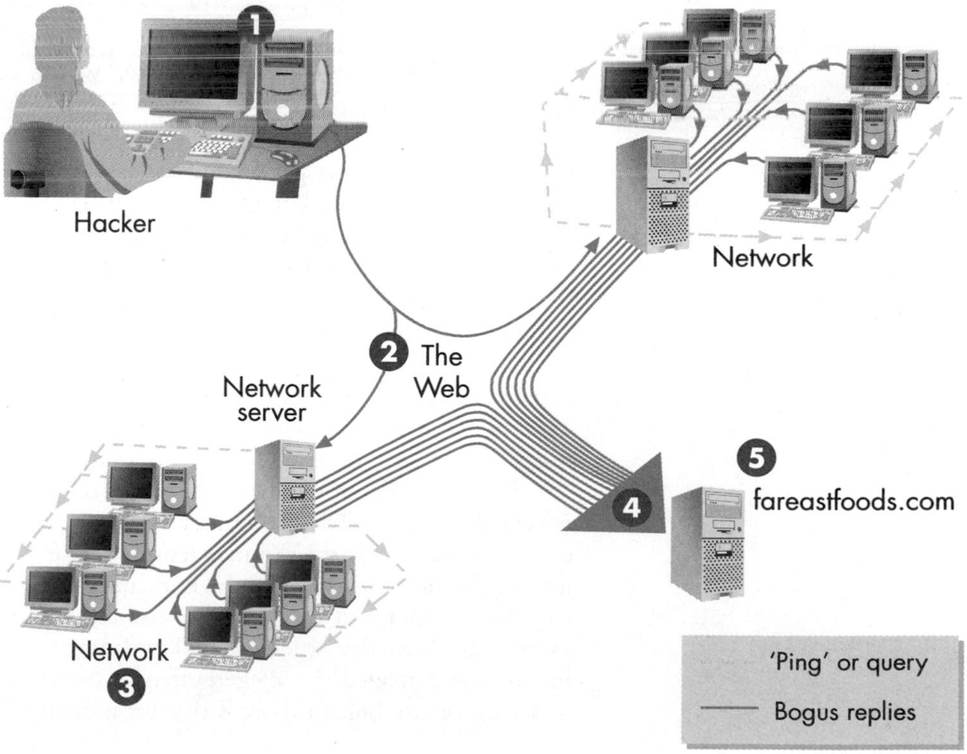

Internet Vulnerabilities

The features that lie at the heart of the tremendous growth of the Internet—its ease of access to the World Wide Web and e-mail—also render computers that communicate over the Internet susceptible to outside attack. In addition, the underlying protocol of the Internet, TCP/IP, and the operating systems on which many Internet servers run have contributed to this susceptibility. When TCP/IP and UNIX were developed in the 1960s and 1970s for use in research and academic environments, security was of little or no concern. Today we use TCP/IP and UNIX for electronic commerce and e-mail, which require a high level of security, and they do not always work as well as we would like. Both TCP/IP and UNIX have a number of well-known holes (that is, ways in which an intruder can gain access to the system) that can allow hackers to break into a computer or network and cause mischief. Hackers have also discovered holes in the server versions of Windows (NT, 2000, and XP) through which they can break into servers. Although vendors and programmers work feverishly to provide fixes for the holes, hackers keep busy finding new ways to break in.

The Web, as shown in the electronic commerce layer model in Chapter 8, runs on top of TCP/IP and uses a series of less than fully secure protocols. As a consequence, hackers can often break into a Web site and change the contents of a Web page. E-mail servers suffer from the same types of problems; they can be breached through security holes and crashed by receiving too many requests for service. When hackers break into an e-mail service, they can wipe out e-mail files or capture passwords for future use. Occasionally, a hacked e-mail system must be shut down so that the security hole can be found and fixed.

Legal Aspects of IT Crime

Prior to 1986, although computer criminals reaped large sums of money from crimes, they rarely spent much time behind bars when convicted, for two reasons. First, few laws actually covered IT crime, and even when they did, prosecutors had difficulty obtaining convictions. In many cases, the only thing stolen was information or computer time, neither of which would constitute a crime under theft laws enacted prior to the age of computers. Often, only mail fraud law came close to applying to IT crime. Even when money was taken, the thief usually got off because he or she was young and clean-cut, and had no previous record. The thief might also be excused if the jury considered the crime victimless because the injured party was a large corporation. In addition, the intricacies of computers were often very difficult to explain to the judge and jury, so the prosecution may have found it difficult to demonstrate how the crime occured.

Second, many computer criminals went unpunished because companies often hesitated to take action against former employees. These companies did not wish to risk the negative publicity of a trial or the appearance of not having adequate security for its computer system. As a result, many white-collar computer criminals were allowed to resign quietly, in some cases moving to a similar position in another company.

To rectify the problem, the federal government and all 50 states have passed laws covering IT crime. The Computer Fraud and Abuse Act of 1986 made it a crime to damage data in any government computer or in any computer used by a federally insured financial institution (which includes most banks). This act also criminalized the use of a computer to view, copy, or damage data across state lines. Under this act, if a victim can show that at least $1000 in damage occurred or that medical records were in any way damaged, the alleged offender can be prosecuted. This act made it a felony to set up or use bulletin boards that list individual or corporate computer passwords.

People convicted under the Computer Fraud and Abuse Act can receive a prison term up to 20 years and a fine of up to $100,000. The author of a particularly malicious software virus called Melissa, for example, was prosecuted under this act. Sections of the USA Patriot Act of 2001, passed in response to the terrorist attacks of September 11 of that year, have since modified this act. The Patriot Act increased the penalties for computer crimes, changed the definition of damage to make it easier to prosecute hackers, and added new penalties for damaging computers used for national security.

IT crime is not restricted to the United States. Other countries, such as Canada and the United Kingdom, have either enacted laws similar to those discussed here or begun to explore legal avenues to combat IT crime. The European Union has been particularly active in working to develop computer-crime-related statutes that would apply to member countries.

1. List the three objects of an attack through the Internet.
2. What do trapdoors and Trojan horses have to do with viruses and worms?

Information Technology Security

Information technology security entails the protection of the IT assets, including hardware, software, data, and information. Given the huge potential losses involved in IT crime, IT security has become a very important topic for individuals and organizations. IT security encompasses a broad group of methods used to protect IT from natural and criminal forces. (Natural forces are included because fire, water, wind, or earthquakes can prove as devastating to IT as criminals or cyberterrorists.) Types of security range from a common-sense approach of locking doors to sophisticated methods of protecting intranets from outsiders.

Threats to Computer Systems

Although unauthorized access by outside parties has received much attention as a threat to a computer system's security, most experts do not consider it to be the most important. More pressing concerns include errors in data and omission of data by employees; the theft of computers and peripheral devices (printers, monitors, and so on); misuse by disgruntled, laid-off, or dishonest employees; damage from fire, water, or natural disasters; or damage from terrorist acts, such as those occurring to the World Trade Center towers and the Pentagon on September 11, 2001.

Although not everyone may agree with a particular ranking of threats to computer hardware, software, and data, no one will deny that organizations must respond to all of these threats. Another way to look at the security problem for computers is to consider physical security, data security, and Internet security. **Physical security** involves the protection of computer hardware from theft or damage, whether caused by nature or humans, in the same way that other office equipment would be protected. **Data security** entails the protection of software and data from manipulation, destruction, or theft. Finally, **Internet security** entails the protection of both the data and information traveling over the Internet and the information technology used to send and receive data and information.

physical security
The protection of computer hardware from theft or damage, whether caused by nature or humans, in the same way that other office equipment would be protected.

data security
The protection of software and data from manipulation, destruction, or theft.

Internet security
The protection of both the data and information traveling over the Internet and the information technology used to send and receive data and information.

As you consider the various types of security discussed in the remainder of this chapter, note that it is important to approach the problem of providing security for an organization in the same step-by-step manner that you use to develop information systems. Recognizing that a security problem exists is only the first stage; you must then analyze the problem, design a solution for it, and then implement that solution. The organizations in and around the World Trade Center that had planned for security were up and running in a few days after the attack. Those that had not may still be trying to rebuild their systems—if those systems even exist now.

Physical Security

Methods of protecting computer hardware from outside forces include employing procedures that exclude unauthorized people from the organizational servers, guarding hardware and software against theft, and protecting the hardware as much as possible from natural disasters and fire and water damage. A key part of any physical security plan relates to the recovery of critical data and programs in case of disaster. This plan should consider the worst-case scenarios facing the organization and set up ways to recover from them. Scenarios should include weather or terrorist disasters as well as problems caused by hackers or malicious employees. Downtime due to loss of equipment or data can also prove very expensive. For example, Figure 11.7 shows the cost of one hour of downtime for various companies based on a survey by Contingency Planning & Strategic Research Corporation. Failure to plan ahead through a comprehensive **business continuity plan** that deals with all possible downtime scenarios could easily result in the organization's going out of business.

business continuity plan
A comprehensive plan that deals with all possible downtime scenarios.

biometrics
A technology that uses thumbprints, eyeprints, or voiceprints to identify individuals.

Organizations can control entry to the areas containing servers, and thereby exclude unauthorized people, by using thumbprints, eyeprints, or voiceprints to identify individuals, via a technology called **biometrics**, and, as simple as it may seem, by keeping doors locked. Protecting computers, especially laptops, from theft represents a growing problem that requires constant vigilance on the part of owners. Keeping serial numbers of computers on file is another important part of physical security because such ID numbers offer the best hope for recovering lost machines. Portable secondary storage devices such as floppy disks, Zip disks, or writable CD-ROMs are even easier to steal than computers when employees leave them lying around or store them in an unlocked desk drawer or cabinet. Keeping doors locked may be even more important for the security of personal computers than for the larger, less portable computers.

Although damage from fire or water appears on everyone's list of dangers, these dangers often are not treated with the respect they deserve. Like any electronic machine, the computer always faces danger from an electrical fire caused by a short circuit. Likewise, nonelectrical fires can cause great damage if water is used to extinguish them, because water can harm the delicate circuits of a computer. For this reason, the

Figure 11.7 Effects of one hour of downtime

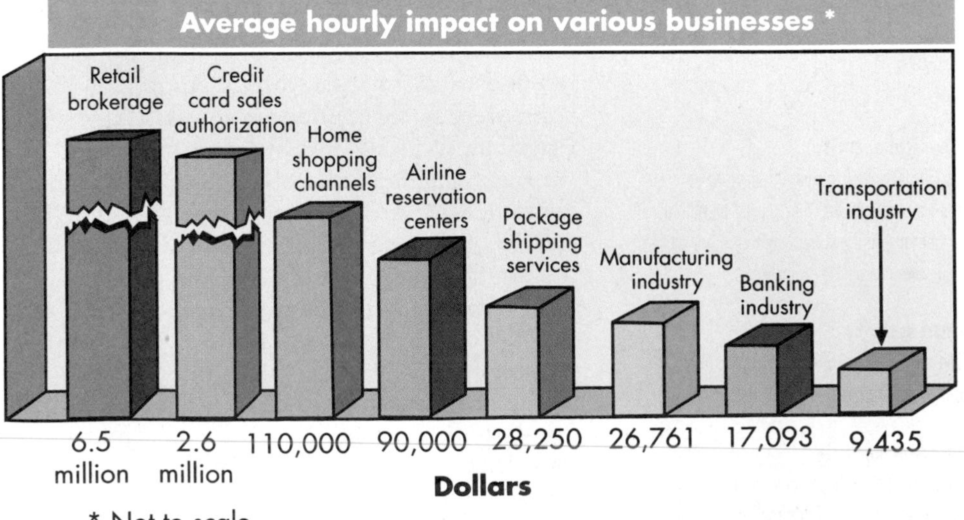

Source: http://www.ontrack.com/datarecovery/cost.asp.

organization must consider the danger of water damage when planning the security system for its computers. Sprinkler systems are a common protection against fire in commercial buildings, but they can cause more damage to a computer than the fire itself. A number of years ago, a government agency's sprinklers went off by accident, soaking the computers and causing many related problems. Newer fire-suppression systems apply combinations of gases to deprive a fire of oxygen, thereby quenching the flames without water. At the very minimum, the organization should own fire extinguishers that are rated to deal with electrical fires.

backup
A second (or even third) copy of a data file on a storage device kept separate from the primary disk storage.

To avoid a complete loss of data in a disaster, the organization should implement a policy of regularly backing up files on the system and storing the backups in an area physically separate from the main computer center. In this context, a **backup** is a second (or even a third) copy of a data file on a storage device separate from the primary disk storage. Unfortunately, personal computer users tend to put this task off as long as possible. To avoid the expensive problems that can occur if the hard disk crashes or an errant command or virus destroys all data on the disk, every user should make regular backups on disk or tape. Although most companies affected by the World Trade Center disaster appear to have instituted business continuity plans involving data storage off-site and backup processing capabilities through such companies as Sungard that enabled them to keep operating, at the time of this writing the amount of data on PC hard disks that were lost remains unknown.

Uninterruptible power supplies, such as this one, provide continuous power in the event of a power loss.

In addition to securing a personal computer against theft, individuals and companies must protect the computer from environmental harm. The normal precautions taken to protect any piece of office equipment against fire, water, dust, and other physical damage apply here. In addition, the user must protect the computer and its data against, of all things, electricity. Too much electricity, too little, the wrong kind, or the right kind applied in the wrong manner can cause problems. A **voltage surge**, or **spike**, in which lightning or some other electrical disturbance causes a sudden increase in the electrical supply, can destroy the delicate chips and other electrical parts of a computer. For this reason, **surge protectors** are a necessity to protect the computer against uneven electrical power. These inexpensive devices plug into a normal wall outlet; the computer and its peripherals, in turn, plug into outlets in the surge protector. When a surge hits the wall outlet, a circuit breaker is thrown in the surge protector, thereby protecting the computer and its peripherals.

voltage surge
A sudden increase in the electrical supply caused by lightning or some other electrical disturbance. Also known as a spike.

surge protector
A device that protects the computer's hardware and memory from a voltage surge.

Just as too much electricity can damage a computer, too little electricity—in the form of a brownout or a blackout—can cause the loss of all data from internal memory. Because RAM is volatile and depends on a constant power source to retain information, loss of power means loss of memory. Devices called **uninterruptible power supplies (UPS)** continue the power to a personal computer when a disruption of electrical current occurs. Although a UPS costs more than a surge protector, it is commonly used by organizations that have a LAN or by people who use a computer on a daily basis. If you do not have a UPS, the best defense against a power outage is frequently saving open files to disk. Figure 11.8 shows the spectrum of electrical power and the effects of too much or too little power.

uninterruptible power supply (UPS)
A device that continues to send power to a computer if the electrical current is disrupted.

Data Security

Protecting software and data from unauthorized access is an entirely different problem from providing physical security. Although computer hardware can almost always be replaced, an organization's data generally represent its most important asset and may be irreplaceable. Even if no destruction of data occurs, having data fall into a competitor's hands can have disastrous implications for private companies and national governments.

password
A sequence of letters and digits, supposedly known only to the user, that must be entered to access a computer system.

The primary tool for protecting access to software and data on computer systems is the password. A **password** is a sequence of letters and digits, supposedly

Figure 11.8 Electrical spectrum

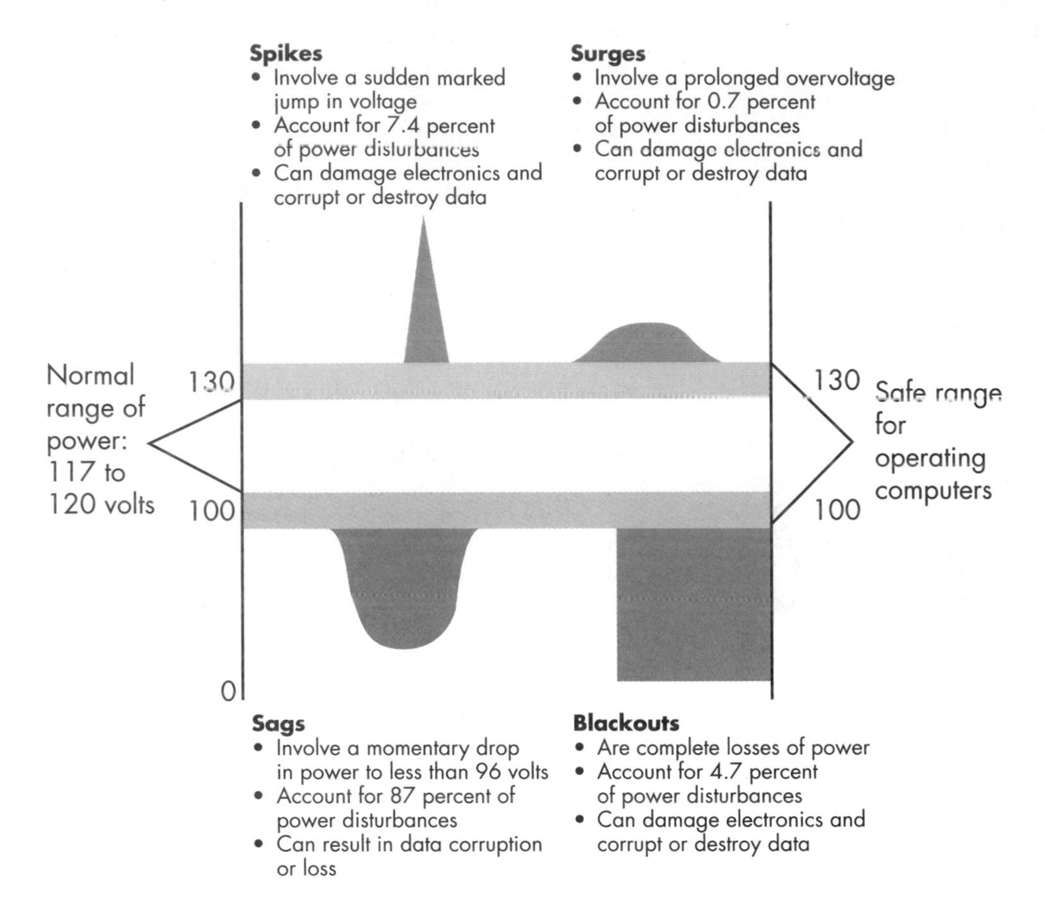
Spikes
• Involve a sudden marked jump in voltage
• Account for 7.4 percent of power disturbances
• Can damage electronics and corrupt or destroy data
Surges
• Involve a prolonged overvoltage
• Account for 0.7 percent of power disturbances
• Can damage electronics and corrupt or destroy data
Normal range of power: 117 to 120 volts
130
100
0
130
100
Safe range for operating computers
Sags
• Involve a momentary drop in power to less than 96 volts
• Account for 87 percent of power disturbances
• Can result in data corruption or loss
Blackouts
• Are complete losses of power
• Account for 4.7 percent of power disturbances
• Can damage electronics and corrupt or destroy data

personal identification number (PIN)
A type of password used to access a bank account from an automatic teller machine.

password policies
Specific company policies designed to protect data and software through responsible use of passwords.

known only to the user, that the user must enter to access the computer system. Most people have become accustomed to using a type of password called a **personal identification number (PIN)** to access their bank accounts from an automatic teller machine (ATM). Unfortunately, many users choose passwords that are easy to remember, such as "test," "system," "password," or the name of the user's significant other, child, or pet. Of course, it is easy to guess such passwords.

Given the importance of passwords in protecting software and data on computer systems, **password policies** that define how a password is assigned, when it is changed, and who should have a password form an important part of any security system. These policies should include the following considerations:

- Users should not be allowed to choose their own passwords, and all passwords should consist of at least four letters or digits.
- Users should have their passwords changed immediately if any evidence suggests a breach of security has occurred.
- All passwords should be changed periodically to make guessing a password more difficult.
- If an employee with supervisory-level network access is fired or leaves under less than pleasant circumstances, *all* passwords to which he or she had access—not just the ex-employee's—should be changed. Failure to change all relevant passwords can leave the system open for intrusion.

Related to password policies are policies dealing with the rights and privileges associated with different levels of authority in an organization. Essentially, the higher up in the IT structure of an organization, the more rights an individual has to access data and information on the network. Typically, higher-level people in the IT department have passwords that give them the priviledge of controlling more of the network. For this reason, the organization must change all passwords when a supervisory-level person leaves the company.

In addition to password policies, other tools used to protect computer software and data include systems audit software, data encryption systems, and antivirus policies and software. **Systems audit software** keeps track of all attempts to log on to the computer, paying particular attention to unsuccessful attempts. An audit of the system log would indicate who has accessed the system at any given time and therefore should reveal whether an unauthorized user has accessed the system. **Data encryption systems** protect the data being transmitted over a network by converting them into an unreadable form. The process transforms **plain text** (a readable form) into **ciphertext** (an indecipherable form) at the source computer and then back to plain text at the destination computer. The quality of an encryption system is measured in terms of the number of bits in the key that is required to convert the ciphertext back to plain text. The higher the number of bits in the key, the more effort required to guess the key: If a key has n bits, then there are 2^n possible combinations in that key. Modern encryption systems use at least 128-bit keys. The Pretty Good Privacy (PGP) system is a popular public-key encryption system (discussed in Chapter 8) that is available for free to users at a variety of sites in the United States and around the world.

As noted earlier in this chapter, viruses and other destructive software pose threats to computer data. **Antivirus policies** are organizational policies that protect the computer system from destructive software. These policies include not allowing downloading of untested software to organizational computers and using **antivirus software** from such companies as McAfee and Symantec to run a virus test every time a computer starts. The system should also check all e-mail messages as they come through the organizational server for attachments containing known malicious software. Note that just having antivirus software installed is not enough; the organization must update this software frequently to add protection against new viruses and worms. Updates are typically free or available for a small subscription fee from the company that originally sold the antivirus software.

systems audit software
Software that keeps track of all attempts to log on to the computer, giving particular attention to unsuccessful attempts.

data encryption systems
Systems that protect data being transmitted over a network by converting them into an unreadable form.

plain text
The readable form of a message.

ciphertext
The encrypted form of a message.

antivirus policies
Organizational policies that protect the computer system from destructive software.

antivirus software
Software used to test for destructive software every time a computer is started, an e-mail is received, or software is downloaded.

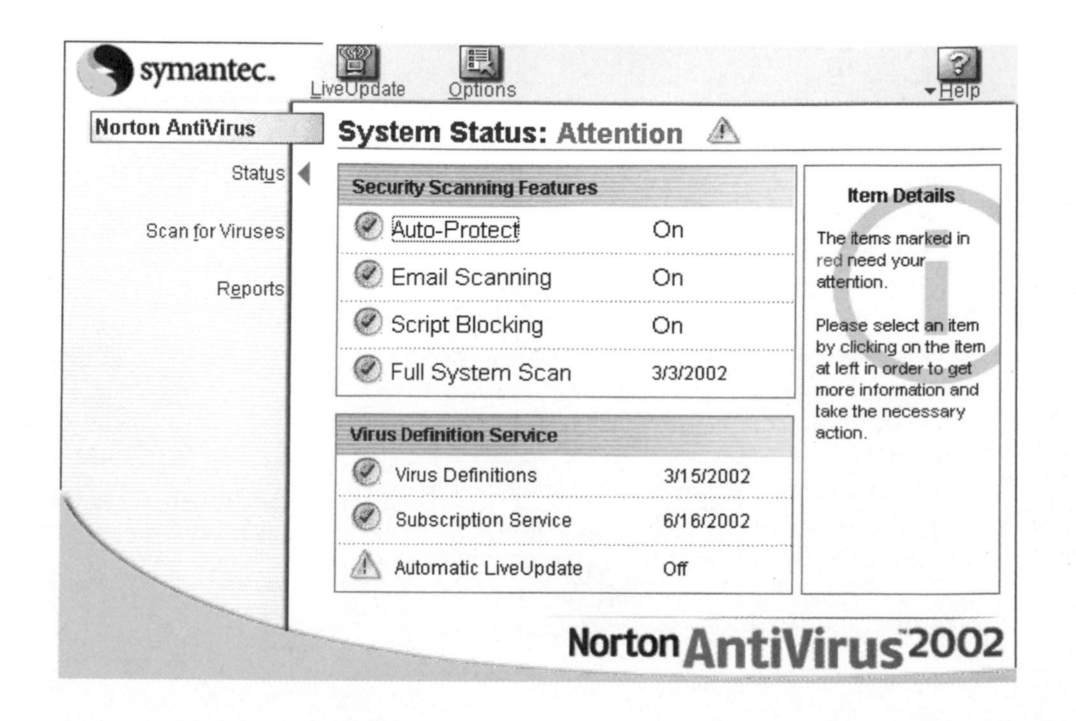

Antivirus software is an important line of defense against malicious software such as viruses.

Security at FarEast Foods

The IT staff at FarEast Foods has implemented measures to provide both physical and data security. From a physical security point of view, the servers at the company reside in a server farm located behind locked doors. A biometric device checks eye-prints to identify and authenticate everyone entering the room. Duplicate servers for each operation exist so that if one goes down, the operation can continue uninterrupted. In addition, the system includes duplicate communication links to the Internet so that if one line becomes cut, the company can continue to operate. All servers are protected by surge protectors, and each has its own UPS that offers two hours of continuous power. A generator can also be used if necessary. The servers are backed up each night to an off-site location to ensure that no data will be lost even if the office buildings are destroyed. All in all, the system includes a good deal of redundancy to protect FarEast Foods' capability to operate.

Data security at FarEast Foods begins with passwords to access the server, which are generated randomly for each person and changed every month. Encryption software encrypts and decrypts order information coming in from browsers over the Internet. In addition, antivirus software loaded and running on all computers is updated as soon as new data become available on the Web, but no less than once a month.

IT INNOVATORS

Phil Zimmermann

One of the best ways to provide security for data and information is to encrypt it. As discussed in Chapter 8, encryption involves converting text from a readable form into an unreadable form; decryption converts it back to the other form. Although the CIA and other government agencies have very sophisticated methods of encryption, almost anyone can use a type of encryption known as Pretty Good Privacy (PGP). PGP is a public key form of encryption in which users exchange digital keys that enable them to encrypt and decrypt messages. A software engineer named Phil Zimmermann invented PGP in 1991. Zimmermann had been interested in encryption since childhood and, after completing a bachelor's degree in computer science, became serious about finding a *workable* encryption. *Workable* means a system that is easy to use while not being easily broken. The result of his work was PGP.

After inventing PGP, rather than trying to sell it, Zimmermann gave it away to anyone requesting it. That policy got him into trouble with U.S. authorities, because it was illegal at that time to export encryption technology. (In July 2000, the Clinton administration significantly reduced encryption export restrictions.) Despite government efforts to stop his work, PGP quickly became the most widely used encryption software in the world. After a three-year effort to prosecute him, the U.S. Justice Department dropped its case and Zimmermann founded PGP, Inc. This company was purchased by Network Associates. Zimmerman now does cryptographic consulting for a number of companies and industry organizations. Although he has received a number of awards for his work, including the 1998 Lifetime Achievement Award from *Secure Computing* magazine, he remains especially proud of the fact that many human rights activists document the atrocities of their governments by encrypting documents with PGP.

Phil Zimmermann is the inventor of the popular PGP public-key encryption system.

Sources: http://www.pgp.com/phil/phil.cgi#about and http://www.animatedsoftware.com/hightech/philspgp.htm.

1. What are the three types of threats addressed by IT security?
2. What is the primary way to protect data on a computer from outside intrusion?

Internet Security

Because the Internet has become the primary method for accessing data and information on computers around the world, Internet security is critical to organizations. Web and e-mail sites often come under attack from destructive software. Protecting Web and other Internet servers from these attacks is, therefore, part of Internet security. The organization must consider the trade-offs between making its Web, e-mail, and other Internet-related servers easily available to its stakeholders and protecting the servers from unwanted intrusions. It must consider how it will secure its network environment.

Securing the Corporate Internet Connection

direct Internet connection
Connecting an enterprise network directly to the Internet with no intervening ISP.

indirect Internet connection
Connecting an enterprise network to the Internet through an ISP.

To secure its connection to the Internet, an organization must first decide how it wishes to connect to the Internet. Two choices exist: direct and indirect connection. In a **direct Internet connection**, the company enterprise network connects to the Internet through its own server and communications link. In an **indirect Internet connection**, the company connects to the Internet through an ISP. Each type of connection has both advantages and disadvantages. A direct connection is faster and more flexible but subjects the company to more risks from outside intrusion. An indirect connection tends to be slower and less flexible but provides the company with more protection from outside forces and, possibly, additional technical expertise.

Once a company decides how it will connect to the Internet to establish a Web presence, it must consider how it will handle the security risks resulting from its Internet connection. Certainly those companies that have their enterprise network directly connected to the Internet will most likely need to install higher levels of security. Three types of security apply to individual computer systems and to the Internet:

- Installing virus protection software to protect against known types of malicious software
- Using passwords to control access to a network
- Using encryption to protect the content of messages being transmitted over the Internet

digital certificate
A piece of digital information indicating that the Web server is trusted by an independent source.

Secure Sockets Layer (SSL)
A protocol used by Internet browsers and Web servers to transmit sensitive information.

Encryption (usually public-key encryption of the type discussed in Chapter 8) is implemented in two ways: digital certificates and Secure Sockets Layer. A **digital certificate** is a piece of digital information indicating that the Web server is trusted by an independent source. It essentially confirms the identity of each computer and provides each computer's public keys to other computers. **Secure Sockets Layer (SSL)** is a protocol used by Internet browsers and Web servers to transmit sensitive information. You can identify a secure link in two ways. First, you will notice that instead of *http*, the address line of your browser contains *https*. Second, you will see a small locked padlock in the status bar at the bottom of the screen. Figure 11.9 shows these two indicators of a secure connection.

Companies with direct connections to the Internet need to consider two other forms of security: fixing known security holes in the UNIX and Windows operating systems and in Web server software, and employing firewalls to protect themselves.

Figure 11.9 Indicators of a secure connection

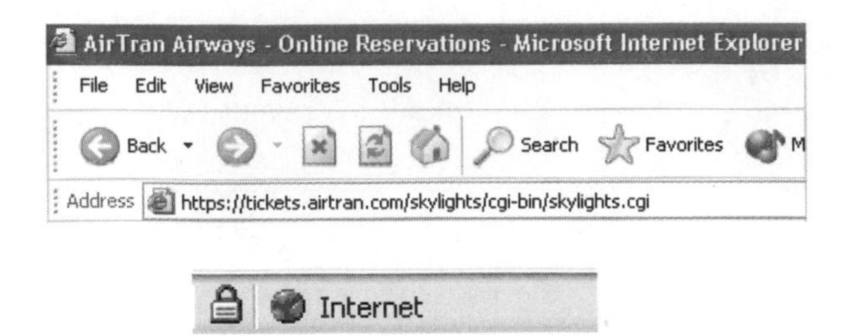

Individuals who use a high-speed connection to the Internet through either cable or DSL face the same security risks as companies with direct connections to the Internet. For that reason, these individuals should also install firewall software to protect their computers from intrusion by hackers or others bent on mischief. In fact, many of the computers used to launch DoS attacks connect to the Internet in this fashion.

Fixing Known Security Holes

As mentioned earlier, the UNIX operating system and Internet protocols were not designed to provide a high level of security. One feature of the UNIX operating system, which runs many Web servers, is that approved users can log on from anywhere, at any time, to administer the system. By gaining access to the main folder of the server, users with an administrator-level password can manipulate files and gain entry to a corporate network. Unfortunately, so can hackers who know how to exploit these features of UNIX. Server versions of the Windows operating system (Windows NT, 2000, and XP) also contain holes that are ripe for exploitation by hackers. Fortunately, users can modify much of the server and operating systems software to greatly improve its level of security.

system administrators
People who manage security and user access to an intranet or LAN. Also known as *network administrators*.

Some well-known holes in the security armor of a company's server software can be fixed by a knowledgeable system administrator. **System administrators**, sometimes called network administrators, manage security and user access to an intranet or LAN. Other holes are less easily fixed, and some remain to be discovered. One of the best ways to protect mission-critical information is to move it onto other servers or networks that are not connected to the Internet. Nevertheless, some critical information may need to be available on the Internet-accessible portion of the corporate network. A system administrator can take several other steps to improve corporate network security, including installing patches, monitoring system logs, running a network scanning program, and staying on top of recent security events in the organization's network.

patch
A piece of computer code that a software publisher releases to fix a problem in its software.

System administrators need to constantly update their Web and e-mail server software by installing patches on their software. A **patch** is a piece of computer code that a software publisher such as Microsoft releases to fix a problem in its software. For example, Microsoft released patches for IIS when Code Red and Nimda exploited problems in that software.

To stay on top of attempts to break into the organization's computer, IT personnel should moitor the server logs. Monitoring software keeps track of all successful and *unsuccessful* attempts to access the server. It is estimated that only 5 percent of all intrusions are detected and, of those detected, only 5 percent are reported. Intruders' latest tricks can be observed and thwarted by monitoring the server logs for any unusual activities. When intrusions or attempts at intrusion appear, the server needs to be fixed quickly to guard against similar attacks. Copycat hackers may attempt to break into a system by exploiting system flaws found by other hackers. Figure 11.10 shows a server log for an educational institution.

Figure 11.10 Server log

Event Viewer - Security Log on \\CLAY

Log View Options Help

Date	Time	Source	Category	Event	User	Computer
3/14/02	10:36:58 AM	Security	Detailed Tracking	592	mellis	CLAY
3/14/02	10:36:58 AM	Security	Detailed Tracking	592	mellis	CLAY
3/14/02	10:36:58 AM	Security	Detailed Tracking	592	mellis	CLAY
3/14/02	10:36:58 AM	Security	Detailed Tracking	592	mellis	CLAY
3/14/02	10:36:58 AM	Security	Detailed Tracking	592	mellis	CLAY
3/14/02	10:36:58 AM	Security	Detailed Tracking	592	mellis	CLAY
3/14/02	10:36:57 AM	Security	Detailed Tracking	593	mellis	CLAY
3/14/02	10:36:57 AM	Security	Detailed Tracking	593	mellis	CLAY
3/14/02	10:36:56 AM	Security	Detailed Tracking	592	mellis	CLAY
3/14/02	10:36:56 AM	Security	Detailed Tracking	592	mellis	CLAY
3/14/02	10:36:56 AM	Security	Detailed Tracking	592	SYSTEM	CLAY
3/14/02	10:36:56 AM	Security	Detailed Tracking	592	SYSTEM	CLAY
3/14/02	10:36:56 AM	Security	Privilege Use	576	mellis	CLAY
3/14/02	10:36:56 AM	Security	Logon/Logoff	528	mellis	CLAY
3/14/02	10:36:56 AM	Security	Privilege Use	576	mellis	CLAY
3/14/02	10:36:56 AM	Security	Logon/Logoff	528	mellis	CLAY
3/14/02	10:36:47 AM	Security	Logon/Logoff	538	mellis	CLAY
3/14/02	10:36:47 AM	Security	Privilege Use	576	mellis	CLAY
3/14/02	10:36:47 AM	Security	Logon/Logoff	528	mellis	CLAY
3/14/02	9:55:06 AM	Security	Logon/Logoff	538	bjhunt	CLAY
3/14/02	9:53:43 AM	Security	Privilege Use	576	bjhunt	CLAY
3/14/02	9:53:43 AM	Security	Logon/Logoff	528	bjhunt	CLAY
3/14/02	9:53:29 AM	Security	Object Access	562	SYSTEM	CLAY
3/14/02	9:53:29 AM	Security	Object Access	560	ANONYMOUS	CLAY
3/14/02	8:55:11 AM	Security	Logon/Logoff	529	SYSTEM	CLAY
3/14/02	8:40:20 AM	Security	Logon/Logoff	538	mellis	CLAY
3/14/02	8:39:17 AM	Security	Privilege Use	576	mellis	CLAY
3/14/02	8:39:17 AM	Security	Logon/Logoff	528	mellis	CLAY
3/14/02	6:51:37 AM	Security	Logon/Logoff	538	bjhunt	CLAY

Rejected login attempt

One way to begin to find the security holes on a network or your server is to run a program that is designed to identify potential security holes. Many of these scanner programs are controversial—hackers can use them to find holes in a network that they can then exploit. Knowing that hackers have these tools at their disposal, it would be wise to use them yourself to close the openings to your server that the hacker can easily identify and crawl through. NMap is one such scanner program.

The nMap software can be used to test the vulnerabilities of an organization's network.

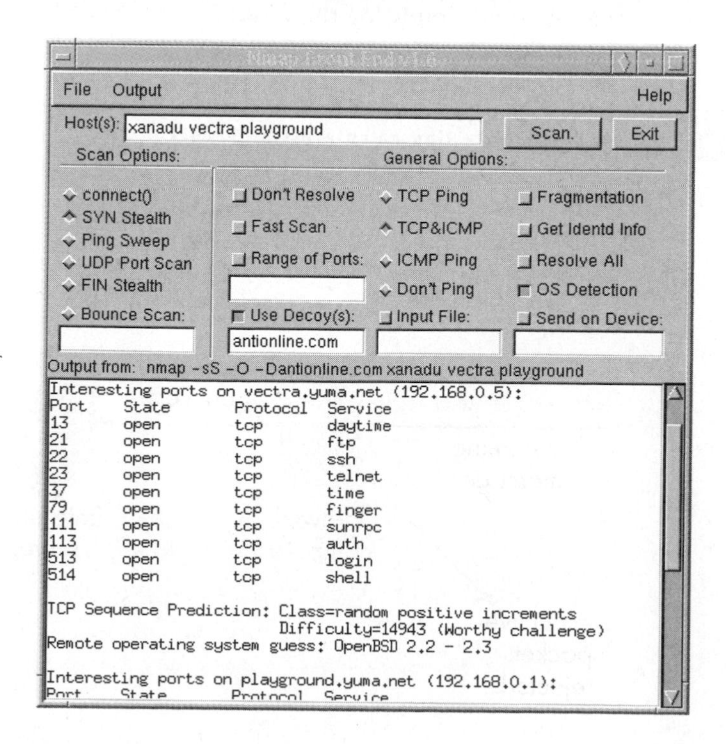

Finally, after fixing the known holes, the network administrator needs to monitor server logs to stay aware of attempts to break into company computers and activities at other sites on the network. To stay abreast of security violations on the Internet as a whole, the network administrator may want to join one of the mailing lists run by Computer Emergency Response Team (CERT) or another mailing list that alerts network administrators to known security problems. The network administrator can also find a wealth of information about the state of security of the Internet by checking the CERT Web site at www.cert.org.

Firewalls

Firewalls are the dominant technology used by businesses to protect their networks from hackers. As mentioned in earlier chapters, a *firewall* protects an organization's network from outside intrusions. Firewalls must guard the enterprise network against outside intruders while simultaneously enabling users inside the firewall to access the Internet. A trade-off is inevitable between user flexibility and the level of security provided for the internal network. Although no firewall balances these two competing demands perfectly many come close. Figure 11.11 shows a conceptual view of a firewall around a corporate network.

Once a corporation decides to put in a firewall, it must program the firewall to support its security needs. A firewall can be restrictive or flexible, depending on the company's goals. For instance, the organization can limit specific services to reduce the probability of break-ins. One of the most common ways for a hacker to break into a server is through the File Transfer Protocol (FTP) software on the server; as a result, many corporations have programmed their firewalls to block outside access to that service.

The firewall looks at every piece of information that is either sent in or out of the internal network. Firewalls act on a message on the basis of user identification, point of origin, file, or other codes or actions. A firewall can take four basic actions when it finds a suspicious piece of information:

1. It can drop (not forward) the piece of information.
2. It can issue an alert to the network administrator.
3. It can return a message to the sender after a failed attempt to send the information.
4. It could simply log the action.

Figure 11.11 Firewall protecting an enterprise network

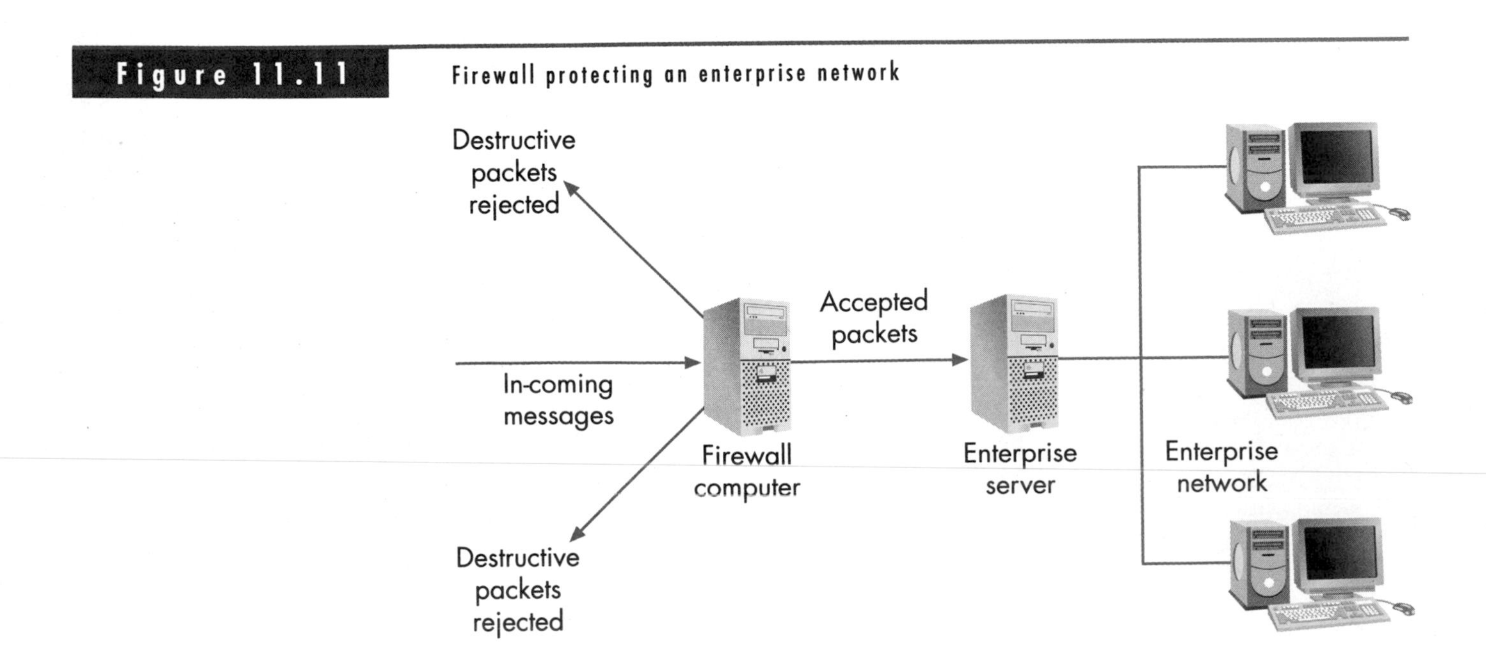

The action taken by the firewall depends on the severity of the perceived attack and the policies set by the system administrator.

IP spoofing
A form of Internet intrusion in which an internal IP address is used to fool the server into believing it is in contact with a trusted computer.

A common way to attack a network behind a firewall is through IP spoofing. **IP spoofing** takes advantage of the fact that the UNIX operating system presumes that anyone who logs in to a server using a previously approved TCP/IP address must be an authorized user, which is not always the case. By altering the source IP address, someone can fool the firewall into believing that a message comes from a trusted source. Once inside the network, the message can wreak havoc. To combat this problem, many firewalls reject all packets that come from outside the firewall but carry an internal source address.

Internet Security at FarEast Foods

In addition to using SET to protect all orders and credit card numbers coming into the fareastfoods.com Web site, FarEast Foods' system administrator runs the latest scanning software periodically to find new holes in the UNIX operating system that runs the company's servers, so as to protect company software and data. FarEast Foods also has a firewall that it keeps constantly updated to keep rogue messages out of the system.

1. How can an organization's enterprise computer network be connected to the Internet?
2. Which two methods are often used to provide security for an enterprise network that connects directly to the Internet?

Security Issues

In addition to the technical issues, legal, insurance, and personnel issues must be considered when creating a security plan. This section will discuss these issues and summarize ways to secure the enterprise network from outside intrusions.

Legal Issues

Many companies overlook the potential legal issues they face when they connect to the Internet. The popular media have focused attention on many of these issues, including employees misusing their Internet connections to download pornography, bootlegged software, or music that violates copyright laws. Other issues have been raised concerning companies that become victims of copyright infringement. Network administrators must be aware of these potential dangers and take measures to protect their company and its employees from lawsuits and loss of valuable copyrighted data.

If the company allows FTP access to its server, it is good policy not to allow external users to place files on the corporate server, or to purge files from it frequently. This stance will guard against unwanted guests using the server as a clearinghouse for pirated software. One well-publicized case occurred at a major university where unknown individuals used a seldom-used computer as a storage facility for pirated software. The owners of a server could potentially be held liable for what resides on their computer, regardless of their knowledge of its contents, and be brought to court for copyright infringement.

To curb access to sexually explicit materials, many companies restrict access to a variety of newsgroups. Although this practice may cut off that particular source, users can always gain access to such materials in other ways. A company's only hope for preventing such actions is to educate employees on those types of behaviors the company will not tolerate and to enforce these policies aggressively. A company cannot constantly monitor the actions of all employees, but by creating an environment that does not condone such behavior it may be able to reduce the likelihood of such incidents.

Employees also need to be educated on copyright laws. Although it is fairly well known that copying commercial, nonshareware computer programs is illegal, other forms of copyright infringement may be less obvious. Downloading a copyrighted music file or distributing an article or picture copied from a Web site without citing it, for example, can violate copyright laws. Furthermore, the company needs to be concerned not only with materials that employees obtain while at work but also with those materials that they may post outside the company. An employee may unwittingly release private company information to others on the Internet, jeopardizing company data or potential profits. Once again, the only way to guard against such situations is through employee education. When in doubt about sharing information or data, employees should contact the company legal department or the network administrator for advice.

Insurance Against IT Crime

Since Lloyds of London entered the market in 1981, many insurance companies have begun offering coverage for IT crime. In many cases, this coverage is part of a more general insurance policy for criminal activities, and most large corporations have some degree of protection. However, these policies tend to include many exclusions related to criminal activities—by either internal employees or hackers. For example, the traditional insurance policy covers only theft from the insured by people outside the firm—not harm to the company from insiders or business interruption. It also usually refers only to money or property. In response to these exclusions, newer policies offer coverage against crimes committed by insiders or hackers, loss of business due to such activities, and activities carried out by a hacker using the company computer. These new policies extend the definition of fraud from "theft when related to the use of any computer" to more than theft, to the extent that they protect a company from going out of business due to a hacking event.

Unfortunately, these new policies carry high price tags (currently with minimum premiums of around $10,000). They may be too expensive for all but the largest corporations, leaving smaller companies uncovered.

Human Aspects of Computer Security

The human role in ensuring the security of a computer system is every bit as important as the technological role. First, education constitutes the primary means of changing people's attitudes toward security. Individual users must understand that unauthorized intrusion, no matter how innocent it may seem, is a crime punishable by a fine, imprisonment, or both. Second, management must be willing to run a complete background check on each prospective employee to determine whether any security problems occurred during that person's previous experiences. It is not uncommon for individuals convicted of some type of IT crime to apply for a similar position once released from incarceration.

On the other side of the coin, a growing trend calls for hiring ex-hackers to help organizations identify weaknesses in their security systems. These individuals are referred to as "white hat hackers" to distinguish them from "black hat hackers" who seek to damage the system. This nomenclature was taken from Western movies, which often put white hats on the good guys and black hats on the bad guys to enable the viewer to tell them apart during a chase scene. These legitimate hackers may use freely available utilties downloaded over the Internet to test the security of the network. The ease with which they can move around the network undetected often surprises even the white hat hacker. However, these surprises can help network supervisors close holes in their security system.

Another way to tighten computer security is to involve users in the design of the security system. Although failure to take this step may still allow a system to be secure, users may end up spending more time devising ways to circumvent the security measures than following appropriate security guidelines. Finally, users of the system need to recognize that people, not machines, bear the real responsibility for computer security. This aspect of computer ethics is growing in importance as the use of computers grows.

Security Summary

Figure 11.12 summarizes the methods that can be used to protect an organization's enterprise network. Note that an employee who wishes to access the intranet server must pass a human guard, satisfy some type of identification system, enter a locked door, and then log on to an intranet client that is linked via a LAN to the server, using a personal identification code or user name. The employee must then provide an acceptable password and may have to go through an additional authentication procedure. At this point, the employee is logged on to the system and passes through an encryption process before actually accessing the intranet server. A user seeking to access the server from the Internet must go through a firewall before following the same process as the internal user.

1. Why is education the only real way for a company to curb employee online access to sexually explicit materials?
2. In what ways are human aspects of security a part of Figure 11.12?

Figure 11.12 Methods used to protect enterprise networks

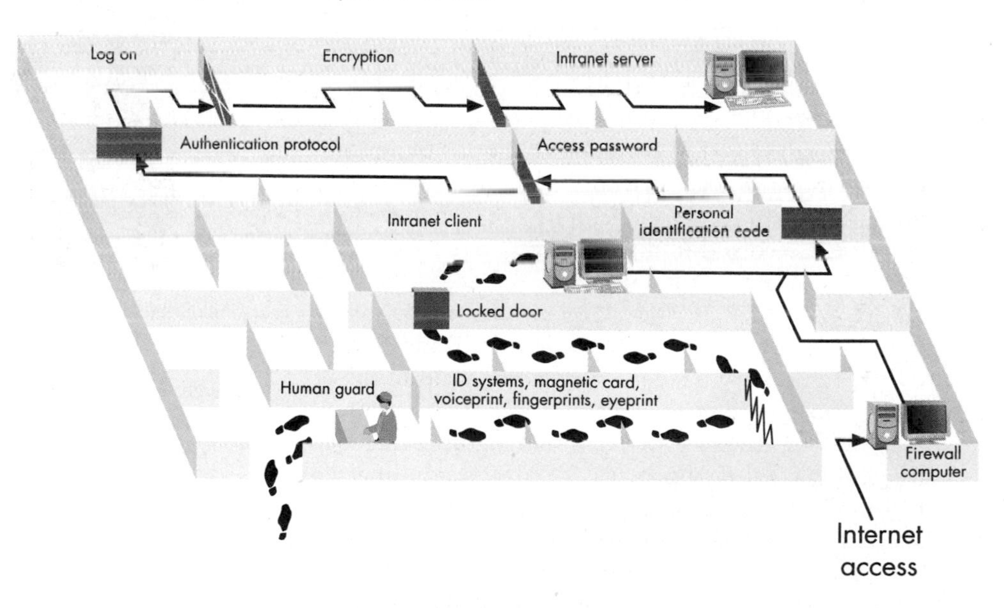

CaseStudy

Empire BlueCross BlueShield

Empire BlueCross BlueShield is the largest health insurance provider in New York state, serving more than 4 million people. Prior to September 11, 2001, Empire's headquarters were located in 10 floors of the World Trade Center, where a total of 1900 people worked. In the terrorist attack on that date, Empire lost nine employees and two consultants. At that time, Kenny Klepper, senior vice president of systems, technology, and infrastructure, was in Bangalore, India, investigating an off-shore code maintenance opportunity there run by IBM. As a result, for the four days after the attack, when no air traffic moved into the United States, Klepper had to direct Empire's IT staff from half a world away with the help of IBM. IBM set up a Bangalore hotel conference room as his office. It was equipped with two around-the-clock open conference-call voice circuits to New York and direct e-mail connections via IBM's global network. From there he was able to determine the extent of the IT loss in terms of data, equipment, and, possibly, people.

As soon as he could arrange a flight to New York, Klepper headed back to continue his work. Once home, he began to rebuild Empire's IT infrastructure. This effort involved finding new office space in midtown Manhattan. Klepper also replaced the 265 servers, more than 2200 desktop systems, and 413 laptops lost in the collapse of the World Trade Center towers. Temporary headquarters were set up in the W Hotel on Lexington Avenue, where significant upgrades to the network infrastructure included installing a gigabit Ethernet LAN that was connected to four T1 lines running at 1.45 Mbps. From the W Hotel, efforts began to rebuild the Web-based interface to the company's systems that had been lost in the disaster—10 days worth of source code. This project was considered critical in serving Empire's customers. Fortunately, the entire object-oriented executable code had been electronically transferred to the company's data center on Staten Island. Finally, Empire converted one of its automated call centers to track the whereabouts of the employees working in the WTC.

Just when it looked like things were starting to turn around six weeks after the attacks, Klepper had to deal with an anthrax scare at its data center. The anthrax report was based on a suspicious piece of mail, but turned out to be a false alarm. Even so, it added to the difficulties Empire faced in conducting business as usual in times that are far from normal.

Empire BlueCross BlueShield was able to set up operations in the W Hotel shortly after the September 11, 2001, disaster.

Source: Bob Brewin, "Empire Blue Cross IT group undaunted by WTC attack, anthrax scare." *Computerworld*, November 1, 2001, http://www.computerworld.com.

Think About It

1. Visit Empire's Web site at http://www.empirehealthcare.com/disaster_recovery/index.shtml and read the press releases from the period after the WTC tower collapse. Comment on the manner in which management communicated with both employees and customers during the crisis.
2. Discuss the ways in which Klepper dealt with rebuilding Empire's IT infrastructure to handle customer needs.
3. Discuss any changes in IT policy that Empire might make in light of its experience with the WTC disaster.

SUMMARY

To summarize this chapter, let's answer the questions posed at the beginning of the chapter.

How has the face of crime changed in the networked economy? In the networked economy, individuals and organizations have found ways to use information technology for illegal purposes. Unfortunately, you can expect the volume of information technology crime to grow right along with the growth of the Internet and electronic commerce. The movement into the networked economy means that criminals will adapt their techniques to this new environment. An IT crime is an illegal act that requires use of information technology. The nature of crime has changed from the industrial economy to the networked economy, becoming more remote, having higher financial implications, using an electronic rather than a physical format, and carrying a lower risk for the criminal. Four types of IT criminals exist: employees, outside people (hackers), organized criminals, and cyberterrorists.

What types of crime are having the greatest effects on individuals and organizations in the networked economy? IT crime can be classified into four categories: theft, fraud, copyright infringement, and attacks on information technology. Theft involves stealing hardware, data, or information from individuals or organizations. Fraud entails use of the Internet to carry out fraudulent activities by deceiving victims. Copyright infringement involves the illegal reproduction of software, music, or videos over computer networks. IT attacks target organizations and individuals over the Internet with a goal of damaging software, data, information, or reputation.

Types of fraud include convincing users to send money for unneeded services, work-at-home schemes, or multilevel marketing schemes, as well as committing identify theft, travel fraud, telephone fraud, credit card fraud, investor fraud, creating illegal IDs, and online auction fraud. Businesses (as opposed to individuals) face the greatest risks from credit card fraud, investor fraud, online auction fraud, and illegal IDs.

In copyright infringement, someone uses copyrighted software, music, or videos (all forms of intellectual property) without paying the copyright holder a fee or obtaining permission to use the copyrighted material. Two types of copyright infringement are installing software on computers without a license to do so (software piracy) and illegally downloading music or video files. Software piracy cost software companies $12 billion in 1998. The Napster music sharing service was shut down because it was violating copyright laws.

What kinds of attacks are being made on information technology? Information technology attacks include use of malicious software to destroy data or software and strikes against Web servers. The most widely used destructive software are viruses and worms, many of which cause a large amount of financial loss. Attacking Web servers with a denial of service (DoS) attack can cause the servers to crash when they become overloaded with too many requests.

The Internet remains vulnerable to outside attack because of the open nature of Internet protocols and the use of the UNIX operating system used to run many Web servers. Hackers have also found holes in Windows operating systems for servers. The primary laws used against IT crime are the Computer Fraud and Abuse Act of 1986, the NET Act of 1997, and certain provisions of the USA Patriot Act of 2001. The European Union has been active in drafting anti-computer crime legislation as well.

How can organizations protect themselves from crime in the networked economy? For information technology, security involves protecting IT assets—including hardware, software, data, and information—from both natural and criminal forces. Types of security range from a common-sense approach of locking doors to sophisticated methods of protecting intranets from outsiders. IT security can be divided into physical security, data security, and Internet security. Physical security protects computer hardware from theft or damage, whether caused by nature or people, in the same way that other office equipment would be protected. It includes always backing up data to a separate location and protecting computers from too much or too little electricity. Data security guards software and data against unauthorized manipulation, destruction, or theft. It includes implementing password policies and using systems audit software, data encryption, and antivirus policies. Antivirus policies include installing antivirus software and updating it frequently.

How can an organization protect its enterprise network from intrusions over the Internet? Internet security involves protecting Web and other Internet servers from attack over the network. An organization can have a direct or an indirect connection to the Internet. An organization with a direct connection must consider the trade-offs between making its Web, e-mail, and other Internet-related servers easily available to its stakeholders and protecting the servers from unwanted intrusions. All organizations should have a plan to secure the network environment. System administrators contribute to network security by fixing known security holes in the UNIX operating system and Web server software via installation of patches in the systems and by setting up firewalls to keep out intruders. Firewalls check each piece of information coming into the organizational computing system and watch for IP spoofing, where an intruder computer masquerades as a computer with a trusted IP address.

What nontechnical issues should you consider in dealing with IT security? Companies must consider legal issues when they connect to the Internet, including the liability associated with downloading child pornography or bootlegged software, and infringement of copyright laws by employees. Insurance against IT loss is available but tends to be quite expensive. To avoid problems, companies must educate their employees to change their attitudes toward IT security as well as run checks on new employees.

REVIEW QUESTIONS

1. Why is it difficult to determine the actual value of IT crime?
2. List the types of crimes involving information technology.
3. List and discuss three types of IT criminals. Which type of criminals comprise the largest group?
4. How did the name *hacker* arise?
5. Why is the risk to IT criminals less today than the risk to criminals 50 years ago?
6. List the various types of IT fraud. Which types primarily affect the merchant but not the individual?
7. How does SET relate to Internet fraud? What is a recent advance in reducing Internet fraud that involves a smart card?
8. Why can you legally make a backup of software but not copy it for a friend?
9. On how many computers can you generally install a single software license? How does a site license change this situation?
10. What is a "one-disk country"? List three of these countries.
11. Why was the free version of Napster shut down?
12. List the various types of destructive software. What is a DoS attack?
13. Why is it relatively easy to break into the Internet? How does the type of e-mail system in use affect this consideration?
14. What did the USA Patriot Act of 2001 add to definitions and penalties for computer crime?
15. What are typical threats to a computer system?
16. What three types of security does a computer need?
17. List three elements that should be included in any password policy that organizations follow to protect their data.
18. Why is just having antivirus software installed on a computer not adequate protection? What must be done on a regular basis to boost security?
19. In addition to those activities used to protect individual computer systems, what else must be done to protect companies with direct connections to the Internet?
20. What four actions can a firewall take when it looks at a piece of information coming into the organization?

DISCUSSION QUESTIONS

1. Discuss the types of IT crime in detail.
2. Research and discuss the No Electronic Theft (NET) Act of 1997. What penalties does this act establish for computer criminals?
3. Research and discuss the USA Patriot Act of 2001. What penalties does this act establish for computer criminals?
4. Research the anti-computer crime actions taken by the European Union and compare them to those taken by the United States.
5. Identify a recent virus or worm, and describe the way in which it attacks the computer and manages to spread itself.

RESEARCH QUESTIONS

1. Research the way in which PayPal tracked down individuals who sought to exploit its payment system, and write a short paper on your findings.
2. Find another company whose IT infrastructure was affected by the September 11, 2001, terrorist attacks and compare its response to that of Empire BlueCross BlueShield in an electronic presentation with at least 10 slides.
3. Investigate the level of losses associated with one of the worms or viruses listed in Table 11.5, and write a short paper on your findings.
4. Investigate the current state of insurance for computer crime or other IT losses, creating an electronic presentation with at least 10 slides on your findings.
5. Research the current state of security for wireless networks, and write a short paper on your findings.

Alex was in the kitchen preparing dinner when he heard what he thought sounded like "gribble nix ... jabber twillig hacked" grumbled from the upstairs office.

"What's the matter dear?" he inquired. "Did your exertions on the trail give you a bad cough?"

"No!" the exasperated voice cried, a little closer now, as Claire descended the stairs to the kitchen. "The Web site has been hacked! Come up and take a look!"

The two rushed upstairs to the computer. Claire reloaded the browser and pulled up the WildOutfitters.com home page. At first, all appeared normal. Alex looked at his wife inquisitively and was about to speak when suddenly the page seemed to burst into flames before his eyes. The flames appeared to incinerate the home page, leaving a "scorched paper" background. In big flaming letters, a message appeared in the center of the screen saying: "Congratulations, you're a victim of the Spoiled Hackers And Miscreants Club!"

Alex and Claire are learning the hard way that the Internet is a difficult environment for maintaining security and the integrity of Wild Outfitters' information. Almost anyone from any country can connect to the Internet and potentially gain unauthorized access to connected systems. In addition, the uncontrolled growth of the Internet has made it difficult to develop technical and legislative remedies for security violations. The Campagnes have given little thought to security so far, thinking that their site would be an unlikely target for IT crime. As they are finding out, in the networked economy, everyone who is connected is a potential victim.

Fortunately, all of the pages for the company's site were stored on backup tape. Claire immediately shut down the site by replacing the home page with a message about technical difficulties. She then began to remove and replace the site's pages with the clean pages from the backup tape. Later that evening, she began to look at the HTML code for the site to see whether she could determine how the hack occurred. The only changes she could find appeared in the home page. Someone had inserted a simple JavaScript program to redirect the browser to the S.H.A.M. page after 15 seconds. Finding that the hackers had made only this small change brought Claire just slight relief. She still did not know how the hackers had put the JavaScript code on the page in the first place.

For the Campagnes, this attack on their site turned out to be just a minor nuisance and disruption. It did, however, bring to light a problem with Wild Outfitters' systems. Claire began to worry about what could happen in the future if the company didn't protect its systems. Someone had already gained unauthorized access to the network and made alterations to the Web pages. What if hackers could also gain access to the database? It contained a lot of valuable information—for the company as well as for its customers. What if the information was destroyed or used to violate the privacy of a customer? Could Wild Outfitters be held liable? Could a hacker enter the system and place bogus orders? What kind of financial problems would that activity cause? Did the Campagnes have insurance to cover the possible losses? All of these questions ran through her mind and frightened her. Claire resolved to call someone familiar with IT security to evaluate Wild Outfitters' system and help make it more secure. The sooner, the better, she thought.

After a few minutes of discussion about what they should do, Claire said, "Scary, isn't it, and so realistic. It's almost as if I can smell the home page burning."

Looking puzzled, Alex sniffed the air and then cried, "Supper!"

He ran downstairs to see what he could salvage while Claire picked up the phone and pressed the speed-dial button for the local Chinese takeout.

Think About It

1. How would you classify the type of IT crime perpetrated against WildOutfitters.com? Which type of IT criminal performed the crime?
2. Do you think that the Campagnes should report this crime? Explain your choice by discussing the pros and cons.
3. What steps should the Campagnes take for securing their system? Should they look for holes in their network? Would utilizing a Web hosting service provide them with greater security?

Hands On

4. Search the news and the Web for recent examples of IT crime. Describe several of these incidents. What categories of IT crime did you find? Who were the victims? Do you notice any trends in the crimes?

Hands On

5. Pretend that you are an IT security specialist. Prepare a presentation for the Campagnes describing the IT security problems to which they might be vulnerable and how they might protect themselves against these problems.

CHAPTER 12

PRIVACY AND ETHICAL ISSUES IN THE NETWORKED ECONOMY

LEARNING OBJECTIVES

After reading this chapter, you will be able to answer the following questions:

- Why has privacy become an important issue in the networked economy?
- What would be the negative effects of stringent privacy laws?
- How is information technology used to collect private data and information?
- What threats to personal privacy exist in the networked economy?
- Why is identity theft such a fast-growing type of financial crime?
- What approaches are used to protect personal data?
- What ethical issues must be dealt with in the networked economy?

Is Your Identity at Risk?

A growing type of crime in the developed world is identity theft, in which criminals use information about the victim to create a virtual person who runs up bills and ruins the credit of the real person. The three following incidents represent just a few of the cases of identity theft that occur every day.

Just the ticket. When David's neighbor, Bill, offered to get him some cheap airline tickets, David throught it odd that Bill asked for his Social Security number, but gave it to him anyway and got the promised airfare. While he was on his trip, Bill called to tell him of a break-in and asked for a copy of his driver's license to fill out a report. When David returned home, he was surprised to receive a $30,000 insurance policy in his name for two watercraft that he did not own, but with Bill's son listed as an authorized operator. After a little checking, David found that the two watercraft had been purchased in his name, but the mailing address had been switched to a post office box. He also determined that his helpful neighbor Bill had used David's name to open three credit card accounts, charging $4375 on one card to make a down payment on the watercraft.

A free ride? Not until the criminal went on a $10,000 shopping spree did Karl realize that someone had stolen blank checks out of his mailbox. As it turns out, this event was just the start of Karl's problems; four years later, someone used his Social Security number from the checks to apply for credit cards, and a man arrested in another state on drug charges used one of the stolen checks to post bail. Karl thought he had resolved the last problem by demonstrating to authorities that he and the other individual were of different races, but he still receives letters stating that his driving priviledges in that state have been revoked. He jokes that he always asks his wife to drive when passsing through that state.

Who wants to be a millionaire? When the financial firm Merrill Lynch received an e-mail request to transfer $10 million from an account belonging to Thomas Siebal, owner of Siebal Systems, it raised enough suspicion that the firm checked with Siebal. He informed them that he had initiated no such transfer, so Merrill Lynch reported the transfer request to authorities. Following a trail of evidence starting with the e-mail message, police found mailboxes in the names of Paul Allen, co-founder of Microsoft, and James Cayne, head of Bear Stearnes. When captured, the perpetrator was picking up equipment to create phony credit cards and carrying the Social Security numbers, home addresses, and birth dates of 217 CEOs and celebrities, which he had found in *Forbes* magazine in an article titled "The 400 Richest People in America." He also had more than 400 stolen credit card numbers.

Sources: Peter Franceschina, "Identity theives growing in number." *News-Press,* May 27, 2001, pp. 1A, 10A; Sheila M. Poole, "Identity theft rising sharply as felons get more devious." *Atlanta Journal-Constitution,* December 16, 2001, pp. D1, D7; and "Tycoons targeted in alleged identity fraud scheme." *Atlanta Journal-Constitution,* March 21, 2001, p. E3.

Information Technology and Privacy

In this chapter, you will learn about how the networked economy affects the individual's personal privacy. As the national and international community has increasingly begun to depend on the Internet and Web for commerce and information, protecting personal information has become a greater concern, as shown in the examples of identity theft described in the opening case. Identity theft is a crime of growing proportions that attacks privacy at its very core, by stealing identities.

The problem of privacy on the Internet is an important area for you to consider, both as an employee in a networked economy organization and as a consumer on the Internet. As an employee, you should know why your company needs to collect data on its customers so as to serve them better while remaining sensitive to their concerns about loss of privacy. You should also be aware of why your company may want to monitor employee computer use and what your privacy rights are. As a consumer, you should recognize threats to your privacy and know how to deal with them. Both considerations require a look at ethics in the networked economy: Although it is often not illegal, invasion of privacy can be unethical.

privacy
Freedom from unauthorized intrusion; on the Internet, the right of users to control personal information and the capability to determine if and how that information should be obtained and used.

Privacy is freedom from unauthorized intrusion.[1] In the case of the Internet, this definition can be expanded to mean that users have the right to control personal information and the capability to determine whether and how that information should be obtained and used. Privacy is often confused with confidentiality, but privacy is broader because confidentiality involves protecting already acquired information from outside intrusion. Many people assume that U.S. citizens have a right to privacy. In reality, the word *privacy* never appears in the U.S. Constitution. Although the Fourth Amendment precludes illegal search and seizure, it does not mention privacy. In fact, the only country with an explicit constitutional guarantee to privacy is Germany.[2]

Privacy usually falls under the broader aegis of *common law*, meaning that it is considered to be a given right that does not need to be guaranteed by a constitution. Advances in technology associated with the Internet, however, have pushed this concept to the limit. Virtually no laws restrict the sharing of personal information between companies, often prompting the concerns raised in this chapter.

How concerned are consumers about their privacy? Various surveys have shown that nearly 60 percent of Web users in the United States are "very concerned" with threats to their online privacy. Respondents were particularly worried about the personal information they provided online being shared with third parties without their consent.[3] Some experts believe that the fact that many prospective customers seem concerned with threats to their privacy represents a major impediment to continued growth of electronic commerce.

Although you may think that this concern for privacy arose with our heavy dependence on information technology (IT), previous technological advances also brought concerns for privacy. For example, the introduction of both the telegraph and telephone in the latter half of the nineteenth century raised privacy questions. Would telegrams be protected by mail privacy laws, or would they be open to reading by the authorities? Could authorities listen in on telephone conversations if protecting the public good seemed to warrant that intrusion?

Information technology brings with it a number of new privacy concerns. First, the growth in the use of the World Wide Web and Internet for research, commerce, and recreation has opened new avenues for organizations to gather personal data on users. Second, tremendous amounts of personal data can be stored on modern database

1. *WWWebster Dictionary*, http://www.m-w.com/cgi-bin/dictionary.
2. Ann Cavoukian, "Go beyond security—Build in privacy: One does not equal the other," Cardtech/Securtech 96 Conference, Atlanta, May 14-16, 1996.
3. Marilyn Geewax, "Poll: Open records, privacy key." *Atlanta Journal-Constitution*, April 4, 2001, p. A11.

Table 12.1 Examples of Databases Containing Personal Data

Category	Database
Government data	Crime databases, immigration records, tax records, Social Security records, welfare records, and armed forces records
Public data	Marriage and divorce proceedings, bankruptcy filings, voting registrations, property taxes, land titles, birth and death records, occupational licenses, hunting and fishing licenses, and court records
Organizational data	Company personnel files, including intelligence, aptitude and personality tests, and supervisor appraisals Political party and club membership lists
Medical data	Hospital stay, doctor visit, and pharmacy records Psychiatric and mental health records Insurance records Workers' compensation records

servers and then matched with data on other databases in a matter of *seconds* to create a very complete picture of groups of consumers or even of individual consumers. Without information technology, such a search would require days, if it were possible at all. Finally, personal data collected and stored for legitimate reasons can become public through software or employee errors, creating problems for the people about whom the data were collected.

The current privacy concerns have their roots in the mid-1960s when, after hearings before the U.S. House of Representatives, Congress rejected a Bureau of the Budget proposal to set up a national database on citizens. People feared that such a database of information on individuals would give the government too much power. Although the United States does not currently have an all-encompassing national database storing data on everyone for every contact individuals have with the federal and state governments, numerous public and private databases containing a great deal of personal information have evolved, and their number increases almost daily. The personal computer has allowed even small organizations to set up databases of client or customer data and communicate with larger database servers storing public data. Many of these public data banks exist for federal and state government use, but numerous private databases have emerged as well. Table 12.1 provides examples of government, public, and private databases that store personal information. It distinguishes between government and public data because most government data are confidential, whereas public data are available from courts or other government sources. Data that are created by information technology are discussed in a separate section.

The Changing Nature of Privacy

To understand the changing nature of privacy concerns, consider Table 12.2, which compares the industrial economy, represented by the year 1952, with the networked economy, represented by the year 2002. This comparison foucuses on three aspects of privacy: searchability, protection, and integration. Searchability refers to the ease with which information can be found about someone. Protection refers to the level of security provided for truly critical information. Integration refers to the degree that data from separate sources can be integrated.

Table 12.2 The Changing Nature of Privacy

Aspect of Privacy	Industrial Economy (1952)	Networked Economy (2002)
Searchability	Difficult	Easy
Protection	Little	Much
Integration	Very little	Moderate

As Table 12.2 shows, searchability was problematic in 1952, but today it is easy. Today, much information is readily available over the Internet using search engines; in 1952, you would have had to visit many offices, go through many books, and ask many questions of people to find a even a fraction of this information. However, because searching was such a difficult task in 1952, less of a need for protection existed. You may have seen movies in which private detectives talk their way into an office to look for information in a filing cabinet. Today, although an avalanche of information is freely available over the Internet, truly critical information is better protected. Finally, almost no integration of personal information took place in 1952 because it was kept in a variety of locations with little or no communication. Today, because software can link widely separated databases—for example, using Social Security numbers to create a fairly complete picture of an individual—moderate integration has become possible. A high level of integration would assume that these databases are automatically linked with no need of additional software. Figure 12.1 shows the difference in approaches between 1952 and 2002 to finding data on individuals.

As Table 12.2 reveals, privacy is at greater risk in the networked economy than it was in the industrial economy because of increased searchability and integration. At the same time, privacy can be better protected if the organizations retaining personal information follow good security practices. Consumers face the greatest risk of having their privacy threatened when organizations choose to cooperate in sharing information, use the information inappropriately, or do not follow proper security procedures.

Figure 12.1 Comparison of approaches to finding data on people

Privacy Trade-offs

Privacy in the networked economy is an evolving issue, with at least two sides to it. On the one hand, no one wants to have his or her privacy intruded upon without permission. On the other hand, a number of studies have shown that implementing stringent privacy restrictions would negatively affect both individuals and the economy as a whole. Strict privacy requirements could result in higher prices, higher mortgage rates and fewer loans, fewer free Web sites, and less shopping convenience. As you know, information technology forms the infrastructure of the networked economy and information is the power that runs this infrastructure. If authorities imposed restrictions on the free flow of information, it could have a cascading detrimental effect on the entire economy.[4]

For example, information enables retailers to direct their advertising to likely customers through such techniques as customer relationship management (CRM), as discussed in Chapter 6. Such narrow advertising campaigns are more cost-effective than broad advertising campaigns, with the customer receiving the benefits of the cost savings.

High mortage rates and fewer loans might result if lenders had less access to credit information. They would have to increase their rates to account for failed loans or refuse more loan applications. A study by Ernst & Young LLP concluded that restrictions on information availability could result in the equivalent of a 3 percent excise tax on loans.[5]

Today, users enjoy visiting many free Web sites that provide infomation, entertainment, and shopping. Without the free flow of information, many of these free sites might have to charge for their services. Currently, these sites can defray their costs by selling the information they glean about visitors' Web browsing and shopping habits.

Finally, although electronic commerce remains only a small fraction of the overall global economy, it is predicted to grow rapidly over the next few years. Business and industry constantly strive to create Web sites that will attract customers and other stakeholders. One technique involves *mass customization*—customizing the Web site to each visitor by learning about the visitor's preferences. They can do so by asking for information about the visitor or by tracking the visitor's purchases and learning about his or her buying habits. In either case, to use mass customization, an organization must create profiles of its Web customers. This information makes the Web site more useful to customers but requires that customers surrender some of their privacy.

This trade-off between the benefits of the free flow of information (such as reduced prices and improved services) and the drawbacks of possible use or misuse of this information (such as the threat to personal privacy) is not new to the networked economy. Indeed, improved service of almost any type has always required additional knowledge about the customer and a resulting loss of privacy. Previously, however, these benefits were available only to the few who could afford to have servants and tailor-made suits. In the networked economy, anyone can take advantage of the benefits, but at the cost of loss of control over personal information.

Another issue that has become important since the terrorist attacks of September 11, 2001, relates to the trade-off between security and privacy. This issue is closely linked to that of cyberterrorism (see Chapter 13). In 2001, the U.S. Congress passed an anti-terrorism act, called the U.S. Patriot Act, that made cyberterrorism a crime and gave the government broad powers under certain circumstances to conduct Internet surveillance using the FBI's Internet surveillance

4. Fred H. Cate, Indiana University Information Law and Commerce Institute, as quoted in "Protection could prove expensive" by Marilyn Geewax, *Atlanta Journal-Constitution*, March 25, 2001, p. D3.
5. Marilyn Geewax, "Protection could prove expensive." *Atlanta Journal-Constitution*, March 25, 2001, p. D3.

system, called Carnivore, *without* a court order. In addition, any ISP must turn over customer information if the FBI claims the records are needed to fight terrorism, and it is not allowed to tell the customer that such a transfer has occurred. These meausres raise concerns among many civil libertarians and highlight the many tensions that exist between privacy and the need to protect citizens from crime.

The following sections outline the types of information that can be collected using information technology and note how this information might be used to threaten privacy.

Quick Review

1. How does the issue of privacy of personal information differ in the networked economy than in the industrial economy?
2. List the arguments against restricting the free flow of information.

Data and Information Collection Using IT

Data and information collection using IT can be classified into three categories: data collected on transactions, data collected from Web visits, and data collected from communications using the Internet. Internet communications data are categorized separately from Web-visit data because different Internet protocols generate them. Table 12.3 summarizes the three types of data that can be collected using information technology.

Transactional Data

transactional data
Data that are created when a transaction takes place that requires the customer to reveal his or her identity.

Transactional data include all data created when a transaction takes place that requires the customer to reveal his or her identity. This type of event usually involves face-to-face point-of-sale, telephone, mail, or Web transactions as well as travel, credit, or communication services. For example, every time you make a purchase with a credit card, make a long-distance or mobile telephone call, rent a car, purchase an airline ticket, or carry out an automatic teller machine transaction, you are creating transactional data that are stored in at least one database. To avoid creating transactional data about yourself as you go through life, you would have to use cash or money orders for all transactions and to have no Internet, telephone, power, water, or cable accounts. In other words, you would have to live like a hermit.

Table 12.3 Types of Data Collected Using IT

Data Type	Source	Contents
Transactional data	All transactions that require buyer to reveal identity	Data on transactions using credit or debit cards, car rentals, airline tickets, lodging reservations, ATM transactions, mobile and land telephone records, prescription drug purchases, magazine subscriptions, hospital or doctor visits, and so on, either face-to-face, by telephone, or over the Web
Web-visit data	Visits to Web sites	Data and information provided to Web sites by the user or collected by the Web server during a visit by the user
Internet communications data	Newsgroups, chat groups, bulletin boards, e-mail	Postings to a newsgroup or bulletin board, conversations in chat rooms, contents of e-mail

Any time you make an airline reservation, you are creating transactional data.

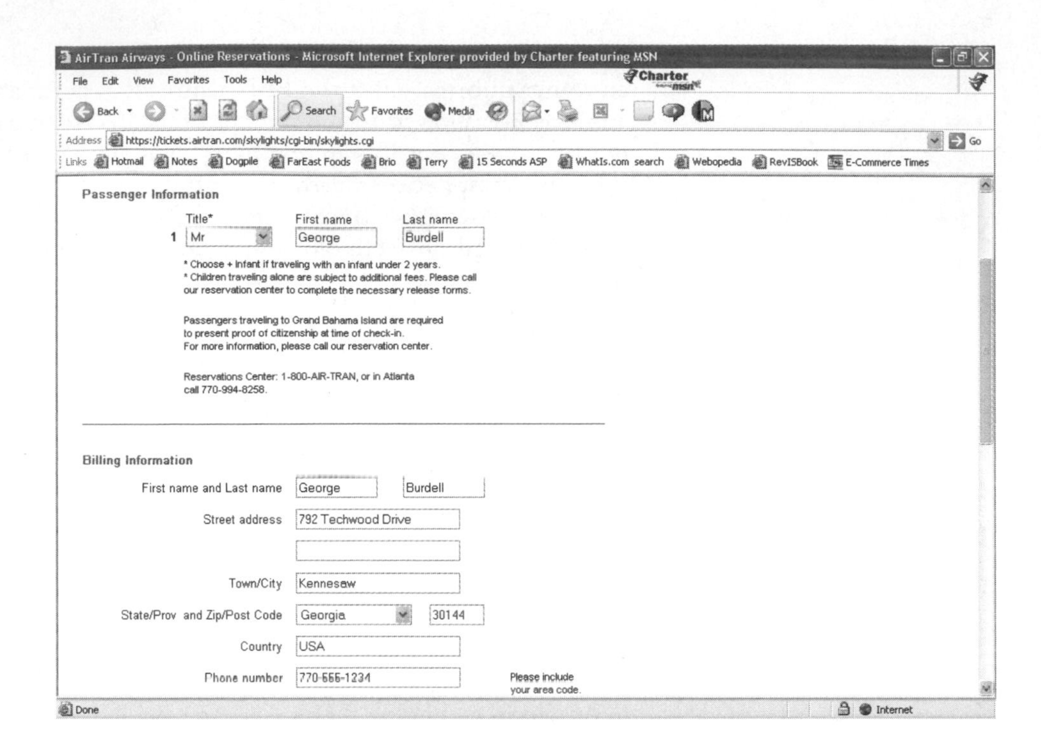

Consumers also create data about themselves and their buying habits whenever they provide product registration or contest information, either on paper or on the Web. This information may be combined with transactional data to create a private database or sold to organizations that collect these data as a sidelight of their main business. For example, when you complete a change-of-address form at the post office, you may not suspect that you are creating data for a database. In fact, the U.S. Postal Service has been known to sell these names and addresses to direct marketers!

Transactional data indicate such facts as what type of clothes you buy, where you eat, where you stay when out of town, where and to whom you make long-distance calls, where and when you take trips, and when you withdraw money from the bank. Together, such data can paint a complete picture of an individual's lifestyle—something that companies seeking to market goods or services will find quite valuable. Table 12.4 lists a number of the ways in which people create transactional data.

No U.S. federal laws prohibit companies from sharing transactional data on their customers. For example, if you subscribe to one magazine, it will often sell your name, along with those of all other subscribers, to other magazines and catalogs, leading to a substantial increase in the amount of postal mail you receive. At one time, banks routinely shared information on their customers with other financial companies, often leading to "preapproved" credit card offers. Recent legislation, however, requires financial institutions to give customers information on their data-sharing practices and offer them the opportunity to opt out of these programs. Other countries, especially those in the European Union, have passed laws regulating the practice of sharing data among companies and other organizations, and many organizations in the United States have begun to call for similar restrictions here. Conversely, the changed attitudes since September 11, 2001, may actually increase the amount of data shared with the government, at least.

Table 12.4 Data Collection Methods

Collection Method	Description
Web browser	Records date, time, source, and amount of all transactions
ATM	Records time, data, amount, and location of transaction; may also capture your image
Prescription drugs	Druggist keeps a record of all drugs purchased, including date and payment method
Employee ID scanner	Records date and time of every entry into the building, including employee name
Mobile telephone	Records date, time, and length of every call; calls can be intercepted
Credit/debit cards	Records date and amount of every transaction
Toll booths	With electronic payment of tolls, records date and time of passage through toll booth
Telephone calls	Record date, time, and duration of all calls
Bar code scanners	With an affinity card, allows store to track every item purchased
Surveillance cameras	Many banks, liquor stores, convenience stores and public buildings now video tape on a 24/7 basis
Product registration	Provides personal information as well as information on buying habits

electronic supervision
Organization managers monitoring the amount of work performed by employees on networked computers for entering data, making reservations, and so on. Also called *computer monitoring.*

In addition to transactional data generated by consumer purchases, data are often collected on employees who handle telephone purchases or deal with customer service questions. In this practice, commonly referred to as **electronic supervision** or **computer monitoring**, managers use software on the network server to monitor the amount of work performed by employees on networked computers in terms of entering data, making reservations, dealing with customer questions, and so on. Typical values measured include the amount of time a telephone operator takes to answer a request, the number of keystrokes a data entry operator enters per second, and the number of errors made by employees. A company can monitor what appears on an employee's screen and how long a computer has been idle. Although many employee organizations believe that this sort of computer monitoring represents an invasion of employee privacy, it is virtually unregulated in the United States at either the state or federal levels. Figure 12.2 depicts the process of computer monitoring.

Figure 12.2 Computer monitoring process

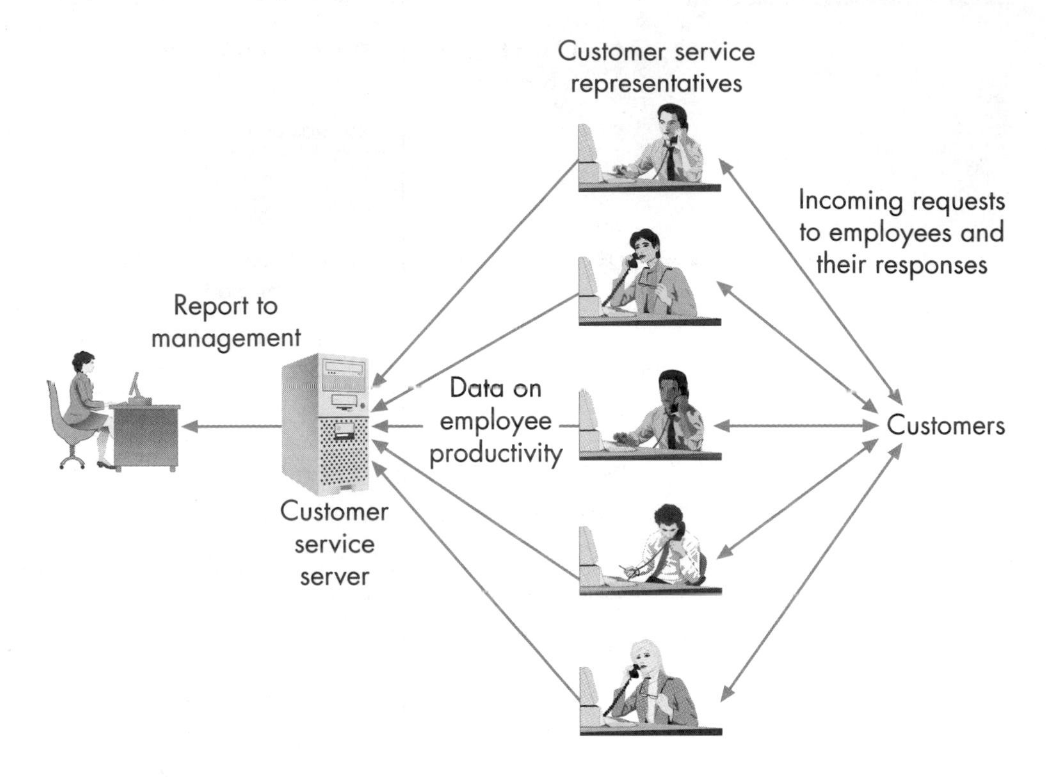

Web-Visit Data

web-visit data
Data generated about a user's computer whenever the user visits a Web site.

Web-visit data include data that are unique to the Web, in that a transaction in the traditional sense of the word does not have to occur. Any visit to a Web site generates a great deal of data about the user's computer. To see what a Web site knows about you, go to www.idzap.com/asurf.php, and click the second link—"Specifically, what can a web site find out about me?" Next, click the link for *This page*. Figure 12.3 shows some of the information gathered by this site on the author's computer. Note that this Web site indicates the operating system in use (Windows NT), the type of browser employed (Netscape Mozilla), and the type of files accepted by the browser. Although not shown here, the Web site also knows the computer's IP address and name. The Web site snoop.anonymizer.com provides even more information on you and your computer.

Figure 12.3 Information on a Web site visitor

Authentication type:
Authenticated user name (for protected pages):
User identity (if server supports RFC931):
Current web page: /cgi-bin/userdata.pl
Previous web page visited: http://www.idzap.com/asurf.php
Cookie:
Browser type: Mozilla/4.0 (compatible; MSIE 6.0; Windows NT 5.1; Charter B1)
Browser acceptable language: en-us
Acceptable data type: image/gif, image/x-xbitmap, image/jpeg, image/pjpeg, application/msword
Acceptable character set:
Acceptable encoding method: gzip, deflate
Cache setting:
Port number of user's software: 1444
Server name: www.idzap.com

This information may or may not be specific to you, depending on how you access the Internet. If you access the Internet through a dial-up connection to an ISP such as Earthlink or AOL, or through a cable modem, you are typically assigned an IP address and computer name dynamically (that is, it persists only during that dial-up session). As a consequence, Web sites know only the name of the ISP through which you access the Internet, making it difficult to track down who actually visits the Web site. On the other hand, if you access the Web from a LAN or through DSL, then your computer has a specific IP address and name assigned to it that do not change unless you reconfigure the network in some way. Thus Web sites know the specific machine you use if you access the Internet over a LAN or cable modem connection, making you more easily traceable. In either case, users who want to surf the Web anonymously can use the www.idzap.com Web site to do so for free.

If law enforcement agencies suspect you of engaging in criminal activities, they can obtain a search warrant that will allow them to view the ISP's logs, which will show which dynamic IP address you used during a dial-up session or through a cable modem. This approach has enabled authorities to track down the creators of malicious software. Hosting companies such as AOL that allow users to assign screen names to themselves are also subject to search warrants or subpoenas to turn over that information to the courts.

In addition to the data that can be collected by a Web site whenever it is visited, many Web sites request information from the visitor such as name and e-mail address, which they then save in a database. As discussed in Chapter 8, such an *adaptable Web site* alters its responses based on its memory of the information you choose to give it. In most cases, if visitors return to that Web site and enter the same information, the Web site will remember their preferences. In addition, *adaptive Web sites* use the information you give them to find and respond to patterns in your interests and purchases. For example, as shown in Figure 12.4, Amazon.com recognizes the author and knows from his previous purchases that he is a Jimmy Buffett fan. Based on this information, the site suggests other CDs by this artist as well as CDs by artists that other Jimmy Buffett fans have purchased. Most electronic commerce sites are becoming more adaptive because this approach does not require users to enter any information when they return to the Web site.

Figure 12.4 An adaptive Web site

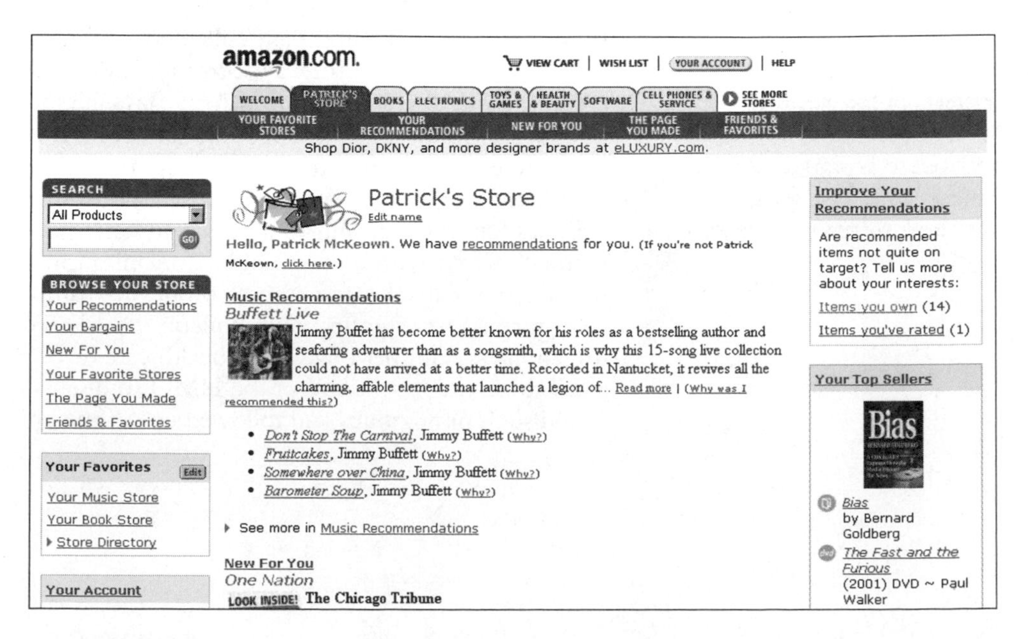

When you visit an adaptive site like the one shown in Figure 12.4, it may welcome you by name without requesting that you sign in. This response is accomplished via a cookie, which the Web server stores on your hard disk in a subfolder of the browser folder. As discussed in Chapter 8, cookies are necessary because the HTTP protocol makes each visit to a Web page independent of all others. Without cookies, a Web server has no memory of what pages it has previously sent to a user. Although you can readily set up your browser to reject the storage of cookie files or to ask you whether to store the files, remember that by so doing, you lose the customized features of adaptive Web sites.

The data collected by Web sites and stored in databases are structured (see Chapter 5), which makes it possible to combine information from multiple databases to create an accurate picture of a customer. An organization running a Web server can query these data to learn more about specific customers or groups of customers.

Many privacy advocates voice concerns about the collection of data by Web sites, either on the server or on cookies. Customers must recognize that a trade-off exists between privacy and customization of the Web site to their needs. If you rejected all cookies, then you would always have to sign in to all Web pages rather than having them recognize you. If the server did not retain any information on Web visitors, then electronic commerce applications would never be customized to users' needs. At one point, the European Union was considering legislation to outlaw the use of cookies on Web sites visited by citizens of the member countries, but finally decided to just require Web sites to alert users to the use of cookies.

Children's Online Privacy Protection Act of 1998 (COPPA)
Legislation requiring commercial Web sites that collect personal information from children younger than age 13 to obtain prior parental consent.

Although collection of data remains an option on most Web sites, it has been restricted on Web sites aimed at children. Implemented as the result of a Federal Trade Commission (FTC) report, the **Children's Online Privacy Protection Act of 1998 (COPPA)** requires commercial Web sites that collect personal information from children younger than age 13 to obtain prior parental consent. The FTC has decided that, for information gathered for internal use by the Web site operator, an e-mail consent is acceptable. For information gathered for sharing with other entities, the Web site operator must have a more verifiable form of consent, such as postal mail, fax, or credit card number. According to a 2001 study conducted by the Annenberg Public Policy Center at the University of Pennsylvania, only a minority of 162 child-oriented Web sites checked followed the FTC rules closely.

Internet Communications Data

Internet communications data
Non-Web data about individuals transmitted over the Internet, including mail messages, conversations in chat rooms, postings to bulletin boards, and messages to newsgroups.

In addition to structured data collected by Web sites, a great deal of unstructured data, sometimes called Internet communications data, exists in other Internet applications. **Internet communications data** include e-mail messages, conversations in chat rooms, postings to bulletin boards, and messages to newsgroups.

These data are unstructured because they are not organized into database form. This fact does not prevent use of the data; it just means that querying the data involves searches for words or phrases and that any integration across data sets must take place manually. Even with these problems, organizations continue to watch Internet communications data closely, because complaints, rumors, and the like are often voiced in chat rooms, newsgroups, and bulletin boards. For example, a woman expressed unhappiness with her HMO in a health care newsgroup and was surprised to receive a call from a representative of the HMO to discuss her complaints. The HMO monitored such newsgroups and followed up on any complaints posted there.

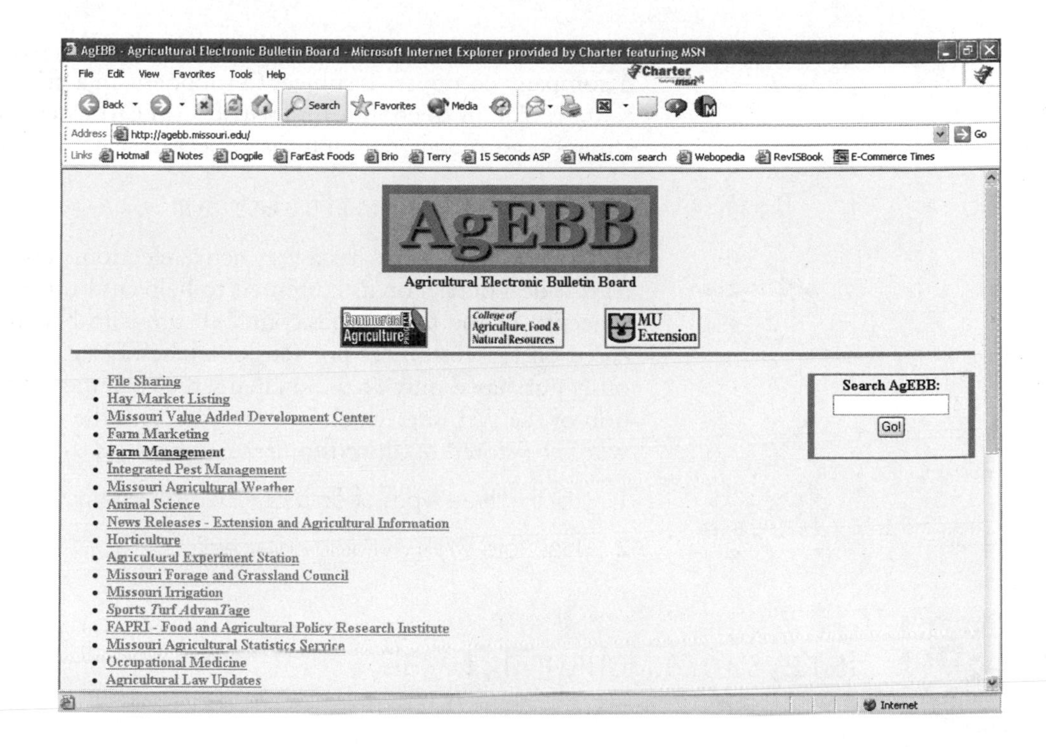

Postings on bulletin boards, like this one at the University of Missouri, are a form of unstructured data and are available for viewing by millions of people—for any purpose.

At the same time that concerns have arisen regarding public and private data banks, the capturing of transactional data, and the use of cookies with Web sites, many people happily tell the world about themselves on newsgroups or chat with people they don't know as if they were old friends. In reality, each piece of correspondence is available for viewing by millions worldwide for years to come and represents a fertile source for cyberdetectives seeking information. Newsgroup postings are even searchable, just like Web sites.

In the networked economy, e-mail has become the method of choice for sending messages. E-mail has the same constitutional guarantees of privacy and similar protections as postal mail has, with one big exception: E-mail sent or received on an employer's computer is the property of the employer. Thus employers have the right to read employees' e-mail at any time to protect themselves from potential liability and to ensure that their computers are used for legitimate business purposes. In fact, an employer's rights concerning computers go even further: The employer can monitor keystrokes for one or all computers or remove a computer or just the hard disk at any time and search it. Although most employers do not snoop into employees' e-mail, if damaging or threatening e-mail messages are sent from a company computer, they have the right to take action to stop it.

In addition to monitoring as transactions discussed earlier, companies typically monitor computer use for two other reasons: to protect the employer from a sexual or racial harassment suit and to protect the company's intellectual property. Companies look at employees' e-mails or Web surfing to identify behavior that might be considered sexually or racially offensive to co-workers and that could lead to a law suit against the company if not stopped. Many companies also monitor computer use because they have trade secrets or vital information that, if sent to a competitor, could lead to very negative results for the company, including the worst case of going out of business.

Also, as Microsoft and many other companies have found, any e-mail saved on a company computer is subject to subpoena in a criminal or civil court action. The same is true in freedom of information (FOI) cases for government agencies; anything on an agency computer is subject to search and release in an FOI case.

Data Collection at Fareastfoods.com

Because FarEast Foods has a very active electronic commerce component, it collects a great deal of data on its customers to help customize its Web site to their needs. The objective of the fareastfoods.com Web site is to become adaptive, customizing itself based on the customer's previous purchases. For example, if the customer had previously purchased only Thai food items, the customer might automatically see that portion of the Web site. Much of these data, plus the customer's login ID and password, would be stored on the customer's computer as a cookie. (It is not now done so.)

1. List the three types of data collected by information technology.
2. How are Web-visit data collected?

TECHNOLOGY ON THE EDGE

Is Keystroke Monitoring Legal?

When the FBI raided a business run by Nicodermo Scarfo, Jr., in January 1999, it used a search warrant to justify copying the contents of his computer. Agents were unable to open one computer file, believed to contain records of Scarfo's illegal gambling (bookmaking) business. It appeared to be protected by a sophisticated program that scrambled its contents when agents attempted to open it without the appropriate password. To obtain access to this file, the FBI obtained a second warrant to learn the password to the scrambling program using **keystroke monitoring**. In this technique, keystrokes from the computer keyboard are monitored either by software installed on the computer or by a hardware device that intercepts the keystrokes as they are sent to the computer. In the Scarfo case, the FBI did not specify which type of monitoring device it used. Using the results of the keystroke monitoring, the FBI was able to obtain the password, access the file, and use its contents to charge Scarfo and another man with gambling and extortion.

Scarfo's lawyers requested to see information on the keystroke monitoring technology to determine whether it went beyond the limits stated in the search warrant by intercepting Internet traffic. If the keystroke monitoring had intercepted any Internet traffic, a much more difficult to obtain wiretap warrant would have been required. Scarfo's lawyers argued that they could not determine whether the technology violated the warrant without having more information about it. To keep the technincal details out of court, the FBI claimed that the technology was protected by the Classified Information Protection Act. In the end, the judge ruled in favor of the government, and Scarfo pled guilty to bookmaking in return for the government dropping charges of loan sharking. Privacy advocates watched this case very closely, as it may be instrumental in determining the extent to which authorities can use technology in fighting crime without violating constitutionally protected rights.

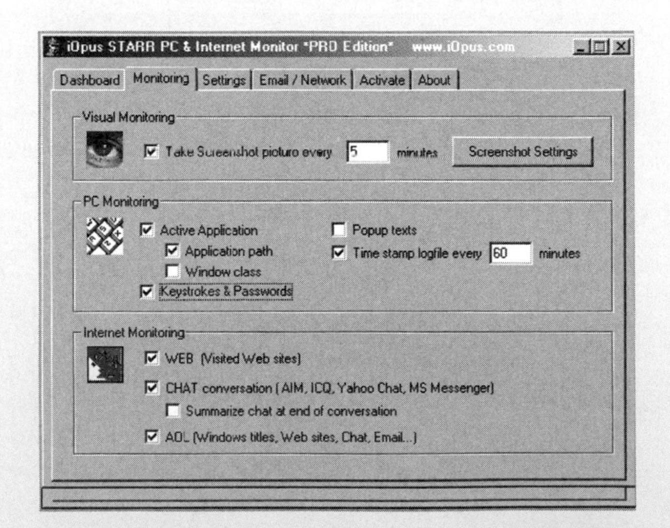

With software such as the iOpus STARR PC and Internet Monitor, you can capture the screen of a PC at set intervals, monitor activities on the computer, or report Internet activities.

Sources: Richard Willing, "FBI technology raises privacy isssues." *USA Today*, July 31, 2001, p. 3A; and George A Chidi, "FBI claims tech is national secret in mob case." *Computerworld*, August 27, 2001, http://www.computerworld.com.

Threats to Privacy

Today, people have become concerned about the threat of organizations and individuals using the Internet to access public and private databases to invade their privacy. This uneasiness has arisen because almost anyone with a personal computer and an Internet connection can search a database and retrieve information about prospective employees, tenants, or other individuals. Table 12.5 lists six general threats to personal privacy, along with the source of the threat and the information at risk. The first five threats are discussed in this section, and the next section will discuss identity theft.

Information Exposure

Exposure of confidential information can pose a threat to an individual's financial and personal well-being, whether such exposure occurs accidentally or intentionally. Accidental exposures are usually caused by gaps in the security of servers on which the information resides. A server administrator may have failed to close security holes, install required software files, or fix known bugs in server software. To prevent accidental exposure, the person or technical group running the servers should fix all security holes and software bugs immediately upon discovery. Organizations face this same problem in providing security for their networks.

Intentional exposure can occur when a hacker breaks into the network, when information is "grabbed" by a Web site that is set up to harvest e-mail address and other data, or through the sale of information. Hackers might search for information on a specific person, seek to expose an entire database to the outside world to embarrass an organization, or use the data for criminal purposes. For example, a hacker was able to download thousands of credit card numbers from a company's database. The hacker tried to blackmail the company by threatening to release the credit card numbers to the public.[6]

Table 12.5 Threats to Privacy

Type of Threat	Source	Information at Risk
Information exposure	Accidental exposure, hackers, Web sites, sale of data	Almost any type of information that is stored in a database connected to a network
Data surveillance	Public or private organizations seeking to identify individuals who fall into a certain group	Information about a person's lifestyle, spending habits, and travel history
Information brokers	Persons or organizations paid to find information about individuals	Current (and past) names, home addresses, telephone numbers, and workplaces
Spyware	Free or commercial software; Web sites	Web site visit data, user name, passwords, and other personal information
Junk e-mail	Persons or organizations that wish to advertise their "services" to anyone with an e-mail address	E-mail inbox is overwhelmed with unwanted e-mail
Identity theft	Criminals	Name, Social Security number, credit rating, and anything else that identifies a person

6. John Markoff, "An online extortion plot results in release of credit card data." *New York Times,* January 10, 2000, http://www.nytimes.com/library/tech/00/01/biztech/articles/10hack.html.

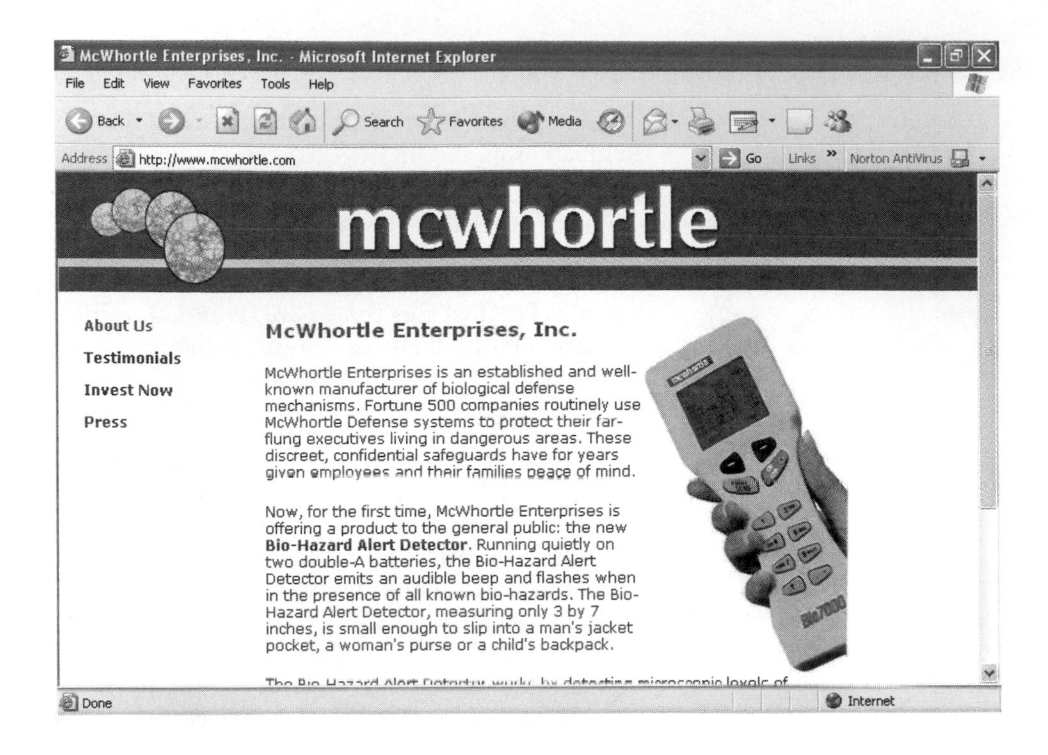

The McWhortle Enterprises Web site at www.mcwhortle.com was created to demonstrate how you can be scammed on the Web.

In the case of exposure of information on Web sites, a Web site may appear to be legitimate but actually exists merely to collect information on visitors. One Web site mimicked the World Trade Organization (WTO) site and was able to collect e-mail addresses of visitors. The hoax was so good that the phony Web site appeared on several search engines.[7]

The final type of information exposure involves the sale of information. When so many dot-com companies were going out of business in 2000-2001, they became tempted to sell one of their few remaining assets—their customer e-mail lists. For example, Toysmart.com tried to sell data on its more than 200,000 customers when it went into bankruptcy. A group of state attorneys stopped this deal, but other failing companies were able to sell their customer lists.

Data Surveillance

As discussed in Chapter 6 and earlier in this chapter, organizations often use customer relationship management (CRM) to find individuals who fall into segments or groups of interest to them. For example, universities seek to find students who will succeed in their academic programs, and retailers look for people who can afford their products and may have an interest in them. Although most uses of information technology to find groups of people who fall into certain categories do not pose problems, governments or private groups that use information technology for surveillance purposes can present a threat to privacy. From a privacy point of view, this process is referred to as **data surveillance** and entails the systematic use of information technology in the investigation or monitoring of the actions or communications of one or more people. This terminology has a much more negative connotation than CRM and highlights the continuing difference of opinion between companies that want to refine their marketing techniques and privacy advocates who view CRM as a form of surveillance.

data surveillance
The systematic use of information technology in the investigation or monitoring of the actions or communications of one or more people.

7. Peter Sayer, "Fake WTO Web site harvests e-mail addresses." *Computerworld*, October 31, 2001, http://www.computerworld.com.

Are Your Records on the Internet?

You might have automatically answered "no" to this question, but think for a minute—is your name in one of the phone books on the Web? Try the Netscape White Pages, Bigfoot (www.bigfoot.com), or Switchboard (www.switchboard.com), and see if you are listed. If you find your name on Switchboard (and you probably will), one more click will show a map of your neighborhood. If you need a more detailed map of your location, go to MapQuest (www.mapquest.com) and have it show you a map with your location pointed out!

Although having your name, phone number, and even a map to your home listed on the Internet may not be so bad, what happens when confidential or embarrassing information appears on the Web? Recent cases have shown just how insecure this medium can be. For example, in addition to the end-of-chapter case involving scholarship data, two other cases of accidental exposure of confidential information came to light in July 2001. In the first case, a service that provides online bookings for people traveling to Chicago, 877Chicago.com, was informed by its conference registration service, RegWeb, that a hole had been found in its system, exposing 300-400 credit card numbers. In the second case, the e-mail addresses of more than 600 people who were taking the antidepressant drug Prozac were released over the Internet. These people had signed up at the Web site for the maker of Prozac, Eli Lilly and Company, to receive an automated e-mail reminding them to take their dose of Prozac. When the company sent a message to all subscribers announcing the end of the service, it mistakenly included all e-mail addresses in the header. In both cases, the firms fixed the problems as soon as they were discovered, but it is unknown what kind of problems were caused before the repairs.

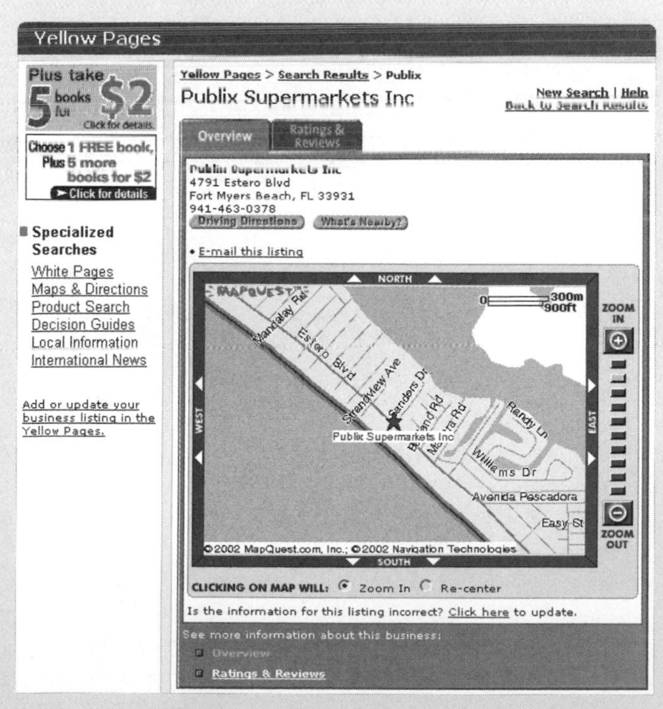

Simply entering an individual's or bussiness's name and city results in an address, telephone number, and map of the location.

Sources: Brian Ploskina, "Hundreds of credit card numbers exposed online." *Interactive Week*, August 1, 2001, http://www.zdnet.com/eweek; and Charles Wilson, "Prozac maker let slip hundreds of e-mail addresses." *Atlanta Journal-Constitution*, July 6, 2001, p. A3.

Regardless of the name (CRM or data surveillance), this process takes advantage of the fact that an individual's record in one database usually shares one or more attributes in common with that same individual's record in other databases. In many cases, this attribute is the Social Security number, which has become virtually a universal identifier for residents of the United States. A credit bureau such as Experian can combine its data on people's income, jobs, bank accounts, purchasing behavior, and credit limits with public data that it draws from motor vehicle and public property records, such as those described in Table 12.1, to create sophisticated lists of consumer information. For example, it might generate a list of all people in Gwinnett County, Georgia, who purchased (and registered) a Mercedes automobile within the last six months, make more than $150,000 per year, and have no more than two children. The process of matching records in two or more databases to determine which records exist in both databases and using the results to create a profile of the individualis is known as **computer matching** or **profiling**. Figure 12.5 graphically depicts this process. Note that after creating a list of raw hits, the organization must filter and edit the list to remove erroneous items and create a list of solid hits, which goes into the organization's database. A danger exists if some of these data are wrong and an individual is refused credit because of the error.

computer matching
The process of matching records in two or more databases to determine which records exist in both databases. The data are often used to create a profile of the individual. Also called *computer profiling.*

The databases used in the computer matching process often contain transactional or Web-visit data because they may yield a very clear picture of a consumer's spending habits. Computer matching can also be an efficient method for alerting consumers to products in which they may have an interest while not bothering

Figure 12.5 Computer matching process

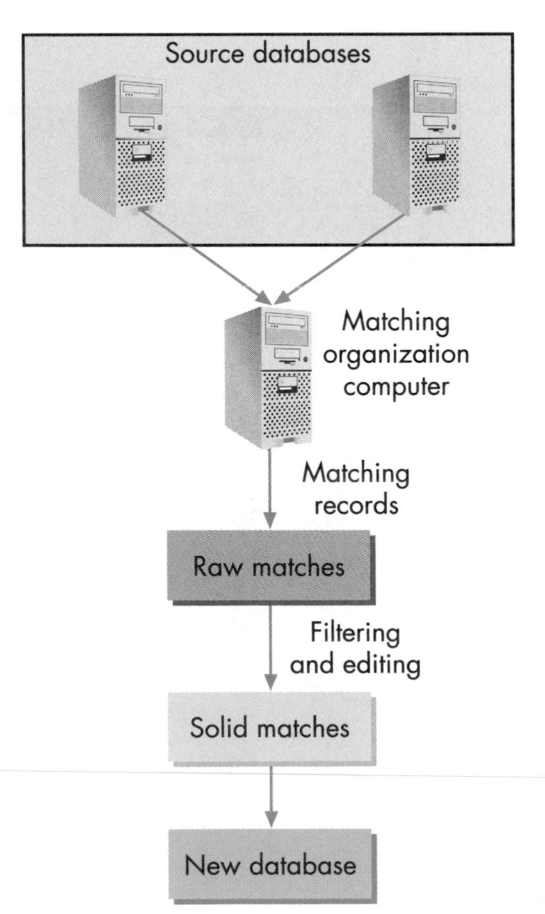

Finding information using a computerized database is much easier than searching through dozens of file cabinets like these.

them with products of little interest. Of course, companies also can prejudge the economic behavior of consumers, thereby limiting some groups' access to information about goods and services. Computer matching may also lead to annoying telephone calls from telemarketers or unexpected and sometimes disconcerting offers from businesses that you have never visited.

At one time, the U.S. government made heavy use of data surveillance techniques to try to increase its efficiency. For example, individuals who had not repaid their government student loans were matched with individuals eligible for tax refunds. This effort enabled the IRS to collect some of the money owed the government (an average of $544.91 per case for a cost of only $3.70). However, in response to complaints that the federal computer matching generated many erroneous hits, the **Computer Matching and Privacy Protection Act of 1988 (CMPPA)** took effect in January 1990. It established a number of fair information practice provisions to apply to data surveillance in the federal government, although it excludes many programs and agencies.

Computer Matching and Privacy Protection Act of 1988 (CMPPA)
A federal law that establishes a number of Fair Information Practice provisions to apply to data surveillance in the federal government but excludes many programs and agencies.

Although there have been numerous calls for regulation to protect the public from the abuses of data surveillance, no new legislation has emerged in recent years. And, as discussed earlier in this chapter, such legislation could have unintended negative effects on consumers and the economy.

Information Brokers

information brokers
Individuals who use databases and other sources to find information on individuals.

With the proliferation of both private and public databases, a new breed of people and organizations called **information brokers** has found ways to profit from these databases. Information brokers will use almost any means—usually legal, but sometimes not so legal—to provide information on an individual of interest to their customers. Typical information provided by these brokers includes credit reports, criminal histories, unlisted phone numbers, and so on. Information brokers find this information in a variety of ways—some the electronic way, and some the old-fashioned way, through personal contact. In the first case, information brokers tap into the growing number of online proprietary databases that store public information. As recently as 10 years ago, information such as death records, marriage licenses, and property deeds was stored in town halls and courthouse basements, requiring a personal visit to review it. Now companies such as Choicepoint make much of this information available online for a fee. Finding information on a person still requires a great deal of research, but now it can be accomplished from the researcher's computer. Figure 12.6 shows the type of report (for a fictitious person) that an information broker can provide.

Figure 12.6 Information broker's report

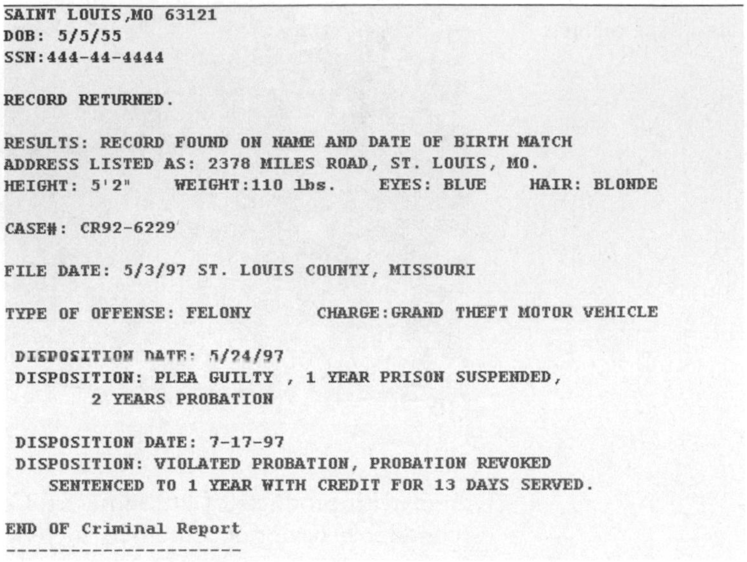

```
SAINT LOUIS,MO 63121
DOB: 5/5/55
SSN:444-44-4444

RECORD RETURNED.

RESULTS: RECORD FOUND ON NAME AND DATE OF BIRTH MATCH
ADDRESS LISTED AS: 2378 MILES ROAD, ST. LOUIS, MO.
HEIGHT: 5'2"     WEIGHT:110 lbs.     EYES: BLUE     HAIR: BLONDE

CASE#: CR92-6229

FILE DATE: 5/3/97 ST. LOUIS COUNTY, MISSOURI

TYPE OF OFFENSE: FELONY        CHARGE:GRAND THEFT MOTOR VEHICLE

 DISPOSITION DATE: 5/24/97
 DISPOSITION: PLEA GUILTY , 1 YEAR PRISON SUSPENDED,
        2 YEARS PROBATION

 DISPOSITION DATE: 7-17-97
 DISPOSITION: VIOLATED PROBATION, PROBATION REVOKED
    SENTENCED TO 1 YEAR WITH CREDIT FOR 13 DAYS SERVED.

END OF Criminal Report
----------------------
```

Gathering information through personal contact is a technique carried over from the industrial economy. One such approach involves a process known as *pretexting*, in which an unscrupulous information broker calls a business or agency under the *pretext* of being the person on whom they are seeking information and asks about that individual's account. If the information broker has the person's Social Security number, he or she can find almost any type of information on the target individual; if not, the information broker must work a little harder, but eventually may find the needed information. Often success occurs when a person at the business or agency unknowingly uses a computer to look up the information for the information broker. Note that many companies engaged in information brokering have spoken out strongly against the use of deception to find information.

Spyware

spyware
Software that gathers information about online activities and transfers it back to a server without your knowledge or permission.

A recent threat to privacy comes from a type of software known as spyware. **Spyware** gathers information about online activities and transfers it back to a server without your knowledge or permission. Three types of spyware exist: spyware that is part of other software, a Web bug, or a virus or worm. Virus/worm software was discussed in Chapter 11, so this section will concentrate on the first two types.

In the first case, spyware can unknowingly be installed on a computer as part of a commercial software package on CD or as a module of a freeware or shareware package that is downloaded over the Internet. In either case, the spyware runs in the background, watching online activities and reporting them back to a server. The parent software can be a game demo, an MP3 player, a utility package, or any number of other software types. For example, in 1999, RealNetworks, which makes the popular RealPlayer media players, was found to be gathering information on users' listening habits, preferred music types, and other information from users who installed its RealJukebox program. Although RealNetworks asserted that it was just gathering information that would help it customize RealJukebox, it quickly released a patch to the software that would disarm the reporting capabilities. Other companies that have collected or continue to collect information using spyware include Comet Cursor, DoubleClick, Gator.com, Radiate, and webHancer.[8] This type of software was a key factor in Tom Clancy's popular novel, *The Bear and the Dragon*.

The best defense against the installation of this type of spyware on your computer is to use a firewall (discussed in Chapter 11) or a software package, such as SpyCop or Lavasoft Ad-aware, that specializes in watching for spyware programs or the data they send back to a server. Figure 12.7 illustrates how spyware works. The spyware checks any data coming into the CPU, and interesting messages are stored on hard disk or possibly sent over the Internet to the person installing the spyware.

Web bug
A tiny image file (usually 1 pixel by 1 pixel) on a Web page that gathers data on the user's online activities.

The second type of spyware is the Web bug. A **Web bug** is a tiny image file (usually 1 pixel by 1 pixel) on a Web page that gathers data on the user's online activities. It can collect information on your IP address, take note of your Web visits, or even access cookies on your hard drive and send that information back to a server. You usually cannot see Web bugs, and cookie and virus filters do not catch them. A variety of "bug-catching" utilities, such as Flow Protector or Bugnosis, can help users avoid this problem.

Figure 12.7 Spyware process

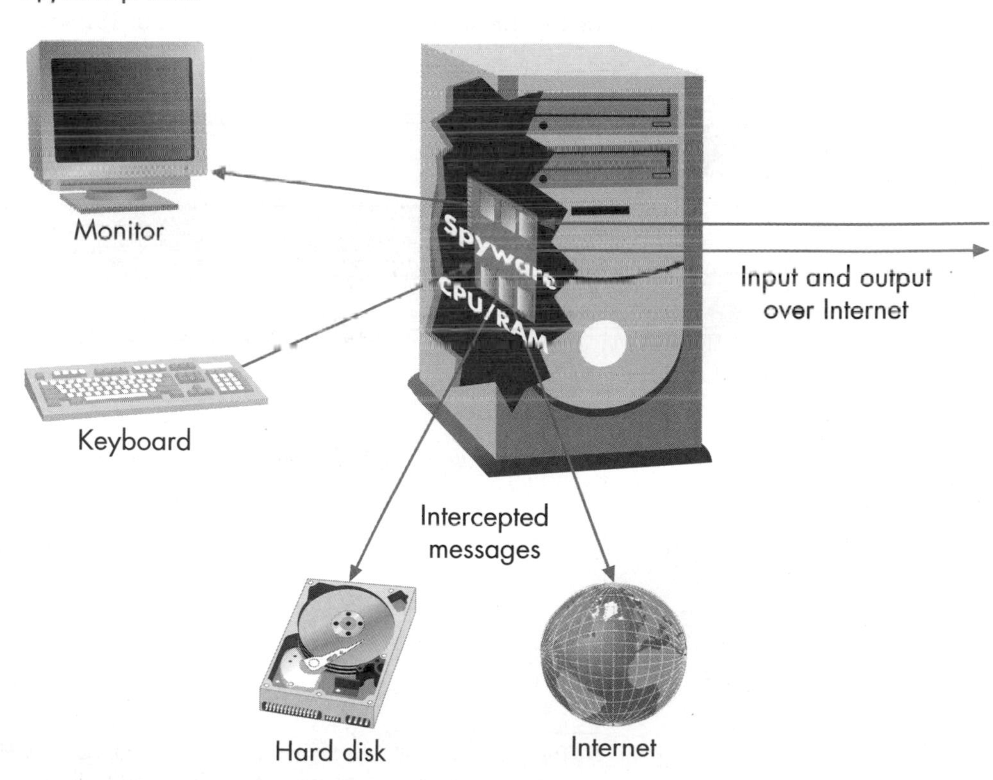

8. Bernard Dy, "They're counting on you." *Smart Computing*, January 2002, pp. 45-47.

Junk E-Mail

Just as junk mail and telemarketing calls were used in the industrial economy to advertise or attempt to sell products and services, e-mail is now being heavily used in the networked economy for similar purposes. This chapter on privacy covers this topic because many people feel that their privacy has been breached when they receive offending junk e-mail.

spam
Junk e-mail.

Commonly referred to as **spam**,[9] junk e-mail is extremely inexpensive to send and (at least in monetary terms) to receive. Because of the low cost, an organization can send spam to enormous numbers of people, and only a small proportion of those receiving the e-mail need to respond for it to be cost-effective. Individuals who post messages to particular newsgroups often receive spam. Others unknowingly sign up for it when they register at a Web site. Because many people replied to spam by *flaming* the sender (that is, sending a nasty reply) or by overloading the spammer's server by sending many messages or very large messages, spammers now use *nonrepliable* messages in which they omit the reply-to address. Instead of replying by e-mail, you must use fax or the postal service to reply to the "wonderful" offer in the e-mail.

In most cases, the best way to respond to spam is simply to delete it before reading the message. Replying in any way confirms to the sender that a person exists behind the e-mail address, and it usually adds your e-mail address to several other spam lists. Figure 12.8 shows an example of spam received by the author in which the "From" and "To" addresses (which have been deleted) were the same and not that of the author. Using filters from your ISP is another way to try to avoid spam.

Figure 12.8 Example of spam

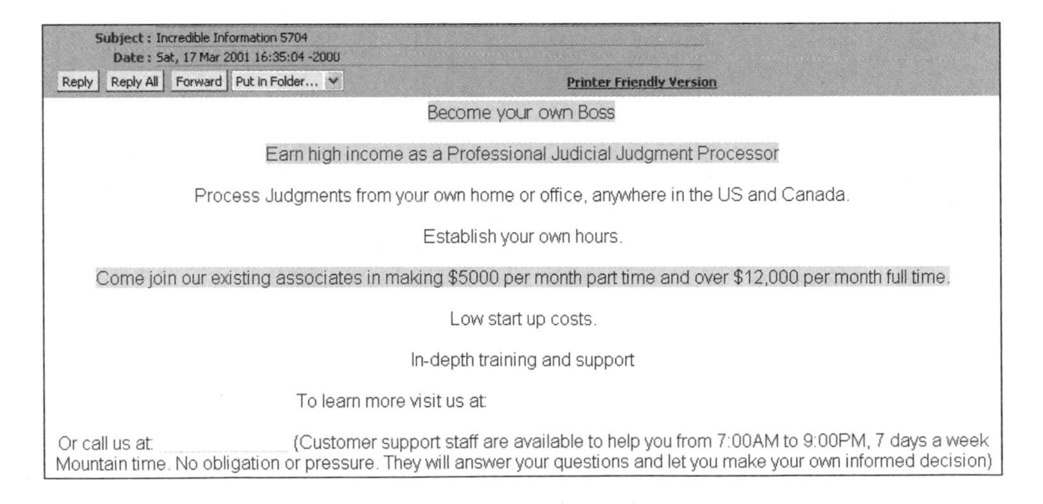
Subject: Incredible Information 5704
Date: Sat, 17 Mar 2001 16:35:04 -2000
Reply | Reply All | Forward | Put in Folder...
Printer Friendly Version

Become your own Boss

Earn high income as a Professional Judicial Judgment Processor

Process Judgments from your own home or office, anywhere in the US and Canada.

Establish your own hours.

Come join our existing associates in making $5000 per month part time and over $12,000 per month full time.

Low start up costs.

In-depth training and support

To learn more visit us at:

Or call us at: (Customer support staff are available to help you from 7:00AM to 9:00PM, 7 days a week Mountain time. No obligation or pressure. They will answer your questions and let you make your own informed decision)

9. The product Spam is a type of lunch meat produced by Hormel. The origin of the term in relationship to unwanted e-mail is shrouded in the myths of the Internet but is thought to have some relationship to the Monty Python skit of the same name.

Although some have attempted to prosecute spammers under state and federal laws, only a few cases have proved successful. Some governments, notably Austria and Australia, have passed laws against sending spam, but these restrictions may be difficult to enforce given the global nature of the Internet. Although no movement has emerged to urge similar federal legislation in the United States, the European Union has considered a ban on spam. As of this writing, no action has been taken.

Pop-up windows are a form of "Web spam" that have become popular with advertisers. As mentioned in Chapter 7, these windows appear during a Web session and users must close them manually. Sometimes, the browser has to be shut down to close all of them. This trend is expected to get worse as advertisers add "pop-under" windows, which appear when the browser is closed, and windows that change shapes and positions within the browser window.

Using Information at Fareastfoods.com

FarEast Foods has been offered a number of opportunities to purchase e-mail lists from target marketers for use in spam campaigns. It also has been solicited to sell information about its customers. In both cases, the company has resisted because employees know that their future success relies on the goodwill of their customers and that using spam could backfire and create negative customer relations.

Quick Review

1. List the four major threats to personal privacy in the networked economy.
2. Who are information brokers, and how do they threaten privacy?

Identity Theft

identity theft
The process of stealing a person's identification for purposes of purchasing goods and services using his or her credit.

As discussed in the opening case, when even a small amount of information about a person falls into the hands of criminals, the worst type of privacy invasion can occur—**identity theft**, in which criminals steal a person's identification for purposes of purchasing goods and services using that individual's credit. Identity theft was introduced in Chapter 11 as a form of IT crime, but because of its implications for privacy, it is discussed in more detail here.

Identity theft is one of the fastest-growing types of financial crime, costing consumers between $2 and $4 billion per year.[10] It has been described as the "signature crime of the twenty-first century," and some estimates expect about 5 to 7 million cases to occur over the next five years.[11] It has been the number one fraud reported to the Federal Trade Commission for the past few years, with four times as many cases reported in 2001 as any other fraud.

Although criminals often view identity theft as a low-risk crime, it brings a large amount of grief to the victim. The Identity Theft Resource Center estimates that the average victim of identity theft spends 175 hours and $808 in out-of-pocket expenses to clear his or her name.[12] Unfortunately, some people continue to receive letters from debt collectors, credit card companies, and even law enforcement agencies long after they believe they have cleared their names.

10. Sheila M. Poole, "Identity theft rising sharply as felons get more devious." *Atlanta Journal-Constitution*, December 16, 2001, p. D1.
11. Lee Shearer, "Identity theft rising threat, expert says." *Athens Banner-Herald*, February 23, 2002, p. A10.
12. www.idtheft.org.

Methods of ID Theft

Credit card numbers, ATM PINs, and Social Security numbers all offer ways for thieves to steal your identity. This information can be obtained in a number of ways, including the following:

- Theft of a pocketbook, purse, wallet, or mail
- Dumpster diving—going into a trash bin to retrieve credit card receipt carbons, medical records, and so on
- Using inside connections to credit agencies to access information about employees or others
- Fraudulent schemes in which the criminal fills out a change-of-address form and has the victim's mail sent to a new address
- Pretexting to fool people into providing information on the victim

In addition, criminals can go to genealogy Web sites to determine a person's mother's maiden name, thereby providing a key answer to a question often asked to confirm identity.

Armed with this information, criminals can use it to apply for driver's licenses, telephone service, and credit cards, or to steal benefits such as pensions and Social Security payments. Often, identity thieves are individuals who have credit problems; sometimes, however, members of organized crime are the culprits. Victims often don't find out that their identity has been stolen until they are denied a loan because of unpaid bills or until bill collectors try to collect on credit card bills the criminal has run up.

Finding out about the identity theft uncovers only the tip of the iceberg. Convincing the creditors and police that your identity has been stolen and restoring your credit rating can be a long and difficult task. To make matters worse, until recently, a person whose identity had been stolen was not even considered a victim; only the credit grantors who suffered a monetary loss were legal victims of identity theft! This oversight was rectified by the **Identity Theft and Assumption Deterrence Act of 1998**, which made stealing someone's identity a crime punishable by up to 15 years in jail. In addition, by 2001, 47 of the 50 states in the United States had passed laws making identity theft a crime, with many of them making it a felony.

Identity Theft and Assumption Deterrence Act of 1998
Legislation under which stealing someone's identity is a crime punishable by up to 15 years in jail.

Social Security cards were never meant to be a national identification system.

Social Security Number Theft

Of all the types of identity theft listed earlier (that is, theft of credit card numbers, ATM PINs, and Social Security numbers), the last catagory accounts for far and away the biggest proportion of identity theft in the United States. Even though Social Security numbers are often used as personal identifiers, they were never meant for that purpose. The original intention of Social Security numbers in the late 1930s was to track earnings so that the Social Security Administration would know who was entitled to benefits. In fact, the original cards were clearly marked as "Not for Identification." Over time, they have evolved into a form of identification, being used for some two dozen purposes, including college IDs, driver's licenses, marriage licenses, blood donor cards, birth certificates, and any application for government benefits such as the Medicare program. Even the federal Parent Locator Service, which tries to track down employees who owe child support, uses Social Security numbers for identification.

In addition to the proliferation of Social Security numbers on many different forms, an even greater problem derives from the fact that private firms can legally sell or reveal Social Security numbers to other firms. Although the U.S. government is restricted in its use of Social Security numbers by the Privacy Act of 1974, no such limitation applies to private firms. Credit bureaus such as Experion and Equifax often sell Social Security numbers to organizations like banks, insurers, credit-granting firms, and so on, because they believe the purchasers need such information. Legislation now before Congress would restrict the sale of Social Security numbers, but at this writing, no action has occurred on it.

Because Social Security numbers serve as personal identifiers, once stolen, they can easily be used for identity theft. Armed with your name and Social Security number, someone can pretend to be you and apply for a credit card in your name. By the time you find out your identity has been stolen, which takes on the average 14 months to occur, your credit record is in tatters. In general, consumers should closely guard their Social Security numbers and not be quick to enter them on a Web site, no matter how secure.

1. Why might criminals consider identity theft to be considered a low-risk crime?
2. Why is the Social Security number considered to be the prime source of identity theft?

Protecting Privacy

Fair Information Practice (FIP) principles
Principles of privacy protection that include notice/awareness, choice/consent, access/participation, integrity/security, and enforcement/redress.

Next, let's take a look at five widely accepted principles of privacy protection and see how they are applied. These principles, known as the **Fair Information Practice (FIP) principles**, were developed by government agencies in the United States, Canada, and Europe. Their development was prompted to a large extent by a 1973 report by the U.S. Department of Health, Education, and Welfare (HEW) on privacy protections in the age of data collection *(Records, Computers, and the Rights of Citizens)*. The five principles are notice/awareness, choice/consent, access/participation, integrity/security, and enforcement/redress. Table 12.6 provides a short description and an example of each principle.

The first principle is key because without notice or awareness, the other principles have no meaning. Also, without choice or consent, consumers cannot "opt out" of having their data collected. Legislation and self-regulation are the two primary approaches employed to ensure that these principles are applied to personal privacy. In the legislative approach, government takes the responsibility for ensuring privacy; in the case of self-regulation, business and industry ensure this privacy. For the best protection of privacy, society must utilize a combination of these approaches.

Table 12.6 Fair Information Practice Principles

Principle	Description	Example
Notice/awareness	Consumer is notified that information is being gathered and with whom it is being shared	Web site informs consumer that information is being gathered and with whom it will be shared
Choice/consent	Consumer is given the option to not have information gathered	Web site allows consumer to opt out of having the information gathered
Access/participation	Consumer is given access to his or her data and has right to contest that data's accuracy and completeness	Web site shows the consumer the information that has been gathered and offers the opportunity to correct errors
Integrity/security	Data collected are accurate and secure	Data collected by Web server are verified before use and kept secure against intrusion
Enforcement/redress	Some mechanism enforces these principles and offers redress for consumers when not enforced	Consumer can report Web site to an industry watchdog group or to a government agency if it fails to follow these principles

Legislative Approach

Fair Credit Reporting Act of 1970
Legislation that regulates some actions of credit bureaus that collect credit information on individuals.

credit check
A inquiry with a credit bureau regarding a person's financial status.

Freedom of Information Act of 1970
Legislation that gave individuals the right to inspect information concerning them held in U.S. government data banks.

Privacy Act of 1974
Legislation that attempted to correct most of the recordkeeping practices of the federal government.

Electronic Communications Privacy Act of 1986
Legislation that extended wiretap laws protecting aural conversations to include communications between computers.

Computer Security Act of 1987
Legislation aimed at ensuring the security of U.S. government computers.

The proposal for a national data bank in the United States in the 1960s, and its subsequent rejection, marked the beginning of a period of legislative action on the privacy issue. These efforts resulted in the following legislation between 1970 and 1987 (in addition to the more recent laws already discussed):

- Fair Credit Reporting Act of 1970
- Freedom of Information Act of 1970
- Privacy Act of 1974
- Electronic Communications Privacy Act of 1986
- Computer Security Act of 1987
- Identity Theft and Assumption Deterrence Act of 1998

The **Fair Credit Reporting Act of 1970** regulates some actions of credit bureaus that collect credit information on individuals. When a person seeks to borrow money (or engage in other activities, including applying for a job), the potential lender runs a **credit check** on the individual by requesting information from a credit bureau. Unfortunately, the tremendous increase in computing power since 1970 led to widespread abuse of this law.

The **Freedom of Information Act of 1970** gave individuals the right to inspect information concerning them held in U.S. government data banks. The **Privacy Act of 1974** attempted to correct most of the recordkeeping practices of the federal government. The **Electronic Communications Privacy Act of 1986** extended wiretap laws protecting aural conversations to include communications between computers. The **Computer Security Act of 1987** was aimed at ensuring the security of U.S. government computers. This act protects public electronic mail, but court decisions have since established that corporate electronic mail messages are the property of the organization. As mentioned earlier, the Identity Theft and Assumption Deterrence Act of 1998 was passed in response to the growing problem of identity theft.

The cornerstones of privacy remain the Fair Information Practice principles. Recent surveys have shown that they are not being widely applied to electronic commerce Web sites, as you might hope. Table 12.7 shows the results of one such survey of 91 of 100 of the most popular Web sites conducted by the FTC in 2000.[13]

13. http://www.ftc.gov/os/2000/05/testimonyprivacy.htm.

Table 12.7 Results from an FTC Study of Electronic Commerce

Item	Percentage of Total
Sites collecting personal information	99
Sites posting any privacy disclosure	100
Sties following all four of the principles	42
Sites participating in a self-regulation program	45

Whereas 99 percent of the sites collected information at that time, only 42 percent followed all four of the Fair Information Practice principles.

The fact that fewer than half of the most popular Web sites were found to participate in a self-regulation program drew the attention of the FTC. Of even more concern was a companion survey of 335 randomly chosen Web sites, which showed that only 8 percent of them participated in a self-regulation program. In its report based on the survey, "Privacy Online: Fair Information Practices in the Electronic Marketplace," the FTC recommended that legislation be enacted at the federal level to enforce the four Fair Information Practice principles. At this writing, no such legislation has been passed.

In addition to these federal laws, a number of states have passed legislation aimed specifically at identity theft. For example, Montana has increased the penalties for identity theft to a $10,000 fine and 10 years in jail. Washington has made it possible for victims to file suit against credit bureaus that continue to give out negative information about them even after they have been sent a police report on the identity theft. Other states are considering increasing their penalties for identity theft as well.

Governments in Canada and Europe have passed strong laws regarding the protection of personal privacy and data. This difference has led to problems regarding the transmission of data between the European Union countries and the United States, for example. These problems were still being negotiated at the time of this book's publication. Interestingly, some representatives of business and industry in the United States point to the lack of such laws in the United States as the reason why electronic commerce is more advanced in the United States than in Europe. On the other hand, European governments believe that their citizens are better protected than those in the United States.

Self-Regulation

The other approach to protection of personal privacy in the networked economy relies on self-regulation. In this approach, members of the electronic commerce industry develop ways to implement the Fair Information Practice principles without government intervention. Efforts at self-regulation have included industry group guidelines, use of privacy seals of approval, advertiser pressure, technology, and user education.

One example of guidelines for industry groups involves the 1998 Online Privacy Alliance (OPA), a coalition of industry groups, which announced its Online Privacy Guidelines to implement the FIP principles. OPA members agree to adopt and implement a posted privacy policy, but no monitoring of member compliance to the guidelines takes place.

privacy seal of approval
An icon that a Web site can display if it has agreed to follow the FIP principles as defined by the seal-granting organization.

A **privacy seal of approval** is an icon that the Web site can display if it has agreed to follow the FIP principles as defined by the seal-granting organization. Such seals of approval enable consumers to identify online businesses that follow specified privacy principles and enable businesses to demonstrate compliance with such principles. By early 2002, a number of privacy seals were available from a variety of organizations, as noted in Table 12.8.

Table 12.8 Privacy Seals of Approval

Name	Stated Purpose	URL
TRUSTe	Build confidence in Internet privacy	www.truste.org
BBBOnline	Accredit Internet merchants with high standards in electronic commerce	www.bbbonline.org
WebWise	Provide independent verification to online businesses under strict criteria	www.webwisecenter.com
Chamber Seal	Provide Web-trust seals verifying the authenticity of companies that own Web sites	www.chamberseal.com
Internet Trade Bureau	Promote integrity and fair business practices on the Internet	www.internettradebureau.com

The most widely used privacy seal at this writing, TRUSTe, claims that its seal assures users that the Web site will disclose what information is being gathered, how it will be used, and with whom it will be shared. Such Web sites are also supposed to offer users choices about how information will be collected, safeguards to protect the information, and ways for users to update or correct information. The privacy seal organizations also claim to monitor the compliance of the online businesses that display the seals. Figure 12.9 shows the seal for TRUSTe at the bottom left-hand corner of the Intuit (makers of Quicken software) Web site.

Figure 12.9 TRUSTe seal on Intuit Web site

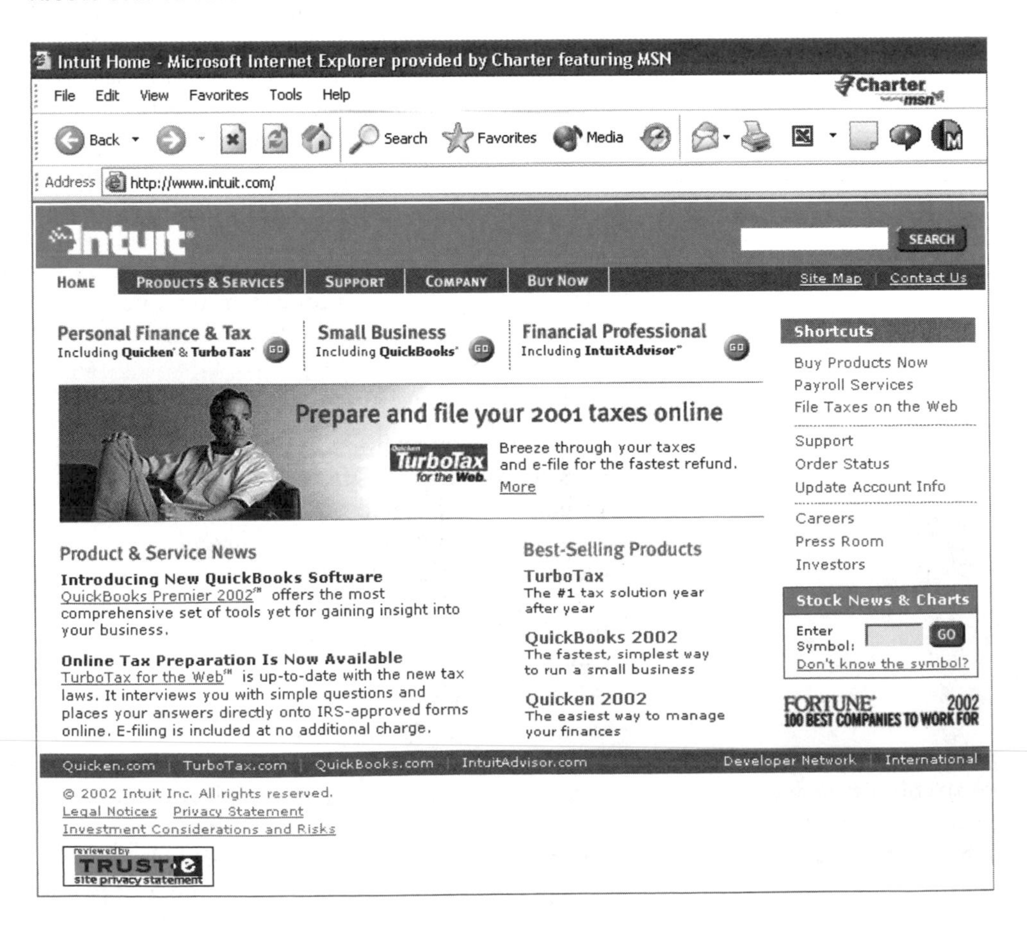

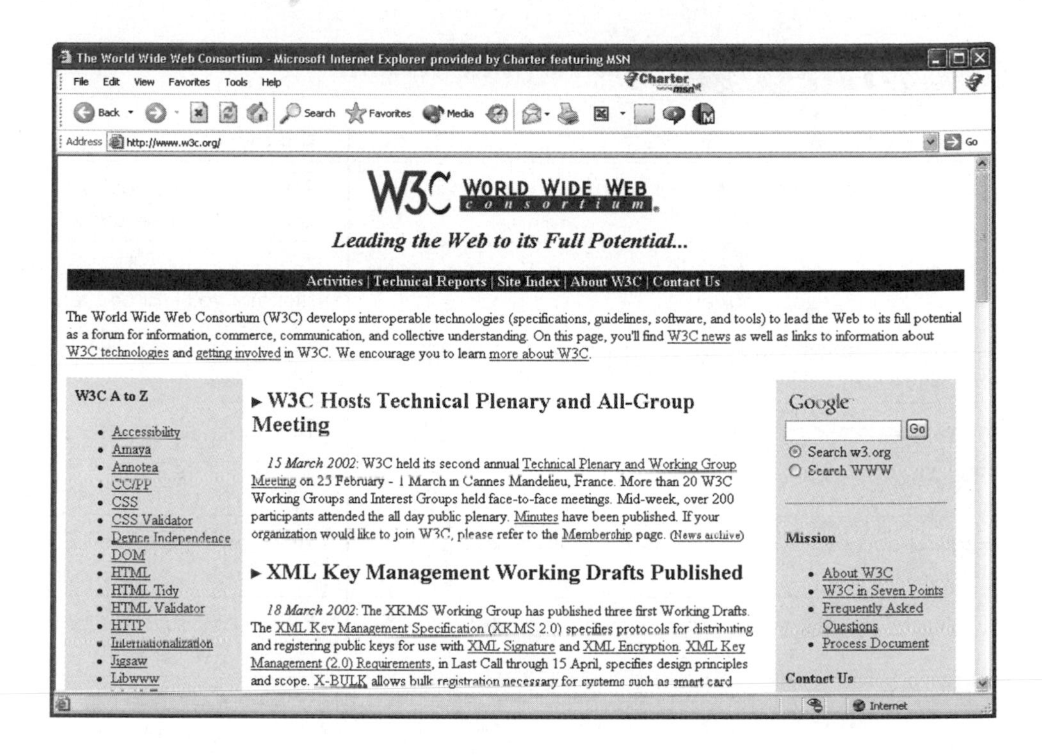

The World Wide Web Consortium is an organization that is aimed at "leading the Web to its full potential."

In terms of pressure from advertisers, two of the largest Web-based advertisers, Microsoft and IBM, have policies that require the Web sites on which they advertise to have clearly defined privacy policies. One obvious way for Web sites to meet this requirement is to obtain a privacy seal from one of the organizations listed in Table 12.8. Unfortunately, the FTC study mentioned earlier showed a low use of these seals.

A group directed by the inventor of the Web, Tim Berners-Lee, known as the **World Wide Web Consortium (W3C)**, has proposed a technology-based solution to the privacy problem. Referred to as the **Platform for Privacy Preferences (P3P)**, this technology would give consumers control over their personal data by building the necessary mechanisms into Web browsers such as Netscape and Internet Explorer. Internet Explorer 6.0 includes a compact version of the P3P technology that applies only to cookies, but it has met with resistance from companies that do not want to rewrite their cookie code to meet P3P requirements.

World Wide Web Consortium (W3C)
A group of more than 500 member organizations aimed at helping the World Wide Web reach its full potential by developing common protocols that promote its evolution and ensure its interoperability.

Platform for Privacy Preferences (P3P)
Proposed technology that would give consumers control over their own personal data by building the necessary mechanisms into Web browsers such as Netscape and Internet Explorer.

User education in the area of online privacy may be the best way to protect personal data. Groups such as the OPA and the privacy seal sponsors are working in this area. Many privacy advocates agree that the most effective way for consumers to protect their privacy is to exert tighter control over their personal data. Table 12.9 offers some suggestions about how you might protect your personal data.

1. What are the five Fair Information Practice principles?
2. What two general approaches are used to protect personal privacy in the networked economy?

Table 12.9 Ways to Protect Your Privacy

Suggestion	Comment
Supply your Social Security number only when required by law	Ask if you can obtain a different identifying number for nongovernment requests.
Find your cookie file and erase it, then set your browser to give you the option to reject cookies	This strategy will allow you to customize only those adaptive Web sites that you trust and refuse to do so for unknown Web sites.
Visit Web sites anonymously by going through sites such as www.idzap.com	This tactic will effectively kill all customization of Web sites to your habits.
Pay cash whenever possible	This approach will reduce generation of transactional data but will require forethought and planning as well as effectively cutting you off from electronic commerce.
Be careful about your use of chat rooms, bulletin boards, and newsgroups	These entities can provide a wealth of information about you, even if you use a pseudonym.
Think twice about filling out warranty cards or contest entries	Receipts are sufficient for most warranties. Have you won any sweepstakes recently?

Roger Clarke

IT INNOVATORS

Although the commercialization of the Internet has drawn a great deal of attention to privacy issues, for some writers and researchers, these issues existed long before the Internet's popularity exploded. One such writer is Roger Clarke, an Australian lecturer, author, and consultant in the field of privacy. Clarke's interests in privacy date back more than 25 years, to when he prepared a series of papers for the New South Wales Privacy Committee. Since 1982, he has written more than 100 papers on the subject, many of which are available at www.anu.edu.au/people/Roger.Clarke/. He divides his works into theory, practice, policy, and papers specific to Australia. Clarke is also a regular participant at the annual Computers, Freedom, and Privacy conferences. His notes on these conferences are linked to his Web page.

Clarke's Web page is a treasure trove of information for anyone interested in privacy. He specializes in an area of privacy known as *dataveillance*, the systematic use of personal data systems in the investigation or monitoring of the actions or communications of one or more people. He introduced this topic in a paper that appeared in the *Communications of the Association for Computing Machinery* and was twice reprinted. Other privacy topics that Clarke explores include identification and anonymity, privacy implications of digital signatures, and person-location and person-tracking technologies.

Roger Clarke has lectured, written, and consulted on privacy issues for over 25 years.

Source: www.anu.edu.au/people/Roger.Clarke/.

Figure 12.10 Ten commandments of Computer Ethics

1. Thou shalt not use a computer to harm other people.
2. Thou shalt not interfere with other people's computer work.
3. Thou shalt not snoop around in other people's files.
4. Thou shalt not use a computer to steal.
5. Thou shalt not use a computer to bear false witness.
6. Thou shalt not use or copy software for which you have not paid.
7. Thou shalt not use other people's computer resources without authorization.
8. Thou shalt not appropriate other people's intellectual output.
9. Thou shalt think about the social consequences of the program you write.
10. Thou shalt use a computer in ways that show consideration and respect.

Ethical Issues in the Networked Economy

ethics
Rules created by cultures and economies about whether certain acts are "good" or "bad," or "right" or "wrong."

All cultures and economies have developed rules about whether certain acts are "good" or "bad," or "right" or "wrong." These rules, known as **ethics**, are inherently value judgments that have resulted from a consensus in society. Such rules are often expressed or supported by laws. The networked economy is no different from previous economies in terms of ethical issues. For example, is it ethical to write virus programs that annoy users without destroying anything? Is it ethical for employers to read their employees' e-mail with no reason (it is certainly legal, but is it *ethical*)? Similarly, is it ethical to compile personal information on customers with or without their knowledge? This section could examine a host of ethical issues, but it will use a code of ethics developed by the Computer Ethics Institute as a framework for the discussion.

Ten Commandments of Computer Ethics
A set of rules that covers many of the ethical issues facing computer users in the networked economy; developed by the Computer Ethics Institute in 1992.

Known as the **Ten Commandments of Computer Ethics**, the list of rules shown in Figure 12.10 covers many of the issues facing computer users in the networked economy. (And, as with any list of general rules, problems can be found with it.[14]) Note that the Computer Ethics Institute developed these commandments in 1992, and they have *not* been modified to reflect the tremendous growth in the use of the Internet since then. However, the rules remain appropriate if, when you see the word *computer,* you include any networks (including the Internet) to which computers are linked. Individual organizations may have their own codes of computer ethics. For example, your university almost certainly has its own code of computer ethics that provides guidance for faculty, staff, and student use of its computers.

Discussion and Application of the Ten Commandments of Computer Ethics

To help you understand the ethical issues facing computer and Internet users in the networked economy, let's look at each of these rules (slightly modified to include networks).

1. *Thou shalt not use a computer (or network) to harm other people.* In this commandment, the key term is *harm,* which you can interpret to mean any type of harm—physical, emotional, monetary, or otherwise. For example, just as it is unethical to harm people by planting a bomb in a public building, according to this rule, it is also unethical to post instructions for bomb making on the

14. http://www.ccsr.cse.dmu.ac.uk/resources/professionalism/codes/cei_command_com.html.

Internet. Hacking into a credit union database and accessing personal financial data would also be considered unethical (as well as illegal) under this commandment, as would purposefully exposing personal data on the Web or collecting personal data without an individual's permission.

2. *Thou shalt not interfere with other people's computer work. Interfere* is the key word in this commandment, because programmers can readily send viruses and other programs over the Internet to interfere with or even destroy other people's computer work. Sending even nondestructive viruses to other computer users violates this commandment if they interfere with work by taking over internal memory. Also, sending an overwhelming number of e-mails or requests for Web pages to a server with the intention of crashing it would be considered unethical, no matter what your feelings are about the purpose of the server.
3. *Thou shalt not snoop around in other people's files.* Files on computers owned by individuals (not organizations) either are the personal property of the individuals or are software that has been licensed to that person. In either case, it is unethical (and, in many cases, illegal) to access these files, including e-mail sent from personally owned computers. On the other hand, files on computers owned by organizations have been ruled to be the property of the organizations, although some employee organizations insist that it is unethical for employers to read their employees' e-mail.
4. *Thou shalt not use a computer to steal.* Using a computer to steal from individuals or organizations is both unethical and illegal. The networked economy has inspired many new types of fraud as well as old schemes repackaged for the Internet.
5. *Thou shalt not use a computer (or network) to bear false witness.* One well-known fact about the Internet is that bad or unflattering news spreads like wildfire. Using a Web page to spread an untruth or inaccurate information and sending an unfounded rumor to a newsgroup are examples of using a computer to bear false witness. The volatility of Internet-related stocks means that good or bad news can cause a large number of investors to gain or lose a great deal of money very quickly before the validity of the news can be determined.
6. *Thou shalt not use or copy proprietary software for which you have not paid.* Although software piracy is clearly illegal, you may think that *borrowing* a copy of a software program from a friend is okay. It is definitely not! The software is *licensed* to a single user (unless a site license has been purchased), and that user is the only person who should use it. Allowing others to use software is both illegal and unethical in the same way as photocopying a copyrighted textbook. Although not always illegal, it is definitely unethical to download shareware and then not pay for it. (**Shareware** is software that can be downloaded from Web sites on the Internet for free, but you are expected to pay a nominal fee to its author. If you choose to use it, then you are "on your honor" to pay for the shareware.)

shareware
Software that can be downloaded from Web sites on the Internet for free but a nominal fee is expected to be paid to its author.

7. *Thou shalt not use other people's computer resources (or network) without authorization or proper compensation.* Your school computer account probably allows you to access your university's computer system, including its e-mail program. You may also have an AOL or local ISP account for which you pay a monthly user fee. In either case, you would not want someone else to break into the computer system and use your account. The same applies to other people's accounts—hacking into them is considered unethical.
8. *Thou shalt not appropriate other people's intellectual output.* Just as copying someone else's math homework or English term paper is cheating, so, too, is copying someone else's computer program. This statement includes copying text, illustrations, or photos from a Web site. Although most would concede that it is appropriate to learn design and programming techniques by investigating a Web site, outright copying of such a site is unethical, and, if the material is copyrighted, illegal as well. You should cite any quoting or paraphrasing of material from a Web site just like any other research source, as this book has done.

9. *Thou shalt think about the social consequences of the program you write or the system you design.* To apply this commandment, to include Web sites under the broad term *program*. Then you should ask yourself if the Web site you have created will in some way harm society. Does it provide information that can be used in a harmful way—say, in the hands of someone with psychological problems? Will the Web site incite anger or other hurtful emotions in those who read it? Will it degrade a group of the population or harm children? As mentioned earlier, several attempts have been made in the United States to control Web site content to protect children, but some have been ruled unconstitutional on freedom of speech grounds.
10. *Thou shalt use a computer (and network) in ways that show consideration and respect for your fellow humans.* E-mail users and others using Internet communication protocols sometimes feel that they can write things that they would not say in a face-to-face conversation. Sending angry e-mail messages to someone or an organization would be considered unethical under this commandment. As with any communications media, politeness and consideration remain the best policy, regardless of how you feel about the other party.

Although these commandments may not cover all possible situations, they provide an overview of ethical use of computers in the networked economy.

Case*Study*

When "HOPE" Spells Problems

Imagine that you are a college student in the state of Georgia who, through hard work in high school, has won a HOPE (Helping Outstanding Students Educationally) scholarship for college tuition and expenses. While surfing the Internet, you find your name, home address, Social Security number, birth date, and amount of money you have received among a listing of HOPE scholarship recipients. Although this scenario may sound ludicrous, it actually happened in May and June of 2001. In the actual case, a HOPE recipient found this information on herself through a search on Google.com. With a little more research, the student was able to find more than 3000 pages of personal information on thousands of students attending private and public colleges in Georgia as well as on students who applied for HOPE scholarhips but attended a college outside the state. Given the increase in identify theft, like the person who actually made this discovery, you would probably be very concerned about it and the potential effect it might have on your future.

How could this supposedly private information end up available to anybody doing a search on the Web? As it turns out, human error was to blame. When technicians installed a software upgrade to the Student Finance Commission's server (the agency that oversees the HOPE program) in April 2001 that would improve access to its files by financial aid officials at colleges and universities in the state, they failed to copy a vital file that would have kept search engines such as Google from finding information on the 527,000 individuals who have applied for HOPE since it initiation in 1993. When Google's Web crawler visited the HOPE site on April 30 of that year, it copied 730 files containing the information on the HOPE applicants to its server. The commission was alerted to the problem by the mother of another student whose name and data appeared on the Web and the security hole was subsequently fixed. Unfortunately, even after the state technicians made the repair, the information remained on the Google site until June 24. The commission did not know if other search engines had picked up this information and saved it on their servers or if users had posted the results on chat rooms or Web pages.

To further compound the problem, HOPE officials failed to disclose this breach of confidentiality until a newspaper reporter called to obtain background information regarding the disclosure in August 2001. Only then did they alert the governor to the problem. He promised to have the Georgia Technology Authority, which oversees state technology spending, investigate the issue. Only after the investigation was complete would the governor consider disciplinary action at the commission. Students and parents were never officially informed of the breach of security.

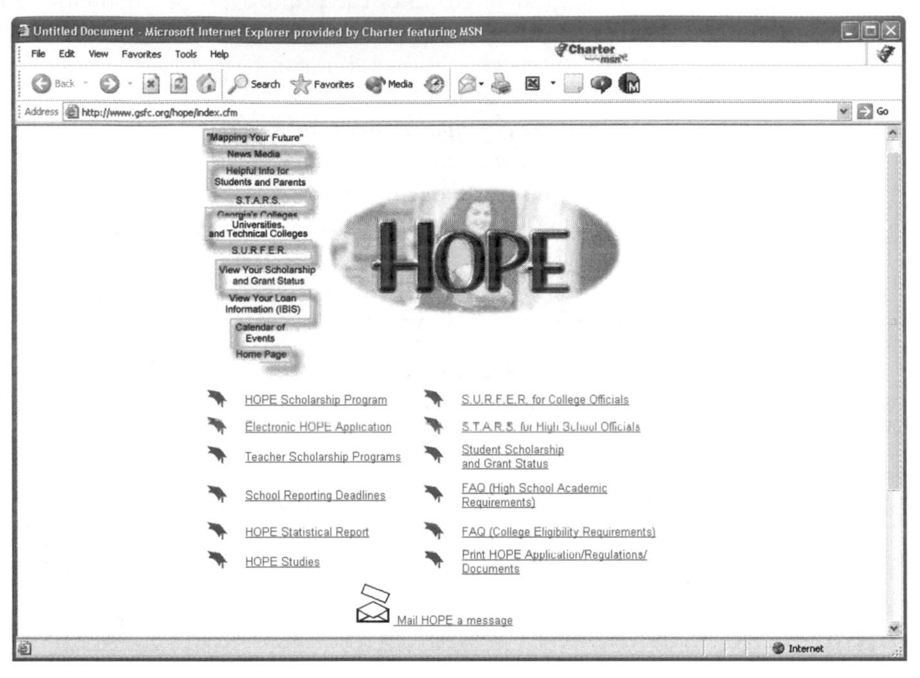

The HOPE program provides scholarships to college students in Georgia, but failed to protect private data on applicants.

Source: Alan Judd and Kathy Brister, "State probing HOPE data exposure." *Atlanta Journal-Constitution*, July 26, 2001, pp. A1, A10.

Think About It

1. Do you agree with HOPE officials' response when they determined that the problem had occurred? Why or why not?
2. Do you think that the governor should take disciplinary action if this case turns out to be just an accident on the part of the technicians?
3. How can the Student Finance Commission ensure that this type of problem never occurs again while still providing access to its data for financial aid officials across the state?

SUMMARY

To summarize this chapter, let's answer the questions posed at the beginning of the chapter.

Why has privacy become an important issue in the networked economy? Personal privacy is an important area for you to consider, both as a consumer in the networked economy and as an employee in a networked economy organization. As a consumer, you should know the threats to your privacy and recognize how to deal with them. As an employee, you should address the privacy concerns of your customers and make sure that your organization balances them against its need for information to serve customers better. Privacy is at greater risk in the networked economy than in the industrial economy because of increased searchability and integration, but the organizations retaining personal information can protect it by implementing good security practices. Consumers face the greatest risk of having their privacy violated when organizations choose to cooperate in sharing information or don't follow security procedures.

People have become concerned about the threat of organizations and individuals using the Internet to access public and private databases to invade their privacy. This worry arises because almost anyone with a personal computer and modem can access the Internet to search a database and retrieve information about prospective employees, tenants, or other individuals.

What would be the negative effects of stringent privacy laws? These negative effects could take the form of higher prices, higher mortgage rates and fewer loans, fewer free Web sites, and less shopping convenience. Less information would cause companies to use broad and less effective advertising and prompt financial companies to charge higher mortgage rates to cover bad loans. Fewer free Web sites would be available because many such sites depend on the sale of Web-visit data to cover their costs. Less information would make it more difficult to customize Web sites by learning about customers' preferences. This information makes the Web site more useful to customers but also requires the customers to give up some of their privacy.

How is information technology used to collect private data and information? Three broad classes of data exist: transactional data, Web-visit data, and Internet communications data. Noncash transactions generate transactional data. Such transactions take place face-to-face, electronically, or by telephone, mail, or the Web and identify the customer. No federal laws in the United States prohibit companies from sharing transactional data on their customers. Data are often also collected on the employees handling the transactions.

Web-visit data about a user's computer are generated whenever a user visits a Web site. Adaptable Web sites request information from the visitor, whereas adaptive Web sites use patterns they find in consumer interests and purchases to match the customer's interests. These Web sites often employ cookie files to save these data on the client computer.

Internet communications data include e-mail messages, "conversations" in chat rooms, postings to bulletin boards, and messages to newsgroups. E-mail sent or received on an employer's computer remains the property of the employer, which means that the employer has the right to read an employee's e-mail at any time.

What threats to personal privacy exist in the networked economy? Threats to privacy involve the use of data or information that has been collected using information technology. Six types of threats to privacy exist in the networked economy: exposure of information, data surveillance, information brokers, spyware, spam, and identity theft. Exposure of information can occur either accidentally or intentionally (by hackers). Data surveillance involves computer matching of databases or profiling of individuals who fall into specific groups. Organizations often use profiling for target marketing. Information brokers find information on an individual for a fee. Spyware is software that monitors Internet usage and reports it back to a server. Spam involves an intrusion on users' personal privacy through the receipt of large amounts of unwanted e-mail. In identity theft, someone steals a person's credit rating or other information specific to that person.

Why is identity theft such a fast-growing type of financial crime? Criminals often view identity theft as a low-risk crime. It brings a large amount of grief to the victims, however, who may continue to receive letters from debt collectors, credit card companies, and even law enforcement agencies long after they believe they have cleared their names. Credit card numbers, ATM PINs, and Social Security numbers all offer ways for thieves to steal your identity, with Social Security numbers being the most prevalent method. This information can be stolen in a number of ways.

What approaches are used to protect personal data? Approaches to protecting personal data involve attempting to follow the Fair Information Practice principles through government legislation and/or self-regulation. In the United States, governments have passed a number of laws to protect privacy. Self-regulation includes industry group guidelines, privacy seals of approval, advertiser pressure, technology, and user education.

What ethical issues must be dealt with in the networked economy? Like any other economy, the networked economy has spawned a variety of ethical issues. These issues can be addressed in the context of the Ten Commandments of Computer Ethics and focus on the ethics of using computers and networks to harm other people, steal from them, spread untruths about them, look into their files, and so on.

REVIEW QUESTIONS

1. What three dimensions are used to compare privacy in the networked and industrial economies?
2. What admendment to the U.S. Constitution has been cited as protecting personal privacy? What problems arise with this approach?
3. What trade-offs are associated with more stringent laws regarding personal privacy? Give an example other than those mentioned in the text.
4. Develop a list of five more databases of personal information that are kept by government, public, or private organizations.

5. List five types of transactional data that you have created in the last week. How would these data create a picture of your likes and dislikes?
6. List the methods that were used to collect the transaction data you mentioned in Question 5.
7. How do employees generate transactional data? How do employers use these data?
8. Why do we say that Internet communications data are unstructured? Does this statement mean they cannot be searched or integrated with other data?
9. What is COPPA? What types of parental consent has the FTC defined under this act?
10. What are the most typical ways in which information exposure occurs?
11. What are the steps in the profiling process? Have you ever been the target of profiling? If so, how?
12. How does profiling relate to CRM? How does it differ from computer matching?
13. What is spyware, and how is it often installed on a computer?
14. What is a Web bug? How does it collect data?
15. Why is spam included in this chapter as a type of threat to personal privacy? Do you agree with this inclusion? Why or why not?
16. How can you avoid being the target of spam? What is one sure way to continue to be the target of even more spam?
17. What is the major identifier used in identity theft? How often have you given it out in the last month?
18. What are four approaches to self-regulation of privacy protection?
19. What is P3P, and how is it related to Microsoft's Internet Explorer browser?
20. Why is it unethical to copy material from other Web sites and use it in yours?

DISCUSSION QUESTIONS

1. Discuss the pros and cons of more stringent privacy legislation. Take one point of view and expand upon it.
2. How is employee e-mail different from personal e-mail? Discuss the pros and cons of a company reading employee e-mail.
3. For a Web site of your choosing, discuss how it uses customization and what type of personal data it collects.
4. Discuss the two approaches to protection of personal privacy. Take one approach and expand upon it.
5. Check with your college computing center about whether it has a code of ethics for computer use. If it does, compare it to the Ten Commandments of Computer Ethics.

RESEARCH QUESTIONS

1. Visit the URL www.idzap.com/asurf.php#findoutaboutme and determine the information being sent from your computer to the Web server. Discuss these data and compare them to that shown in Figure 12.3 in a one-page paper.
2. Go to a child-oriented Web site and determine how it addresses the requirements of COPPA. Prepare a presentation on your findings to give to the rest of the class.
3. Use the Web to research recent occurrences of information exposure or identity theft. Write a two-page paper comparing the occurrences and drawing conclusions about similarities and differences.

CaseStudy

WildOutfitters.com

Claire was awakened when she felt Alex stirring beside her. Groggily, she asked, "What's the matter, honey?"

"Nothing, I just had a strange dream," Alex replied. Before Claire could settle back to sleep, he continued. "I was climbing on a high mountain, when I came upon an eagle's nest. I was happy to see the eagle and began to ask questions to get to know her better. Each time that I asked her a question, I gave her a worm. The questions kept getting more personal and soon I knew almost everything about the eagle and her family."

"I don't think eagles eat worms," Claire said, not quite fully awake.

"Give me a break," Alex responded. "It's a dream."

"Okay then, but wasn't the eagle nervous about answering your questions?"

"No, it seemed that as long as I had nice, juicy worms to give her she was willing to tell me anything."

Alex's dream probably can be attributed to the latest ideas that the Campagnes had been discussing for the Web site. One reason for the success of their bricks-and-mortar store has been their ability to give personalized service to their customers. Admittedly, the number of regular customers to the store has been relatively small but this approach has allowed the Campagnes to know these customers well. This knowledge has served them on a number of occasions by enabling them to recommend trips or gear that were a custom fit with the customers' interests. Claire and Alex would like to extend this ability to Wild Outfitters' Web site, but doing so would involve collecting and using more personal information from their Web customers.

Customization is not entirely new to their site. The Campagnes already collect some personal information from users who register for their customer incentive program and track their sales levels. This information, plus the use of a cookie, allows registered users to log in to the site quickly and check how close they are to winning a prize. Also, for customers who want it, Wild Outfitters has been sending targeted e-mail about special deals. The Campagnes are also more confident in their ability to protect information due to the improved security of their systems.

Lately, they have been struggling with ideas on how to customize the site further. One alternative is to make their site adaptable. With this option, they would use information provided by each user to present more customized views on the Web pages. Although still offering the main options and maintaingin a fairly standardized look and feel, they could incorporate some objects on the pages that would vary according to the user. They might even go so far as to allow users to create their own WildOutfitters.com-related home page on the site.

Another alternative is to make the Web site adaptive. Instead of requesting specific information from the customers, the Campagnes could enhance the cookie to monitor customers' usage of the site. Then, using this information, they can make more subtle changes to the site when the customer visits.

For each alternative, Alex and Claire must answer several questions: What information will they need to collect, and how will they collect it? What uses for this information are acceptable? At what point would their questions become too personal and obtrusive? Could this approach turn some customers off? What additional responsibilities would they assume to protect the privacy of their customers? The Campagnes are well aware of the trade-offs in regard to privacy brought by the networked economy. Because of the possible risks to privacy, they have decided to think very carefully about how they may use the personal information of their customers in a responsible manner.

"Strange dream, bird brain," Claire said. "What made you wake up?"

"I didn't realize that she'd had enough. I tried to give her unwanted worms with little ads tattooed on the sides," he replied. "She got mad and started to peck at my hand. That's when I woke up."

Claire giggled, "I've heard of snail-mail before, but worm-spam, that's a new one on me." She was barely able to dodge the pillow that flew her way.

Think About It

1. How would you classify the data that would be collected with the system discussed in the case?

2. Alex and Claire are trying to decide between making their Web site adaptive or adaptable. What are the advantages and disadvantages of each choice? Which would you choose for the WildOutfitters.com site? If the Campagnes decide to make their site adaptive, should they notify their customers about the changes in the information that is stored in the cookie? Explain.

3. As Wildoutfitters.com begins to collect personal information from its Web site customers, what rights do the Campagnes have in terms of use of this information? What additional responsibilities do they have in regard to this information and their customers?

Hands On

4. Browse the Web for various sites that collect personal information from customers. How many of these sites have a privacy statement? Read a few of these privacy statements to see what they contain. Do they make you feel secure about entrusting your personal information to the site? Using what you learn, write a privacy statement for WildOutfitters.com.

SOCIAL ISSUES IN THE NETWORKED ECONOMY

LEARNING OBJECTIVES

After reading this chapter, you will be able to answer the following questions:

- How has the widespread use of computers and networked communications changed our concepts of time, distance, and borders?
- What economic issues do we face in the networked economy?
- How has the virtual office changed the lives of many employees in the networked economy?
- What health issues has the networked economy brought on?
- How has online content caused social problems?
- What international issues do we face in the networked economy?
- What kinds of changes might we see in the future of the networked economy?

The U.S. Department of the Interior: Working Without the Internet

Have you ever thought about trying to carry out your everyday activities without a link to the Internet? For more than 70,000 U.S. Department of the Interior employees, that was the case beginning in early December 2001 because of a federal judge's concerns over the security of financial accounts held in trust for 300,000 Native Americans. The judge's order came after a study revealed severe lapses in the trust fund system security. As a part of a test, a computer specialist hacked into the system, created a false account, and altered existing ones. Because the Department of the Interior was not able to determine what sort of connections linked its various Web sites, e-mail servers, and the trust fund, it chose to shut down all of its Web sites and to deny Internet access to all of its employees. Although Interior employees could communicate via e-mail over their LAN with other employees in the same office, they had to use the telephone or fax machine to communicate with employees in other agencies within the department. To make sure that no problems exist, firewalls and intrusion detection software were installed, but, as of March 2002, only 40 percent of the Web sites were back in service.

Shutting down the Department of the Interior Web sites and shutting off employee Internet e-mail had some ripple effects, including the following:

- The ParkNet reservation service run by the National Park Service had to be replaced by a toll-free telephone number.
- Time cards for hourly employees, which had been transacted electronically, had to be filled by hand and sent by express mail service to Denver, Colorado, for processing.
- Oil and gas companies in Wyoming could no longer electronically submit drilling permits to the Bureau of Land Management.
- Conservation groups could not easily access environmental-impact studies and had to submit their comments by postal mail.

Most ironically, the disruption affected 40,000 Native Americans, who normally received electronic royalty checks from the Interior Department for leases on their land. For at least a month after the computer blackout, these individuals did not receive their money. Their attorneys noted that, while Interior Department employees continued to be paid, the Native Americans were not despite the judge's order to ensure that this problem did not occur.

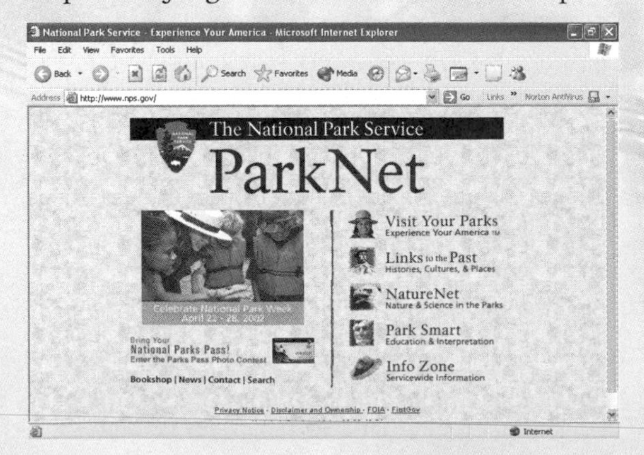

Sources: Tom Kenworthy, "Phone, fax will have to do for offline Interior Dept." *USA Today*, January 4, 2002, p. 9A; and Robert Gehrke, "Interior Department computers still shut down 1 month after court order." *Nando Times*, January 4, 2002, http://www.nandotimes.com/politics/story/211184p-2038870c.html.

The Net as a Part of Our Lives

It seems impossible to believe that it was only a little more than 10 years ago, on January 15, 1992, that Tim Berners-Lee first demonstrated his invention—the World Wide Web—to a roomful of physicists. The world has not been the same since that momentous event. The Internet or, as the popular media often refer to it, the Net, has become an integral part of day-to-day activities for many people. It's difficult to imagine how, as described in the opening case, U.S. Department of Interior employees managed to do their jobs without access to e-mail or the Web for more than a month. As it turned out, this blackout affected not just the Interior Department, but many other people and organizations that had previously relied on the agency's servers.

As described throughout this book, the Internet and its two most popular applications, the Web and e-mail, have created a networked economy that extends around the world. By February 2002, 57 percent of the U.S. and Canadian populations were online—almost 181 million people—and this number has most likely increased since that survey.[1] Although most work can go on without access to the Internet, our lives would certainly be less interesting and less productive without it. And, experts predict, in the future the Internet and the networked economy will only become a more significant part of our lives, both personally and professionally.

Social Cornerstones of the Networked Economy

From a societal point of view, the networked economy is built on three cornerstones: education, free movement of ideas and trade, and the widespread use of computers and communications technology. Education and the free movement of ideas and trade were also cornerstones of the industrial economy. For example, you could argue that the growth of Western economies began with the invention of the printing press. It made knowledge in printed form available to the majority of citizens, whereas before its invention only an elite few had access to extremely scarce, hand-lettered books. The access to knowledge facilitated by the printing press, coupled with the freedom to express ideas and to trade goods freely, led to more inventions and new ways of doing things that spurred economic growth. The key difference between the industrial economy of the last 200 years and the newly emerging networked economy involves the addition of another cornerstone—widespread use of computers and communications technology, as shown in Figure 13.1.

The Gutenburg press changed civilization by making education widely available.

1. Janet Kornblum, "After 10 years on the Web, impact keeps unfolding." *USA Today*, December 27, 2001, p. 3D.

Figure 13.1 Social cornerstones of the networked economy

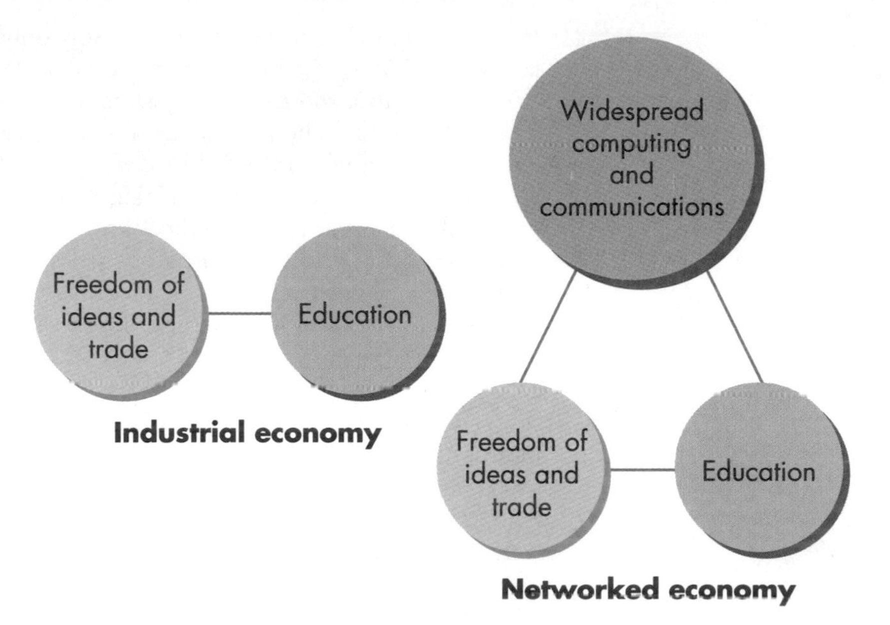

The combination of education and free movement of ideas and trade with widespread computing and communications results in a new type of economy. Instead of requiring employees with physical strength and dexterity to labor in manufacturing plants, power stations, and so on, the networked economy needs employees who know how to use information technology. It generally requires a higher level of education than was required in the industrial economy for an individual to be successful.

Consequences of Internet Use

The shift to the networked economy, with its heavy use of the Internet and networked communications, has resulted in three consequences that you have probably already noticed or been affected by:

- Death of distance
- Homogenization of time
- Disintegration of borders

death of distance
The idea that people are no longer restricted by geography.

The **death of distance** means that geography no longer restricts human activities. That is, you can work from just about anywhere and share your results with co-workers around the world in a matter of seconds. The death of distance has made it possible for many people to work from home using the Internet or private networks, thereby avoiding long commutes to their places of employment.

For example, consider the creation of this textbook. The author lives in Georgia, and the editor lives in Massachusetts. Through use of the Internet, we have been able to develop this textbook in much less time than we would have needed without it. After the author wrote chapters, he converted them to Adobe Portable Document Format (pdf) files and posted them to the Web. Reviewers at educational institutions around the United States downloaded the pdf files, then sent comments on them to the editor via e-mail. In turn, the editor sent the reviews and edited chapters to the author, again via e-mail. Finally, the author e-mailed a second draft to the editor, who then submitted final chapters via e-mail to the publisher for subsequent production. This electronic process can save months of time compared to the earlier method of sending manuscripts via express or postal mail, and it is also much more

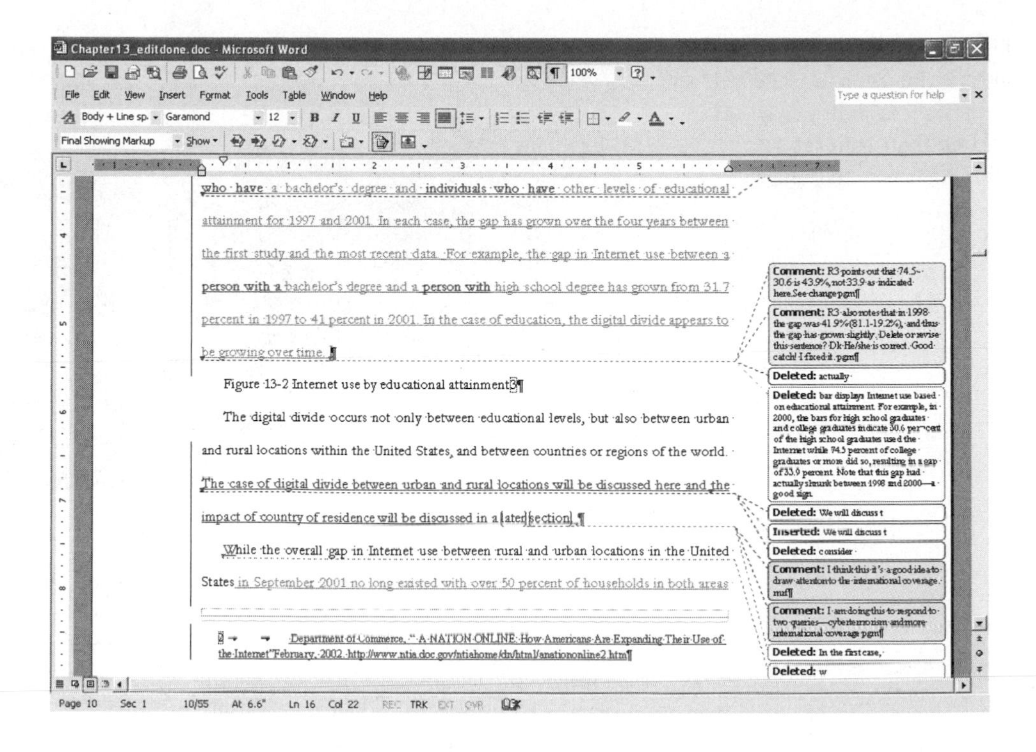

Editing features in word processing packages enable multiple authors and editors—all in different locations—to work on a document, with each contributor's changes and comments being identified by color.

cost-effective. Submitting and processing manuscripts electronically has become the norm. In addition, authors and other professionals can now be more mobile, taking work with them wherever they go, and working from many time zones away if they so choose. For example, Arthur C. Clarke, author of *2001: A Space Odyssey* and many other books, works from Sri Lanka with a U.S.-based publisher.

The **homogenization of time** means that we live in a 24/7 world where business continues somewhere all of the time. Because network servers and software are designed to always remain available, people can work or conduct commerce around the clock, an effect that blurs the boundaries between work and home. For example, as part of an information systems development project, development teams in widely separated time zones might collaborate, so that work on the project could progress continuously. The homogenization of time also means that workers who work at home must be careful to separate their professional and home lives; otherwise, they may have a tendency to work all the time.

homogenization of time
The idea that people live in a 24/7 world, where business continues somewhere all of the time.

Anthropologists and sociologists who study changes in lifestyle have noticed some trends associated with the homogenization of time. In **personal outsourcing**, individuals outsource many of their nonprofessional responsibilities (such as automobile or yard maintenance, transporting children, or even grocery shopping and cooking) to someone else so they can keep working. In **reach creep**, individuals take on more and more work because technology allows it. For example, an author might engage in reach creep by choosing to self-publish using desktop publishing software rather than engaging the services of a publishing house. Both personal outsourcing and reach creep can have decidedly negative effects on your personal life if taken to the extreme.

personal outsourcing
A situation in which individuals outsource many of their nonprofessional responsibilities, such as automobile or yard maintenance, transporting children, or even grocery shopping and cooking, to someone else so they can keep working.

reach creep
A situation in which individuals take on more and more work because technology allows it.

Finally, the **disintegration of borders** means that ideas and electronic goods can flow freely into countries around the world without being subject to search or duties. The rise of the Internet and Web has made it possible to access information, buy goods and services at lower prices with no sales tax, and engage in interactive activities from virtually anywhere in the world. In fact, many social researchers have predicted that the disintegration of borders will lead to a global democratization of society.

disintegration of borders
A situation in which ideas and electronic goods can flow freely into countries around the world without being subject to search or duties.

Kids Kab provides a personal outsourcing service to parents in the Silicon Valley by transporting their children to and from various activities.

In response, a number of states within the United States and countries around the world are attempting to reinstitute their borders through legislation or technology. For example, states have sought ways to tax electronic commerce sales, and countries that fear the import of ideas perceived as dangerous to the government have attempted to institute content controls. In addition, a number of countries now censor gambling and adult-oriented material. Of the three consequences of the Internet, the disintegration of borders appears to be the one that may not continue to grow along with the Net. A later section on international issues in the networked economy will address this topic in more detail.

Social Issues in the Networked Economy

Accompanying the growth of the networked economy are a host of societal issues, including economic issues, workplace-related issues, health concerns, online content issues, international issues, and concerns about the future direction of information technology. No cut-and-dry responses to these issues are possible because, as noted in the opening case, restricting one set of actions may have additional undesirable effects. For example, the emergence of widespread computing and communications has led to a blurring of work and home life. A company might state that employees should never use the Internet for nonwork activities, but if work and home life become blended, how do you define nonwork activities? Also, how do you know whether restricting Internet use might cause the firm to miss some new opportunity that only Web surfers have identified? All of these social issues will be discussed later in this chapter.

1. What are the cornerstones of the networked economy?
2. How are countries trying to reinstitute borders?

Economic Issues in the Networked Economy

The networked economy has already brought about many changes in our lives. One dramatic change involves the definition of *scarcity.* Like land, labor, and capital, consumer attention span has become a scarce commodity. Companies seeking to do business in the networked economy must take this change into account. In addition to this new version of scarcity, the networked economy has created or amplified other economic issues, including the division between those who have access to the Internet and electronic communication and those who do not, and whether and how to tax electronic commerce.

Development of Digital Divides

Because the number of manufacturing jobs is declining and the demand for workers is shifting toward software and service, most new hires in the future will be knowledge workers. Recall that a *knowledge worker* is someone who works with information in an organization. In the networked economy, knowledge workers will be in great demand, as the very nature of work changes from requiring physical strength and dexterity to requiring knowledge. In September 2001, a U.S. Department of Commerce survey found that more than 80 percent of managerial and professional workers use a computer at work; this percentage has certainly gone up since then.[2] The same survey found that college graduates are three times more likely to use the Internet than high school graduates, and ten times more likely than non-high school graduates. Based on these rates, which will assuredly increase throughout the twenty-first century, it will be increasingly more difficult to enter the workplace without some understanding of information technology. Even positions such as automobile mechanics have undergone restructuring to reflect the higher level of information technology required to service today's automobiles.

2. Department of Commerce, "A Nation online: How Americans are expanding their use of the Internet," February 2002, http://www.ntia.doc.gov/ntiahome/dn/html/anationonline2.htm.

Bringing Wireless Broadband to Native Americans

INTERNET IN ACTION

Like many groups in rural America, Native American tribal leaders in the San Diego area faced a problem in bringing broadband Internet to their reservations. They either had to upgrade the existing telecommunications infrastructure or find a way to deploy new technologies that would not depend so heavily on existing systems. A collaborative effort between the University of California-San Diego (UCSD) and the Southern California Tribal Chairman Association (SCTCA) has resulted in "technology leapfrogging" to put such a high-speed network in place. In this case, a wireless broadband approach called the High Performance Research and Education Network (HPWREN) was implemented on three reservations as a very successful pilot project. The effort was funded by a $2.3 million National Science Foundation grant and involved deployment of a wireless backbone consisting of 45 Mbps point-to-point links with high-speed access to individual users with low-cost "Wi-Fi" connections.

This approach cost much less and took a shorter period of time to implement as compared to wired options. Installation took months instead of years and cost a few hundred thousand dollars instead of a few million. In fact, the incremental equipment cost of a tribal connection is estimated to be in the range of $10,000 as a fixed, one-time cost. This cost drops even further when one considers that the Wi-Fi receivers can be purchased for around $100 and provide broadband capability comparable to multiple T-1 lines. The main problem involved in setting up the network was achieving clear lines-of-sight between the various wireless connection points in and around the mountainous territory of two of the reservations. The pilot project has proved so successful in providing the three reservations with high-speed Internet connections that the SCTCA plans to build on the HPWREN model to connect all 18 reservations in the San Diego area in a wireless backbone to create a network of virtual tribal villages.

The Southern California Tribal Chairman Association (SCTCA) is installing a wireless network to bring connectivity to native American reservations near San Diego.

Source: Kade L. Twist, "Native networking trends: wireless broadband networks." *Digital Divide Network*, September 24, 2001, http://www.digitaldividenet.org.

digital divide
A separation of the world population based on people's knowledge of and access to the Internet.

The need for workers who can use information technology has prompted concerns that those who lack knowledge of or access to information technology and the Internet might miss out on participating in the networked economy. This widening gap between the digital haves and have-nots of the world is called the **digital divide**. To measure it, studies have begun to identify trends in Internet use by various characteristics, including education, race/national origin, rural versus urban location, and country of residence.

Although race/national origin has and continues to be a concern in terms of the digital divide, studies show that education is actually a more important factor. This finding should not surprise anyone: Education is essential to managing information technology and adapting to the changing work requirements in the networked economy. Without education, workers will struggle to find positions that will enable them to share in the wealth generated by the networked economy.

Figure 13.2 shows education trends affecting the digital divide. In the figure, each bar shows the gap between Internet use for individuals with a bachelor's degree and other levels of educational attainment for 1997 and 2001. In each case, the gap has grown over the four years between the first study and the most recent data. For example, the gap in Internet use between people holding a bachelor's degree and people holding a high school degree grew from 31.7 percent in 1997 to 41 percent in 2001. In the case of education, the digital divide appears to be expanding over time.

The digital divide occurs not only between educational levels, but also between urban and rural locations within the United States and between countries or regions of the world. The case of digital divide between urban and rural locations is discussed here; a later section will focus on the implications of country of residence.

In September 2001, the overall gap in Internet use between people living in rural and urban locations in the United States no longer existed, as more than 50 percent of households in both areas had Internet access. However, the Department of Commerce's Nation Online study reveals that the gap in access to high-speed broadband Internet through either cable or DSL remains significant. In 2000, only 7.3 percent of rural areas had access to some form of broadband as compared to 12.2 percent for central cities.

Figure 13.2 Internet use by educational attainment

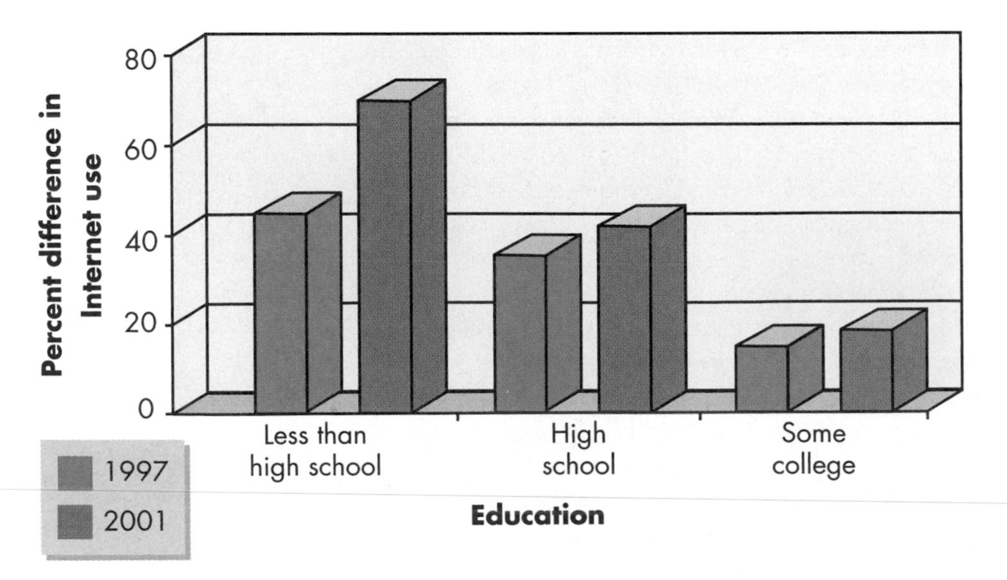

Source: Department of Commerce, "A nation online: How Americans are expanding their use of the Internet," February 2002, http://www.ntia.doc.gov/ntiahome/dn/html/anationonline2.htm.

In 2001, although the percentage of rural households had grown to 12.2 percent, the percentage of central city households had grown to 22 percent, increasing the gap between urban and rural broadband access from 5.2 percent in 2000 to 9.8 percent in 2001. Clearly, rural areas remain largely restricted to lower-speed telephone connections and thus will enjoy less access to the many features of broadband access.

Taxation of Electronic Commerce

With the rapid growth of electronic commerce, there is pressure in the United States and globally to subject that commerce to some form of sales tax. In the United States, because sales made through electronic commerce often go across state borders, many states lose sales tax revenue that they would otherwise collect on sales made at stores within their borders. Because they rely on sales taxes for more than one-third of their revenues, states and counties are definitely hurt by Web sales across state lines. In fact, state and local jurisdictions will lose an estimated $13 billion in sales taxes by 2004 from the expected $155 billion in e-commerce sales.

Internet Tax Freedom Act of 1998
Legislation that established a three-year waiting period on state and local Internet taxes in the United States and created the Advisory Commission on Electronic Commerce to study this issue. It was extended in 2002 until 2003.

Because the long-term results of a tax on electronic commerce were unclear, the **Internet Tax Freedom Act of 1998** placed a three-year waiting period on state and local Internet taxes in the United States and created the Advisory Commission on Electronic Commerce to study this issue and make recommendations to Congress by April 2000 on taxing electronic commerce. The commission completed its work by simply recommending an extension of the waiting period. Some of the issues it encountered in its work included the following:

- Electronic commerce businesses hesitate to collect sales taxes in all 50 states because they fear the states will begin imposing corporate income and franchise taxes on them. Currently, a business is subject to such taxes only if it has a presence in the state in the form of an office or store.
- With 7600 state and local taxing jurisdictions, electronic commerce firms want to avoid having to collect and remit taxes to all of them.
- Local governments do not want to lose their ability to set sales tax rates for fear of losing tax revenue. Many counties and cities collect local-option sales taxes on top of the sales taxes collected by the state, which provide revenue to fund operations or new projects.

Shortly after the three-year waiting period for the imposition of Internet taxes expired, a two-year extension was passed that will continue the moratorium until late 2003. This extension also gives the states five years to come up with a simplified tax-collection plan with one national tax rate or one tax rate per state.

1. How does education relate to the digital divide?
2. Why do electronic commerce companies want to avoid collecting and remitting sales taxes to the various political entities in the United States?

The Virtual Workplace

virtual workplace
A concept relating the capability of a worker to work at any place and any time.

Related to the death of distance and homogenization of time is the concept of the **virtual workplace**, in which a worker can work at any place and any time. The increased availability of wireless networking has enhanced the attractiveness of the virtual workplace by no longer requiring that the worker use a wired connection. The most popular approach to the virtual office is teleworking from home, temporary or shared office space, or a mobile office.

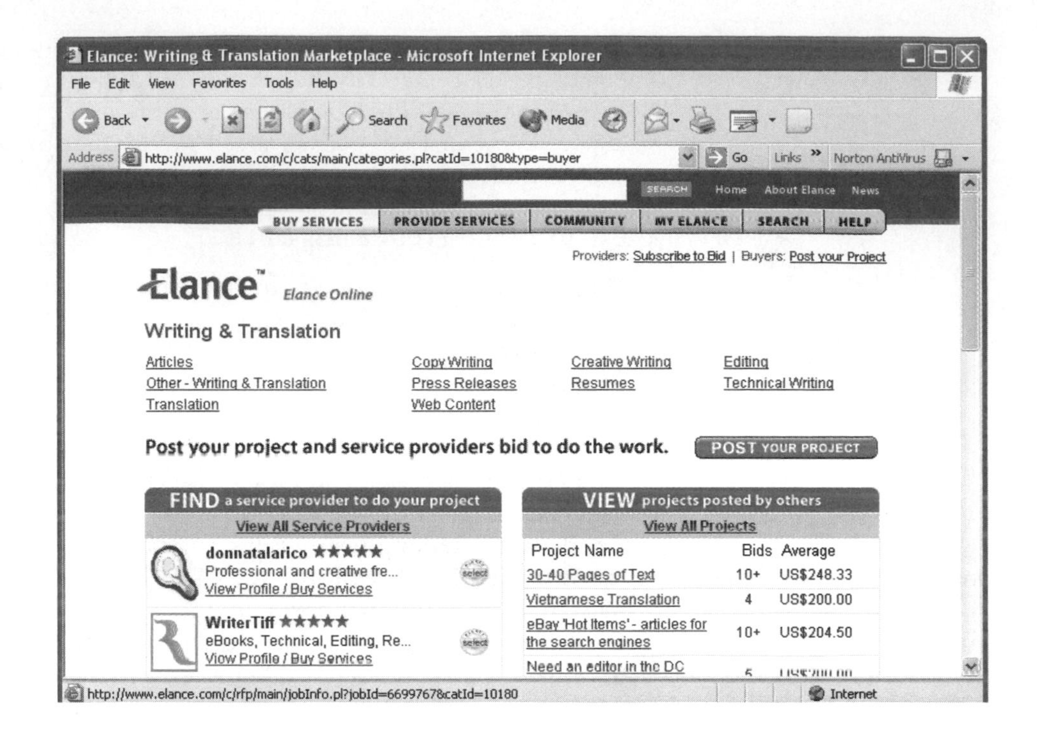

The elance.com Web site enables e-lancers to learn about jobs and bid on them.

e-lancing
Electronically connected freelancers working together on project teams for a week, a month, or however long it takes to complete a project.

virtual assistant
A person who works as an assistant on an as-needed basis over the Internet for a number of employers.

In addition, a new concept called e-lancing is becoming more popular. With **e-lancing**, electronically connected freelancers work together on project teams for a week, a month, or however long it takes to complete a project. Several Web sites operate to connect e-lancers with companies looking for people with particular skills to work on a project including www.elance.com and www.FreeAgent.com. Another version of this idea is the **virtual assistant**, a person who works as an assistant on as-needed basis over the Internet for a number of employers. This section concentrates on teleworking and its application in the networked economy.

Teleworking

teleworking
The use of networks to engage in work outside the traditional workplace. Also called telecommuting.

Teleworking (also known as **telecommuting**), the use of networks to engage in work outside the traditional workplace, is an increasingly popular approach to work. As shown in Figure 13.3, surveys conducted by the International Telework Association and Council (ITAC) revealed that teleworking grew dramatically between 1990 and 2001, the last year for which data were available, increasing by more than 20 million workers to 28.8 million. Forecasters expect teleworking to continue to grow for the foreseeable future.

Many companies have established policies for teleworking, including AT&T, IBM, and Lucent Technologies. AT&T adopted a corporate teleworking policy in 1992. By 2001, the percentage of U.S.-based AT&T managers who teleworked at least one day per month had grown to 56 percent, with 11 percent teleworking 100 percent of the time.[3] The most recent ITAC survey determined that 24.1 percent of teleworkers preform their duties in some type of mobile environment and 21.7 percent work from home. A smaller proportion, 11.7 percent, work from some type of temporary office, but most of them, 42.4 percent, combine working from home with some other type of teleworking.

3. http://www.att.com/telework/docs/congressional_testimony.pdf.

Figure 13.3 Growth of teleworking

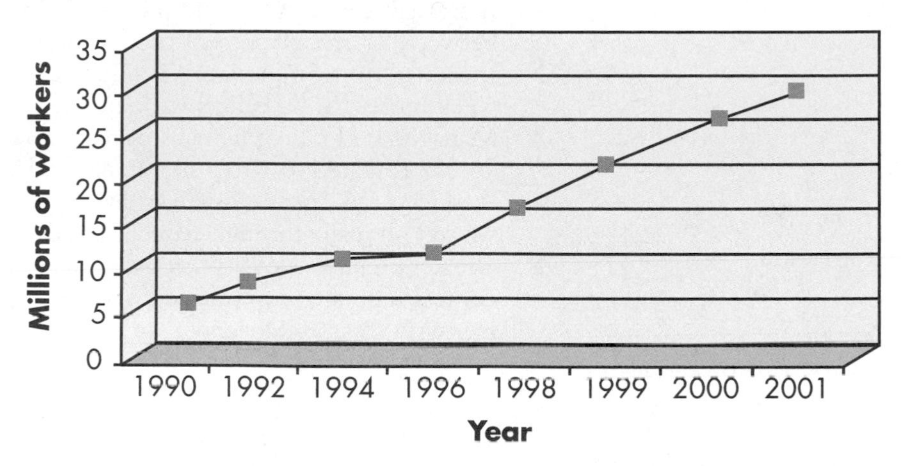

Source: http://www.telecommute.org.

Clean Air Act of 1990
Legislation aimed at reducing the amount of air pollution, which included a recommendation for increased teleworking.

Family and Medical Leave Act of 1992
Legislation that encouraged teleworking as a way of employees remaining at home to care for family members.

Two federal laws particularly encourage teleworking: amendments to the Clean Air Act of 1990 and the Family and Medical Leave Act of 1992. The **Clean Air Act of 1990** aimed to reduce the amount of air pollution and included a recommendation for increased teleworking. The **Family and Medical Leave Act of 1992** encouraged teleworking as a way for employees to remain at home and care for family members. Teleworking was widely used in Atlanta during the 1996 Olympics, for example, and is being encouraged again to combat the area's severe air quality problems during the summer months. Much of this problem is believed to stem from automobiles used by Atlanta's commuters who face a longer ride to work than people in any other metropolitan area in the United States.

A variety of surveys have identified many benefits to teleworking for the employee, the company, and the community. Many employees, like the increased opportunities that this approach gives them to be with their family. They also save, on average, more than 50 minutes of commute time by teleworking, equivalent to more than five working weeks over the course of a year.

The level of commuting in Atlanta is so great that it is causing high levels of air pollution during the summer months.

For the company, teleworking can provide considerable benefits as well:[4]

- Increased productivity: JD Edwards, Inc., found that its teleworkers were between 20 and 25 percent more productive than its office workers.
- Increased time on the job: AT&T's teleworkers work an average of 5 hours per week more than its office workers.
- Manager/staff ratio: The ratio of managers to staff in a virtual workplace is 1 to 40, as compared to 1 to 4 in a traditional workplace.
- Reduced real estate costs: On average, a company can save $8000 per teleworker annually through reduced need for office space.
- Greater employee loyalty: An overwhelming majority (80 percent) of teleworkers feel a greater commitment to the organization and most say they plan to stay with their employer.

Finally, for the community, teleworking leads to less traffic and air pollution. Once again, consider the average 50-minute commute that is saved by teleworking, resulting in far fewer cars on the road and much less air pollution. In fact, in Atlanta, teleworking represents an integral part of the effort to reduce air pollution problem (see www.cleanaircampaign.org).

On the downside, teleworkers require a different type of management because they are not present in the traditional workplace. In addition, teleworking is not for every worker because it requires a large degree of self-motivation and discipline. Finally, it can erode the quality of home life if the worker cannot separate his or her work life from home life, because it can lead to too much homogenization of time.

Teleworking Locations

As noted earlier, most teleworkers work either from the home or in the home in combination with some other location. Home offices created out of a spare bedroom or a space in the living area are now common, with almost half of all teleworkers reporting having a home office. Many employers provide their teleworking employees with office furniture in addition to technology such as computers, Internet connections, and fax and copy machines. Interestingly enough, more teleworkers report that they work during times other than the traditional work day than do office workers, but also experience less conflict and an enhanced quality of life.

In addition to working from the home, four other options for teleworking exist: the virtual/mobile office, hoteling, the satellite office, and the telework center.

virtual/mobile office
Equipping employees with the communications tools and technology they need to perform their jobs from wherever they need to be—home, office, customer location, airport, and so on.

hoteling
The use of temporary office space by employees from many companies for short-term periods.

satellite office
A fully equipped office set up by a firm where its employees can reserve space and work one or more days per week closer to their homes, thereby reducing commute times and helping ease traffic congestion.

In a **virtual/mobile office**, employees are equipped with the communications tools and technology needed to perform their jobs from wherever they need to be—home, office, customer location, airport, and so on.

With **hoteling**, an office building contains temporary office space for drop-in use by employees from many companies. This office space, which is typically offered by a third party, comes equipped with standard office technology—phones, PCs, faxes, printers, copiers, e-mail, Internet access, and so on—and employees either reserve space in advance or drop in to use a cubicle on an as-needed basis. Employees or their companies are charged for the amount of time that the space is used.

A **satellite office** is a fully equipped office set up by a firm where its employees can reserve space and work one or more days per week closer to their homes, thereby reducing commute times and helping ease traffic congestion. Satellite offices differ from hoteling in that they are part of a company's office space, whereas hoteling office space is not.

4. http://www.telecommute.org/resources/facts.shtm.

At telework centers, employees can carry out their work closer to home.

telework center
A fully equipped office used by employees from different organizations, with employers being charged for the space and services utilized by each employee per day.

Finally, a **telework center** resembles a satellite office except that employees from different organizations use the office space, with employers being charged for the space and services utilized by each employee per day. These centers are typically located closer to employees' homes than their regular office locations. A telework center represents a permanent version of hoteling where the company rents space on a long-term basis.

Virtual Workplaces at FarEast Foods

FarEast Foods has always encouraged its employees to telework if they feel it would benefit both the company and them. Many of the firm's programmers work at home and submit their programs over the Internet to the company Web server for testing. The company provides them with the latest technology and gives them a $1500 allowance for home office furniture. Teleworkers come into the office once a week to meet with other team members and their supervisor; they use the telephone for audio teleconferencing at other times.

1. What is a virtual assistant?
2. What are the five types of teleworking?

Health Issues in the Networked Economy

In general, computers and the Internet have proved beneficial to society and have allowed us to do many things not otherwise possible. For example, they have created virtual communities of people who otherwise would never have met. However, working at the speed of the Internet can cause health problems associated with long-term use of computers, especially the keyboard and mouse. These problems cost employers large sums in worker's compensation costs and lost productivity. In addition, some psychological health problems are associated with use of the Internet. This section covers the various health problems and possible solutions.

Repetitive Stress Injuries

In the early part of the twentieth century, workers in many blue-collar occupations that required repetitive motions often complained of a variety of aches and pains that they accepted as a part of the job, such as "stitcher's wrist," "brick layer's shoulder," "meat cutter's wrist," and "cotton twister's hand." Today, with the widespread use of computers, workers in computer-related occupations are experiencing similar types of

repetitive stress injury (RSI)
A condition in which workers suffer from moderate to severe muscle and joint problems in the hand, wrist, arm, and shoulder.

aches and pains. These **repetitive stress injuries (RSI)**, in which workers suffer from moderate to severe muscle and joint problems in the hand, wrist, arm, and shoulder, are becoming epidemic in computer-related jobs. In fact, the use of the computer has created a dimension in occupational health and safety unique to computer users. A National Academy of Science panel looking into the problem found that 1 million U.S. workdays were lost annually due to RSI. Although not all RSIs stem directly from computer use, a significant percentage undoubtedly originate from this source.

RSIs have increased concurrently with growth in the numbers of heavy computer users (that is, users who spend long, uninterrupted sessions working at the computer). That long hours at the computer causes problems is not surprising when you consider that a typical keyboard operator's hands travel about 16 miles across the keyboard during an eight-hour workday. This effort involves striking 115,200 keys, a daily workload equivalent of lifting 1.25 tons!

At one time, RSIs were restricted to professions that made heavy use of the keyboard; today, that is no longer the case. The growth in Internet use through such activities as sending, receiving, and processing e-mail, Web surfing, participating in chat rooms, and so on has brought RSIs home to many people who would not consider themselves to work in a profession involving heavy use of the keyboard.

RSI is also called *cumulative trauma disorder (CTD)* and *typing injury (TI)*. Regardless of the exact name used, RSI always involves numbness and tingling in the hands as well as pain and edema (swelling) of the hands, arms, shoulders, neck, and/or back. It occurs when muscles, tendons, and nerves become damaged by irritation resulting from prolonged use of a keyboard with the body in an unnatural, unrelaxed position. Often the injury results from using an improperly designed workstation (in this context, a **workstation** comprises the computer and supporting furniture). RSI can put a worker out of work for weeks or even months, and be very expensive to treat.

workstation
A computer and supporting furniture.

carpal tunnel syndrome (CTS)
An ailment in which, because of excessive keyboarding, the median nerve in the arm becomes compressed when swollen, inflamed tendons exert pressure on a nerve.

tendonitis
A general inflammation and swelling of the tendons in the hands, wrists, or arms.

RSI can take several forms, including carpal tunnel syndrome and tendonitis. **Carpal tunnel syndrome (CTS)** results when the median nerve in the arm becomes compressed because swollen, inflamed tendons exert pressure on a nerve. **Tendonitis** entails a general inflammation and swelling of the tendons in the hands, wrists, or arms. An irritation of the tendons connecting the forearm to the elbow joint is known as tennis elbow. Table 13.1 compares tendonitis and CTS.

Carpal Tunnel Syndrome

Carpal tunnel syndrome is a very common form of RSI. If friction occurs when the tendons connecting the fingers in the hand slide back and forth in their sheaths, inflammation and pain may occur. If the swollen tendons squeeze the arm's median nerve at the wrist, where the median nerve passes to the fingers through a narrow passage called the carpal tunnel, then carpal tunnel syndrome can occur. Figure 13.4 depicts this process. When the nerve becomes compressed, severe pain, numbness, and loss of strength result. These sensations are more common at night, often awakening the victim. Symptoms may recede if the repetitive activity stops, but can worsen quickly if the activity continues. As the nerve compression continues, damage to the nerve can occur.

To diagnose carpal tunnel syndrome, a physician takes a history of the worker's activity and symptoms. Also, electrical impulse procedures are undertaken to assess the nerve's ability to fully handle the impulse—a delay indicates an obstruction in the carpal tunnel. Nonsurgical treatment is, of course, preferred. It includes immobilizing the hand by splinting the hand and wrist for three to six weeks and prescribing anti-inflammatory drugs. If no improvement results in two to three months, then surgical intervention becomes necessary. The surgical procedure, most often done on an outpatient basis, involves severing the transverse carpal ligament, thereby releasing the pressure on the median nerve. Complete healing takes about two months. When the syndrome is recognized and treated early, 80 percent of all cases can be reversed.

Table 13.1 Comparing Tendonitis and Carpal Tunnel Syndrome

	Tendonitis	Carpal Tunnel Syndrome
What is it?	Inflammation in and around the tendon	Pain in the wrists brought on by repetitive flexing or extension
Who gets it?	Almost anyone due to overuse, disease, or injuries	Athletes, manual laborers, office workers, and those who use keyboards or small handtools
Symptoms	Pain or tenderness near a joint; numbness or tingling might appear	Painful tingling and loss of muscle power in one or both hands
Treatment	Anti-inflammatory drugs, physical therapy, or, in the worst cases, surgery	Anti-inflammatory drugs, rest, immobilization, or, in the worst cases, surgery

Source: USA Today Research, "High court raises bar for ADA." *USA Today*, January 9, 2002, p. 3A.

Causes of Repetitive Stress Injuries

Research into RSI has identified a number of risk factors. Heavy computer users were recognized as the most likely subjects to develop RSI. Improper workstation design is another key risk factor. Clearly, if the worker uses the keyboard for extended periods of time in a strained or unnatural position, injury is much more likely. Surveys by labor-related organizations reveal that many workers do not understand how the workstation should be designed to avoid RSI.

Other RSI risk factors include the worker's psychological state relating to production and deadline pressures, fear of losing the job, boredom with repetitive tasks, or isolation or discouragement of socializing with co-workers. Also, certain physical factors have been identified as risks, including diabetes, pregnancy, and thyroid disorders. These tendencies are not fully understood but have become the subject of study.

Figure 13.4 Carpal tunnel syndrome

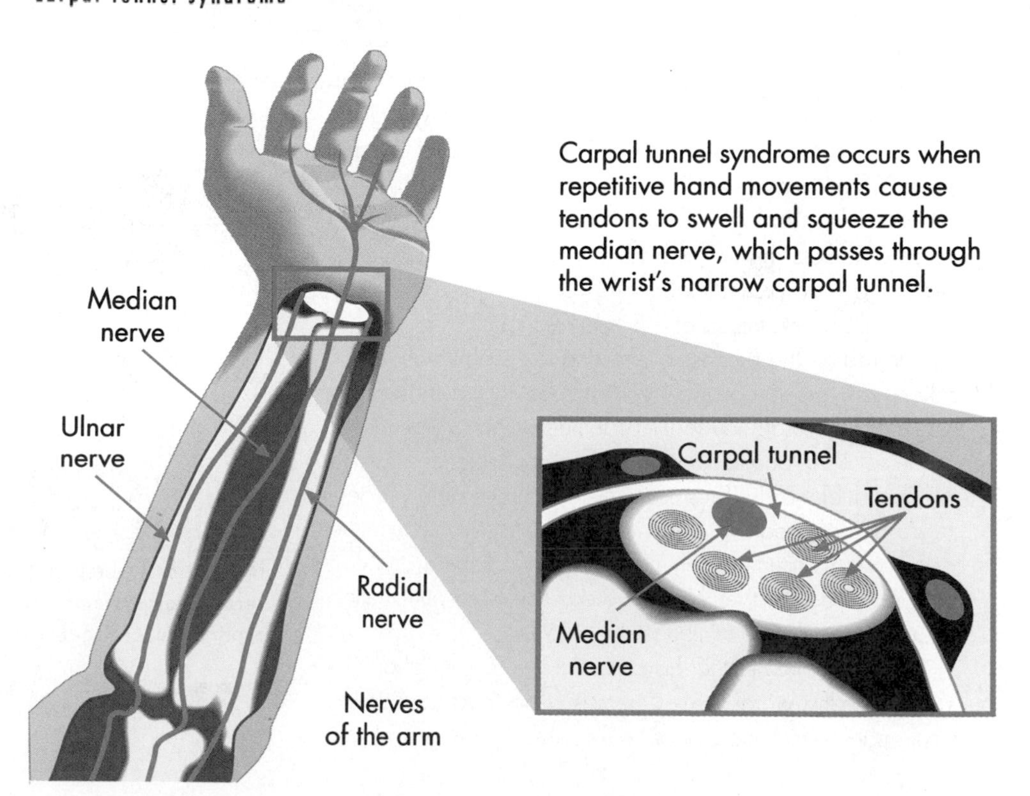
Carpal tunnel syndrome occurs when repetitive hand movements cause tendons to swell and squeeze the median nerve, which passes through the wrist's narrow carpal tunnel.
Median nerve
Ulnar nerve
Radial nerve
Nerves of the arm
Carpal tunnel
Tendons
Median nerve

Researchers have found that a comprehensive examination of the causes of RSI and other health-related problems must consider the following issues:

- The work motion required to complete the task
- The equipment design of the workstation
- The education provided for the worker about how to use the workstation equipment
- The attitude of management toward the worker
- The physical characteristics that may predispose the worker to repetitive stress injury

This comprehensive approach may provide greater understanding of the causes of RSI.

Mouse Problems

Although not nearly as problematic as keyboard use, improper use of a mouse can also cause shoulder, back, and arm pain. People who use a mouse positioned to the side of their computer keyboard often have greater muscle tension in the upper shoulder, back, and arm than those who use a centrally located trackball pointer. Because the mouse is used 30 percent of the time in word processing and 80 percent of the time in graphics work, proper positioning can be critical to avoiding "mouse shoulder" or "mouse arm." Training and short breaks can also cut this tension in half.

Solving RSI Problems with Voice Recognition

TECHNOLOGY ON THE EDGE

When his doctor told him in 1993 to give up typing due to problems with tenderness and swelling in the tendons in his wrist, David Pogue knew he had to find another solution to the problem. After all, writing was his profession—changing careers was not an option. Pogue tried to use a hired typist and underwent months of hand therapy, but neither solved his problem. Finally, he spent $2500 for an early version of a speech recognition program. Although it had its problems (for example, it required a pause between each word), the program did allow Pogue to continue his writing career while giving his wrists time to heal. Today, the much higher processing capability of PCs allows Pogue to dictate at a rate of 130 words per minute while speaking conversationally. People coming into his office and watching his latest voice recognition software pour text into a word processing program or e-mail client think they have walked onto the set of a *Star Wars* movie.

In reality, the process of entering text with voice recognition software is nothing like that seen in the movies. To maximize accuracy by looking at the content, voice recognition software analyzes entire phrases or sentences instead of individual words. Thus users must formulate an entire sentence before speaking. They must also explicitly include the punctuation, inserting commas and semicolons in the correct locations, and let the software know what punctuation to use in ending the sentence.

Pogue estimates that his current software gets about 98 percent of the words correct, which is not bad considering how many errors most of us make while working from the keyboard. Although the voice recognition software never misspells a word, it does misinterpret words, leading to some very funny mistakes that Pogue refers to as "wordos"; unfortunately, spell checkers cannot pick up these errors. For example, he has seen "inscrutable" transcribed as "in screw double" and "or take a shower" come out as "Ortega shower." Anyone who uses voice recognition software must be especially careful to proofread the final document—a small price to pay to continue working at your chosen profession.

In addition to helping people solve RSI problems, voice recognition systems have become an important part of how medical workers generate information for medical records.

Source: David Pogue, "If typing won't do, speak up." *New York Times*, April 19, 2001, http://www.nytimes.com.

Figure 13.5 Thumb-based trackball

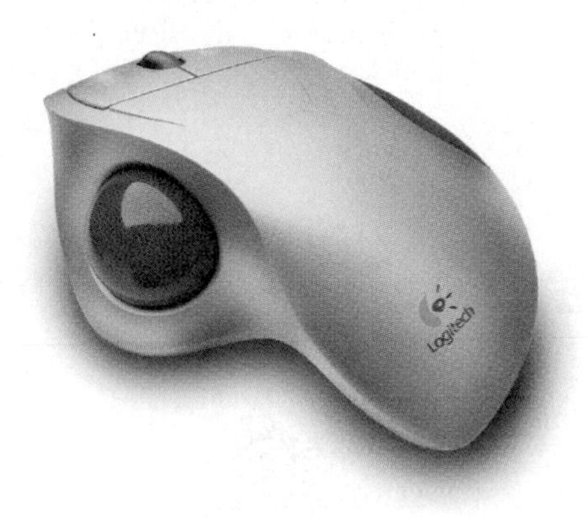

Today's keyboards are much wider than the versions used when the mouse was first introduced in 1984, and the mouse has been moved farther to the side of the user. Users tend to keep their arms straight out, causing shoulder problems. With a centrally located trackball pointer, the pointing motion moves right in front of the user. The author has switched from a mouse to a trackball for this reason.

Even for recreational computer users, extended use of the keyboard and mouse carries a risk. Researchers have found that Web surfing can be hazardous to your health, particularly because it is a leisure activity and people tend to be less aware of their posture and the way they use the mouse and keyboard. By leaning away from the keyboard, users put extra stress on wrists and elbows. They also tend to keep constant pressure on their mouse-click finger while scrolling through long Web pages and don't take advantage of natural waiting periods during downloads, instead keeping their hands on the mouse in a position called "mouse freeze." Figure 13.5 shows a trackball designed for users who find it easier to use their thumb to move the cursor.

Ergonomics

Ergonomic workstations help workers to avoid physical problems associated with computer use.

Ergonomics is the science of designing the workplace in such a way as to keep people healthy while they work, resulting in higher morale and more productivity. The increase in heavy computer usage and rising numbers of RSIs have made ergonomics essential. Ergonomics combines the knowledge of engineers, architects, physiologists, behavioral scientists, environmental scientists, physicians, and furniture designers and manufacturers to determine the best design of tools, tasks, and environments.

The goal of ergonomics is to create an optimal balance between productivity and well-being. An ergonomically designed workstation allows a worker to work in a comfortable posture, thereby reducing the risk of developing a repetitive stress injury. Ergonomic design includes consideration of the height and position of the monitor, the height and angle of the keyboard, the task chair, proper indirect and task lighting, proper ventilation, noise reduction, and a footrest if the user's feet do not rest squarely on the floor. Of these factors, the task chair or seat is probably the most important element. It should have an adjustable back and arm supports and an adjustable height. Figure 13.6 shows an ergonomically designed workstation.

Although the workstation equipment may be ergonomically adjustable, the worker must know how to make adjustments to the workstation and recognize the importance of using them. Many employers approach the RSI problem as a team effort; including the information system manager as part of the team often proves beneficial. This team approach makes using the equipment properly as important as efficiency and productivity. The team approach also improves the working relationship

Figure 13.6 Ergonomically designed workstation

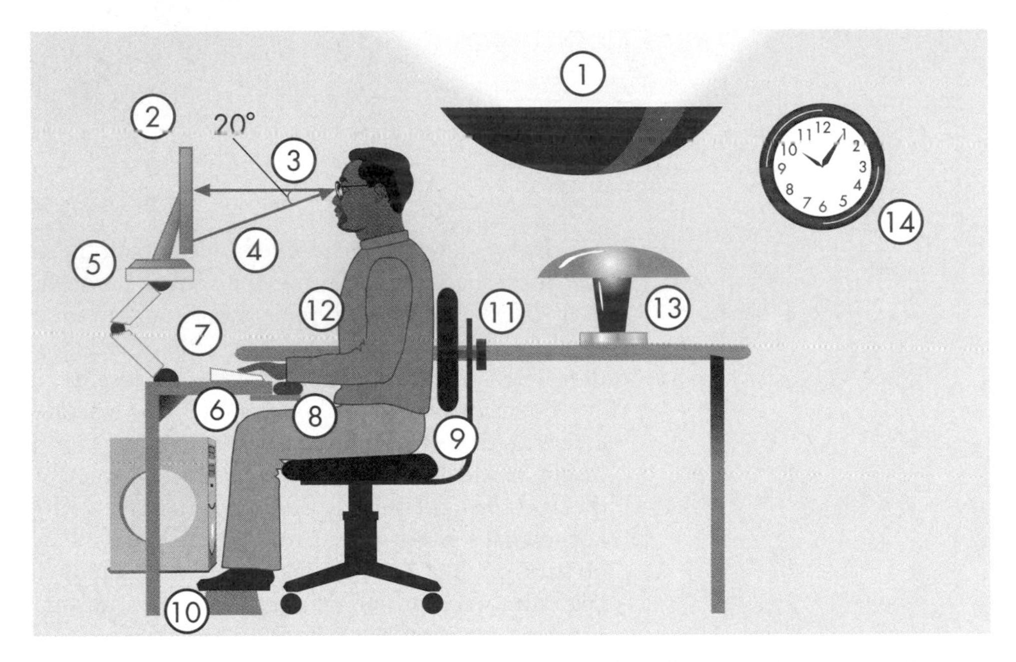

1. Indirect Lighting—Fixtures that bounce light off of ceilings or walls provide a softer light that is less likely to reflect off of monitor screens and create glare.
2. Monitor Height—The top of the monitor should be no higher than eye level.
3. Monitor Distance—16 to 22 inches is recommended for visual acuity; 28 inches or more is recommended if there are emission concerns.
4. Monitor Display—High resolution, screen with dark letters on a light background with an anti-glare screen coating.
5. Adjustable Monitor Support—The monitor should move up and down forward and back, and fill on its axis.
6. Keyboard support—Adjustable between 23 and 28 inches in height. Operator's arm should hang straight down from the shoulder and bend 90 degrees at the elbow and enable the operator to type without flexing or hyperextending the wrist.
7. Keyboard—Adjustable tilt to enable typist to keep hands in a straight line with the wrists and forearms.
8. Wrist Rest—Rounded, padded, adjustable support for the heel of the hand or forearm without constricting the wrist.
9. Seat Height—Adjustable between 16 and 19 inches. Users should be able to bend their hips and knees at 90 degrees and sit with their feet flat on the floor.
10. Foot Rest—Enables typists with shorter legs to rest feet flat while working.
11. Back Support—Back rest should adjust up, down, forward, and backward to support the lumbar portion of the spine in the small of the back.
12. Work Surface Height—A comfortable height for reading, handwriting, drawing, and other non-keyboard work.
13. Reading Light—An independent light source for reading letters, reports, books, etc.
14. Clock—Schedule regular breaks, preferably 5 minutes per hour.

Figure 13.7 Ergonomic keyboard

between management and the worker. A positive mutual attitude has been shown to be productive both in work quality and injury reduction.

Because the keyboard lies at the root of many repetitive stress injuries, several keyboard designs have become available, each claiming that users may use it with their hands in a natural position. Figure 13.7 shows one of these ergonomic keyboards. High-speed scanners and bar code wands can also help by reducing repetitive key data entry.

Psychological Problems

The psychologically healthy person uses the computer in healthy ways but may also suffer some negative effects. As a tool, the computer extends the expectation of work productivity. When results fall short of expectations and they frequently do—a slow process of frustration and anxiety may result in a gradual erosion of self-esteem, leading to feelings of inadequacy and depression. After all, self-esteem is enhanced by success. When perceived failures, no matter how minor, occur, disappointments and depression may ensue. Pressures to produce may take time from the worker that was previously available to family, which can create problems in his or her relationships with a spouse, children, and significant others.

Like any other tool, the computer can exacerbate existing pathologies by presenting another avenue for expressing potentially negative behavior. Typical types of behavior facilitated by the computer and the Internet include "Web addiction," introversion, depression, and pedophilia. A widely quoted 1999 survey of Web use appeared to show that 6 percent of all Internet users were addicted and 10 percent were abusers whose Internet use impinged on the rest of their lives. However, if one looks closely at the survey, which had 17,000 respondents, it should be noted that this data were self-reported and that the users who demonstrated Web addiction appeared to have other psychological problems. Addiction can take many forms, including tobacco, alcohol, and gambling, so heavy use of the Web could be considered simply another form of overall addictive behavior.

Health Problems at FarEast Foods

FarEast Foods provides its employees, whether working in the office or teleworking, with the most ergonomic workstations it can find and encourages its employees to report any work-related health problems immediately. It also runs periodic educational programs to train its employees in the proper use of a keyboard and ways to make ergonomic adjustments to their workstations.

1. What does RSI stand for? Why is RSI on the rise in the networked economy?
2. List three features of an ergonomic workstation.

Online Content Issues

The Internet has brought a tremendous advance in the availability of information, resources, and activities. Unfortunately, some of the content made available online raises issues about defamatory messages, adult-oriented Web sites, and Web-based gambling. This section discusses these online content issues and the ways in which they are being handled.

Defamatory Content

The Internet has developed into a combination postal service, dorm bulletin board, local newspaper, and town crier. Through the Internet you can send mail, post messages on electronic bulletin boards or chat rooms, and create your own Web page with news, editorials, gossip, and classified advertisements. Because it is possible to send a large number of messages at little or no cost or to easily republish someone else's negative information, the Internet also provides a ready avenue for spreading negative information about others. In some cases, groups and individuals use the Internet to spread information that may harm other individuals, groups, and companies. For example, a historian at Emory University was attacked by what he refers to as "Web stalkers" for a book he wrote on the use of guns prior to 1877. He was vilified in reviews at Amazon.com and eventually driven off the Internet after a large number of viruses were anonymously sent to him. As he and many others have found out, because news moves so fast over the Internet, the person being attacked has little chance to correct the erroneous assertions before they get out of control.

Court cases have yielded conflicting results regarding what kinds of messages are legal on the Internet and who is liable for them. In one case, a judge dismissed a libel suit brought by two physicians against a person who reposted libelous claims about them. In another case, however, a physician was awarded a large libel settlement against a person who made a defamatory anonymous posting on an Internet message board. Both rulings were based on the Communications Decency Act of 1996, which does not hold Internet service providers liable for what is posted on their servers, but does hold the original creator liable for it.

In two other cases, former employees were held responsible for sending highly negative e-mails to current employees or posting negative messages on bulletin boards about their former employers. In both cases, the defendants claimed that the ruling would hamper free speech. Clearly, a great deal of litigation will focus on defining what is appropriate and legal information to send or post on the Internet.

Adult-Oriented Web Content

Like virtually every other form of communication through human history, the Web has become a popular source of adult-oriented content, with a large number of such sites being advertised widely via spam. Although few would argue that adult-oriented Web sites are inappropriate for children, a great deal of discussion has centered on how to ensure that children don't visit these sites. Various approaches have been tried, including the following:

- Passing legislation restricting the content of Web sites
- Using filtering software on browsers
- Using filtering software on servers

Communications Decency Act (CDA) of 1996
Legislation, since ruled unconstitutional, that aimed at protecting citizens from pornography on the Internet.

Under the first approach, the United States has enacted several laws attempting to restrict Web content that is indecent or patently offensive from all online systems accessible to minors. The first of these laws, the **Communications Decency Act (CDA) of 1996**, was struck down by the U.S. Supreme Court in 1998. Later that year, Congress passed the Children's Online Privacy Protection Act of 1998 (COPPA).

CyberPatrol is a filtering system that works at the browser level to restrict access to Web sites that contain objectionable material.

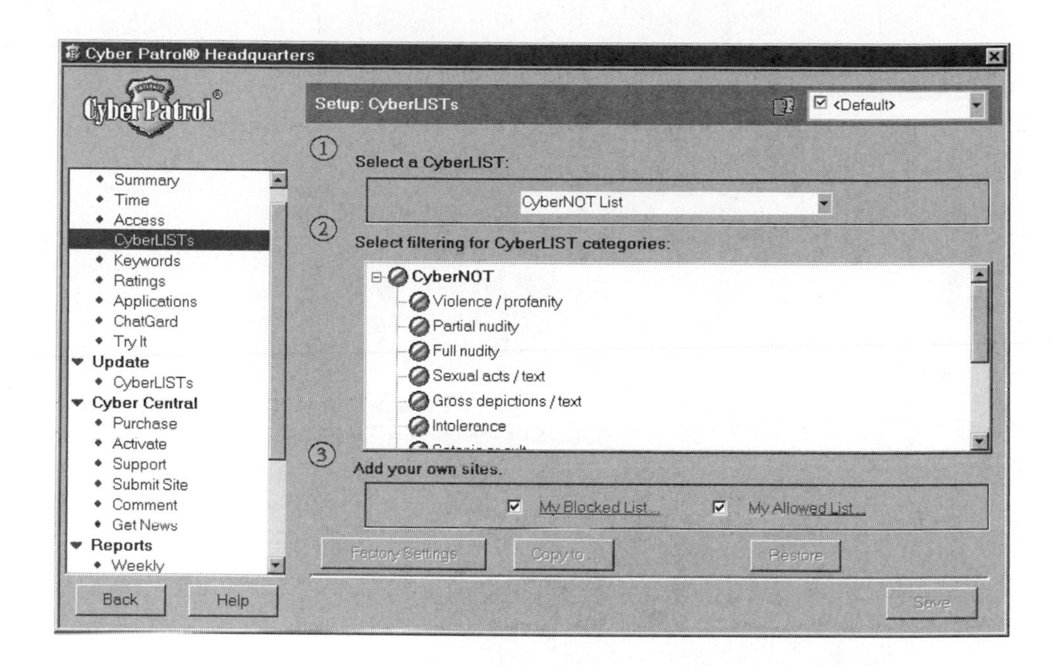

These two attempts to restrict online content and the corresponding legal rulings invalidating the first act point out the problems associated with censoring the Internet. In the United States, the First Amendment to the Constitution guarantees freedom of speech, and the courts have interpreted it as including the Internet.

browser-based filters
Software that works with Web browsers to restrict access to Web sites that contain objectionable material.

Instead of depending on legislation to protect their families from inappropriate material on the Internet, many parents are adding **browser-based filters** to their computers to restrict access to Web sites that contain objectionable material. Browser-based filtering software, such as Cyber Patrol and Net Nanny, works at the browser level to choose which sites will be blocked. These programs range from a simple system based on words found in the Web sites to more complex systems that can determine the pictorial content. Filtering based on words has problems because sites such as that of the Girl Scouts of America could be restricted if *girl* is used as a keyword on which to base Web site blocking. Filtering software also has another problem: Much of it can be bypassed by using instructions available from a quick Web search.

Children's Internet Protection Act (CIPA)
Legislation that places restrictions on the funding to libraries that do not have technology in place to filter out certain material from being accessed through the Internet.

The issue of using filtering software on computers in libraries connected to the Internet in the United States has been heavily debated. The **Children's Internet Protection Act (CIPA)**, which went into effect on April 20, 2001, placed restrictions on the use of funding to libraries that do not have technology in place to filter out certain material from being accessed through the Internet. At this writing, this legislation remained in effect but was being appealed in the courts by various groups. Figure 13.8 shows the browser-based filtering approach to controlling objectionable Web content.

server-based filtration
A filtering system installed on the Web server of the Internet service provider that allows the browser to access only those Web sites or chat rooms that are approved by the ISP.

The third option, **server-based filtration**, installs the filtering system on the Web server of the Internet service provider (ISP) so that the browser can access only those Web sites or chat rooms that are approved by the ISP. The server filters out all other sites. Although this system has the best chance of keeping children away from adult-oriented sites, it depends on the ISP to select those sites that pass through the filter and may be restrictive on others in the household. An example of server-based filtration is a national ISP, MayberryUSA, at www.mbusa.net. This ISP indicates it is dedicated to filtering pornographic and hate Web sites for its users.

Figure 13.8 Browser-based filtering

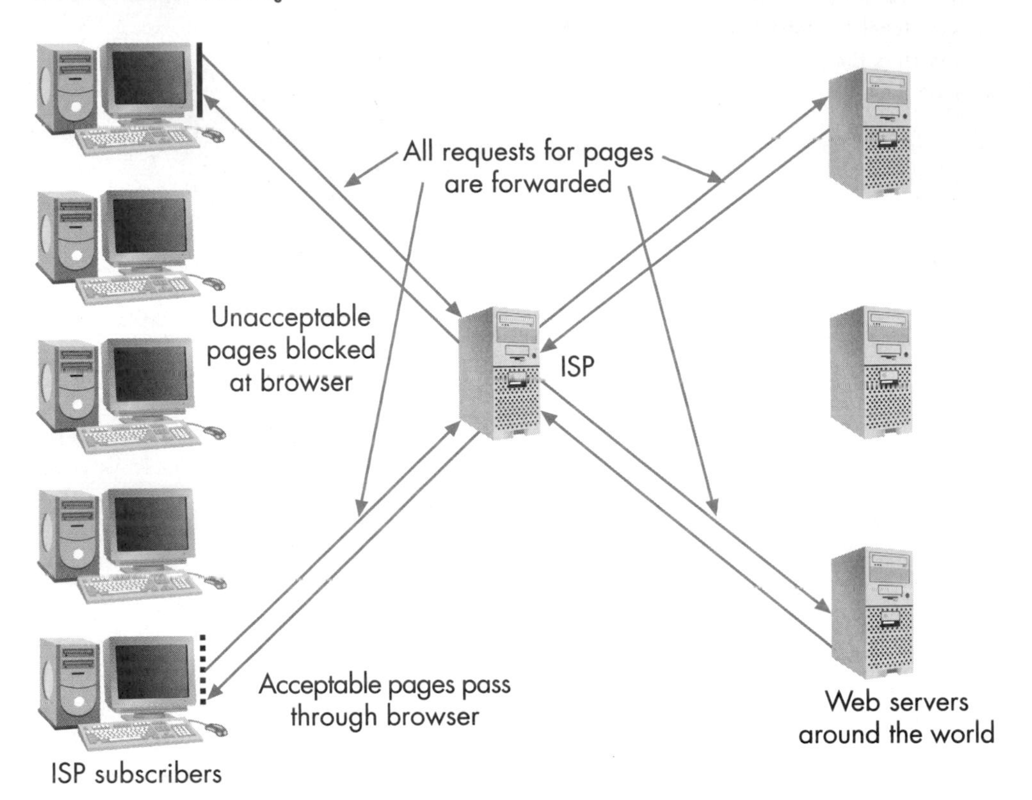

MayberryUSA is an ISP that filters all material sent to subscribers.

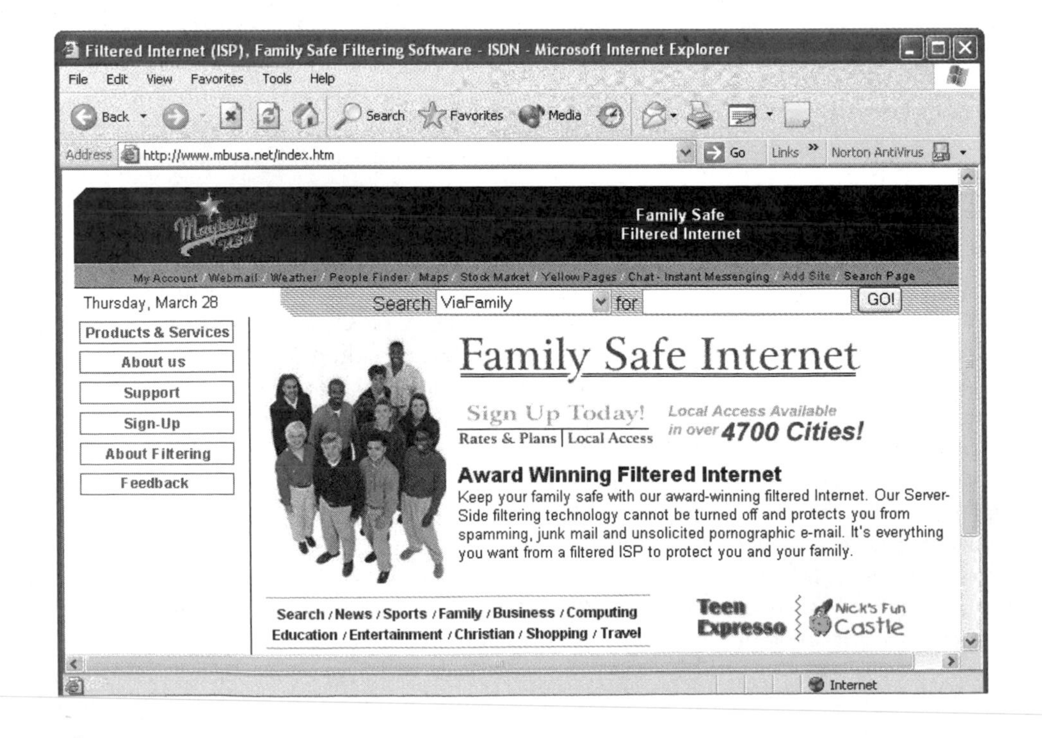

Web-Based Gambling Sites

In 2001, worldwide online gambling revenue totaled an estimated $6.7 billion and some 2.9 million people engaged in this activity. These numbers are expected to grow to almost $21 billion and 7.4 million people by 2005. Online gambling sites had doubled between 2000 and 2001, when they reached an estimated 1400 Web sites. Obviously, this online activity remains very popular, despite the fact that several governments are working to put a stop to it.

In 1999, a U.S. commission on gambling recommended banning Internet wagering not already approved in the United States. In another case, a New York State Supreme Court judge ruled that even though a casino was located in the country of Antigua, it was still subject to New York laws. In that case, the judge ruled that the Internet site created a virtual casino in the user's computer and therefore broke state and federal gambling laws. Conversely, gambling proponents argued that because the server was located in a foreign country, it should not be subject to U.S. laws. Currently, no online casinos are located in the United States. Instead, they operate out of places such as Antigua, Barbados, Costa Rica, and Australia. In 2001, Australia moved to ban online gambling by its citizens but not by noncitizens on Australian gambling Web sites.

Steve Jobs

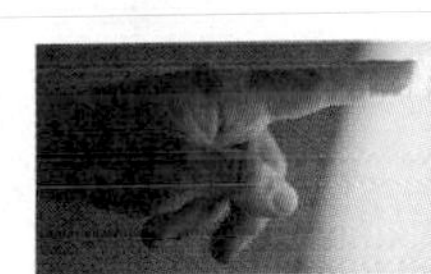

IT INNOVATORS

On April Fool's Day 1976, Steve Jobs and Steve Wozniak founded the Apple Computer Company to sell the computers that they were building in their garage. For the next nine years, Jobs served in various positions with the company, including president and chairman of the board, as it grew into a giant in the computer field. During this time, he was instrumental in developing the Apple II and Macintosh lines of computers. The Apple II enjoyed wide use in education, while the Macintosh pioneered the use of the graphical user interface that has since become the standard for virtually all computers.

In 1985, Jobs lost a power struggle with John Sculley and left Apple to found a new company, NeXT Computers. When NeXT failed to be a success, Jobs sold it to Apple. He then became involved with Pixar, Inc., the company that gave us the *Toy Story* computer-animation movies, as its CEO and president.

In 1997, after a period of falling sales and stock prices, Apple chose to bring Steve Jobs back as the interim CEO. Under his leadership, Apple introduced the extremely popular iMac, iBook, and G4 lines of computers and released several much-needed upgrades to the operating system including the UNIX-based OS X. Even though its speed rating in megahertz is no higher than that of competing chips, the PowerPC chip, on which the G4 computers rely, achieves a much faster speed by processing more data at a time. It has been able to perform more than 1 *billion* operations per second (1 *gigaflop*)—fast enough for the U.S. Department of Defense to classify it as a supercomputer and restrict its export to certain countries. Another innovation from Apple, under Jobs, is the IPod, which enables the storage and replay of more than 1000 songs in MP3 format, yet weighs less than seven ounces. Jobs is now the full-time CEO of both Apple and Pixar. Jobs still lives with three of his four children in the part of Silicon Valley where he grew up.

Steve Jobs was the co-founder of Apple Computers and is currently the CEO of both Apple and Pixar Studios.

Figure 13.9 Gambling Web site

At this writing, the future of Internet gambling in the United States and other countries remained uncertain, with its fate depending on the outcome of legislation and rulings by various courts. Even if such gambling is outlawed in the United States and countries such as Australia, authorities may find it very difficult to prosecute offshore gambling sites if their nations have no extradition treaties with the countries hosting the sites. Just as putting Napster out of business had very little effect on the distribution of music on the Internet, outlawing gambling in a particular country probably will not deter its citizens from visiting offshore Web sites. Figure 13.9 shows the Web page for one of these gambling sites.

1. What three approaches have been used to protect children from adult-oriented content?
2. Why is online gambling so difficult to control?

International Issues in the Networked Economy

The Internet and the networked economy have created a number of thorny international issues for individuals and governments. Although the Internet brings definite benefits to some developed areas, in less developed parts of the world, which have little or no access to the Internet, it is creating another form of digital divide and causing environmental problems. In other parts of the world, the disintegration of borders has brought problems for governments by introducing unwelcome content. This same disintegration of borders has made possible a new form of terrorism based on use of the Internet and other networks to attack governmental and industrial infrastructure.

International Digital Divide

An earlier section discussed two types of digital divides, based on education and on urban versus rural locations. A third type of digital divide involves the gap between the countries of the world. A 1999 United Nations study[5] suggested that the planet is splitting into two very different worlds: one inhabited by a minority that uses information technology to improve its standard of living, and another inhabited by a majority of poverty-stricken citizens in low-tech countries. For example, for approximately one-half of the people in the world, their next telephone call will be their first.

5. Marilyn Geewax, "Chasm between the 'haves' and 'have nots' is widening." *Atlanta Journal-Constitution.* July 25, 1999, p. F3.

Discarded computers often end up in less developed countries like China.

One reason that this difference arises is that the less developed countries do not have an adequate telecommunications infrastructure with which to provide Internet access. In addition, their citizens cannot afford to keep up with the rapid changes in information technology. The same study showed that the average American pays less than one month's wage for a computer, whereas the average Bangladeshi would spend eight *years'* income to purchase the same machine. The Bangladeshi citizen would then have to learn English to use the Internet, because 80 percent of all Web sites are in English. Although some have suggested taxing e-mail messages in the developed countries to support the creation of the needed infrastructure in underdeveloped countries, this event appears highly unlikely to occur.

Another type of regional digital divide arises when developed countries send some of the more than 15 million computers discarded annually to less developed countries for reclaiming of precious metals. That practice causes severe environmental problems for those doing the reclaiming. Computers contain a number of toxic chemicals that can poison the ground water when the computers are recycled. For example, in Guiyu, China, about 100,000 people—many of them children—work to retrieve usable materials from the computers. The resulting pollution is causing severe problems with the ground water supply in the area around Guiyu.[6]

Politically Objectionable Content

As noted at the beginning of this chapter, many countries are attempting to reinstitute their borders so as to keep certain Web content out. A number of countries have screened Web site content for political or religious reasons, including China, Vietnam, Iraq, Iran, and Saudi Arabia. Prior to late 2001, the former Taliban regime of Afghanistan had completely banned the use of the Internet. In Saudi Arabia, the number of banned Web sites reached 200,000 in 2001.

To screen content, in some countries all connections must go through a single computer that filters out Web sites that are deemed objectionable. This approach to filtering objectionable Web content at the national level uses the server based filtration systems discussed earlier. One main difference separates the two systems: In filtering for adult content, the subscriber *chooses* to have content censored; in the case of politically objectionable content, *governments* make that choice for their citizens. Figure 13.10 shows the use of server-based filtration to screen content coming into a country.

In other countries, service providers monitor what their clients view. For example, Australia has enacted legislation requiring its ISPs to prevent pornographic or other content deemed indecent from reaching their clients. ISPs must remove objectionable content from their servers within 24 hours of being directed to do so by national authorities and must provide a means to block access to objectionable overseas sites as well.

Cyberterrorism

cyberterrorists
Individuals who would attempt to destroy others' ability to use the Internet and other computer networks.

With the terrible events of September 11, 2001, came the realization that not only is the country's physical infrastructure at risk, but the electronic infrastructure may also be at risk from so-called **cyberterrorists** who might attempt to destroy the ability to use the Internet and other computer networks. In some ways, precursors of such attacks have already emerged in the form of denial of service attacks on commercial Web sites such as eBay, CNN.com, and Amazon.com.

6. "Activists expose computer dumps," http://www.msnbc.com/local/pencilnews/320230.asp.

Figure 13.10 Use of server-based filtration to screen content at the country level

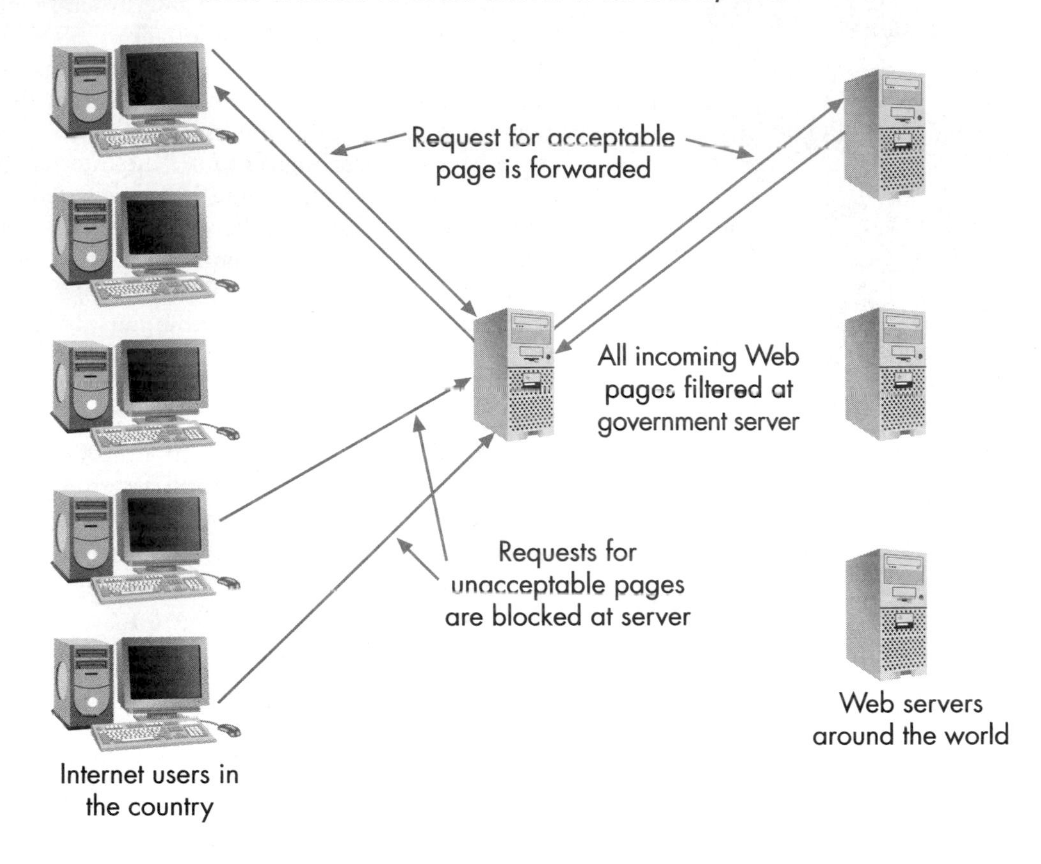
Request for acceptable page is forwarded
All incoming Web pages filtered at government server
Requests for unacceptable pages are blocked at server
Web servers around the world
Internet users in the country

Information technology is the underpinning of modern society and, as discussed in Chapter 11, is based on protocols that were never meant to be secure. With this recognition, the danger from cyberterrorism becomes evident. The long list of possible targets includes, to name but a few, electric power distribution networks, banking and finance networks, communication networks, civil aviation networks, and the Internet itself. For example, if a hacker launched a denial of service attack on the routers that control the flow of messages across the Internet, it might be possible to bring the entire network to a halt.

In summer 2001, the Central Intelligence Agency (CIA) warned the U.S. Congress that the United States' heavy reliance on computer systems could give adversaries the potential to circumvent the country's advantages in conventional military power. The day after the Pentagon and World Trade Center attacks, the FBI sent out a warning about the potential for attacks on the electronic infrastructure.

A study by the Institute for Security Technology Studies at Dartmouth College found that lessons learned from recent acts of cyberterrorism include that cyberattacks often accompany physical attacks; they are increasing in volume, sophistication, and coordination; and cyberterrorists are attracted to high-value targets. [7] The study went on to say that, during the current war on terrorism, cyberterrorists might attempt to deface Web sites, use denial of service attacks, and commit unauthorized intrusions into networks, resulting in infrastructure outages and corruption of important data. Finally, it suggested that security against

7. Institute for Security Technology Studies at Dartmouth College, "Cyber attacks during the war on terrorism: A predictive analysis," September 22, 2001.

cyberterrorism should include developing increased levels of alert, reporting suspicious activity to authorities, installing and updating systems to prevent intrusion, and using filters to avoid denial of service attacks.

In a move to protect the Internet and computer networks from cyberterrorists, Congress passed the Patriot Act of 2001. This act, among other things, greatly increases the government's capability to view e-mail messages, to use roving wiretaps to track suspected terrorists as they move from computer to computer, and to request information from ISPs about individuals' use of the Internet. It also enlarges the scope of the Computer Fraud and Abuse Act of 1986 and increases the penalties associated with it.

Although not wishing to protect terrorists in any way, civil libertarians, especially the Electronic Freedom Foundation, remain very concerned about the dangers to civil liberties associated with this act. In the future, you will probably read about cases focusing on its constitutionality.

1. How is the international version of the digital divide similar to or different from the other two types of digital divides discussed earlier?
2. What three things does the Dartmouth study suggests may occur during the current war on terrorism?

The Future of the Networked Economy

Predicting the future of the networked economy is a risky business, at best. However, one safe prediction is that a continuing movement will occur toward a "u-world" of computing and connectivity—that is, more ubiquity, universality, uniqueness, and unison both in computing and connectivity.[8] These concepts were discussed in Chapter 7 as they related to electronic commerce, but are reviewed here as they relate to the future.

Ubiquity will occur as computing and communications devices become smaller, cheaper, and faster. As bandwidth increases, the number of electronic devices other than computers connected to the Internet will continue to grow. They will include not only mobile phones and personal digital assistants (PDAs) but everyday devices such as ovens, washing machines, refrigerators, and so on.

Universality will emerge as the PDA, mobile phone, pager, smart cards, and credit cards merge into a single mobile device that also identifies its owner and pays his or her bills. This device might use an advanced form of the GSM digital mobile telephone standard now employed in Europe to handle wireless broadband communications, which will eventually enable the transmission of full-motion video. In the trend toward universality, many devices and communication standards will converge into a single mobile device.

Uniqueness will occur as you become able to customize virtually all information sent to you to your own tastes. Each day, you can receive just the news you want over your mobile device as either audio, video, text, or graphics, depending on your personal tastes.

Finally, in terms of *unison*, you will be able to synchronize all of the devices you use in the office, at home, and on the road. Thus what your refrigerator knows about its contents will also be available on your mobile device, or what your office computer has on its to-do list will also show up on your mobile device—all without you having to even think about synchronizing them.

These trends toward ubiquity, universality, uniqueness, and unison will be manifested in a number of ways. Consider the following examples:

---

8. Richard T. Watson, "U-Commerce: The Ultimate." *Ubiquity*, http://www.acm.org/ubiquity/views/r_watson_1.html.

XM Radio receives signals from satellites rather than from transmitters using antennas.

- Connection of consumer appliances to the Internet gives homeowners the capability to control these devices from a distance as well as enables them to communicate with one another or, if necessary, a repair service.
- Expanded wireless Internet connections give people the capability to connect to the Internet from any location—car, boat, hiking trail, and so on. Satellite radio already covers the entire United States, so moving to wireless connectivity everywhere is a logical progression.
- Wireless computer networks in the home enable families to communicate via e-mail, instant messaging, or short message systems.
- Client computers in hotel rooms or on the backs of airline seats access data stored on a network server, eliminating the need to lug around a laptop.
- Time is becoming the scarcest resource of all, leading to increased use of virtual meetings through videoconferencing so as to avoid long business trips.

As a student of information systems, you must find ways to become knowledgeable about these trends so that you can take advantage of them. Failure to do so will put you at a severe disadvantage. Remember that the networked economy will include many new jobs plus old jobs recast in new forms. Understanding the trends and pitfalls of the networked economy will enable you to take advantage of these new opportunities.

A Day in the Life of Mary: 2007

To help you envision this future, consider the following day in the life of a college student named Mary in the year 2007. Mary attends the University of Georgia on a HOPE Scholarship. Although all of these changes may not occur by that year, many of them likely will—some long before 2007.[9]

7:00 A.M. Mary is awakened at home by her computer playing some new Celtic music videos, which had been downloaded in compressed format overnight and automatically charged to her credit card account. The videos are displayed on a flat-screen monitor hung on the wall (it doubles as a watercolor landscape when not in use). After five minutes, the display switches to a Web page showing news customized to Mary's interests, including scores from the latest Georgia Bulldogs sporting events. After a quick breakfast, Mary leaves for campus. As she drives in, her automobile connects to the Internet and reads her e-mail to her through its sound system; it then switches to a satellite radio system that enables her to listen to a live concert of a new Fado group from Lisbon, Portugal. Her automobile also reminds Mary of the classes she is scheduled to attend that day.

8:00 A.M.–9:15 A.M. Mary's first class is Globalization, Regionalism, and Information Technology Systems (commonly known as GRITS), an elective course examining how IT can be used to solve global problems. Today the class is considering ways in which wireless capabilities help to overcome the digital divide in

9. The author wishes to thank Professor Richard Watson of the University of Georgia for writing the original version of this section.

some rural portions of sub-Saharan Africa; it features speakers from the School of Business Leadership at the University of South Africa as well as speakers from Botswana and Kenya. In addition to the students in Georgia, students in Singapore, Norway, Brazil, and South Africa view the live broadcast. All are taking the same course, which has to start at 8:00 A.M. in Georgia because of the time differences between the various locations.

9:15 A.M.–10:30 A.M. After class, Mary visits the combination computer lab-coffee shop, where she purchases a bottle of fruit juice and a muffin. As she leaves the food area, she checks the wall-mounted LCD panel to verify that the correct amount was deducted from her account by the mobile device in her backpack that communicates automatically with the checkout device. Cash practically disappeared from her university four years previously, following the introduction of contactless smart cards that did not require swiping. Now all you need to do is walk through the food area exit. Mary's mobile device handles all such transactions in addition to other chores. In fact, Mary can program her mobile device to display selected information every time she approaches a wireless access point, which is almost all the time! For example, for her Finance class, Mary's team is managing a portfolio of mature Internet stocks, and she has programmed her mobile device to display the portfolio's latest value every 15 minutes. The bottom line of the LCD window shows that the portfolio is down 1.5 percent for the day based on the activities on a number of stock exchanges around the world.

Choosing a seat at a table with a flat-screen display device, Mary tells the hidden microphone to log her onto her network account. To validate her access to the account, she presses her ring finger on an area of screen and is instantly connected to the wireless local area network. She checks her Web-based to-do list and is reminded that she must take a quiz for her Networked Economy class and finish a report on Mississippi.com for her Strategic Management course. Before working on her assignments, Mary checks her integrated mailbox, answers three e-mails and one voice mail, and deletes video spam. The quiz takes about 20 minutes, and she is relieved to immediately find out that she scored 92 on it.

Next, to finish her Mississippi.com assignment, Mary consults an online collection of databases—introduced at her university in the late 1990s and checks a few Web sites. When she finishes the electronic report, she submits it via e-mail to the professor. Even though the professor is currently working with an MBA team on a consulting assignment in Portugal, she knows that she will grade the report within a couple of days and return it with attached audio and text comments. Mary thinks this mix of classroom lectures and independent learning provides good preparation for her business career, because she is learning how to learn by herself. She could have done all of the work on her mobile device using audio output, but she likes to see the graphics available on the flat-screen display.

10:30 A.M.–11:45 A.M. Mary attends her Networked Economy class and, via the Web, participates in an interesting class discussion that includes the use of voice over the Internet. Whereas some of her fellow students are in her classroom, others are at home or in offices as many as five time zones away. All work from the same Web page and wear a special headset-microphone combination device that allows them to hear and respond to other class members' comments.

1:00 P.M.–2:30 P.M. After lunch, Mary's Data Management class team (Mary, Eduardo from Brazil, and Tore from Norway) meets to review its database design. The team holds an audio conference using a shared screen, so all team members see the same data model and can take turns amending it until they agree. This high-fidelity model focuses on the timetable for the São Paulo subway. The Data Management class is simultaneously taught with the partner business schools in Brazil and Norway, and students learn how to design and query databases while honing their skills in working on cross-cultural teams. After completing the project, Mary catches a bus to the recreation center to play racquetball.

5:00 P.M.–6:00 P.M. Mary's team in her Strategic Management class meets at the video booth in another lab. An alumnus working in New York has agreed to critique the team's presentation prior to its final meeting with the client. It takes only a moment to set up the camera and check the video connection, and then the team delivers its video presentation. As he watches the presentation on his network computer in New York, the alumnus makes electronic notes whenever he spots something of concern. Presentation analysis software tags the comment to the exact point in the presentation so that, during the playback, he can go directly to the portions of the presentation that need more work and point out the problems. After a few retakes of portions of the presentation and some more comments by the alumnus, the team declares victory.

6:30 P.M.–7:30 P.M. Mary's boyfriend brings some Chinese food for dinner. During the meal, her sound system stops playing the latest REM DVD and announces that a priority voice mail has arrived. Mary uses the remote to instruct the system to play it for her; it is the alumnus who viewed her team's presentation that afternoon. He was impressed by her role in the presentation and wondered if she could cut and paste her section of the presentation and mail it to the company's recruiter. It takes Mary about five minutes to locate the video on the college server, edit it, and e-mail it to the recruiter. After dinner, she celebrates her new job opportunity by ordering a digital copy of the old video *There's Something about Mary* from the Web-based video library. Her computer automatically transfers funds from her online account through her mobile device to pay for the video.

Case *Study*

Teleworking for the Federal Government

The U.S. government has supported teleworking since 1991, when the Federal Flexible Workplace Pilot Project was established. It was followed in 1993 by the funding of a number of satellite work locations across the country and a Presidential memo in 1994 directing agencies to establish flexible work arrangements. To assess the effectiveness of these efforts, the Office of Personnel Management undertook a study to find successful teleworkers and their supervisors. Along the way, it found five common themes among the teleworking success stories: increased productivity on the part of the teleworker, reduced stress for the employee, a saving of time, equivalent access to e-mail and information through information technology, and an increased need on the part of the employee to maintain relationships with co-workers and supervisors. The following information from that report focuses on two employees who have been teleworking one day per week for three years at the U.S. Department of Agriculture National Agricultural Statistics Service (NASS). The employees are at the GS 13/14 level and are classified as information technology specialists. NASS's mission is to provide accurate and current statistical information to those who make informed agricultural decisions.

From the supervisor's perspective, teleworking has been a "perk" that has enabled him to retain highly knowledgeable staff. Because they work on the cutting edge of information technology, his staff are in demand in the private as well as the public sector. He permits his staff to telework one day a week, but asks that everybody be in the office on Tuesday. He has seen no downsides to telework and there have been no problems with the staff or with his customers. It appears that the quality of the service provided is the same whether the specialist is working in the office or at another location.

From the employees' perspectives, telework has had many advantages. One works from home and the other from a telecenter 10 minutes from his home. Both believe that they are more productive when teleworking because there are fewer distractions. They did report that teleworking took some getting used to and required a conscious communication effort to make sure that other members of their team were kept up-to-date on projects. Although there was some resistance to teleworking from other managers who felt that working from home was not "working," the teleworkers credited it with being an important factor in their job satisfaction.

Source: U.S. Office of Personnel Management, Office of Merit Systems Oversight and Effectiveness, "Telework works: a compendium of success stories," May 2001.

Think About It

1. In reading this success story of two teleworkers at NASS, speculate on what downside the two individuals may face in teleworking.
2. Why do you think that some managers did not equate teleworking with really "working"?
3. Compare the teleworking approaches taken by the two employees—home and telecenter—in terms of reasons why one approach might be chosen over the other.

SUMMARY

To summarize this chapter, let's answer the questions posed at the beginning of the chapter.

How has the widespread use of computers and networked communications changed our concepts of time, distance, and borders? The networked economy depends on the Internet for communications. From a social point of view, the networked economy is built on three key elements: widespread computing and communications technology, education, and free movement of ideas and trade. The widespread use of networked computers has changed humans' perceptions of time and distance. Three of its effects include the death of distance, the homogenization of time, and the disintegration of borders. The death of distance means that geography no long restricts activities, making it possible for many workers to avoid long commutes to and from work by teleworking using the Internet or private networks. The homogenization of time means that people live in a 24/7 world where business continues somewhere all of the time. This trend has led to the blurring of lines between work and home, as people find that they can work around the clock if necessary. Finally, the disintegration of borders means that information can flow freely across international borders through the Internet. In response, many countries today are attempting to reimpose their boundaries (limit or control the flow of information) through legislation and technology.

What economic issues do we face in the networked economy? The primary economic issues in the networked economy are the digital divide and taxation policies. In the digital divide, people become separated into two groups—those who have access to information technology, and those who do not—based on education, rural versus urban location, and regional location. Many individuals with the lowest educational attainment will likely use the Internet the least and therefore miss out on economic opportunities. Rural locations are falling behind urban locations in terms of access to broadband Internet capability. A great deal of discussion has focused on taxation of electronic commerce sales. The U.S. government decided to extend an existing moratorium on taxing electronic commerce until late 2003, with a requirement that the states work out a simplified sales tax system.

How has the virtual office changed the lives of many employees in the networked economy? The capability to work anywhere at any time has created the virtual office. The most widely used form of the virtual office involves teleworking—that is, the use of information technology to perform one's job from some place other than the primary office. Teleworking offers numerous advantages for the company, the employee, and the community. For the company, the advantages include increased productivity, more time on the job, lower manager-to-staff ratio, less need for office space, and greater employee loyalty. For the employee, teleworking leads to a less stressful job, a shorter (or no) commute to work, and a better family life. For the community, teleworking produces less traffic and less air pollution. On the other hand, teleworkers require a different type of management because they are not in the traditional workplace. In addition, teleworking is not for every worker because it requires a large degree of self-motivation and discipline. Finally, teleworking can cause problems when workers cannot separate their work life from home life, and find themselves working all the time.

What health issues has the networked economy brought on? Working at the speed of the Internet can cause health problems associated with long-term use of computers, especially the keyboard and mouse. These problems cost employers large sums of money in worker's compensation and lost productivity. The primary problem is repetitive stress injury (RSI) due to the overuse of a keyboard in work and leisure activities involving the computer and the Internet. Carpal tunnel syndrome (CTS) is a common form of RSI. Ergonomic workstations represent one way to reduce the risk of RSIs. Ergonomics seeks to create the optimal balance between productivity and well-being. An ergonomically designed workstation allows the worker to work in a comfortable posture. Other health problems linked to the networked economy involve users who become pyschologically dependent on the Internet.

How has online content caused social problems? Online content has caused problems in three ways: defamatory messages, adult-oriented Web sites, and gambling sites. People can use the Internet to spread negative information about another person or group, and this information can prove very difficult to correct. In the second case, adult-oriented Web sites should be screened from young users. A variety of approaches are employed to protect children from adult-oriented material, including passing laws against such content, using filters on Web browsers or servers to screen the content, and filtering all Web content coming into a country. In the third case, governments are also attempting to keep their citizens from participating in off shore gambling activities over the Internet.

What international issues do we face in the networked economy? International issues include creation of a regional digital divide, politically objectionable content, and cyberterrorism. Many underdeveloped parts of the world lack access to the technology needed to participate in the networked economy. This shortcoming puts them in danger of falling further behind more developed parts of the world. Just as they have made efforts to protect children from adult Web content, some governments are attempting to legislate against certain content or to use server-based filtration systems to block their citizens from viewing certain content. Along with physical acts of terror come the danger of acts of cyberterrorism, in which groups attempt to destroy others' ability to use the Internet and other computer networks as well as destroy or control other elements of the country's infrastructure.

What kinds of changes might we see in the future of the networked economy? A continuing movement will occur toward a "u-world" of computing and communications—that is, ubiquity, universality, uniqueness, and unison will increase. This trend will bring about both opportunities and challenges for society as people learn to deal with these changes. Opportunities will be available to those who learn how to take advantage of these changes, whereas those who don't will be at a severe disadvantage in the networked economy.

REVIEW QUESTIONS

1. Why is education one of the social cornerstones of the networked economy?
2. What social issues have arisen from widespread use of computers and networked communications?
3. What is *reach creep*, and how does it relate to the homogenization of time?
4. What is the *death of distance*, and how does it relate to teleworking?
5. How have educational requirements for working in the networked economy changed compared to those for the industrial economy?
6. What three types of digital divides are mentioned in the text?
7. What is required of states as part of the extension of the U.S. moratorium on imposition of sales taxes on electronic commerce?
8. What are the two points of view taken by companies and state/local governments regarding taxation of electronic commerce?
9. What is e-lancing? How does it relate to teleworking?
10. What are some advantages of teleworking for an organization? Why is a different type of management necessary for teleworkers?
11. What must you consider in any analysis of the causes of RSI?
12. What is ergonomics? How does it relate to RSI?
13. What kinds of psychological issues are related to the Internet?
14. Why are defamatory messages sent over the Internet so difficult to deal with?
15. What is server-based filtering? How does it differ from browser-based filtering?
16. Which type of filtering do governments use to block their citizens from accessing objectionable Web sites? How does this differ from the filtering used to protect children from adult-oriented Web sites?
17. How does legislative control of online gambling relate to the disintegration of boundaries?
18. What targets might a cyberterrorist seek to strike?
19. What are the four u's in the future of computing and communications?

DISCUSSION QUESTIONS

1. Talk with your parents or adults who are at least 20 years older than you to compare communications when they were your age to how you communicate today. What differences do you find? What similarities do you find?
2. Discuss the differences between the various types of digital divides. Which do you think will disappear the most quickly? Why?
3. Discuss why computer health problems are no longer restricted to people working in computing-intensive positions.
4. Discuss the various types of teleworking. Which do you think you would choose if you were to become a teleworker? Why?
5. Discuss why the disintegration of boundaries is a concept that may not reach its full potential.

RESEARCH QUESTIONS

1. Go to the Web site for the Digital Divide Network (www.digitaldividenet.org), and look at some of the articles posted there on efforts to bridge the digital divide. Summarize one of those articles in a two-page paper.
2. At least 26 states are working together in the Streamlined Sales Tax Project in an effort to meet the requirements of the extension on the U.S. moratorium on Internet sales taxes. Visit the Web site for this project at http://www.geocities.com/streamlined2000/, and write a two-page paper on its progress.
3. Research the current status of the Australian law requiring ISPs within the country to censor Web content. Write a two-page paper on your findings.
4. Research known occurrences of cyberterrorism. Create an electronic presentation of your findings containing at least 10 slides.
5. A new use of wireless devices is radio frequency identification (RFID) to identify a wide variety of items. Research the current use of RFID, and write a two-page paper on your findings.

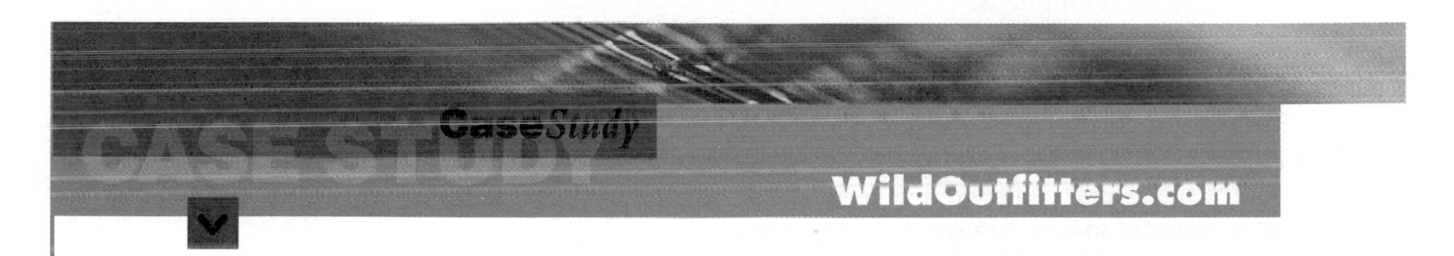

Alex was pulling an extension cord out of the front window of the store to hook up the laptop when Claire pulled into the drive. The laptop was sitting on the collapsible camp table that he had set up on the front lawn of the store.

Claire greeted him with a kiss and asked, "Working outside to enjoy the nice weather?"

"That's only a fringe benefit," Alex replied. "Actually, I need to write some product reviews but Jill's at my desk working on some invoices. You might want to join me out here, because Bob's doing some system upgrades at your desk."

"It seems that we need to do something about our lack of space. Things didn't seem so crowded in the old days," Claire sighed.

"I've been thinking that we might have an alternative to adding onto the store," said Alex. "One word—teleworking!"

The Campagnes have had to expand their staff because of the increased business they have experienced since starting their Web site. With success came new problems, such as a shortage of space in their store to accommodate the new employees. The couple has been thinking about adding more office space to the store. As Alex mentioned, they might not need an addition or could at least delay it if they provided teleworking for their employees.

Several advantages linked to teleworking make the idea attractive to the Campagnes. First, it could provide savings in terms of facility costs. They could realize a very real savings if they did not need to build additional space. Second, teleworking programs can yield increases in productivity from 10 percent to 40 percent. Third, remote access to important functions such as network maintenance and customer service would make the company more resilient to disruptions. Fourth, and possibly the most attractive feature to the Campagnes, teleworking might provide an improved quality of life for their staff. This idea is dear to their hearts, because the Campagne's major motivation for starting the Wild Outfitters store was to improve their own quality of life.

To set up a teleworking program, the pair will have to consider additions to their information system. Information system resources for teleworking will include home office equipment, central site equipment at Wild Outfitters, telecommunications access for both the home offices and the central site, and software for security and network management. The technology available for these components varies in expense, access speeds, bandwidth, and reliability.

Perhaps more importantly, they must examine their management policies to understand how they should change for a teleworking program. The Campagnes will need to identify the qualities that make a good teleworker and incorporate them into their hiring and training practices. Other issues to examine include the hours of duty, pay and leave policies, and overtime guidelines. It will be important to establish clear work policies for the teleworkers and to ensure that these policies are communicated effectively. Such a program will take a lot of thought and planning, but it seems to be a good next step in the evolution of Wild Outfitters.

"You know, if we do this teleworking right," Claire said as she stirred sugar into the coffee that she had prepared on the camp stove, "we could set up an office like this one anywhere."

Alex leaned back against a pile of sleeping bags and smiled. "Only a little while ago it seemed that we were just a couple of small store owners. Now look at how far we've come," he said as he waved his hand over the makeshift office.

"I can see the headlines now: Campagnes Office-less Due to Cyber-Success." Claire and Alex beamed with pride over the new challenges that their networked business had provided.

Think About It

1. What benefits and costs could the Campagnes derive from starting a teleworking program at Wild Outfitters?
2. What characteristics should a person have to be a good teleworker? Explain.
3. The Campagnes seem to have a good grasp of the advantages of teleworking. Are they overlooking any disadvantages of teleworking? Describe any disadvantages that you can imagine, and state how Wild Outfitters might overcome these disadvantages.

Hands On

4. A teleworking program creates a nontraditional working environment. Old methods of management and evaluation may not work well in this new environment. Search the Web for suggestions about how to manage teleworkers. Use your findings to develop a teleworking agreement that may be used to communicate the expectations between a supervisor and a teleworking employee at Wild Outfitters.

Hands On

5. Alex and Claire would like to analyze the costs and benefits of developing a teleworking program at Wild Outfitters. They have found that several options exist for providing the resources discussed in the case. A little research has allowed them to further develop several equipment options. The information related to the costs and benefits of these options are stored in an Excel file on the textbook's Web site. Use this information to perform a cost/benefit analysis for each option. Based on your analysis, discuss the options that you think would work best for Wild Outfitters.

3G device
A mobile device that adheres to third-generation protocols leading to higher transmission speeds and consistent digital standards.

access control
Techniques for controlling access to stored data or computer resources.

Active Server Pages (ASP)
Web pages that include one or more embedded programs, which are processed on a Web server to generate and return a Web page to the user.

adaptable site
A Web site that can be customized by the visitor.

adaptive site
A site that learns from the visitor's behavior and determines what should be presented.

ad hoc programming process
A process in which an individual or a group that needs a new or revised system meets with a programmer, and together they decide what should be done.

analog devices
Devices that convert conditions, such as movement, temperature, and sound, into analogous electronic or mechanical patterns.

anonymous FTP site
An FTP site that does not require users to have user IDs and passwords.

ANSI X 12
An EDI protocol used in the United States.

antivirus policies
Organizational policies that protect the computer system from destructive software.

antivirus software
Software used to test for destructive software every time a computer is started, an e-mail is received, or software is downloaded.

Apache
An open-source Web server.

applets
Self-contained computer programs on Web pages written in Java.

application service provider
An IT company that runs one or more application servers, which a customer can access to process data, to query data warehouses, or for a host of other purposes.

artificial intelligence (AI)
Hardware and software systems that exhibit the same type of intelligence-related activities as humans, including listening, reading, speaking, solving problems, and making inferences.

ASCII
(pronounced "as-key") An acronym for *American Standard Code for Information Interchange*.

asynchronous
A type of communication in which only one of the parties can send messages at a time.

attractor
A Web site that continually attracts a high number of visitors.

attribute
A field or column in a relational database.

backbone
A transmission medium created to connect networks.

back office server
Software that processes data from transactions and uses it to track inventory and send purchase orders when inventory levels fall below a designated level.

backup
A second (or even third) copy of a data file on a storage device kept separate from the primary disk storage.

bandwidth
A measure of the capacity of a communication channel, expressed in bits per second.

bar codes
Combinations of light and dark bars that are coded to contain information.

baseband
A classification of digital transmission in which the full capacity of the transmission medium is used and multiple sets of data are transmitted by mixing them on a single channel.

batch processing system
A system that combines data from multiple users or time periods and submits them to the computer for processing together.

benchmark test
A test in which the development team compares competing products using programs and data typical of the actual conditions under which the proposed system will operate.

best practices
A list of the best ways that have been found to carry out operations as discovered by individuals or groups.

beta testing
Engaging the services of users to test prerelease versions of software.

binary number system
Base-2 number system based on zeros and ones.

biometrics
A technology that uses thumbprints, eyeprints, or voiceprints to identify individuals.

bit
The basic unit of measure in a computer; contraction of **B**Inary and dig**iT**.

bits per second (bps)
A measure of the data rate.

bridge
A combination of hardware and software that connects two similar networks.

broadband
Simultaneous analog transmission of large amounts and types of data, including audio, video, and other multimedia, using different frequencies.

broad customization
A strategy in which a Web site attempts to communicate with several types of stakeholders or many of the people in one stakeholder category.

brochureware
Web sites that reproduce print publicity and advertising documents.

browser
Client software used on the Web to fetch and read documents on-screen and print them, jump to other documents via hypertext, view images, listen to audio files, and view video files.

browser-based filters
Software that works with Web browsers to restrict access to Web sites that contain objectionable material.

bugs
Errors in a computer program.

bus
A primary cable to which other network devices are connected.

business continuity plan
A comprehensive plan that deals with all possible downtime scenarios.

business-critical application
A software application that is critical to the continued existence of the organization.

business intelligence
Information systems aimed at helping an organization prepare for the future by making good decisions. Also called decision support systems.

business services infrastructure layer
The software layer of electronic commerce that handles the services required to support business transactions (for example, encryption).

business strategy
The long-term plans that describe how a firm will achieve its desired goals.

bus network
A computer network in which computers are tied into a main cable, or bus.

byte
A group of eight bits—equivalent to a single character.

card-based digital cash
The storage of value on a plastic card, such as a prepaid telephone card or a smart card, that can have value added to or removed from it.

carpal tunnel syndrome (CTS)
An ailment in which, because of excessive keyboarding, the median nerve in the arm becomes compressed when swollen, inflamed tendons exert pressure on a nerve.

CASE repository
A database of metadata about the project that is used to automate much of the paper flow typically associated with structured development.

CASE tools
CASE software packages.

CD-ROM disks
A form of read-only optical storage using compact disks.

CD-RW drives
Optical disk drives that can write to a CD-ROM as well as read from it.

central processing unit (CPU)
The part of the computer that handles the actual processing of data into information.

character
A byte, or group of eight bits; equivalent to a single character.

chat rooms
Reserved areas that allow users to carry on group conversations on the Internet.

Children's Internet Protection Act (CIPA)
Legislation that places restrictions on the funding to libraries that do not have technology in place to filter out certain material from being accessed through the Internet.

Children's Online Privacy Protection Act of 1998 (COPPA)
Legislation requiring commercial Web sites that collect personal information from children younger than age 13 to obtain prior parental consent.

chip
A tiny piece of silicon consisting of millions of electronic elements that can carry out processing activities.

ciphertext
The encrypted form of a message.

Clean Air Act of 1990
Legislation aimed at reducing the amount of air pollution, which included a recommendation for increased teleworking.

click stream data
Data that are captured about users' activities when they visit a Web site.

client
A computer running an application that can access and display information from a server.

client/server computing
A combination of clients and servers that provides the framework for distributing files and applications across a network.

client-side processing
Processing that takes place on the browser itself before data are actually sent to the server.

coaxial cable
A medium for data transfer composed of a center wire, an insulating material, and an outer set of wires. Similar to that used to transmit cable television signals into your home.

codification
The process of writing knowledge down in some fashion.

command-driven interface
An interface in which the software responds when the user enters the appropriate command or data.

commercial off-the-shelf (COTS) software
Commercially prepared software on disk(s), with instructions and documentation all wrapped together.

common gateway interface (CGI)
A method of communication between the Web server software and a computer program that processes data sent from the user.

Communications Decency Act (CDA) of 1996
Legislation, since ruled unconstitutional, that aimed at protecting citizens from pornography on the Internet.

communications flip-flop
The reversal of the seller controlling the flow of information through print and media advertising to the customer controlling the flow of information by first finding Web sites and then choosing which ones to visit.

computer
A device that accepts data and manipulates it into information based on a sequence of instructions.

computer-aided software engineering (CASE)
Software used to help in all phases of the systems development process so as to improve the productivity of systems development.

computer-based digital cash
The storage of value on a computer, usually linked to the Internet, allowing for payment directly between the customer and merchant computers or for a transfer of funds between individuals.

computer language
A language used by humans to give instructions to computers.

computer matching
The process of matching records in two or more databases to determine which records exist in both databases. The data are often used to create a profile of the individual. Also called *computer profiling*.

Computer Matching and Privacy Protection Act of 1988 (CMPPA)
A federal law that establishes a number of Fair Information Practice provisions to apply to data surveillance in the federal government but excludes many programs and agencies.

computer network
A combination of two or more computers with a communications system that allows exchange of data, information, and resources between the computers.

Computer Security Act of 1987
Legislation aimed at ensuring the security of U.S. government computers.

connectivity
The availability of high-speed communications links that enable the transmission of data and information between computers and conversations between people.

conversion process
Transactions associated with the production of goods and services.

cookie
A small data file containing data about the user that is kept on the user's computer and read by the browser when the user visits a particular Web site.

creative destruction
A concept emphasizing that the most important part of the change process for a business is not what remains after the change but rather what has been destroyed.

credit check
A inquiry with a credit bureau regarding a person's financial status.

customer convergence
The Web marketing concept stating that firms must describe their products and services in such a way that potential customers converge on the relevant Web pages.

customer relationship management (CRM)
Techniques applied to organizational data with the goal of segmenting customers in order to serve them better.

cyberterrorists
Organized groups of individuals who attempt to damage the IT infrastructure of a country or culture.

data
Facts, numbers, or symbols that can be processed by humans or computers into information.

database
A collection of information that is arranged for easy manipulation and retrieval.

data-based DSS
A decision support system that is aimed at exploring databases and data warehouses to analyze data found there so as to answer questions from decision makers.

database management system (DBMS)
Software used to organize, manage, and retrieve data stored in a database.

data dependence
A relationship between data and the software used to store it.

data encryption systems
Systems that protect data being transmitted over a network by converting them into an unreadable form.

data flow diagram (DFD)
A pictorial representation of the flow of data into and out of the system.

data hierarchy
The order in which data are organized in a database.

data integrity
The process of ensuring that data are accurate and reliable.

data management system
The part of a decision support system that retrieves information from a database as needed.

data mart
A scaled-down version of a data warehouse designed to suit the needs of a specialized group of knowledge workers.

data mining
A search for relationships within the data in a database or data warehouse.

data model
One of several models specifying how data will be represented in a database management system.

data rate
The number of bits per second transmitted between computers.

data redundancy
The repetition of data in multiple files.

data security
The protection of software and data from manipulation, destruction, or theft.

data sharing
A function of a database query language that coordinates the sharing of database information by multiple end users.

data surveillance
The systematic use of information technology in the investigation or monitoring of the actions or communications of one or more people.

data types
Specification of the type of data that will be stored in a database field.

data warehouse
A subject-oriented snapshot of an organization at a particular point in time.

death of distance
The idea that people are no longer restricted by geography.

decision support systems
Information systems aimed at helping an organization prepare for the future by making good decisions.

decryption
The conversion of an encrypted, seemingly senseless character string into the original message.

dedicated server network
A network in which at least one of the computers linked to the network acts as a server.

demand report
A report generated by information-based DSS upon a request by a manager.

demonstration prototype
A prototype that is used instead of a written proposal or a slide presentation to demonstrate what the final system will accomplish.

denial of service (DoS) attack
A destructive use of software in which a Web site is bombarded with thousands of requests for Web pages, rendering the Web server unusable.

digital cash
The storage of value in a digital format in one of two broad forms: card-based or computer-based.

digital certificate
A piece of digital information indicating that the Web server is trusted by an independent source.

digital devices
Devices that process and store data in a binary (0-1) form.

digital divide
A separation of the world population based on people's knowledge of and access to the Internet.

digital signature
A digital code that is attached to an electronically transmitted message and that uniquely identifies the sender.

digital subscriber line (DSL)
A digital method of data transmission using existing telephone lines.

direct Internet connection
Connecting an enterprise network directly to the Internet with no intervening ISP.

disintegration of borders
A situation in which ideas and electronic goods can flow freely into countries around the world without being subject to search or duties.

disintermediation
The process of eliminating intermediaries.

disk drive
A device that writes information to or reads information from a magnetic disk.

document database
A database that, instead of storing tables, stores related documents.

document management system
Technologies used to store and manage information in a digital format.

domain database
The part of an expert system knowledge base that contains the facts about the subject being considered by the expert system.

domain name
Another name for the server address.

dot con
A fraud carried out over the Internet.

dumb terminal
A computer with no CPU or secondary storage. Its sole purpose is to serve as an input/output device for a mainframe.

DVDs
Digital versatile disks used for storing data as well as audio and video programs.

economic feasibility
An indication that solving the problem or developing a system will offer financial benefits to the organization.

EDIFACT
An EDI protocol used in Europe.

e-lancing
Electronically connected freelancers working together on project teams for a week, a month, or however long it takes to complete a project.

electronic commerce
The activity of carrying out business transactions over computer networks.

electronic commerce strategy
The manner in which electronic commerce is used to further the goals and aims of the business or organization.

Electronic Communications Privacy Act of 1986
Legislation that extended wiretap laws protecting aural conversations to include communications between computers.

Electronic Data Interchange (EDI)
A communication protocol that allows computers to exchange data and information electronically, thereby automating routine business between retail stores, distributors, and manufacturers.

electronic funds transfer (EFT)
Any transfer of funds from one bank account to another without paper money changing hands; also, the transfer of payments

between consumers or between organizations engaged in business-to-business electronic commerce and businesses.

electronic publishing infrastructure layer
The layer that permits organizations to publish a full range of text and multimedia over the message distribution infrastructure.

electronic shopping cart
Software on a Web server that enables the customer to keep shopping without having to check out after selecting each item.

electronic supervision
Organization managers monitoring the amount of work performed by employees on networked computers for entering data, making reservations, and so on. Also called *computer monitoring*.

electronic wallets
A digital form of storage that enables the user to electronically store multiple credit cards in a combination of software and data.

encrypt
To convert readable text into characters that disguise the original meaning of the text.

encryption
The conversion of readable text into characters that disguise the original meaning of the text.

end-user development
Development of an information system by an end user.

end users
Non-IT professionals who use computers to solve problems associated with their jobs.

enterprise information portal (EIP)
A Web site on an intranet that allows the individual seeking help with a decision to use all three types of information systems without having to worry about which one is being used.

Enterprise resource planning (ERP)
A multi-module application software that helps an organization manage the important parts of its business, including managing the supply chain, maintaining inventories, providing customer service, and tracking orders.

Ethernet protocol
The most popular protocol for controlling LANs; runs on a bus network and uses collision avoidance methodology.

ethics
Rules created by cultures and economies about whether certain acts are "good" or "bad," or "right" or "wrong."

evolutionary prototype
A prototype that evolves into a working application.

exception report
A report generated by an information-based DSS when some condition falls outside a previously defined acceptable range.

executive information system (EIS)
A personalized, easy-to-use system for executives, providing both internal and external data, often in a graphical format.

expenditure process
The process that handles the transactions associated with the payment of expenses associated with running the organization.

expert
An individual who, because of his or her knowledge in a specific area, can provide solutions to problems in that area.

expert system
A computer-based system that uses knowledge, facts, and reasoning techniques to solve problems that normally require the abilities of human experts.

explicit knowledge
Knowledge that is codified and transferable.

Extensible Markup Language (XML)
A markup language designed to make information in a document self-describing on the World Wide Web, intranets, and elsewhere.

extranet
A wide area network using the TCP/IP protocol to connect trading partners.

facial recognition technology
Software that uses biometrics to translate the image of a person's face into a numerical code that can be compared to faces stored in a database.

Fair Credit Reporting Act of 1970
Legislation that regulates some actions of credit bureaus that collect credit information on individuals.

Fair Information Practice (FIP) principles
Principles of privacy protection that include notice/awareness, choice/consent, access/participation, integrity/security, and enforcement/redress.

Family and Medical Leave Act of 1992
Legislation that encouraged teleworking as a way of employees remaining at home to care for family members.

fat client
A client computer, usually a PC, that can also be used as a stand-alone computer.

fax conversion
The use of optical character recognition to convert incoming fax documents into ASCII format.

fax modem
A hardware component that combines the capabilities of a modem and a facsimile machine.

feature creep
A process in which users continually want to add more features to the system as they use it.

feedback
Information about output that may cause a system to change its operation.

fiber-optic cable
The newest type of data transfer medium that consists of thousands of glass fiber strands that transmit information over networks.

field
Part of a database file that stores specific information such as a name, a Social Security number, or a profit value.

field name
An identifier given to a field in a database file.

file management system
Database software that can work with only one file at a time.

files
Programs, data, or information to which the user or software assigns a name.

file server
A server computer with a large amount of secondary storage that provides users of a network with access to files.

File Transfer Protocol (FTP)
A protocol that supports file transfers over the Internet.

financial process
A summary in accounting terms of all the transactions generated by the revenue, expenditure, and conversion processes.

firewall
A device placed between an organization's network and the Internet to control access to data and systems.

firmware
Instructions on a ROM chip.

forecasting model
A process that uses currently available information to predict future occurrences.

foreign key
A primary key for another table placed in the current table.

form page
A type of Web page that sends data to a Web server.

Freedom of Information Act of 1970
Legislation that gave individuals the right to inspect information concerning them held in U.S. government data banks.

frequently asked questions (FAQs)
A form of structured knowledge in which answers are provided for the questions that are most often asked about a subject.

Gantt chart
A graphical project management tool used for developing workplans and schedules.

gateway
A combination of hardware and software that connects two dissimilar computer networks.

gigabyte (GB)
A commonly used measure of computer storage, approximately equal to approximately 1 billion bytes of storage or 500,000 pages of text.

global information infrastructure (GII) layer
The infrastructure layer composed of various national information infrastructures, in which some components may differ depending on the country.

Global System for Mobile communication (GSM) protocol
The most widely used standard mobile telephone protocol in the world. A digital communication system, it is only now being adopted in the United States.

graphical applications development environment
A software package that allows end users to create menus, boxes, and so on, and then to write just the instructions needed for a specific menu or box.

graphical user interface (GUI)
An interface that uses pictures and graphic symbols to represent commands, choices, or actions.

hackers
Individuals who seek unauthorized and illegal access to computers and computer networks for a variety of reasons, including "for the fun of it."

hard disk
A type of magnetic disk that is fixed in the computer.

hardware
The electronic part of the computer that stores and manipulates symbols under the direction of the computer software.

homogenization of time
The idea that people live in a 24/7 world, where business continues somewhere all of the time.

host computer
A computer in a network that is connected to the Internet and has a unique Internet address.

hoteling
The use of temporary office space by employees from many companies for short-term periods.

hub
A device for concentrating connections to multiple network devices.

human capital
Those individuals in an organization who have knowledge about the social networks and problem solving.

hypertext
A method of linking related information in which there is no hierarchy or menu system.

Hypertext Markup Language (HTML)
A computer language used to create Web pages consisting of text, hypertext links, and multimedia elements.

Hypertext Transfer Protocol (HTTP)
The communication protocol for moving hypertext files across the Internet.

icons
Graphical figures that represent operations in a GUI.

identity theft
The process of stealing a person's identification for purposes of purchasing goods and services using his or her credit.

Identity Theft and Assumption Deterrence Act of 1998
Legislation under which stealing someone's identity is a crime punishable by up to 15 years in jail.

IF-THEN rule
The rule used in an expert system that, together with facts, create the knowledge base.

imaging
The process of converting paper versions of documents to a digital form using some type of scanner and saving them to optical or magnetic secondary storage.

implementation stage
The stage of the structured systems development process in which the information system design is built, tested, installed, and maintained. Training also takes place during this stage.

indexing
The process of using data values or descriptors to search through documents.

indirect Internet connection
Connecting an enterprise network to the Internet through an ISP.

inference engine
The deductive part of an expert system that uses the information in the knowledge base to make suggestions or ask additional questions.

influence filter
A Web site feature that makes the site more attractive to a specific stakeholder group.

information
Data that have been processed into a form that is useful to the user.

information-based DSS
A generic name for many types of information systems that have the same goal: to provide decision makers with the information they need in an appropriate form.

information brokers
Individuals who use databases and other sources to find information on individuals.

information system (IS) life cycle
The various phases in the life of an information system—systems development, operational use, and decline in usefulness.

information systems (IS)
Systems that develop the information that managers and other employees combine with knowledge to make decisions.

information technology (IT)
Technology that is used to create, store, exchange, and use information in its various forms.

infrastructure
The underlying foundation or basic framework of a system or organization.

input
Receiving the data to be manipulated and the instructions for performing that manipulation.

input/processing/output (IPO)
A table showing the inputs to a process, the required outputs for that process, and the logic needed to convert the inputs into the desired outputs.

instant messaging (IM)
A form of IRC in which users carry on private conversations.

integrated data management
The storage of all data for an organization in a single database.

intellectual property
The ownership of any creation of the mind, including inventions, literary and artistic works, and symbols, names, images, and designs used in commerce.

intelligent agents
Computer programs that can be trained to carry out a search over the Internet for needed information. Also called *bots*.

interactivity
The capability of the Web site to interact with the user in some way.

interface
The design of the screen that users will see when they access an information system.

Internet
A worldwide network of computer networks in private organizations, government institutions, and universities, over which people share files, send electronic messages, and have access to vast quantities of information.

Internet bill payment/presentment (IBPP)
A system through which both the bill itself and the payment of the bill are presented in an electronic fashion over the Web.

Internet communications data
Non-Web data about individuals transmitted over the Internet, including mail messages, conversations in chat rooms, postings to bulletin boards, and messages to newsgroups.

Internet effects
Somewhat unexpected effects that the Internet can have on manufacturers and retailers, including price transparency, communications flip-flop, and customer demand for perfect choice. Also called *Internet threats*.

Internet Relay Chat (IRC)
An Internet protocol that enables a user to carry on a conversation by typing messages.

Internet security
The protection of both the data and information traveling over the Internet and the information technology used to send and receive data and information.

Internet Service Provider (ISP)
A company that provides access to the Internet to individuals and organizations.

Internet Tax Freedom Act of 1998
Legislation that established a three-year waiting period on state and local Internet taxes in the United States and created the Advisory Commission on Electronic Commerce to study this issue. It was extended in 2002 until 2005.

interorganizational system (IOS)
A networked information system used by two or more separate organizations to perform a joint business function.

intranet
An intraorganizational network based on using Internet technology; it enables people within the organization to communicate and cooperate with one another.

IP address
A numeric address for a server on the Internet consisting of four groups of four digits.

IP spoofing
A form of Internet intrusion in which an internal IP address is used to fool the server into believing it is in contact with a trusted computer.

IS cycle
Information systems for handling the present, remembering the past, and preparing for the future.

IS steering committee
An internal group composed of management, users, and developers that reviews proposals for information system development.

IT crime
An illegal act that requires the use of information technology.

iterative development
A development process in which the user tries out the latest version of the information system and provides feedback to the development team.

IT security
The methods used to protect hardware, software, data, and users from both natural and criminal damage.

Java Virtual Machine (JVM)
Software that is built into operating systems or can be downloaded and added to them; it interprets the code in a Java applet and executes it within the browser.

joint application development (JAD)
A process in which the development team meets with the project sponsor and the users to discuss the project at all stages of development.

key
In encryption an algorithm used to encode and decode messages.

keyboard
An input device made up of keys that allow for input of alphanumeric and punctuation characters.

keystroke monitoring
A methodology in which keystrokes from the computer keyboard are monitored either by software installed on the computer or by a hardware device that intercepts the keystrokes as they are sent to the computer.

kilobyte (KB)
A measure of computer storage equal to 1024 bytes or approximately one-half page of text.

knowledge
A human capacity to request, structure, and use information.

knowledge base
In an expert system, the facts, judgments, rules, intuition, and experience provided by the group of experts.

knowledge workers
Workers in organizations who use their knowledge to work with information.

legacy system
Another name for a mainframe computer system.

listserv
A group e-mail function available on the Internet; it enables and users to subscribe to special-interest mailing lists.

local area network (LAN)
A computer network composed of at least one client and one server that is restricted to a single geographical area.

logic trace
A trace of the line of reasoning used by an expert system to reach a conclusion.

macro language
A computer language built into personal productivity software, such as spreadsheets, that extends their capabilities.

mainframe
A very large and fast computer that requires a special support staff and a special physical environment.

maintenance
The ongoing process of keeping a system up-to-date by making necessary changes.

management information system (MIS)
A system for providing information to support operations, management, and decision-making functions in an organization.

many-to-many communication
A form of communication in which many people can communicate with many other people.

many-to-many relationship
In a data model, the situation in which multiple fields are related to one another.

mass customization
The process of making each visitor to a Web site feel that the site has been customized to his or her particular needs.

megabyte (MB)
A measure of computer storage equal to approximately 1 million bytes or 500 pages of text.

message distribution infrastructure layer
The software layer of electronic commerce that sends and receives messages.

metadata
Data about the data in a database, including the type of database being used, the names of the tables and the fields in the tables, the primary key for each table, and the foreign keys in each table.

meta tag
An invisible HTML tag that describes the contents of the Web page.

metric
The measurement of a particular characteristic of a program's performance or efficiency.

microwaves
High-frequency radio transmissions that can be transmitted between two earth stations or between earth stations and communications satellites, which are commonly used to transmit television signals.

middleware
The software that converts requests for data from a client into queries that are sent to a database server.

minicomputer
A computer that is between a mainframe and a personal computer in size.

mobile commerce
The use of laptops, mobile telephones, and personal digital assistants to connect to the Internet and Web to conduct many of the activities normally associated with electronic commerce.

model
A simplified version of a system that allows an analyst to understand the system's important parts.

model base
A collection of models that can be used to analyze the data from a database.

model-based DSS
A decision support system that combines data from the database or data warehouse with mathematical models to answer questions asked by management.

model management system
The part of a decision support system used to select a model that can help find a solution to a problem.

modem
A communications device that converts computer characters into outgoing signals and converts incoming signals into computer characters.

module
A separate program that performs a specific task and shares data with other modules to lead to an integrated system.

monitor
A cathode ray tube or flat-panel output device that displays output.

mouse
An input device—about the size of a mouse and connected to the computer by a long cord or wireless signal—that allows input through movement over a flat surface.

MP3
An acronym for a method of compressing a music file into a size that can be transmitted over a network.

multidimensional databases
A database with two or more dimensions in which each dimension represents one parameter that can be varied to determine an effect on a variable of interest.

multimedia
An interactive combination of text, graphics, animation, images, audio, and video displayed by and under the control of a personal computer.

multimedia files
Digitized images, videos, and sound that can be retrieved and converted into appropriate human-recognizable information by a client.

multimedia messaging system (MMS)
The transmission of richer content types, including photographs, images, voice clips, and eventually video clips, over mobile devices.

nanosecond
One-billionth of a second.

national information infrastructure (NII)
Communication networks and protocols, including satellite and cable television networks, telephone networks, mobile communication systems, computer networks, EDI, and Internet protocols (TCP/IP).

network cabling
The actual physical wire over which computers communicate.

network computer (NC)
A computer that can be used only when connected to a client/server network.

networked economy
Enhanced, transformed, or new economic relationships based on computer networks and human knowledge.

network interface card (NIC)
A card in a PC that connects the PC to the network and that handles all electronic functions of network access.

Network News Transfer Protocol (NNTP)
The protocol used for newsgroups that makes all messages sent to the specific newsgroup available to all subscribers.

network operating system (NOS)
The software that controls a computer network.

neural networks
Computer hardware, using multiple processors, and software systems that seek to operate in a manner modeled on the human brain.

newsgroup
One of a vast set of discussion lists that can be accessed through the Internet.

object
A self-contained programming module that combines data and instructions and that cooperates with other objects in a program by passing strictly defined messages to the other objects.

object-oriented database
A database that contains a data type called an object, which incorporates both data and the rules for processing that data.

object-oriented languages
Computer languages that use objects to carry out the required logic of a program.

one-to-many relationship
In a data model, the situation in which one field is related to multiple other fields.

online analytical processing (OLAP)
A software tool that enables an analyst to extract and view data from a variety of points of view.

online transaction processing (OLTP)
A process in which each transaction is processed at the time of its entry rather than being held for later processing.

open-source software
Software that is created and supported by volunteers who make it freely available to users who can then add any features desired to it.

operating system software
The software that manages the many tasks going on concurrently within a computer.

operational decisions
Operations that control the day-to-day operation of the organization.

optical character recognition (OCR)
The use of a scanner to convert a document to digital form, followed by use of software to determine the letters and symbols present.

optimization
The use of a mathematical technique to find the best solution to a model.

organizational feasibility
An indication that the problem can be solved or a system developed within the limits of the current organization.

organizational memory
Remembering the past through data, information, and knowledge management.

output
The result of processing as displayed or printed for the user.

output unit
A unit that provides the result of processing for the user.

outsourcing
A process that involves turning over some or all of the responsibility for the development or maintenance of an information system to an outside group.

packets
Data that have been grouped for transmission over a network.

packet switching
In a wide area network, dividing long messages into smaller data units that can be transmitted more easily through a network.

parallel processing
Processing that uses multiple CPU chips to perform multiple processing operations at the same time.

password
A sequence of letters and digits, supposedly known only to the user, that must be entered to access a computer system.

password policies
Specific company policies designed to protect data and software through responsible use of passwords.

patch
A piece of computer code that a software publisher releases to fix a problem in its software.

path
A portion of the URL that includes the name of the home page file plus any directories or folders in which it is located.

peer-to-peer network
A network configuration in which each computer can function as both a server and a workstation.

perfect choice
The demands by customers for a wide range of products and a choice in how to buy them at a variable price.

peripherals
Devices such as printers and speakers that are attached to a PC or a computer network.

persistent cookie
A file that exists indefinitely on the user's hard disk and that the browser uses to identify the user to the corresponding Web site.

personal computer (PC)
A small, one-user computer that is relatively inexpensive to own and does not require a special environment or special knowledge for its use.

personal digital assistant (PDA)
A form of handheld computer that does not require a separate keyboard or monitor.

personal identification number (PIN)
A type of password used to access a bank account from an automatic teller machine.

personalization
Personal sharing of knowledge.

personalized Web site
A Web site that creates an interactive relationship with customers.

personal outsourcing
A situation in which individuals outsource many of their nonprofessional responsibilities, such as automobile or yard maintenance, transporting children, or even grocery shopping and cooking, to someone else so they can keep working.

physical security
The protection of computer hardware from theft or damage, whether caused by nature or humans, in the same way that other office equipment would be protected.

plain text
The readable form of a message.

Platform for Privacy Preferences (P3P)
Proposed technology that would give consumers control over their own personal data by building the necessary mechanisms into Web browsers such as Netscape and Internet Explorer.

point-and-click operations
A method that involves using a mouse or other pointing device to position the pointer over a hypertext link or the menu bar, toolbar, location window, or directory buttons and then click a button to retrieve a Web page or execute a corresponding command.

point-of-sale (POS) transaction processing
An input system used to store and process important information that is obtained at the time a sale is made.

pop-up windows
Browser windows that are generated automatically when you visit certain Web sites and that must be closed separately from the browser window you originally opened.

portable document format (.pdf)
A form of electronic document created with Adobe's Acrobat Exchange that can be easily shared with anyone who has an Acrobat reader.

price transparency
The capability of customers to use the Web to learn about the prices of products from a variety of sellers.

primary key
A field or combination of fields that uniquely identifies each record in a table.

printer
An output device that places words, symbols, and graphics on paper.

privacy
Freedom from unauthorized intrusion; on the Internet, the right of users to control personal information and the capability to determine if and how that information should be obtained and used.

Privacy Act of 1974
Legislation that attempted to correct most of the recordkeeping practices of the federal government.

privacy seal of approval
An icon that a Web site can display if it has agreed to follow the FIP principles as defined by the seal-granting organization.

private key
In a public-key encryption system, the only key that can decrypt the message.

private-key encryption
A form of encryption in which a single key is used to both encrypt and decrypt the message.

processing
Converting data into information.

program
A series of instructions to the computer.

programming
The process of writing a series of instructions for the computer to follow in performing some specific task.

project manager
A person who ensures that the project is completed on time and within budget and who takes responsibility for bringing other people onto the team.

project plan
A document that describes the desired information system, including a clear statement of the scope of the desired project.

project sponsor
A person who has an interest in seeing the system succeed and who will provide necessary business expertise to the project.

prompt
An indicator on the computer screen that data or commands should be entered.

protocol
A formal set of rules for specifying the format and relationships when exchanging information between communicating devices.

prototype
A version of the system that contains the bare essentials and that can be used on a trial basis.

prototyping
A software development process in which the development team creates a quick-and-dirty version of the final product, often using special languages or software tools.

pseudocode
A way of expressing the logic of processing in structured English rather than in a computer language.

public key
In a public-key encryption system, a key that is freely distributed to encrypt messages.

public-key encryption
An encryption system with two keys—one private and one public—where the public key is used to encrypt a message and the private key is used to decrypt it.

query
A formalized request for data sent to a database.

random access memory (RAM)
The section of memory that is available for storing the instructions to the computer and the data to be manipulated.

rapid application development (RAD)
Methods and tools that allow for faster development of application software.

reach creep
A situation in which individuals take on more and more work because technology allows it.

read-only memory (ROM)
The section of memory that is placed in the computer during the manufacturing process and remains there even after the computer is turned off.

record
A collection of fields with information that usually pertains to only one subject (such as a person, place, or event).

reintermediation
The process of creating new intermediaries in electronic commerce.

relational database management system (RDBMS)
A database system in which elements are represented as being parts of tables, which are then related through common elements.

relations
Data organized as a table and used in a relational database.

repetitive stress injury (RSI)
A condition in which workers suffer from moderate to severe muscle and joint problems in the hand, wrist, arm, and shoulder.

request for proposals (RFP)
A complete list of specifications used by vendor or contractors to prepare a bid on project.

remote job entry (RJE)
A site where data are stored locally on a PC and then submitted to the mainframe or supercomputer for manipulation.

revenue management
The application of disciplined tactics that predict customer demand at the micro-market level and optimize product price and availability so as to maximize revenue growth.

revenue process
The receipt of revenue in exchange for goods and services.

router
A computer that determines the path that a message will take from the sending computer to the receiving computer.

rule database
The part of an expert system knowledge base that contains the rules to be used by the reasoning element of the expert system.

satellite office
A fully equipped office set up by a firm where its employees can reserve space and work one or more days per week closer to their homes, thereby reducing commute times and helping ease traffic congestion.

satellite transmission
The use of direct broadcast, which uses microwaves for one-way downloads of data to homes and offices.

scanner
A device used to translate a page of a document into an electronic form that OCR software can understand.

scheduled report
A report generated by an information-based DSS on a regular basis, containing summary

reports of the results of the data processing operation.

search engines
Web sites that use technology to find as many Web sites as possible that match the user's request.

secondary storage
Storage area that is used to save instructions, data, and information when the computer is turned off.

secure electronic transaction (SET)
A protocol that provides a way for buyers to transfer credit card information to the credit card issuer over the Internet without the seller seeing the credit card information.

Secure Sockets Layer (SSL)
A protocol used by Internet browsers and Web servers to transmit sensitive information.

server
A computer on a network running an application that provides services to client computers.

server address
The address of a server computer on the Internet.

server-based filtration
A filtering system installed on the Web server of the Internet service provider that allows the browser to access only those Web sites or chat rooms that are approved by the ISP.

server farm
A group of servers that work together to handle processing chores.

server includes
A way of processing data from the user that involves integrating the programming code into the Web page that is being sent to the server, but in such a way that users do not see the code.

server-side processing
Accepting data from a user's browser, processing it, and generating and returning a Web page to the user.

service resource
Another name for a protocol on the Web.

session
A client/server protocol in which a continuous sequence of transactions occurs between client and server.

session cookie
A cookie that exists only during the current series of interactions between the browser and Web server.

shareware
Software that can be downloaded from Web sites on the Internet for free but a nominal fee is expected to be paid to its author.

Short Message Service (SMS)
A system that enables mobile telephones to send and receive text messages up to 160 characters in length.

shrink-wrapped software
See *commercial off-the-shelf software.*

siftware
Software used in data mining to sift through the data looking for elements that match the search criteria.

signal type
The type of signal—digital or analog—being used to transmit bits between computers.

Simple Mail Transfer Protocol (SMTP)
A communication protocol for transferring mail messages over the Internet.

simulation
The use of probability-based models to imitate a real phenomenon.

site license
A license to use software that covers all of the computers in an organization.

slicing and dicing
A technique that enables the user to extract portions of the aggregated data and study them in detail.

smart card
A card, containing memory and a microprocessor, that can serve as personal identification, a credit card, an ATM card, a telephone credit card, a critical medical information record, and cash for small transactions.

smart object
A chip-based device with an antenna that provides information about the object in which it is embedded.

sniffer
A computer program on an intermediate Internet computer that will briefly intercept and read a message.

software
One or more programs that direct the activity of the computer.

spam
Junk e-mail.

speakers
Devices designed to broadcast sounds.

spiders
Computer programs that crawl the Web by following links from one page to the next, then send information back to a database of visited pages.

spyware
Software that gathers information about online activities and transfers it back to a server without your knowledge or permission.

stakeholder analysis
An analysis to determine the effect of a new or revised system on a particular person or group.

stakeholders
The people who determine the future of the organization, such as stockholders, employees, and customers.

stateless protocol
A client/server protocol in which the server has no memory of an interaction with the client other than logging some information about it.

statistical model
A model in which the objective is to learn about tendencies within the data set or to prove that differences exist between parts of the data.

steganography
A form of encryption that hides messages within graphic or audio files.

strategic decisions
Decisions that determine the long-term direction of the organization by creating policies.

structured decisions
Decisions made by following a set of rules and usually made on a repetitive basis; decisions that can be programmed in advance. Also called programmed decisions.

Structured Query Language (SQL)
A computer language for manipulating data in a relational database.

structured systems development
An approach to systems development in which each stage must be completed in a specific order after certain objectives are achieved in the previous stage, including a specified set of deliverables and management approval.

supercomputer
The biggest, fastest computer used today.

supply chain
The oversight of materials, information, and finances as they move in a process from supplier to manufacturer to wholesaler to retailer to consumer.

surge protector
A device that protects the computer's hardware and memory from a voltage surge.

switching costs
Costs associated with an individual or an organization changing to a new supplier.

synchronous
A type of communication in which more than one of the parties can send messages at a time.

system
A group of elements (people, machines, cells, and so forth) organized for the purpose of achieving a particular goal.

system administrators
People who manage security and user access to an intranet or LAN. Also known as *network administrators.*

system conversion
The process of changing over from an old system to a new one.

system degradation
The point in the IS life cycle in which the performance of the system drops off markedly and the quality of information provided by the system suffers.

system proposal
A document that describes what the new system should look like.

system request
A document that lists the business need for a systems development project, its expected functionality, and the benefits that would likely result from its completion.

systems analysis and design (SAD)
Another name for the systems development process.

systems analyst
A person who carries out the systems analysis and design process.

systems audit software
Software that keeps track of all attempts to log on to the computer, giving particular attention to unsuccessful attempts.

systems development
The process of developing a system design to meet a new need or to solve a problem in an existing system.

systems development life cycle (SDLC)
Another name for the structured approach to systems development.

system specification
(1) A document that is used in the implementation stage to develop the new system internally, outsource it, or acquire it, depending on the decision made at the beginning of the design stage. (2) A complete and detailed group of deliverables, including the physical data model, physical models of each process in the process model created in the analysis stage, and interface screens.

system unit
The main case of the PC housing the processing unit, internal memory, secondary storage devices, and modem.

table
A database model composed of rows and columns, with rows specifying a particular person, place, or thing, and columns giving the specific details about each person, place, or thing.

tacit knowledge
Personal knowledge, experience, and judgment that is difficult to codify.

tactical decisions
Decisions made to implement the policies created by strategic decisions.

target refractor
A method for customizing a Web site to meet the needs of stakeholders.

T-carrier circuit
A digital method of data transmission over dedicated telephone lines.

technical feasibility
An indication that the technology exists to solve the problem or develop the system.

telework center
A fully equipped office used by employees from different organizations, with employers being charged for the space and services utilized by each employee per day.

teleworking
The use of networks to engage in work outside the traditional workplace. Also called telecommuting.

Telnet protocol
The capability to use the Internet to log on to a computer other than your local computer.

Ten Commandments of Computer Ethics
A set of rules that covers many of the ethical issues facing computer users in the networked economy; developed by the Computer Ethics Institute in 1992.

tendonitis
A general inflammation and swelling of the tendons in the hands, wrists, or arms.

terabyte
A measure of computer storage equal to approximately 1 trillion bytes or 500 million pages of text.

thin client
A client computer on a network that cannot be used in a stand-alone mode.

threaded
An organization of questions and answers in which answers or comments that relate to a previous question or comment are linked to it.

three-tiered client/server architecture
A client/server architecture in which an intermediate computer exists between the server and the client.

throw-away prototype
A prototype that developers use to carry out exploratory work on critical factors in the system. It is discarded after being developed.

tiered pricing
Goods and services being offered at different price points to meet different customers' needs.

top-level domain
One of the 14 domain names that define the type of service or area of interest of the server.

touchpoint system
A contact point through which a company interacts with the customer, including the Web, e-mail, personal sales, direct mail, call centers, and so on.

touch screen
A type of input device that enables the user to touch parts of the computer screen to select commands.

transaction
Any event that involves the digital transfer of money or information between entities.

transactional data
Data that are created when a transaction takes place that requires the customer to reveal his or her identity.

transaction processing system (TPS)
A system for converting raw data produced by transactions into a usable, electronic form.

Transmission Control Protocol/ Internet Protocol (TCP/IP)
The basic communication language or protocol of the Internet.

trapdoor
A secret entrance into a computer through which criminals can access the computer or network.

tuple
A row in a relation used in a relational database.

twisted pair
A medium for data transfer that is made of pairs of copper wire twisted together.

U-commerce
An extension of mobile commerce in which the *U* stands for a number of things, including ubiquitous, universal, or unison.

Unicode Worldwide Character Standard (Unicode)
An international coding scheme that can process and display written texts in many languages.

uninterruptible power supply (UPS)
A device that continues to send power to a computer if the electrical current is disrupted.

unstructured decisions
Decisions that involve complex situations and often must be made on a once-only or ad hoc basis using whatever information is available. Also called *unprogrammed* or *ad hoc decisions*.

update
In a database, to make additions, deletions, or changes to one or more columns for a particular row.

URL (uniform resource locator)
A standard means of consistently locating Web pages or other resources, no matter where they are stored on the Internet.

user
In a model-based DSS, the person working with the models and data to generate alternative solutions to a problem.

user interface
What the user sees on the screen.

user-interface prototype
A prototype that demonstrates an example interface for the information system.

user interface system
The part of a decision support system that handles the interactions between the analyst and the computer.

value-added network (VAN)
A public network, available by subscription, that provides data communications facilities beyond standard services; often used to support EDI.

value chain model
The chain of business activities in which each activity adds value to the end product or service.

value system
The linkage of the value chains of two organizations.

videoconferencing
A way of enabling groups or individuals in

different locations to meet at the same time through real-time transmission of audio and video signals between the different locations.

virtual assistant
A person who works as an assistant on an as-needed basis over the Internet for a number of employers.

virtual/mobile office
employees with the communications tools and technology they need to perform their jobs from wherever they need to be—home, office, customer location, airport, and so on.

virtual team
A team of people who attempt to use information systems to help structure, focus, and facilitate the transfer of information and knowledge among themselves.

virtual workplace
A concept relating the capability of a worker to work at any place and any time.

virus
Malicious or destructive software that damages resources on a target computer.

voltage surge
A sudden increase in the electrical supply caused by lighting or some other electrical disturbance. Also known as a spike.

waterfall approach
Another name for the structured approach to systems development.

Web bug
A tiny image file (usually 1 pixel by 1 pixel) on a Web page that gathers data on the user's online activities.

Web conferencing
A combination of telephone conferencing and visual interaction over the Web.

Web page
A special type of document that contains hypertext links to other documents or to various multimedia elements.

Web services
Distributed computer applications that can be easily located, accessed, and used over the Internet.

Web site
An Internet server on which Web pages are stored.

web-visit data
Data generated about a user's computer whenever the user visits a Web site.

wide area network (WAN)
A network covering more than a single geographic area.

Wintel
A combination of Windows and Intel, in which a PC with an Intel chip runs a version of the Windows operating system.

Wireless Application Protocol (WAP)
A protocol designed to enable mobile telephones to access the Internet and the Web.

workplan
A document that lists the tasks that must be accomplished to complete a project, along with information about each task, the number of persons required, and an estimated time to complete each task.

workstation
(1) A client computer that allows the use of specialized applications requiring high-speed processing of data into information. (2) A computer and supporting furniture.

World Wide Web Consortium (W3C)
A group of more than 500 member organizations aimed at helping the World Wide Web reach its full potential by developing common protocols that promote its evolution and ensure its interoperability.

worm
Malicious or destructive software that uses up resources on a target computer.

PHOTO CREDITS

p. xvi Tori Bauer/University of Georgia
p. 6 AP/Wide World Photos
p. 7 © Donovan Reese/Stone
p. 8 © Bettmann/CORBIS
p. 13 Courtesy of Intel Corporation
p. 16 Courtesy of Microsoft Corporation
p. 17 © Alan Levenson/TimePix
p. 20 © 2002 PhotoDisc
p. 26 Courtesy of General Electric Company
p. 32 AP/Wide World Photos
p. 34 Courtesy of IBM Corporation
p. 35 Courtesy of American Express Company
p. 36 Copyright © Nokia, 2002
p. 37 Courtesy of Intel Corporation
p. 38 Courtesy of Western Digital Corporation
p. 41 © 2002 PhotoDisc
p. 42 AP/Wide World Photos
p. 46 © Tony Freeman/PhotoEdit
p. 47 © James D. Wilson/Liaison International
p. 50 Courtesy of IBM Archives
p. 55 Courtesy of Sun Microsystems, Inc.
p. 66 Courtesy of Corrugated Supplies Company, LLC
p. 73 © 2002 PhotoDisc
p. 74 Courtesy of Linksys Group, Inc.
p. 76 Courtesy of Asanté Technologies, Inc.
p. 77 Courtesy of EDIdEv LLC
p. 80 Courtesy of D-Link System, Inc.
p. 80 Courtesy of D-Link System, Inc
p. 80 Courtesy of 3Com Corporation
p. 83 Courtesy of Motorola
p. 84 Courtesy of Ericsson
p. 85 Courtesy of 3Com Corporation
p. 85 © 2002 PhotoDisc
p. 87 Courtesy of Genuity, 2002
p. 95 © Reuters NewMedia Inc./CORBIS
p. 95 Courtesy of Dan Bricklin with permission from Bob Metcalfe
p. 111 AP/Wide World Photo
p. 113 Courtesy of Peachtree Software
p. 116 © David J. Sams/Stone
p. 118 Courtesy of IBM Corporation
p. 120 Courtesy of Symbol Technologies, Inc.
p. 122 Courtesy of Alien Technology
p. 127 © 2002 PhotoDisc
p. 131 © Charlie Westerman/Stone
p. 135 AP/Wide World Photos
p. 136 Courtesy of 1 EDI Source, Inc.
p. 156 Courtesy of Teradata, a division of NCR Corporation
p. 169 © Ron Chapple/FPG
p. 171 Courtesy of Lotus Development Corp.
p. 174 © Zigy Kaluzny/Stone
p. 179 Courtesy of Red Robin Gourmet Burgers
p. 186 Courtesy of Brio Software, Inc.
p. 188 Courtesy of MicroStrategy, Inc.
p. 194 Courtesy of Balanced Scorecard Collaborative
p. 199 Photo Courtesy of Pine Cone Systems, Inc.
p. 204 © Charles Gupton/corbisstockmarket.com
p. 209 © Lawrence Manning/CORBIS
p. 215 © Giansanti Gianni/CORBIS SYGMA
p. 221 © Michael Newman/PhotoEdit
p. 226 Courtesy of Edmunds.com
p. 231 © Bill Bachmann/PhotoEdit
p. 234 AP/Wide World Photos
p. 237 AP/Wide World Photos
p. 246 © Reuters NewMedia Inc./CORBIS
p. 255 Courtesy of SendMail, Inc.
p. 274 Images provided by Neil F. Johnson, Ph.D. All Rights Reserved
p. 278 Courtesy of QWallet
p. 282 Courtesy of Akamai Technologies, Inc.
p. 289 Photo courtesy of Mannheim Auctions
p. 296 © Da Silva Peter/CORBIS SYGMA
p. 312 © Eyewire
p. 319 Courtesy of Corrugated Supplies Company, LLC
p. 324 © Alan Schein/corbisstockmarket.com
p. 331 Courtesy of Visible Systems Corporation
p. 337 Courtesy of Ed Yourdon
p. 343 Courtesy of InCert Software (www.incert.com)
p. 356 Courtesy of Putnam Lovell Securities Inc.
p. 363 AP/Wide World Photos
p. 365 © Jim Bounds/CORBIS SYGMA
p. 366 AP/Wide World Photos
p. 371 AP/Wide World Photos
p. 372 © 2002 PhotoDisc
p. 381 Courtesy of American Power Conversion
p. 384 Courtesy of Phil Zimmermann
p. 392 Courtesy of Starwood Hotels
p. 415 © Michael Goldman/FPG
p. 420 © Bettmann/CORBIS
p. 426 Courtesy of Roger Clarke
p. 437 © Bettmann/CORBIS
p. 440 © Brad Mangin
p. 441 Courtesy of the National Science Foundation Grant Number ANI-0087344, and the University of California, San Diego (http://hpwren.ucsd.edu/)
p. 445 © Jim Richardson/CORBIS
p. 447 Courtesy of Lockheed Martin
p. 450 Courtesy of ScanSoft
p. 451 Courtesy of Logitech
p. 451 Courtesy of GWS Systems
p. 453 © Eyewire
p. 457 AP/Wide World Photos
p. 459 AP/Wide World Photos
p. 462 Courtesy of XM Satellite Radio

INDEX